Dynamical Aspects of Nuclear Fission

4th International Conference on

Dynamical Aspects of Nuclear Fission

Častá-Papiernička, Slovak Republic 19–23 October 1998

Editors

Yu. Ts. Oganessian

Joint Institute for Nuclear Research, Russia

J. Kliman

Joint Institute for Nuclear Research, Russia

Š. Gmuca

Slovak Academy of Science, Slovakia

World Scientific

Singapore • New Jersey • London • Hong Kong

Published by

World Scientific Publishing Co. Pte. Ltd.

P O Box 128, Farrer Road, Singapore 912805

USA office: Suite 1B, 1060 Main Street, River Edge, NJ 07661

UK office: 57 Shelton Street, Covent Garden, London WC2H 9HE

British Library Cataloguing-in-Publication Data
A catalogue record for this book is available from the British Library.

Proceedings of the 4th International Conference on
DYNAMICAL ASPECTS OF NUCLEAR FISSION

ISBN 981-02-4187-9

Printed in Singapore.

This book is dedicated to the memory of Boris Pustylnik who organized and edited all DANF conferences, a great physicist and wonderful personality

PREFACE

The international conference DANF 98 was the fourth of the series of conferences on Dynamical Aspects of Nuclear Fission, which started in 1991. It took place from 12–17 October 1998 in the surroundings of castle Červený Kameň placed in Častá – Papiernička, Slovak Republic and was organized traditionally with international cooperation from the Institute of Physics in Bratislava with the JINR Dubna, and supported by Comenius University in Bratislava. Thirty-seven scientists participated in this conference.

The aim of the DANF conference series is to present, interchange and consult brand new theoretical and experimental results in the area of dynamics of nuclear fission as well as to create good conditions for giving rise to new international collaborations.

As usual, the program of this conference was designed to cover a wide range of physical phenomena – from low to intermediate energies for spontaneous and induced fission and also for the fragmentation of hot nuclear systems. The program of DANF'98 was slightly modified, due to recent discoveries and tremendous development of theories, as well as experiments, in the field of superheavy elements, and the fusion – fission processes area was added. This proceedings covers new results on the synthesis of superheavy elements and decay of complex nuclear systems, binary and ternary fission, fragmentation and decay of hot nuclei and nuclear structure of neutron-rich nuclei. It also covers some questions of incineration and transmutation of actinides of nuclear waste. Except for the main stream of the conference a place was also given for reviews on recent progress and to explore future directions in developing radioactive ion beam facilities.

We wish to give thanks to members of the International Advisory Committee for their valuable comments and powerful assistance in preparation of the conference. We would like to thank the Organizing Committee and personnel at the conference site for their tireless work and technical support. Especially we would like to thank to Dr. Emil Běták, who fast edited and printed in Acta Physica Slovaca Vol. 49 (1999) pp.1–124 some contributions from the conference.

However, we feel particularly obliged to give thanks to all participants for their friendly approach. They promoted a creative atmosphere and made DANF'98 a pleasant and successful meeting.

Yu. Ts. Oganessian
J. Kliman
Š. Gmuca

Editors

Organizers:

Flerov Laboratory of Nuclear Reactions, JINR, Dubna
Institute of Physics, Slovak Academy of Sciences, Bratislava
Faculty of Mathematics and Physics, Comenius University, Bratislava

International advisory board:

Yu. G. Abov (Moscow)
N. Carjean (Bordeaux)
H. Faust (Grenoble)
G. Giardina (Messina)
F. Gönnenwein (Tübingen)
W. Greiner (Frankfurt)
F. Hanappe (Brussels)
M. G. Itkis (Dubna)
G. Münzenberg (Darmstadt)
M. Mutterer (Darmstadt)
A.M. Stefanini (Legnaro)
I. Wilhelm (Prague)

Organizing committee:

Yu. Ts. Oganessian (Dubna,chairperson)
W. I. Furman (Dubna)
Š. Gmuca (Bratislava)
J. Kliman (Dubna, Bratislava)
A. N. Mezentsev (Dubna)
B. I. Pustylnik (Dubna)

Local organizing committee:

P. Bandžuch
M. Benovič
E. Běták
A. Duka-Zolyomi
Š. Gmuca (chairperson)
M. Jandel
J. Leja
M. Morháč
V. Matoušek
M. Veselský

CONTENTS

SCIENTIFIC PROSPECTS AND FURTHER DEVELOPMENT OF THE ACCELERATOR COMPLEX OF FLNR. PROJECT "DRIBS"

YU.TS. OGANESSIAN

The present status

After putting into operation the U-400 and U-400M accelerators and equipping them with the new ERC-type sources of multicharged ions as well as with new equipment for the bunching system and the system for extracting beams, the record intensities of ion beams with energies of up to 50 MeV/n have been achieved.

Experiments aiming at the synthesis of superheavy elements using a ^{48}Ca beam, the intensity of which today is about 10^{13} 1/s, have been started at the U-400+ECR-4M cyclotron complex. Within a 20-day experiment in May 1998 at the electrostatic separator of recoil nuclei VASSILISSA, the new nuclide – isotope of element 112 with A=283 – has been synthesized in the reaction ^{48}Ca+^{238}U.

In the bombardment of a ^{244}Pu target with ^{48}Ca ions at the gas-filled separator of recoil nuclei a decay sequence consisting of an implanted heavy atom, three subsequent α- decays, and a spontaneous fission, all correlated in time and position, was detected. The large alpha-particle energies together with the long decay times and spontaneous fission terminating the chain offer evidence of the decay of nuclei with high atomic numbers. They are good candidates for originating from the α-decay of the parent nucleus ^{289}114, produced in the 3n-evaporation channel with a cross section of about 1 pb.

Within this field of research, also the chemical properties of transactinide elements are being investigated with the help of fast radiochemistry methods realized at the set-up KHIPTI. The experiments are carried out in co-operation with radiochemical groups of PSI (Villigen, Switzerland), FZ (Rossendorf, Germany) and INP (Krakow, Poland).

At the accelerator complex U-400M+DECRIS-14 employing the separator ACCULINNA, intensive beams of ^{7}Li and ^{11}B ions provide obtaining high-quality secondary beams of ^{6}He and ^{8}He. The beam spot of the secondary beam at the target is about 0.25 cm^2, the energy spread - not more than 2.5%; the ^{6}He and ^{8}He beams intensity - about $5\cdot10^5$ 1/s and $5\cdot10^3$ 1/s, respectively.

Using the ^{6}He ion scattering reaction at H_2 and ^{4}He nuclei, the structure of nuclei in which a strongly bound "nucleus – core" interacts with 2-4 outermost neutrons has been studied. In the ^{6}He nucleus a "di-neutron" structure of outermost – "valent" neutrons has been revealed. Experiments on searching for neutron correlations in the ^{8}He nucleus are now underway. Physicists of GANIL (France) are participating in the experiments.

Experiments with the use of beams of radioactive nuclei also include the study of properties of proton-rich nuclei, the structure of the ^{8}B nucleus in particular (the MULTI set-up). The mechanism of fragmentation reactions and characteristics of radioactive decay of the most exotic nuclei will be studied in accordance with the accepted program using a wide-aperture high-resolution separator COMBAS that is being prepared now for the experiment.

At the beams of the U-400 cyclotron, the process of heavy and superheavy nuclear fission is being investigated in a wide range of accelerated ion masses. It has been shown that in a collective motion of a nucleus undergoing a strong change in shape and fissioning into two fragments, different fission modes are realized. These modes differ greatly in their characteristics. With the help of a tandem - the CORSET (FLNR) set-up and the European multimodal neutron detector DEMON – the dynamics of different fission modes of the whole number of nuclei from Th to $^{286}112$ has been investigated.

The experiments are carried out within a wide collaboration with IReS (France) and Texas A&M University (USA).

Upon obtaining first results at the 4π- detector FOBOS, a new series of experiments has been prepared intended for the work of the detector with a complete set of detector modules during a long period of time. Some preliminary measurements have beam made which confirm the detector's readiness for the work in such operating conditions.

Considering the work results of the past years it is possible to make the following conclusions:

1. The major part of work on creation and putting into operation new equipment for the accelerator technique has been completed at the U-400 and U-400M accelerator complex. The record intensities have been achieved, including the beams of the most rare ^{36}S, ^{48}Ca and other isotopes. It has been experimentally shown that 2.5% of neutral atoms, injected into the camera of an ECR-source, can be accelerated and delivered to a target.
2. Experiments are started and first results have been obtained in the field of synthesis and study of properties of superheavy elements, heavy nuclei fission, structure of exotic nuclei of the lightest elements etc.
3. The prospects for such experiments have been shown and wide collaborations have been created attracting material and human resources of the biggest world centers into participation in the JINR experimental program in the field of heavy ion physics.

Prospects for development of the research basis in the field of nuclear physics at the JINR

While considering the prospects for the major lines of investigation it would be natural to analyze investigations carried out lately so that one could estimate the place of development in this field. Note that practically all the mentioned above investigations have been performed within the past 1.5-2 years.

It is not very difficult to imagine the development of events of evolutionary character: irradiation of all "promising" targets with ions of ^{48}Ca and synthesizing elements with Z=114-118; study of scattering and nucleon transfer reactions with $^{6,8}He$ ions using targets of different isotopes of hydrogen (including tritium targets), helium and lithium; experiments on searching for the proton halo in the ^{8}B nucleus; study of the cluster structure of Be isotopes etc.

In other words, "harvesting" the results of invested ideas and funds.

At the same time one can also clearly imagine what goals cannot be reached due to principal restrictions imposed already in preparation of the experiments.

Production of exotic nucleus beams

In the study of the interaction between extremely neutron-rich nuclei of the $^{8}He+^{3}H$ type, when 3 protons and 8 neutrons take part in the interaction and even more neutron-rich systems (of the ^{10}He+p or the $^{4}H+^{7}H$ type) can be formed in the entrance channel, it is of utmost importance to obtain high precision in determining the energy of reaction products in missing mass experiments. On the other hand, formation of super neutron-rich systems happens at a low probability and requires considerable increase in the intensity of an exotic nucleus beam. The necessity of variation of the beam energy (from 4 to 20 MeV/n) depending on the reaction type and target matter must be added to that.

In other words, the intensity of beams of ions such as $^{6,8}He$, ^{8}B and ^{11}B must be increased tens and hundreds times with an essential improvement of emittance and energy resolution of a beam (almost by an order of magnitude).

It is practically impossible to obtain such characteristics in the systems of direct production of a secondary (exotic) beam, as it is done today at the separator ACCULINNA, since a rigid approach to the reaction product selection will require increasing the intensity of a primary beam by several orders of magnitude.

However, such results still can be achieved, as is shown in many original approaches and projects, if the functions of isotope production and isotope acceleration are separated using the so-called system ISOL. Such a scheme suggests the use of two accelerator set-ups: the primary beam accelerator inducing the reaction at a production target, and the accelerator of radioactive nuclei transported into its ion source or directly into the center of an accelerator chamber. Two acting FLNR accelerators of the U-400 class can separate these two functions.

Calculations show that the highest yield of exotic nuclei may be achieved in the fragmentation reactions at ion energies $E \geq 30$ MeV/n. This is the role of the U-400M cyclotron as an accelerator for the primary beam production.

Of all radioactive isotopes used as secondary beams, the nuclei having exotic structure and formed with a relatively high cross section are in our opinion of particular interest. They are the nuclei ^{11}Be (14 s) and ^{6}He (0.8 s), both having the structure in a form of a core and a halo, consisting of one and two neutrons, respectively; the ^{8}He nucleus (0.12 s) which evidently possesses a neutron skin-layer; the ^{8}B nucleus whose structure may have a proton halo.

The choice of target nuclei ^{4}He, ^{10}Be and ^{7}Be, playing the role of "cores", and the use of hydrogen isotopes ^{1,2,3}H open up a new area of research of nuclear interactions and nuclear forces under completely unique conditions (Fig. 1).

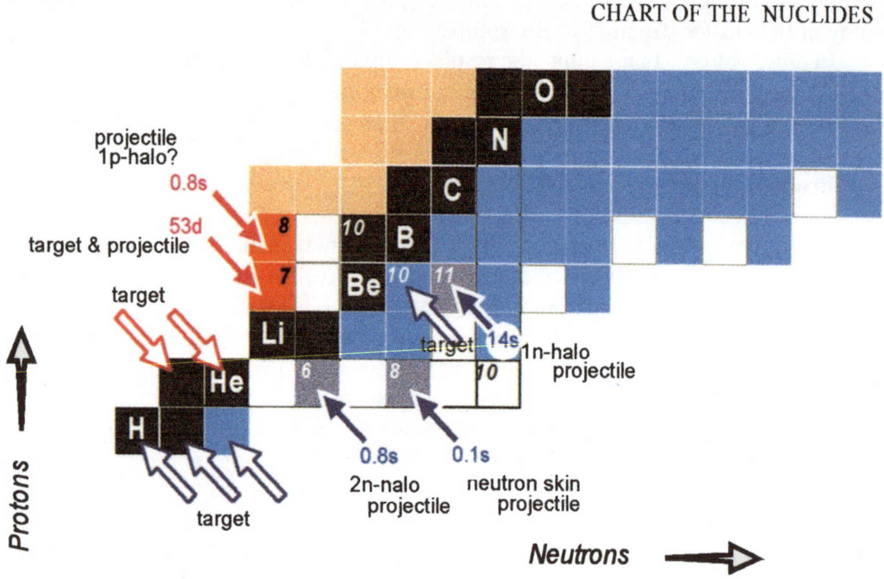

Fig 1. Isotopes of light elements used as beams and targets in the study of the structure and reactions between exotic nuclei.

Table I presents energies and intensities of the primary ion beam, the majority of which have been already achieved at the U-400M cyclotron, and the intensities of the secondary ion beam which can be achieved at external targets of the U-400 accelerator.

It is supposed that reaction products formed in the target will be transported in a gaseous state into an ion source and then after mass-separation – being in a single-charge state – into the accelerator hall via a gallery. They can be injected directly into the U-400 accelerator chamber bypassing the ECR-4M ion source and extracted from the final radius by means of stripping on a thin foil with an almost 100% efficiency (Fig.2).

As is seen from Table I, the beam intensity increases approximately by two-three orders of magnitude (depending on an ion) in comparison with the record results which are obtained today using the direct method of the radioactive nucleus beam production at the U-400M cyclotron.

Table I U-400M + U-400 RIB facility

Isotope	$T_{1/2}$ /sec.	Primary beam		Target	Production rate $1/10^{13}$ pps	Charge state for acceleration	Efficiency	RIB [pps] intensity
		Ion	Energy MeV/n					
^{6}He	0.81	^{7}Li	35	Be	$1\cdot10^{10}$	1^+	$3\cdot10^{-3}$	$3\cdot10^8$
^{8}He	0.12	^{11}B	35	---	$2\cdot10^7$	1^+	$3\cdot10^{-3}$	$5\cdot10^5$
^{9}Li	0.18	^{11}B	35	---	$7\cdot10^8$	1^+	$3\cdot10^{-3}$	$2\cdot10^6$
^{7}Be	$4.6\cdot10^6$	^{10}B	37	---	$1\cdot10^{10}$	1^+	$3\cdot10^{-3}$	$3\cdot10^8$
^{11}Be	13.8	^{13}C	45	---	$1\cdot10^9$	$1^+;2^+$	$3\cdot10^{-2};6\cdot10^{-3}$	$3\cdot10^7;6\cdot10^6$
^{8}B	0.77	^{10}B	37	---	$1\cdot10^9$	1^+	$3\cdot10^{-2}$	$3\cdot10^7$
^{10}C	19.3	^{12}C	45	---	$2\cdot10^9$	$1^+;2^+$	$3\cdot10^{-2};6\cdot10^{-3}$	$6\cdot10^7;1\cdot10^7$
^{14}O	70.6	^{16}O	40	---	$1\cdot10^9$	2^+	$6\cdot10^{-3}$	$6\cdot10^6$

The energy spread of the beam may be obtained at a level of 100-150 keV ($\Delta E/E$=0.2-0.3%) at a minimal cross section of 8-10 mm^2.

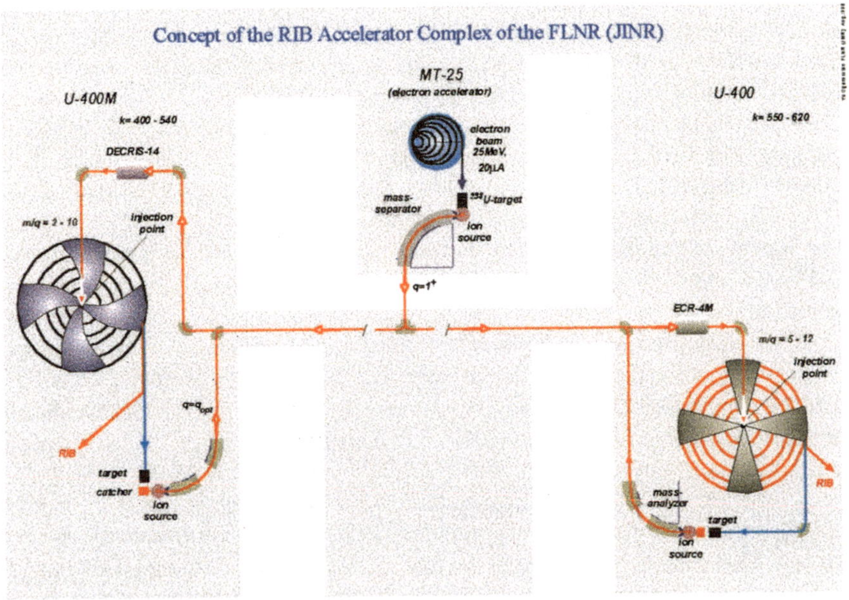

Fig 2. The scheme of location of the MT-25 microtron and the U-400 and U-400 M cyclotrons connected by a transport line. Different variants of radioactive beam injection into the cyclotrons are shown.

Acceleration of fission fragments

In the field of superheavy element synthesis and study of heavy nucleus fission (collective dynamics of many-body systems) in any target-ion combination we don't reach the closed neutron shell N=184, where gigantic changes in the structure are expected, and hence, changes in the heavy nucleus stability. Even when using the most heavy targets of ^{244}Pu, ^{248}Cm or ^{249}Cf and ^{48}Ca beams, some nuclei are formed which in their ground states are 7-9 a.m.u. away from the enhanced stability region. The use of a radioactive beam of ions with neutron excess higher than in the stable ^{48}Ca nucleus could solve the problem.

Here it is important to pay attention to the inverse fission reactions, and first of all to producing beams of neutron-rich magic nuclei, i.e. fission fragments ^{132}Sn ($T_{1/2}$=40s); ^{133}Sb (2.5 min) and ^{134}Te (42 min), which determine the fission fragment mass distribution and thus the fission mode of a weakly excited nucleus. Acceleration of fission fragments is perhaps the most ambitious part of the project.

The idea and the dream to produce a beam of neutron-rich isotopes being formed in the fission of uranium and thorium nuclei have existed during many years. New generation accelerator complexes of radioactive nuclei use the reaction of ^{238}U fission with high-energy protons (E_p=0.6÷50 GeV; CERN, KEK), with fast neutrons generated by a deuterium beam with E_d=20 MeV (the PARNE project, Orsay- GANIL), with thermal neutrons in the reaction ^{235}U(n_{th},f) - the PIAFFE project (Grenoble) and the new PIAFFE' project (Munich).

To make the picture complete, note that it is possible to use extremely expensive, but very convenient spontaneous fission source – the isotope ^{252}Cf, produced in high-flux reactors (Oak-Ridge, Dimitrovgrad), as well as photofission of ^{238}U. The latter deserves separate consideration.

At an energy of gamma-quanta in the range of the ^{238}U nucleus giant resonance (E_γ~13.5-14 MeV), the fission cross section, according to literature, is approximately 160

At these energies, as in the fast neutron fission, massive ^{238}U target of up to 100 g and heavier can be used. Experiments carried out at the Flerov Laboratory at the MT-25 microtron (E_e=25 MeV, I_e=20 μA) with a 100 g ^{238}U block yielded a neutron flow which corresponded to the ^{238}U fission frequency of $\sim 1.3 \cdot 10^{11}$ 1/s (see Table II).

It is not very difficult to estimate that the yield of every heavy Xe isotopes with masses from 137 to 140, corresponding to a given fission frequency, is about $5 \cdot 10^9$ 1/s. This value can be taken in the first approximation for comparison with the other reactions of uranium fission. Table III presents parameters of the set-ups producing Xe isotopes comparable with those parameters expected at an electron beam of only 0.5 kWt.

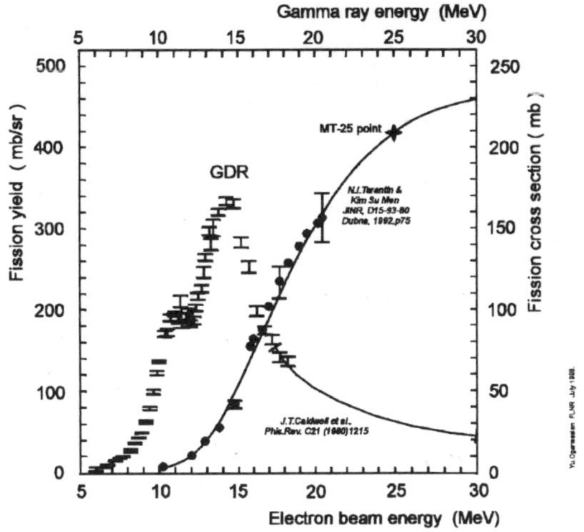

Fig 3. Dependence of the ^{238}U fission cross section on the gamma-quanta energy (red points and axes), and the fission fragment yield (the black curve and dots) taken from the works indicated in the figure.

The yelds of Kr and Xe isotopes produced with an electron beam from the microtron MT-25

Table II

Microtron MT-25

Electron beam energy: 25 MeV

beam intensity: 20 μA

Reaction: $^{238}U(\gamma,f)$ at the energy $E_\gamma = 15$ MeV

fission probability $P_f = 0.38$

neutron number $\bar{\nu} = 3.6$

Neutron yeld:

$Y_n = \sigma_{\gamma n} \cdot \nu_{\gamma n} + \sigma_{\gamma f} \cdot \nu_f = 0.62 \cdot 1.5 + 0.38 \cdot 3.6 = 2.29$

So, **40.5%** of Y_n from **γn-reaction**
and 59.5**%** of Y_n from **fission**

Experimental values:

Target size; 30 x 15 mm^2 = 100g
thickness: 35g/cm^2
neutron rate (in 4π sr) 8x10^{11}/s

Then: fission rate is;8x10^{11}/s x 0.165 = **1.3x10^{11}/s**

According to D.De Frenne et al. Phys. Rev. C (1982)1356

for **Kr(Z=36)** isotopes:

A	$T_{1/2}$ s	Y(A)%	P(Kr)	P(Kr)	Y(A,Z=36)s^{-1}
90	32.3	4.5	35.8	0.77	4.5x10^9
91	8.6	5.0	36.2	0.68	4.4x10^9
92	1.8	5.5	36.6	0.50	3.6x10^9
93	1.3	5.8	37.1	0.30	2.4x10^9
94	0.2	6.0	37.4	0.12	1.0x10^9
95	0.8	5.9	37.8	0.03	2.5x10^8
for **Xe(Z=54)** isotopes					
140	13.6	5.7	54.2	0.59	4.4x10^9
141	1.7	5.8	54.6	0.52	3.9x10^9
142	1.2	6.0	54.9	0.40	3.1x10^9
143	0.3	5.9	55.3	0.18	1.4x10^9
144	1.2	5.8	55.7	0.05	3.8x10^8
145	0.9	5.7	56.1	0.01	0.7x10^8

Table III: Fission sources used for the production of ^{137}Xe with rate $5 \cdot 10^9$ 1/s

Projectiles / Source	Target	Beam intensity
Protons (600 MeV)	^{238}U (100 g.)	50 μA
Spontaneous Fission	^{252}Cf (0.2 g.)	---
Thermal neutrons	^{235}U (0.4 g.)	$1.6 \cdot 10^9$ n/cm ^{2} s
Fast neutrons (deutrons 20 MeV)	^{238}U (100 g.)	1000 μA
γ - quanta (electrons 25 MeV)	^{238}U (100 g.)	20 μA

Some calculations showed that the yields of isotopes, i.e. magic nuclei with Z=50÷52 and N=82 which are of interest to us, amounted to $2 \cdot 10^9$ 1/s and $5 \cdot 10^9$ 1/s, the initial parameters (intensity, electron beam energy and the ^{238}U target weight) having been the same (Fig.4).

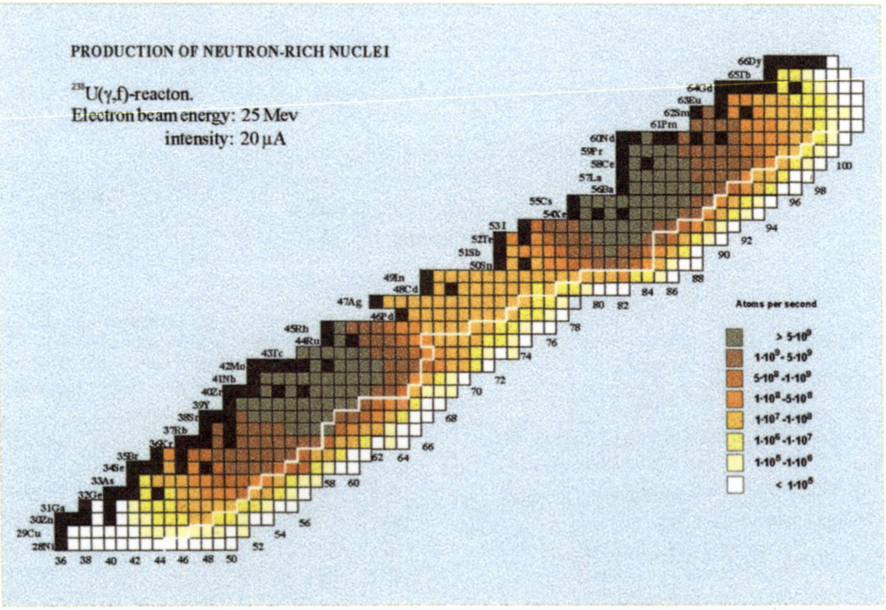

Fig 4. Calculated yields of different isotopes formed as photofission fragments of ^{238}U at the beam of the MT-25 microtron. The white line separates the zones according to the half-lives: the left-hand side – for the nuclei with $T_{1/2}$>1 s, the right-hand side – for those with $T_{1/2} \leq 1$ s.

Within the time comparable with half-lives of nuclei of interest (40 s $\leq T_{1/2} \leq$ 40 min) , isotopes Sn, Sb and Te will be separated and delivered into the ion source of a mass-separator for separating according to the mass number. The low-energy beam of single-charged ions (E_I~20 keV) will be transported to a distance of ~50 m and injected into the ion source ECR-4M of the U-400 accelerator.

In a high-temperature plasma of the ECR-source the ion charge will increase up to $7^+ \div 15^+$ for nuclei with A~80, which corresponds to the accelerated ion beam energy of 3.5÷16 MeV/n and up to $13^+ \div 25^+$, which corresponds to the accelerated ion beam energy of 5÷18 MeV/n at the U-400 cyclotron. The range of energies and beam intensities (without taking into account the losses due to decay in the process of the primary product extraction) are illustrated by Table IV. With the intensity of the radioactive beam of ~10^6 1/s and with the use of spectroscopic targets (~500 μg/cm^2) it is possible to carry out detailed research of reactions between fission fragments including complete fusion of extremely neutron-rich and magic nuclei.

The advantage of this variant consists in the use of a compact electron accelerator (the complete consumed power is ~20 kWt) and one of the U-400 accelerators. Simultaneously, the other accelerator may be switched off or used in an individual mode of operation.

Radioactive beams at the JINR

The project of radioactive beam production at the JINR is based on the acting accelerators of heavy ions of the U-400 class and an electron accelerator of the microtron.

Table IV Acceleration of nuclei with A~80 and A~130 on the U-400 cyclotron

Process	Time, sec	Efficiency A~80			Efficiency A~130		
		A^{7+}	A^{10+}	A^{15+}	A^{13+}	A^{20+}	A^{25+}
ε_1 – diffusion from catcher	1÷10	1	1	1	1	1	1
ε_2 – transport to ion source	0.1÷1	1	1	1	1	1	1
ε_3 – ionization 1^+	$1\cdot10^{-3}$	0.5	0.5	0.5	0.5	0.5	0.5
ε_4 – transport of 1^+ beam	$5\cdot10^{-4}$	0.8	0.8	0.8	0.8	0.8	0.8
ε_5 – capture in ECR	0	0.35	0.35	0.35	0.35	0.35	0.35
ε_6 – ionization n^+	$40\cdot10^{-3}$	0.03	0.06	0.015	0.06	0.06	0.01
ε_7 – acceleration capture	0	0.5	0.5	0.5	0.5	0.5	0.5
ε_8 – acceleration	$5\cdot10^{-5}$	0.8	0.8	0.8	0.8	0.8	0.8
ε_9 – extraction	0	0.25	0.25	0.25	0.25	0.25	0.25
ε_Σ - total	1÷10	$4\cdot10^{-4}$	$8\cdot10^{-4}$	$2\cdot10^{-4}$	$8\cdot10^{-4}$	$8\cdot10^{-4}$	$1.5\cdot10^{-4}$
Q/A		~0.08	~0.12	~0.18	~0.1	~0.15	~0.19
E MeV/n		3.5	7	16	5	11	18
Intensity [pps]		$3\cdot10^6$			$3\cdot10^6$		

The project is based on results of experiments on determination of optimal regimes of the working matter consumption (acceleration of ^{48}Ca ions), on obtaining maximal intensities of ions by means of choosing the optimal working regime of a ECR-ion source and creating beam bunching systems, increasing the efficiency of the beam extraction from the U-400M cyclotron up to 95%, etc. These parameters of the FLNR accelerators are at the highest technical level.

Realization of the RIB project assumes some modernization of the accelerators, i.e. decreasing the magnetic induction with the aim of saving energy supplies, creation of an electrostatic beam extraction system at the U-400M cyclotron providing variation of the beam energy in a wide range (from 4 to 80 MeV/n), creation of new beam channels, etc. All this will definitely modernize the whole complex and will make it more mobile and attractive for carrying out elaborate experiments.

The following is provided by the project:

a) For producing beams of lightest nuclei: R&D of systems of collection and extraction of reaction products, and their express transportation into the ion-source of the separator. Moderate requirements concerning mass resolution ($m/\Delta m \leq 100$) are imposed upon separation of the reaction products. The system on the whole should be fast (~0.1 s), which will provide ^{8}He production without any considerable losses.

b) For producing fission fragment beams it is necessary to create several types of ion sources, operating with elements in gaseous states (isotopes of Kr and Xe), alkaline elements (Rb and Cs isotopes), as well as with mentioned above elements of the Sn, Sb and Te group. In the latter case, separation of fission fragments not only according to their masses but also according to their atomic numbers is necessary. A number of ideas for solving the problem are connected with the use of methods of thermochromatography of volatilies of these elements.

For mass-separation of fission fragments it is necessary to create a mass-spectrometer with a mass-resolution of $m/\Delta m \sim 500$ and a target set-up containing uranium in the metal or melted forms, uranium carbide with a filler etc. Specialists of the JINR working in the field of nuclear mass-spectrometry and mass separation are supposed to take part in solving the problem together with physicists of CERN (ISOLDA), CNSMS (ORSAY) and GANIL (CAEN).

c) For transportation of low-energy beams of light nuclei to a distance of about 120 meters via the existing gallery and their injection into the U-400 cyclotron, it is necessary to create (or modernize) a transport line with the vacuum level of up to 10^{-7} torr. It will demand a change in the pumping system along the whole trajectory of the beam. The transport line should be equipped with the diagnostics elements measuring beam intensities in a wide range from 10^{3} to 10^{10} 1/s. The existing systems of focusing and deflecting the beam may be substituted by more simple ones (less energy-consuming and more compact). For the transportation of a fission fragment beam it is necessary to create an additional line connecting the MT-25 microtron hall with the main transport gallery (Fig. 5).

d) On the second floor of the building where the MT-25 microtron is situated it is planned to create a laboratory for research in the field of atomic physics, nuclear spectroscopy, solid state physics, methods of applied character. The beam of fission fragments goes through the hall of ion-sources No.2. It will allow elaboration of working regimes for the ion sources for tuning single-charged ions into multi-charged ones without injecting the beam into the accelerator hall. This permits the use of the U-400 cyclotron for solving current problems.

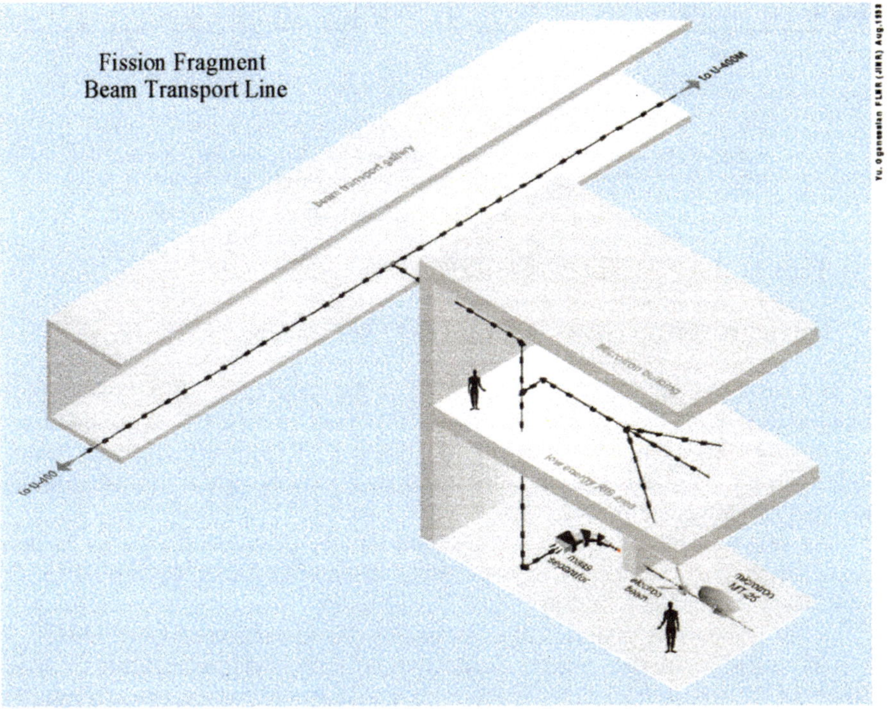

Fig 5. Location of the MT-25 set-up and the fission fragment separator in the microtron hall. The line of transportation of a low-energy beam into the gallery is shown.

The known NUPECC yields of isotopes are compared in Fig.6.

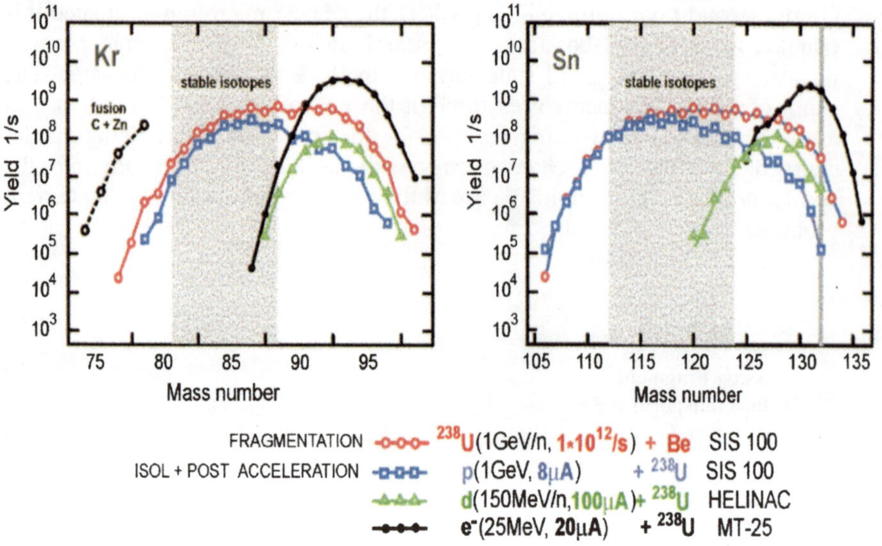

Fig 6. Projected NUPECC yields at GSI-GANIL-FLNR

The status, cost and dates of the project's realization

It is supposed that the RIB project will have the status of an all-institute project in Dubna as a basis for development of nuclear physics in the JINR. It should unite, in a possibly maximal way, interests of all the JINR scientific research teams working in the field of nuclear physics, solid state physics, physics of accelerators and other experimental lines of research.

The real cost of the project's realization will be determined after a detailed consideration and a number of measurements which are going to be carried out in immediate future at the beams of U-400M and MT-25.

In our opinion, the total sum of expenditures should not exceed 4.0-4.5 M$.

The duration of the project should not be more than three years without disturbing the strained scientific program proposed for the period extending up to 2000.

The project can be realized in stages. Every stage of the project should produce some concrete results in the form of radioactive beams, which can be immediately used for performing the experiments.

It is supposed that the project will be financed within the frames of the funds allocated to the development of the JINR theme "Nuclear physics" and the grants of the interested institutes and ministries.

The experimental basis created at the JINR during the past years as well as methods and nuclear physics set-ups of collaborating institutes of the JINR member and non-member states, will be used in the work with radioactive beams.

DEVELOPMENTS IN FISSION, FUSION, CLUSTER RADIOACTIVITY AND THE EXTENSION OF THE PERIODIC SYSTEM OF ELEMENTS

W. GREINER

Institut für Theoretische Physik, J.W. Goethe-Universität, D-60054 Frankfurt, Germany

E-mail: greiner@th.physik.uni-frankfurt.de

The extension of the periodic system into various new areas is investigated. Experiments for the synthesis of superheavy elements and the predictions of magic numbers are reviewed. Different ways of nuclear decay are discussed like cluster radioactivity, cold fission and cold multifragmentation, including the recent discovery of the triple fission of ^{252}Cf and long–living three–cluster–nuclear molecules. Furthermore, the possibility of production of clusters of hypermatter and antimatter in high energy heavy–ion collisions is envisaged. A perspective for the future is given.

1 Introduction

There are fundamental questions in science, like e. g. "how did life emerge" or "how does our brain work" and others. However, the most fundamental of those questions is "how did the world originate?". The material world has to exist before life and thinking can develop. Of particular importance are the substances themselves, i. e. the elements. On this I want to report today.

The elements existing in nature are ordered according to their atomic (chemical) properties in the **periodic system** which was developed by Mendeleev and Lothar Meyer. The heaviest element of natural origin is uranium. Its nucleus is composed of $Z = 92$ protons and a certain number of neutrons ($N = 128 - 150$). They are called the different uranium isotopes. The transuranium elements reach from neptunium ($Z = 93$) via californium ($Z = 98$) and fermium ($Z = 100$) up to Lawrencium ($Z = 103$). The heavier the elements are, the larger are their radii and their number of protons. Thus, the Coulomb repulsion in their interior increases, and they undergo fission. In other words: the transuranium elements become more instable as they get bigger.

In the late sixties the dream of the superheavy elements arose. Theoretical nuclear physicists around S.G. Nilsson (Lund) [1] and from the Frankfurt school [2,3,4] predicted that so-called closed proton and neutron shells should counteract the repelling Coulomb forces. Atomic nuclei with these special **"magic" proton and neutron numbers** and their neighbours could again

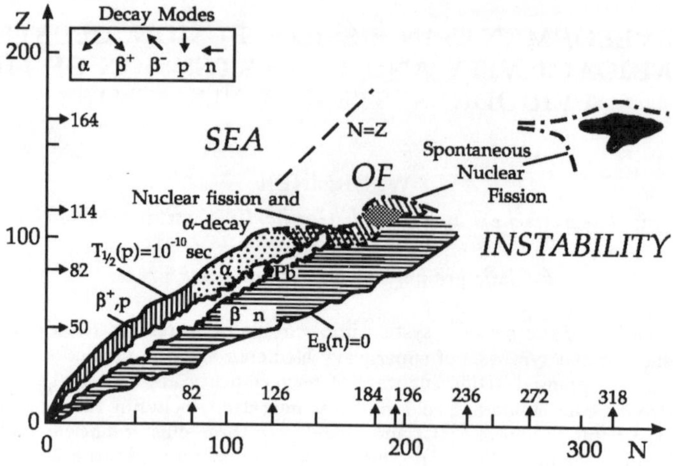

Figure 1. The periodic system of elements as conceived by the Frankfurt school in the late sixties. The islands of superheavy elements ($Z = 114$, $N = 184$, 196 and $Z = 164$, $N = 318$) are shown as dark hatched areas.

be rather stable. These magic proton (Z) and neutron (N) numbers were thought to be $Z = 114$ and $N = 184$ or 196. Typical predictions of their life times varied between seconds and many thousand years. Fig. 1 summarizes the expectations at the time. One can see the islands of superheavy elements around $Z = 114$, $N = 184$ and 196, respectively, and the one around $Z = 164$, $N = 318$.

2 Cold valleys in the potential

The important question was how to produce these superheavy nuclei. There were many attempts, but only little progress was made. It was not until the middle of the seventies that the Frankfurt school of theoretical physics together with foreign guests (R.K. Gupta (India), A. Sandulescu (Romania)) [5] theoretically understood and substantiated the concept of bombarding of double magic lead nuclei with suitable projectiles, which had been proposed intuitively by the Russian nuclear physicist Y. Oganessian [6]. The two-center shell model, which is essential for the description of fission, fusion and nuclear molecules, was developed in 1969-1972 together with my then students U. Mosel and J. Maruhn [7]. It showed that the shell structure of the two final fragments was visible far beyond the barrier into the fusioning nucleus. The

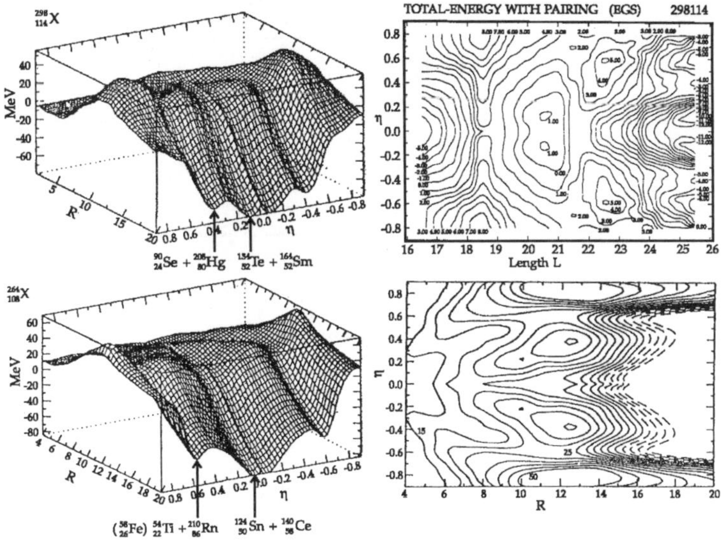

Figure 2. The collective potential energy surface of $^{264}108$ and $^{184}114$, calculated within the two center shell model by J. Maruhn et al., shows clearly the cold valleys which reach up to the barrier and beyond. Here R is the distance between the fragments and $\eta = (A_1 - A_2)/(A_1 + A_2)$ denotes the mass asymmetry: $\eta = 0$ corresponds to a symmetric, $\eta = \pm 1$ to an extremely asymmetric division of the nucleus into projectile and target. If projectile and target approach through a cold valley, they do not "constantly slide off" as it would be the case if they approach along the slopes at the sides of the valley. Constant sliding causes heating, so that the compound nucleus heats up and gets unstable. In the cold valley, on the other hand, the created heat is minimized. The colleagues from Freiburg should be familiar with that: they approach Titisee (in the Black Forest) most elegantly through the Höllental and not by climbing its slopes along the sides.

collective potential energy surfaces of heavy nuclei, as they were calculated in the framework of the two-center shell model, exhibit pronounced valleys, such that these valleys provide promising doorways to the fusion of superheavy nuclei for certain projectile-target combinations (Fig. 2). If projectile and target approach each other through those **"cold" valleys**, they get only minimally excited and the barrier which has to be overcome (fusion barrier) is lowest (as compared to neighbouring projectile-target combinations). In this way the correct projectile- and target-combinations for fusion could be predicted. Indeed, Gottfried Münzenberg and Sigurd Hofmann and their group at GSI (P. Armbruster, ...) [8] have followed this approach with the help of the SHIP mass-separator and the position sensitive detectors, which were especially de-

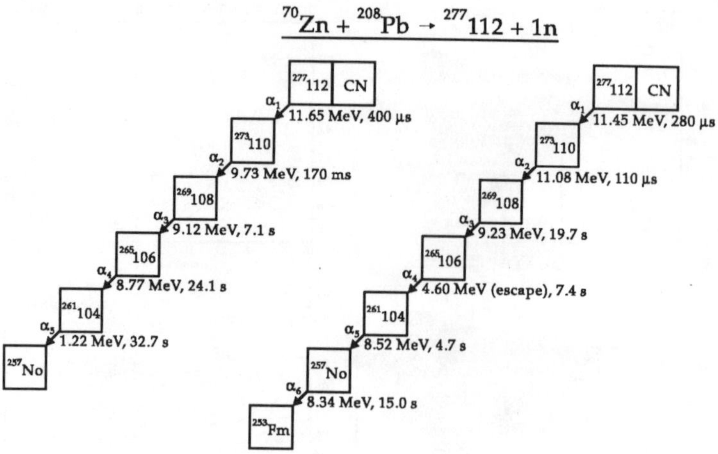

Figure 3. The fusion of element 112 with ^{70}Zn as projectile and ^{208}Pb as target nucleus has been accomplished for the first time in 1995/96 by S. Hofmann, G. Münzenberg and their collaborators. The colliding nuclei determine an entrance to a "cold valley" as predicted as early as 1976 by Gupta, Sandulescu and Greiner. The fused nucleus 112 decays successively via α emission until finally the quasi-stable nucleus ^{253}Fm is reached. The α particles as well as the final nucleus have been observed. Combined, this renders the definite proof of the existence of a $Z = 112$ nucleus.

veloped by them, they produced the pre-superheavy elements $Z = 106, 107, \ldots 112$, each of them with the theoretically predicted projectile-target combinations, and only with these. Everything else failed. This is an impressing success, which crowned the laborious construction work of many years. The last example of this success, the discovery of element 112 and its long α-decay chain, is shown in Fig. 3.

3 Shell structure in the superheavy region

Studies of the shell structure of superheavy elements in the framework of the meson field theory and the Skyrme-Hartree-Fock approach have recently shown that the magic shells in the superheavy region are very isotope dependent [9] (see Fig. 4). **According to these investigations $Z = 120$ being a magic proton number seems to be as probable as $Z = 114$.** Additionally, recent investigations in a chirally symmetric mean–field theory (see also below) result also in the prediction of these two magic numbers [31,32]. Thus, this region provides an open field of research. R.A. Gherghescu et al. have

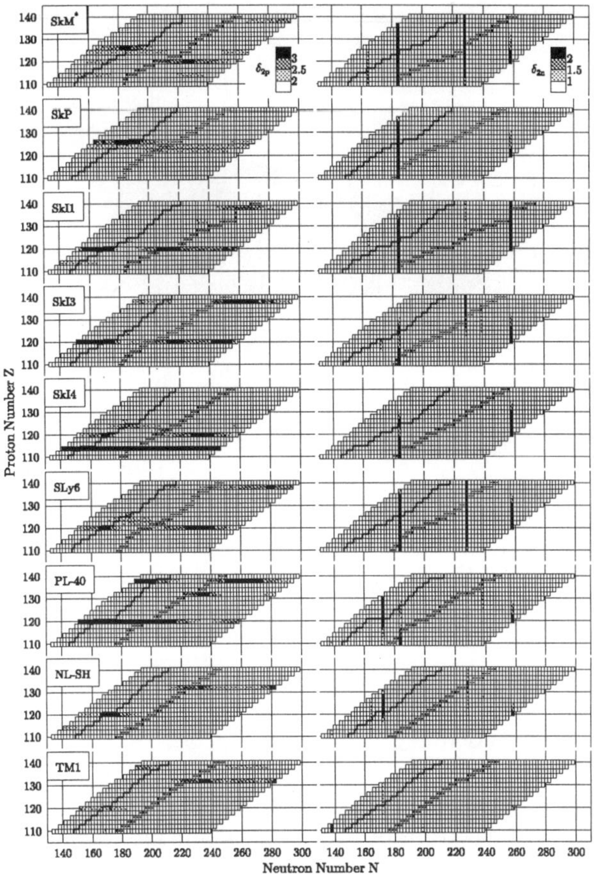

Figure 4. Grey scale plots of proton gaps (left column) and neutron gaps (right column) in the N-Z plane for spherical calculations with the forces as indicated. The assignment of scales differs for protons and neutrons, see the uppermost boxes where the scales are indicated in units of MeV. Nuclei that are stable with respect to β decay and the two-proton dripline are emphasized. The forces with parameter sets SkI4 and PL-40 reproduce the binding energy of ${}^{264}_{156}108$ (Hassium) best, i.e. $|\delta E/E| < 0.0024$. Thus one might assume that these parameter sets could give the best predictions for the superheavies. Nevertheless, it is noticed that PL-40 predicts only $Z = 120$ as a magic number while SkI4 predicts both $Z = 114$ and $Z = 120$ as magic numbers. The magicity depends — sometimes quite strongly — on the neutron number. These studies are due to Bender, Rutz, Bürvenich, Maruhn, P.G. Reinhard et al. [9].

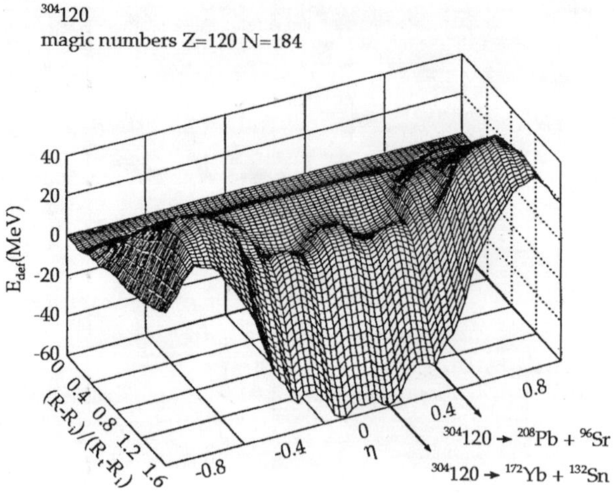

Figure 5. Potential energy surface as a function of reduced elongation $(R - R_i)/(R_t - R_i)$ and mass asymmetry η for the double magic nucleus $^{304}120$.

calculated the potential energy surface of the $Z = 120$ nucleus. It utilizes interesting isomeric and valley structures (Fig. 5).

The determination of the chemistry of superheavy elements, i. e. the calculation of the atomic structure — which is in the case of element 112 the shell structure of 112 electrons due to the Coulomb interaction of the electrons and in particular the calculation of the orbitals of the outer (valence) electrons — has been carried out as early as 1970 by B. Fricke and W. Greiner [10].

4 Asymmetric and superasymmetric fission - Cluster radioactivity

The potential energy surfaces, which are shown prototypically for $Z = 114$ in Fig. 2, contain even more remarkable information that I want to mention cursorily: if a given nucleus, e. g. uranium, undergoes fission, it moves in its potential mountains from the interior to the outside. Of course, this happens quantum mechanically. The wave function of such a nucleus, which decays

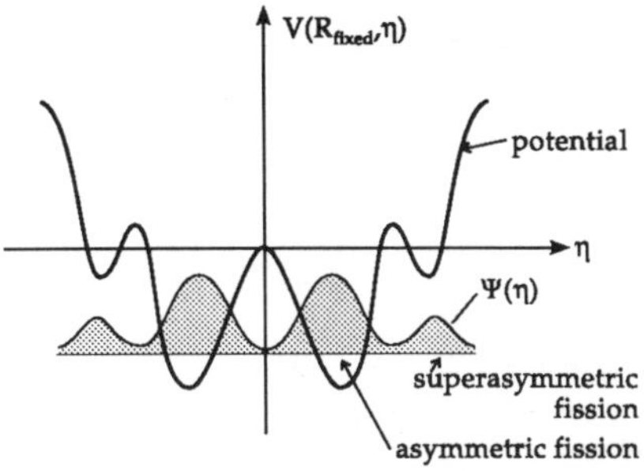

Figure 6. The collective potential as a function of the mass asymmetry $\eta = (A_1 - A_2)/(A_1 + A_2)$. A_i denotes the nucleon number in fragment i. This qualitative potential $V(R_{\text{fixed}}, \eta)$ corresponds to a cut through the potential landscape at $R = R_{\text{fixed}}$ close to the scission configuration. The wave function is drawn schematically. It has maxima where the potential is minimal and vice versa.

by tunneling through the barrier, has maxima where the potential is minimal and minima where it has maxima. This is depicted in Fig. 6.

The probability for finding a certain mass asymmetry $\eta = (A_1 - A_2)/(A_1 + A_2)$ of the fission is proportional to $\psi^*(\eta)\psi(\eta)\mathrm{d}\eta$. Generally, this is complemented by a coordinate dependent scale factor for the volume element in this (curved) space, which I omit for the sake of clarity. Now it becomes clear how the so-called **asymmetric** and **superasymmetric** fission processes come into being. They result from the enhancement of the collective wave function in the cold valleys. And that is indeed, what one observes. Fig. 7 gives an impression of it.

For a pronounced mass asymmetry ($\eta \approx 0.8 - 0.9$) there exist very narrow valleys. They are not as clearly visible in Fig. 2, but they have interesting consequences. Through these narrow valleys nuclei can emit spontaneously not only α-particles (helium nuclei) but also ^{14}C, ^{20}O, ^{24}Ne, ^{28}Mg, and other nuclei. Thus, we are lead to the **cluster radioactivity** (Poenaru, Sandulescu, Greiner [11]).

By now this process has been verified experimentally by research groups in Oxford, Moscow, Berkeley, Milan and other places. Accordingly, one has to revise what is learned in school: there are not only 3 types of radioactivity (α-, β-, γ-radioactivity), but many more. Atomic nuclei can also decay through spontaneous cluster emission (that is the "spitting out" of smaller nuclei like carbon, oxygen,...). Fig. 8 depicts some nice examples of these processes.

The knowledge of the collective potential energy surface and the collective masses $B_{ij}(R, \eta)$, all calculated within the Two-Center-Shell-Model (TCSM),

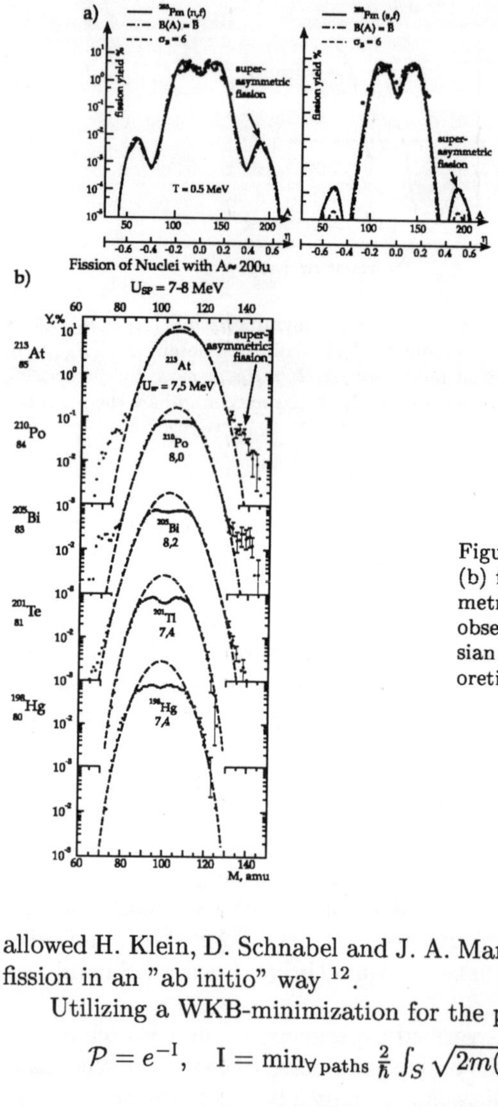

Figure 7. Asymmetric (a) and symmetric (b) fission. For the latter, also superasymmetric fission is recognizable, as it has been observed only a few years ago by the Russian physicist Itkis — just as expected theoretically.

allowed H. Klein, D. Schnabel and J. A. Maruhn to calculate lifetimes against fission in an "ab initio" way [12].

Utilizing a WKB-minimization for the penetrability integral

$$\mathcal{P} = e^{-\mathrm{I}}, \quad \mathrm{I} = \min_{\forall\,\mathrm{paths}} \frac{2}{\hbar} \int_S \sqrt{2m(V(R,\eta) - E)}\, \mathrm{d}s$$

$$= \min_{\forall\,\mathrm{paths}} \frac{2}{\hbar} \int_0^1 \sqrt{\underbrace{2mg_{ij}}_{B_{ij}}(V(x_i(t) - E)\frac{\mathrm{d}x_i}{\mathrm{d}t}\frac{\mathrm{d}x_j}{\mathrm{d}t}}\, \mathrm{d}t \qquad (1)$$

where $\mathrm{d}s^2 = g_{ij}\mathrm{d}x_i\mathrm{d}x_j$ and g_{ij} — the metric tensor — is in the well-known fashion related to the collective masses $B_{ij} = 2mg_{ij}$, one explores the minimal

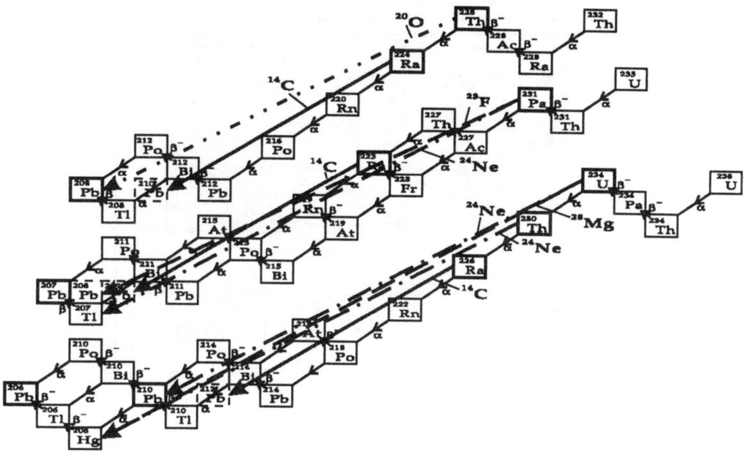

Figure 8. Cluster radioactivity of actinide nuclei. By emission of ^{14}C, ^{20}O, ... "big leaps" in the periodic system can occur, just contrary to the known α, β, γ radioactivities, which are also partly shown in the figure.

paths from the nuclear ground state configuration through the multidimensional fission barrier (see Fig. 9).

The thus obtained fission half lives are depicted in the lower part of Fig. 9. Their distribution as a function of the fragment mass A_2 resembles quite well the asymmetric mass distribution. Cluster radioactive decays correspond to the broad peaks around A_2 = 20, 30 (200, 210). The confrontation of the calculated fission half lives with experiments is depicted in Fig. 10. One notices "nearly quantitative" agreement over 20 orders of magnitude, which is – for an ab-initio calculation – remarkable!

Finally, in Fig. 11, we compare the lifetime calculation discussed above with one based on the Preformation Cluster Model by D. Poenaru et al. [13] and recognize an amazing degree of similarity and agreement.

The systematics for the average total kinetic energy release for spontaneously fissioning isotopes of Cm and No is following the Viola trend, but ^{258}Fm and ^{259}Fm are clearly outside. The situation is similar also for ^{260}Md, where two components of fission products (one with lower and one with higher kinetic energy) were observed by Hulet et al. [14]. The explanation of these interesting observations lies in two different paths through the collective potential. One reaches the scission point in a stretched neck position (i.e. at a lower point of the Coulomb barrier - thus lower kinetic energy for the fragments)

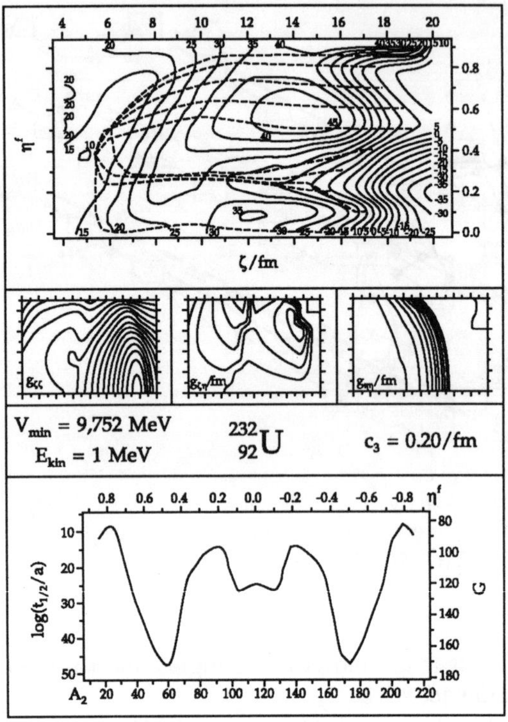

Figure 9. The upper part of the figure shows the collective potential energy surface for $^{232}_{92}U$ with the ground state position and various fission paths through the barrier. The middle part shows various collective masses, all calculated in the TCSM. In the lower part the calculated fission half lives are depicted.

while the other one reaches the scission point practically in a touching-spheres-position (i.e. higher up on the Coulomb barrier and therefore highly energetic fragments) [15]. The latter process is cold fission; i.e., the fission fragments are in or close to their ground state (cold fragments) and all the available energy is released as kinetic energy. Cold fission is, in fact, typically a cluster decay. The side-by-side occurrence of cold and normal (hot) fission has been named bi-modal fission [15].

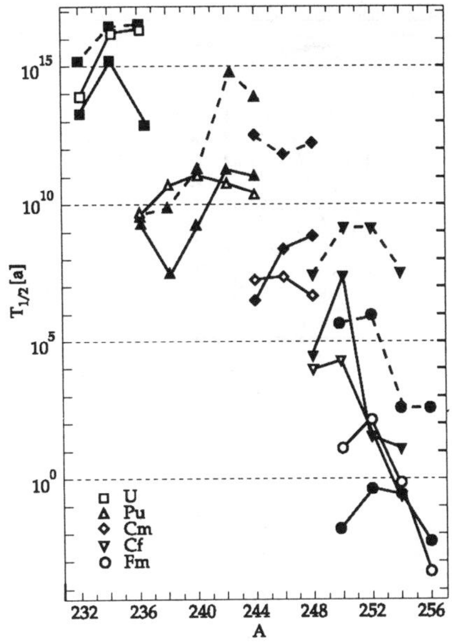

Figure 10. Fission half lives for various isotopes of $Z = 92$ (), $Z = 94$ (Δ), $Z = 96$ (), $Z = 98$ (∇) and $Z = 100$ ($\circ$). The open symbols represent the experimental values. The full symbols with full and dashed lines correspond to calculations with different choices of the barrier parameter in the Two Center Shell Model ($c_3 \approx 0.2$ and 0.1 respectively).

5 Fission observed with the Gamma-sphere: Long living nuclear molecules

There has now been put forward a fantastic idea [16] in order to study cold fission (cluster decays) and other exotic fission processes (ternary-, multiple fission in general) very elegantly: By measuring with e. g. the Gamma–sphere characteristic γ-transitions of individual fragments in coincidence, one can identify all these processes in a direct and simple way (Fig. 12). First confirmation of this method by J. Hamilton, V. Ramaya et al. worked out excellently [17]. This method has high potential for revolutionizing fission physics! With some physical intuition one can imagine that **triple** - and **quadruple** fission processes and even the process of **cold multifragmentation** will be discovered - absolutely fascinating! We have thus seen that fission physics (cold fission, cluster radioactivity, ...) and fusion physics (especially the production of superheavy elements) are intimately connected.
Indeed, very recently, triple fission of

$$
\begin{aligned}
{}^{252}\mathrm{Cf} &\rightarrow {}^{146}\mathrm{Ba} + {}^{96}\mathrm{Sr} + {}^{10}\mathrm{Be} \\
&\rightarrow {}^{112}\mathrm{Ry} + {}^{130}\mathrm{Sn} + {}^{10}\mathrm{Be} \\
&\rightarrow \ldots
\end{aligned}
$$

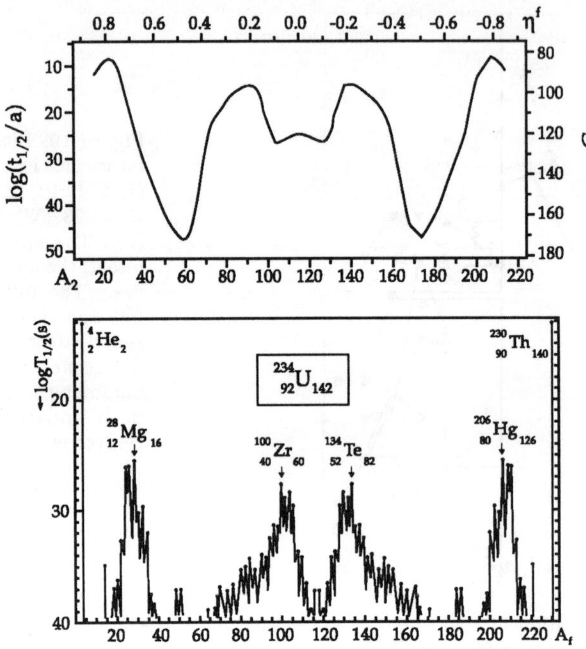

Figure 11. Comparison of the fission half lives calculated in the fission model (upper figure – see also Fig. 8) and in the Preformation Cluster Model [13]. In both models the deformation of the fission fragments is not included completely.

has been identified by measuring the various γ–transitions of these nuclei in coincidence (see Fig. 13). Even though the statistical evidence for the ^{10}Be line is small ($\approx$ 50 events) the various coincidences seem to proof that spontaneous triple fission out of the ground state of ^{252}Cf with the heavy cluster ^{10}Be as a third fragment exists. Also other triple fragmentations can be expected. One of those is also denoted above. In fact, there are first indications, that this break–up is also observed. The most amazing observation is, however, the following: The cross coincidences seem to suggest that one deals with a simultaneous three–body breakup and not with a cascade process. For that one expects a configuration as shown in Fig. 14. Consequently the ^{10}Be will obtain kinetic energy while running down the combined Coulomb barrier of ^{146}Ba and ^{96}Sr and, therefore, the 3368 keV line of ^{10}Be should be Doppler-broadened. Amazingly, however, it is not and, moreover, it seems to be about 6 keV smaller than the free ^{10}Be γ–transition. If this turns out to be true, the

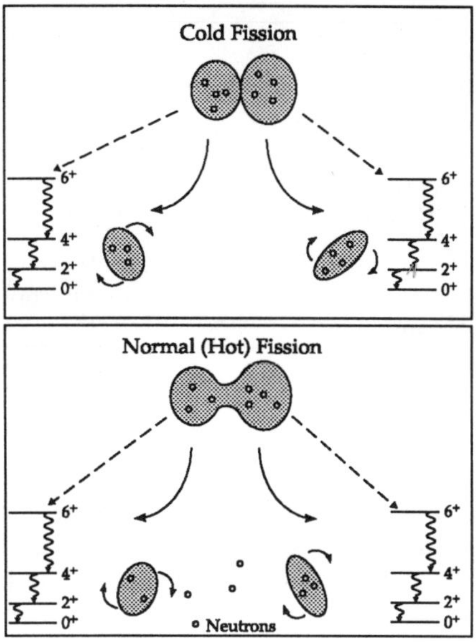

Figure 12. Ilustration of cold and hot (normal) fission identification through multiple γ-coincidences of photons from the fragments. The photons serve to identify the fragments.

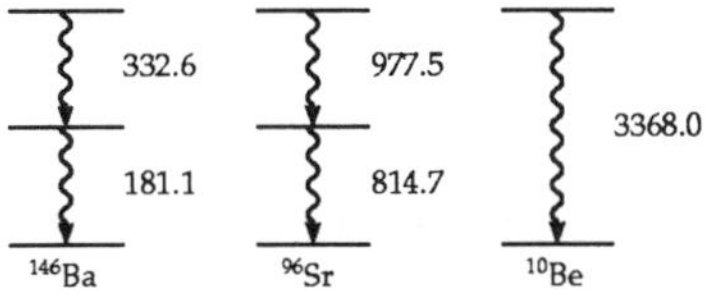

Figure 13. The γ-transitions of the three fission products of ^{252}Cf measured in coincidence. Various combinations of the coincidences were studied.

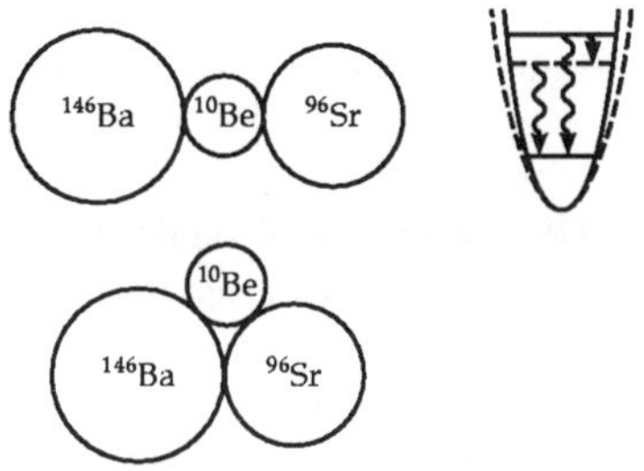

Figure 14. Typical linear cluster configuration leading to triple fission of ^{252}Cf. The influence of both clusters leads to a softening of the ^{10}Be potential and thus to a somewhat smaller transition energy. Recent theoretical investigations indicate that the axial symmetry of this configuration is broken (lower left hand figure).

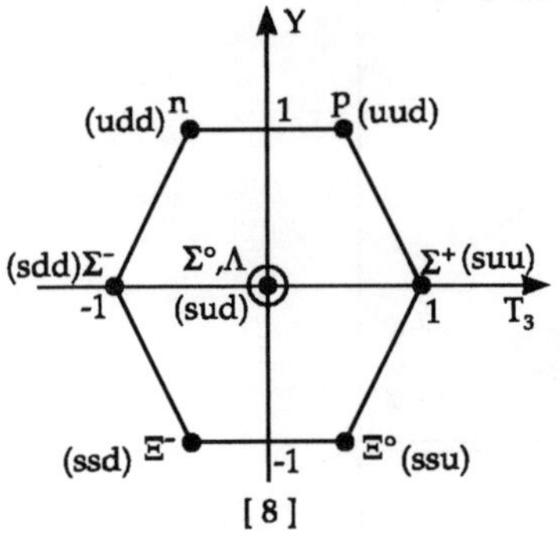

Figure 15. Important baryons are ordered in this octet. The quark content is depicted. Protons (p) and neutrons (n), most important for our known world, contain only u and d quarks. Hyperons contain also an s quark. The number of s quarks is a measure for the strangeness.

only explanation will be that the gamma is emitted while the nuclear molecule of the type shown in Fig. 14. holds. The molecule has to live longer than about 10^{-12} sec. The nuclear forces from the ^{146}Bo and ^{96}Sn cluster to the left and right from ^{10}Be lead to a softening of its potential and therefore to a somewhat smaller transition energy. Thus, if experimental results hold, one has discovered long living ($\approx 10^{-12}$ sec) complex nuclear molecules. This is fantastic! Of course, I do immediately wonder whether such configurations do also exist in e.g. U + Cm soft encounters directly at the Coulomb barrier. This would have tremendous importance for the observation of the spontaneous vacuum decay [19], for which "sticking giant molecules" with a lifetime of the order of 10^{-19} sec are needed. The nuclear physics of such heavy ion collisions at the Coulomb barrier (giant nuclear molecules) should indeed be investigated!
The "cold valleys" in the collective potential energy surface are basic for understanding this exciting area of nuclear physics! It is a master example for understanding the **structure of elementary matter**, which is so important for other fields, especially astrophysics, but even more so for enriching our "Weltbild", i.e. the status of our understanding of the world around us.

6 Extension of the periodic system into the sections of hyper- and antimatter

Nuclei that are found in nature consist of nucleons (protons and neutrons) which themselves are made of u (up) and d (down) quarks. However, there also exist s (strange) quarks and even heavier flavors, called charm, bottom, top. The latter has just recently been discovered. Let us stick to the s quarks. They are found in the "strange" relatives of the nucleons, the so-

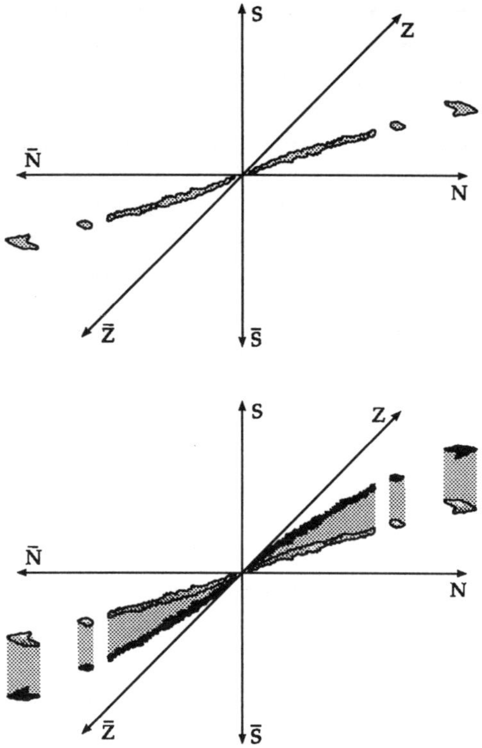

Figure 16. The extension of the periodic system into the sectors of strangeness (S, $\bar{S}$) and antimatter ($\bar{Z}$, $\bar{N}$). The stable valley winds out of the known proton (Z) and neutron (N) plane into the S and $\bar{S}$ sector, respectively. The same can be observed for the antimatter sector. In the upper part of the figure only the stable valley in the usual proton (Z) and neutron (N) plane is plotted, however, extended into the sector of antiprotons and antineutrons. In the second part of the figure it has been indicated, how the stable valley winds out of the Z-N-plane into the strangeness sector.

called hyperons (Λ, Σ, Ξ, Ω). The Λ-particle, e. g., consists of one u, d and s quark, the Ξ-particle even of an u and two s quarks, while the Ω (sss) contains strange quarks only. Fig. 15 gives an overview of the baryons, which are of interest here, and their quark content.

If such a hyperon is taken up by a nucleus, a **hyper-nucleus** is created. Hyper-nuclei with one hyperon have been known for 20 years now, and were extensively studied by B. Povh (Heidelberg) [20]. Several years ago, Carsten Greiner, Jürgen Schaffner and Horst Stöcker [21] theoretically investigated nuclei with many hyperons, **hypermatter**, and found that the binding energy per baryon of strange matter is in many cases even higher than that of ordinary matter (composed only of u and d quarks). This leads to the idea of extending the periodic system of elements in the direction of strangeness.

One can also ask for the possibility of building atomic nuclei out of **antimatter**, that means searching e. g. for anti-helium, anti-carbon, anti-oxygen. Fig. 16 depicts this idea. Due to the charge conjugation symmetry antinuclei should have the same magic numbers and the same spectra as ordinary nuclei. However, as soon as they get in touch with ordinary matter, they annihilate with it and the system explodes.

Now the important question arises how these strange matter and antimatter clusters can be produced. First, one thinks of collisions of heavy nuclei, e.g. lead on lead, at high energies (energy per nucleon $\geq$ 200 GeV). Calculations with the URQMD-model of the Frankfurt school show that through **nuclear shock waves** [22–24] nuclear matter gets compressed to 5–10 times of its usual value, $\rho_0 \approx 0.17\,\text{fm}^3$, and heated up to temperatures of $kT \approx 200$ MeV. As a consequence about 10000 pions, 100 Λ's, 40 Σ's and Ξ's and about as many antiprotons and many other particles are created in a single collision. It seems conceivable that it is possible in such a scenario for some Λ's to get captured by a nuclear cluster. This happens indeed rather frequently for one or two Λ-particles; however, more of them get built into nuclei with rapidly decreasing probability only. This is due to the low probability for finding the right conditions for such a capture in the phase space of the particles: the numerous particles travel with every possible momenta (velocities) in all directions. The chances for hyperons and antibaryons to meet gets rapidly worse with increasing number. In order to produce multi-Λ-nuclei and antimatter nuclei, one has to look for a different source.

In the framework of meson field theory the energy spectrum of baryons has a peculiar structure, depicted in Fig. 17. It consists of an upper and a lower continuum, as it is known from the electrons (see e. g. [19]). Of special interest in the case of the baryon spectrum is the potential well, built of the scalar and the vector potential, which rises from the lower continuum. It is known since P.A.M. Dirac (1930) that the negative energy states of the lower continuum have to be occupied by particles (electrons or, in our case, baryons). Otherwise our world would be unstable, because the "ordinary" particles are found in the upper states which can decay through the emission of photons into lower lying states. However, if the "underworld" is occupied, the Pauli-principle will prevent this decay. Holes in the occupied "underworld" (Dirac sea) are antiparticles.

The occupied states of this underworld including up to 40000 occupied

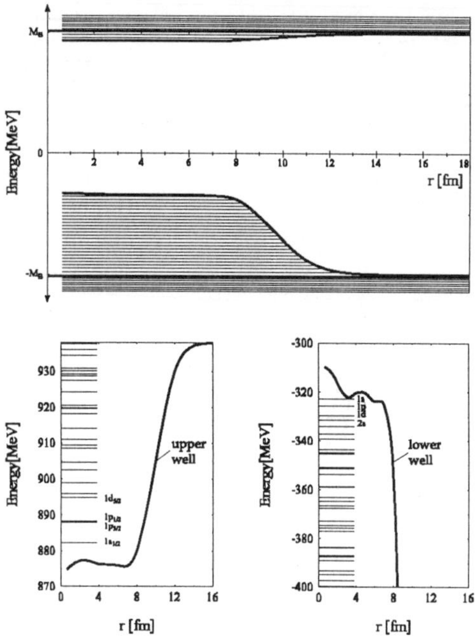

Figure 17. Baryon spectrum in a nucleus. Below the positive energy continuum exists the potential well of real nucleons. It has a depth of 50-60 MeV and shows the correct shell structure. The shell model of nuclei is realized here. However, from the negative continuum another potential well arises, in which about 40000 bound particles are found, belonging to the vacuum. A part of the shell structure of the upper well and the lower (vacuum) well is depicted in the lower figures.

bound states of the lower potential well represent the **vacuum**. The peculiarity of this strongly correlated vacuum structure in the region of atomic nuclei is that — depending on the size of the nucleus — more than 20000 up to 40000 (occupied) bound nucleon states contribute to this polarization effect. Obviously, we are dealing here with a **highly correlated vacuum**. A pronounced shell structure can be recognized [25,26]. Holes in these states have to be interpreted as bound antinucleons (antiprotons, antineutrons). If the primary nuclear density rises due to compression, the lower well increases while the upper decreases and soon is converted into a repulsive barrier (Fig. 18). This compression of nuclear matter can only be carried out in relativistic nucleus-nucleus collision with the help of shock waves, which have been proposed by the Frankfurt school [22,23] and which have since then been confirmed

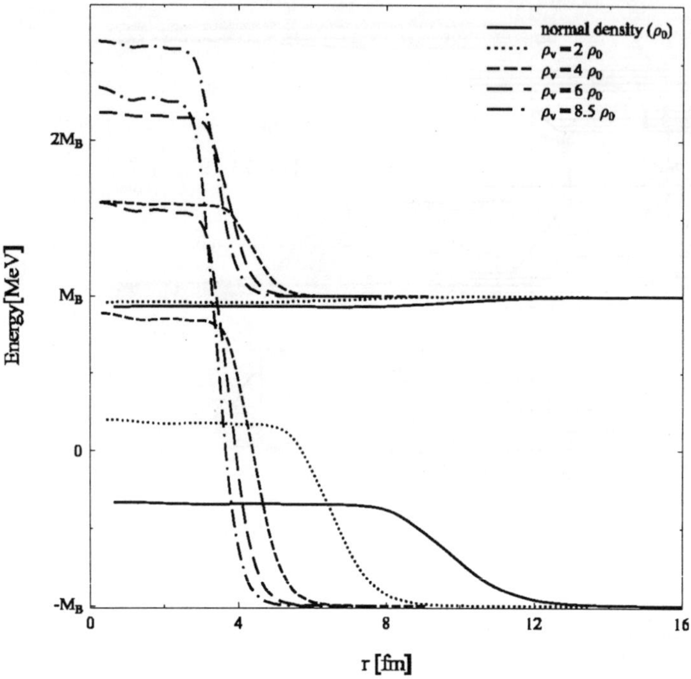

Figure 18. The lower well rises strongly with increasing primary nucleon density, and even gets supercritical (spontaneous nucleon emission and creation of bound antinucleons). Supercriticality denotes the situation, when the lower well enters the upper continuum.

extensively (for references see e.g. [27]). These **nuclear shock waves** are accompanied by heating of the nuclear matter. Indeed, density and temperature are intimately coupled in terms of the hydrodynamic Rankine-Hugoniot-equations. Heating as well as the violent dynamics cause the creation of many holes in the very deep (measured from $-M_Bc^2$) vacuum well. These numerous bound holes resemble antimatter clusters which are bound in the medium; their wave functions have large overlap with antimatter clusters. When the primary matter density decreases during the expansion stage of the heavy ion collision, the potential wells, in particular the lower one, disappear.

The bound antinucleons are then pulled down into the (lower) continuum. In this way antimatter clusters may be set free. Of course, a large part of the antimatter will annihilate on ordinary matter present in the course of the expansion. However, it is important that this mechanism for the production of antimatter clusters out of the highly correlated vacuum does not proceed

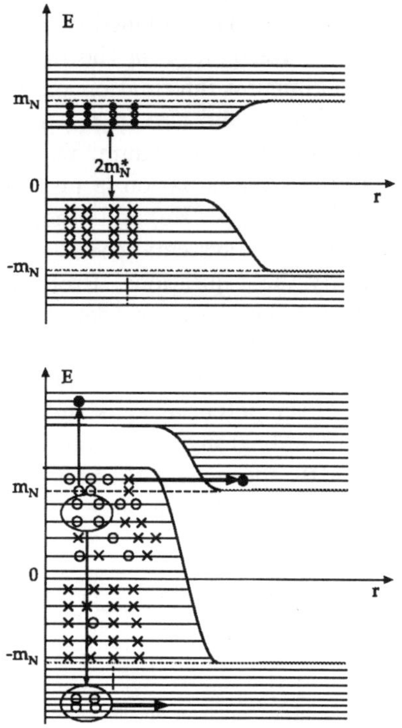

Figure 19.
Due to the high temperature and the violent dynamics, many bound holes (antinucleon clusters) are created in the highly correlated vacuum, which can be set free during the expansion stage into the lower continuum. In this way, antimatter clusters can be produced directly from the vacuum. The horizontal arrow in the lower part of the figure denotes the spontaneous creation of baryon-antibaryon pairs, while the antibaryons occupy bound states in the lower potential well. Such a situation, where the lower potential well reaches into the upper continuum, is called supercritical. Four of the bound holes states (bound antinucleons) are encircled to illustrate a "quasi-antihelium" formed. It may be set free (driven into the lower continuum) by the violent nuclear dynamics.

via the phase space. The required coalescence of many particles in phase space suppresses the production of clusters, while it is favoured by the direct production out of the highly correlated vacuum. In a certain sense, the highly correlated vacuum is a kind of cluster vacuum (vacuum with cluster structure). The shell structure of the vacuum levels (see Fig. 17) supports this latter suggestion. Fig. 19illustrates this idea.

The mechanism is similar for the production of multi-hyper nuclei (Λ, Σ, Ξ, Ω). Meson field theory predicts also for the Λ energy spectrum at finite primary nucleon density the existence of upper and lower wells. The lower well belongs to the vacuum and is fully occupied by Λ's.

Dynamics and temperature then induce transitions ($\Lambda\bar{\Lambda}$ creation) and deposit many Λ's in the upper well. These numerous bound Λ's are sitting close to the primary baryons: in a certain sense a giant multi-Λ hypernucleus has been created. When the system disintegrates (expansion stage) the Λ's distribute over the nucleon clusters (which are most abundant in peripheral collisions). In this way multi-Λ hypernuclei can be formed.

Of course this vision has to be worked out and probably refined in many respects. This means much more and thorough investigation in the future. It is particularly important to gain more experimental information on the properties of the lower well by (e, e' p) or (e, e' p p') and also ($\bar{p}_c p_b$, $p_c \bar{p}_b$) reactions ($\bar{p}_c$ denotes an incident antiproton from the continuum, p_b is a proton in a bound state; for the reaction products the situation is just the opposite) [28]. Also the reaction (p, p' d), (p, p' ^{3}He), (p, p' ^{4}He) and others of similar type need to be investigated in this context. The systematic scattering of antiprotons on nuclei can contribute to clarify these questions. Problems of the meson field theory (e. g. Landau poles) can then be reconsidered. An effective meson field theory has to be constructed. Various effective theories, e. g. of Walecka-type on the one side and theories with chiral invariance on the other side, seem to give different strengths of the potential wells and also different dependence on the baryon density [29]. The Lagrangians of the Dürr-Teller-Walecka-type and of the chirally symmetric mean field theories look quite differently. We exhibit them — without further discussion — in the following equations:

$$\mathcal{L} = \mathcal{L}_{kin} + \mathcal{L}_{BM} + \mathcal{L}_{vec} + \mathcal{L}_0 + \mathcal{L}_{SB}$$

Non-chiral Lagrangian:

$$\begin{aligned}
\mathcal{L}_{kin} &= \frac{1}{2}\partial_\mu s \partial^\mu s + \frac{1}{2}\partial_\mu z \partial^\mu z - \frac{1}{4}B_{\mu\nu}B^{\mu\nu} - \frac{1}{4}G_{\mu\nu}G^{\mu\nu} - \frac{1}{4}F_{\mu\nu}F^{\mu\nu} \\
\mathcal{L}_{BM} &= \sum_B \overline{\psi}_B [i\gamma_\mu \partial^\mu - g_{\omega B}\gamma_\mu \omega^\mu - g_{\phi B}\gamma_\mu \phi^\mu - g_{\rho B}\gamma_\mu \vec{\tau}_B \vec{\rho}^\mu \\
&\quad - \frac{1}{2} e\gamma_\mu (1+\tau_B) A^\mu - m_B^*]\psi_B \\
\mathcal{L}_{vec} &= \frac{1}{2}m_\omega^2 \omega_\mu \omega^\mu + \frac{1}{2}m_\rho^2 \rho_\mu \rho^\mu + \frac{1}{2}m_\phi^2 \phi_\mu \phi^\mu \\
\mathcal{L}_0 &= -\frac{1}{2}m_s^2 s^2 - \frac{1}{2}m_z^2 z^2 - \frac{1}{3}bs^3 - \frac{1}{4}cs^4
\end{aligned}$$

Chiral Lagrangian:

$$\begin{aligned}
\mathcal{L}_{kin} &= \frac{1}{2}\partial_\mu \sigma \partial^\mu \sigma + \frac{1}{2}\partial_\mu \zeta \partial^\mu \zeta + \frac{1}{2}\partial_\mu \chi \partial^\mu \chi - \frac{1}{4}B_{\mu\nu}B^{\mu\nu} \\
&\quad - \frac{1}{4}G_{\mu\nu}G^{\mu\nu} - \frac{1}{4}F_{\mu\nu}F^{\mu\nu} \\
\mathcal{L}_{BM} &= \sum_B \overline{\psi}_B [i\gamma_\mu \partial^\mu - g_{\omega B}\gamma_\mu \omega^\mu - g_{\phi B}\gamma_\mu \phi^\mu - g_{\rho B}\gamma_\mu \vec{\tau}_B \vec{\rho}^\mu \\
&\quad - \frac{1}{2} e\gamma_\mu (1+\tau_B) A^\mu - m_B^*]\psi_B
\end{aligned}$$

$$\mathcal{L}_{vec} = \frac{1}{2}m_\omega^2\frac{\chi^2}{\chi_0^2}\omega_\mu\omega^\mu + \frac{1}{2}m_\rho^2\frac{\chi^2}{\chi_0^2}\rho_\mu\rho^\mu + \frac{1}{2}m_\phi^2\frac{\chi^2}{\chi_0^2}\phi_\mu\phi^\mu$$

$$\mathcal{L}_0 = -\frac{1}{2}k_0\chi^2(\sigma^2+\zeta^2) + k_1(\sigma^2+\zeta^2)^2 + k_2(\frac{\sigma^4}{2}+\zeta^4) + k_3\chi\sigma^2\zeta$$

$$- k_4\chi^4 + \frac{1}{4}\chi^4\ln\frac{\chi^4}{\chi_0^4} + \frac{\delta}{3}\ln\frac{\sigma^2\zeta}{\sigma_0^2\zeta_0^2}$$

$$\mathcal{L}_{SB} = -\left(\frac{\chi}{\chi_0}\right)^2\left[m_\pi^2 f_\pi\sigma + (\sqrt{2}m_K^2 f_K - \frac{1}{\sqrt{2}}m_\pi^2 f_\pi)\zeta\right]$$

The non-chiral model contains the scalar-isoscalar field s and its strange counterpart z, the vector-isoscalar fields ω_μ and ϕ_μ, and the the ρ-meson ρ_μ as well as the photon A_μ. For more details see [29]. In contrast to the non-chiral model, the $SU(3)_L \times SU(3)_R$ Lagrangian contains the dilation field χ introduced to mimic the trace anomaly of QCD in an effective Lagrangian at tree level (For an explanation of the chiral model see [29,31]).

The connection of the chiral Lagrangian with the Walecka-type can be established by the substitution $\sigma = \sigma_0 - s$ (and similarly for the strange condensate ζ). Then, e.g. the difference in the definition of the effective nucleon mass in both models (non-chiral:$m_N^* = m_N - g_s s$, chiral:$m_N^* = g_s\sigma$) can be removed, yielding:

$$m_N^* = g_s\sigma_0 - g_s s \equiv m_N - g_s s \tag{2}$$

for the nucleon mass in the chiral model.

Nevertheless, if the parameters in both cases are adjusted such that ordinary nuclei (binding energies, radii, shell structure,...) and properties of infinite nuclear matter (equilibrium density, compression constant K, binding energy) are well reproduced, the prediction of both effective Lagrangians for the dependence of the properties of the correlated vacuum on density and temperature is remarkably different. This is illustrated to some extent in Fig. 20. Accordingly, the chirally symmetric meson field theory predicts much higher primary densities (and temperatures) until the effects of the correlated vacuum are strong enough so that the mechanisms described here become effective. In other words, according to chirally symmetric meson field theories the antimatter-cluster-production and multi-hypermatter-cluster production out of the highly correlated vacuum takes place at considerably higher heavy ion energies as compared to the predictions of the Dürr-Teller-Walecka-type meson field theories. This in itself is a most interesting, quasi-fundamental question to be clarified. Moreover, the question of the nucleonic substructure (form factors, quarks, gluons) and its influence on the highly correlated

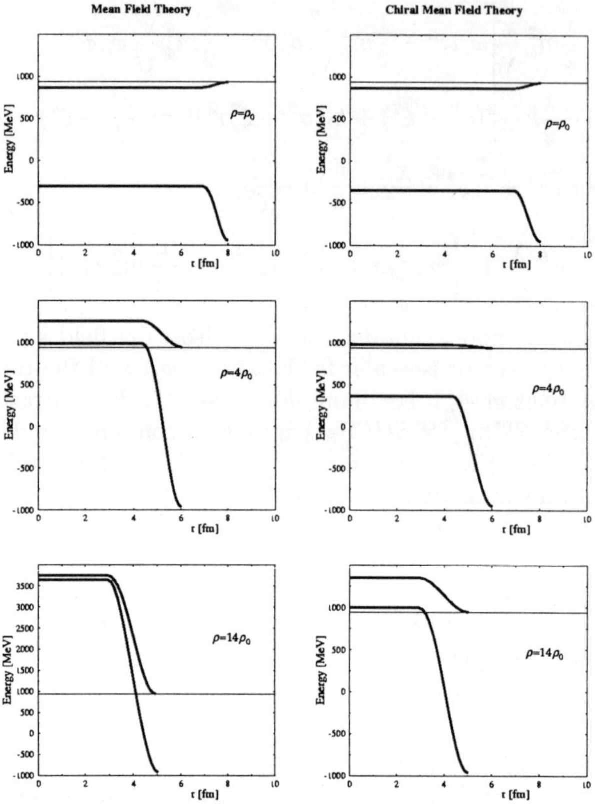

Figure 20. The potential structure of the shell model and the vacuum for various primary densities $\rho = \rho_0$, $4\rho_0$, $14\rho_0$. At left the predictions of ordinary Dürr-Teller-Walecka-type theories are shown; at right those for a chirally symmetric meson field theory as developed by P. Papazoglu [29,31].

vacuum structure has to be studied. The nucleons are possibly strongly polarized in the correlated vacuum: the Δ resonance correlations in the vacuum are probably important. Is this highly correlated vacuum state, especially during the compression, a preliminary stage to the quark-gluon cluster plasma? To which extent is it similar or perhaps even identical with it? It is well known for more than 10 years that meson field theories predict a phase transition qualitatively and quantitatively similar to that of the quark-gluon plasma [30] — see Fig. 21.

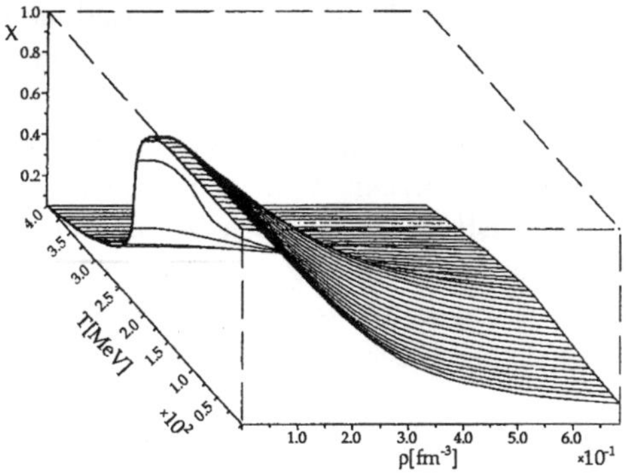

Figure 21. The strong phase transition inherent in Dürr-Teller-Walecka-type meson field theories, as predicted by J. Theis et al. [30]. Note that there is a first order transition along the ρ-axis (i.e. with density), but a simple transition along the temperature T-axis. Note also that this is very similar to the phase transition obtained recently from the Nambu-Jona-Lasinio-approximation of QCD [33].

7 Concluding remarks - outlook

The extension of the periodic system into the sectors hypermatter (strangeness) and antimatter is of general and astrophysical importance. Indeed, microseconds after the big bang the new dimensions of the periodic system, we have touched upon, certainly have been populated in the course of the baryo- and nucleo-genesis. Of course, for the creation of the universe, even higher dimensional extensions (charm, bottom, top) come into play, which we did not pursue here. It is an open question, how the de-population (the decay) of these sectors influences the distribution of elements of our world today. Our conception of the world will certainly gain a lot through the clarification of these questions.

For the Gesellschaft für Schwerionenforschung (GSI), which I helped initiating in the sixties, the questions raised here could point to the way ahead. Working groups have been instructed by the board of directors of GSI, to think about the future of the laboratory. On that occasion, very concrete (almost too concrete) suggestions are discussed — as far as it has been presented to the public. What is missing, as it seems, is a **vision on a long term basis**. The ideas proposed here, the verification of which will need the **commitment for 2–4 decades of research**, could be such a vision with considerable attraction for the best young physicists. The new dimensions of the periodic system made of hyper- and antimatter cannot be examined in the "stand-by" mode at CERN (Geneva); a dedicated facility is necessary for this field of research, which can in future serve as a home for the universities. The GSI — which has unfortunately become much too self-sufficient — could be such a home for new generations of physicists, who are interested in the **structure of elementary matter**. GSI would then not develop just into a detector laboratory for CERN, and as such become obsolete. I can already see the enthusiasm in the eyes of young scientists, when I unfold these ideas to them — similarly as it was 30 years ago, when the nuclear physicists in the state of Hessen initiated the construction of GSI.

Acknowledgments

I am grateful to Dipl. Phys. T. Bürvenich for helping preparing the manuscript.

References

1. S.G. Nilsson et al. *Phys. Lett. B* **29**, 458 (1969); *Nucl. Phys. A* **131** (1969) 1; *Nucl. Phys. A* **115** (1968) 545
2. U. Mosel, B. Fink, W. Greiner: *Contribution to "Memorandum Hessischer Kernphysiker"* Darmstadt, Frankfurt, Marburg (1966)
3. U. Mosel, W. Greiner *Z. Phys* **217**, 256 (1968); *ibid.* **222** (1969) 261
4. a) J. Grumann, U. Mosel, B. Fink, W. Greiner: *Z. Phys.* **228** (1969) 371; b) J. Grumann, Th. Morovic, W. Greiner: *Z. Naturforschung* **26a** (1971) 643
5. A. Sandulescu, R.K. Gupta, W. Scheid, W. Greiner *Phys. Lett. B* **60**, 225 (1976);
   R.K. Gupta, A. Sandulescu, W. Greiner *Z. Naturforschung* **32a**, 704 (1977);
   R.K. Gupta, A.Sandulescu, W. Greiner *Phys. Lett. B* **64**, 257 (1977);
   R.K. Gupta, C. Parrulescu, A. Sandulescu, W. Greiner *Z. Phys. A* **283**, 217 (1977)
6. G.M. Ter-Akopian et al. *Nucl. Phys. A* **255**, 509 (1975);
   Yu.Ts. Oganessian et al. *Nucl. Phys. A 239*, 353 (1975) and 157
7. D. Scharnweber, U. Mosel, W. Greiner *Phys. Rev. Lett.* **24**, 601 (1970);
   U. Mosel, J. Maruhn, W. Greiner *Phys. Lett. B* **34**, 587 (1971)
8. G. Münzenberg et al. *Z. Phys. A* **309**, 89 (1992);
   S. Hofmann et al. *Z. Phys A* **350**, 277 (1995) and 288
9. K. Rutz, M. Bender, T. Bürvenich, T. Schilling, P.-G. Reinhard, J.A. Maruhn, W. Greiner *Phys. Rev. C* **56**, 238 (1997)

10. B. Fricke, W. Greiner *Phys. Lett. B* **30**, 317 (1969);
B. Fricke, W. Greiner, J.T. Waber *Theoret. Chimica Acta (Berlin)* **21**, 235 (1971)
11. A. Sandulescu, D.N. Poenaru, W. Greiner *Sov. J. Part. Nucl.* **11**, 528 (1980)
12. H. Klein: *Thesis.* Inst. für Theoret. Physik, J.W. Goethe-Universität Frankfurt a. M. 1992;
D. Schnabel: *Thesis.* Inst. für Theoret. Physik, J.W. Goethe-Universität Frankfurt a.M. 1992
13. D. Poenaru, J.A. Maruhn, W. Greiner, M. Ivascu, D. Mazilu, R. Gherghescu *Z. Phys. A* **328**, 309 (1987); *ibid.* **332** (1989) 291
14. E.K. Hulet et al. *Phys. Rev. Lett.* **56**, 313 (1986)
15. K. Depta, W. Greiner, J. Maruhn, H.J. Wang, A. Sandulescu, R. Hermann *Intern. J. of Mod. Phys. A* **5**, 3901 (1990);
K. Depta, R. Hermann, J.A. Maruhn, W. Greiner: *in* Dynamics of Collective Phenomena *(Ed. P. David).* World Sci., Singapore 1987, p. 29
16. A. Sandulescu, W. Greiner: *in discussions at Frankfurt with J. Hamilton* (1992/1993)
17. J.H. Hamilton, A.V. Ramaya et al. *J. Phys. G* **20**, L85 (1994)
18. E.K. Hulet et al. *Phys Rev C* **40**, 770 (1989)
19. W. Greiner, B. Müller, J. Rafelski: *QED of Strong Fields.* Springer Verlag, Heidelberg 1985;
For a more recent review see W. Greiner, J. Reinhardt: *Supercritical Fields in Heavy–Ion Physics*, Proc. 15th Adv. ICFA Beam Dynamics Workshop Quantum Aspects of Beam Physics, World Scientific 1998
20. B. Povh *Rep. Progr. Phys.* **39**, 823 (1976);
Ann. Rev. Nucl. Part. Sci. **28** (1978) 1; *Nucl. Phys. A* **335** (1980) 233; *Progr. Part. Nucl. Phys.* **5** (1981) 245; *Phys. Blätter* **40** (1984) 315
21. J. Schaffner, C. Greiner, H. Stöcker *Phys. Rev. C* **45**, 322 (1992); *Nucl. Phys. B* **24B** (1991) 246;
J. Schaffner, C.B. Dover, A. Gal, D.J. Millener, C. Greiner, H. Stöcker: *Ann. Phys.* **235** (1994) 35
22. W. Scheid, W. Greiner *Ann. Phys.* **48**, 493 (1968); *Z. Phys.* **226** (1969) 364
23. W. Scheid, H. Müller, W. Greiner *Phys. Rev. Lett.* **13**, 741 (1974)
24. H. Stöcker, W. Greiner, W. Scheid *Z. Phys. A* **286**, 121 (1978)
25. I. Mishustin, L.M. Satarov, J. Schaffner, H. Stöcker, W.Greiner *J. Phys. G* **19**, 1303 (1993)
26. P.K. Panda, S.K. Patra, J. Reinhardt, J. Maruhn, H. Stöcker, W. Greiner *Int. J. Mod. Phys. E* **6**, 307 (1997)

27. H. Stöcker, W. Greiner *Phys. Rep.* **137**, 279 (1986)
28. J. Reinhardt, W. Greiner: *to be published*
29. P. Papazoglou, D. Zschiesche, S. Schramm, H. Stöcker, W. Greiner *J. Phys. G* **23**, 2081 (1997);
P. Papazoglou, S. Schramm, J. Schaffner-Bielich, H. Stöcker, W. Greiner *Phys. Rev. C* **57**, 2576 (1998)
30. J. Theis, G. Graebner, G. Buchwald, J. Maruhn, W. Greiner, H. Stöcker, J. Polonyi *Phys. Rev. D* **28**, 2286 (1983)
31. P. Papazoglou, D. Zschiesche, S. Schramm, J. Schaffner-Bielich, H. Stöcker, W. Greiner: *nucl-th/9806087 (accepted for publication in Phys. Rev. C)*
32. P. Papazoglou: PhD Thesis, University of Frankfurt, 1998;
C. Beckmann et al.: *in preparation*
33. S. Klimt, M. Lutz, W. Weise *Phys. Lett. B* **249**, 386 (1990)

EXPERIMENTS ON THE SYNTHESIS OF NEW SUPERHEAVY ELEMENTS

F.P. HEẞBERGER
Gesellschaft für Schwerionenforschung mbH, Planckstrasse 1, 64291 Darmstadt, Germany
E-mail: hess@clri6a.gsi.de

Evaporation residue production in cold fusion reactions of Pb- and Bi- target nuclei with projectiles of elements between Ti ($Z = 22$) and Zn ($Z = 30$) leading to compound nuclei $Z_{CN} = 104 - 113$ was investigated at SHIP. The result was the first unambiguous identification of isotopes of elements $Z = 110$, $Z = 111$ and $Z = 112$ in bombardments of ^{208}Pb ,^{209}Bi with ^{62,64}Ni, ^{70}Zn. We further report on the identification of ^{256}Db, ^{252}Lr and on new, improved decay data of ^{253}Lr, ^{255}Rf, ^{257}Db, allowing a first deeper insight in the nuclear structure of these isotopes.

1 Introduction

More than thirty years ago early extensions of the nuclear shell model to regions of nuclei having proton and neutron numbers considerably higher than the heaviest ones known at that time led to the prediction of closed spherical nuclear shells at $Z = 114$ and $N = 184$ [1]. Nuclei close to the crossing point of both shells were predicted to be stabilized against spontaneous fission by large shell correction energies and consequently long half-lives up to values comparable to the age of the universe were expected. Much effort to produce and to detect these 'superheavy' nuclei in the laboratory as well as to search for them in nature has been brought up since these days - however, unsuccessfully up to now. Nevertheless theoretical models have been refined considerably and experimental work has produced heavier and heavier nuclei, slowly approaching the magic number of $Z = 114$. The present situation is illustrated in Fig. 1, where ground-state shell correction energies obtained by Smolanczuk et al. [2] using a Nilsson-Strutinsky approach are plotted as function of N and Z. Two regions of enhanced nuclear stability are visible: one around $Z = 114$ and $N = 184$, representing the 'classical' superheavy element region and a second one around $Z = 108$ and $N = 162$ with almost equal maximum shell correction energies but β_2 - values of up to $\approx$0.24, representing 'deformed' nuclear shells.
The search for superheavy elements was a substantial motivation for the construction of the UNILAC and the velocity filter SHIP [3] at GSI in Darmstadt, and research work in this field has been carried out since about twenty years

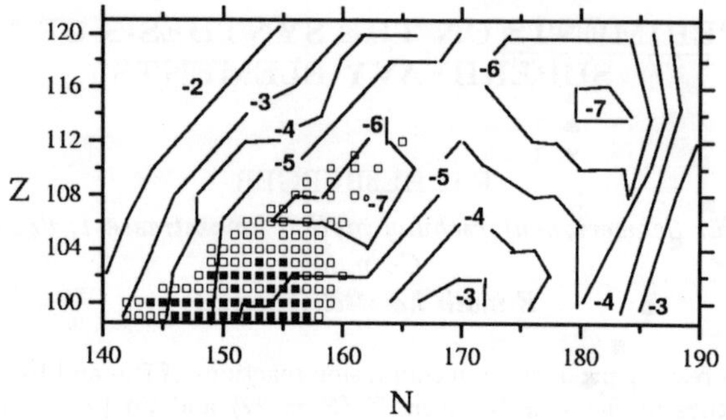

Figure 1. Ground state shell correction energies according to [2]. The squares represent the presently known nuclei, the full squares those already known at early days of superheavy element predictions (ca. 1966).

at these facilities. In the present paper we will report on the results of a series of experiments performed in the period 1994 - 1998 aiming at evaporation residue production in reactions leading to compound nuclei with atomic numbers $Z = 104 - 113$. For details on the experimental set-up we refer to [4].

2 Synthesis of new elements $Z > 109$

2.1 Excitation functions

Although 'cold' fusion of lead or bismuth targets with medium heavy projectiles like ^{50}Ti or ^{64}Ni has proven to be up to now the most successful method for the production of transactinide nuclei, we are faced with the fact that maximum cross sections typically decrease by a factor of ≈ 3.5 when the atomic number of the compound nucleus is increased by one (Fig. 2). Since, on the other hand, the formation cross section of a specific nuclide in a given reaction is strongly dependent on the excitation energy E^* of the compound nucleus the choice of the optimum value is crucial for the successful production of heaviest nuclei. We decided to determine it empirically by extrapolating the results from measured excitation functions.

Excitation functions, recently measured at SHIP for the systems ^{50}Ti, ^{58}Fe, and ^{64}Ni + ^{208}Pb are shown in Fig. 3a-c. Excitation energies, referring to the center of the target, were calculated using experimental mass excesses published by Wapstra and Audi [5] and values predicted by Myers and Swiatecki [6]. The data in Fig. 3 indicate a clear trend of decreasing excitation energy for

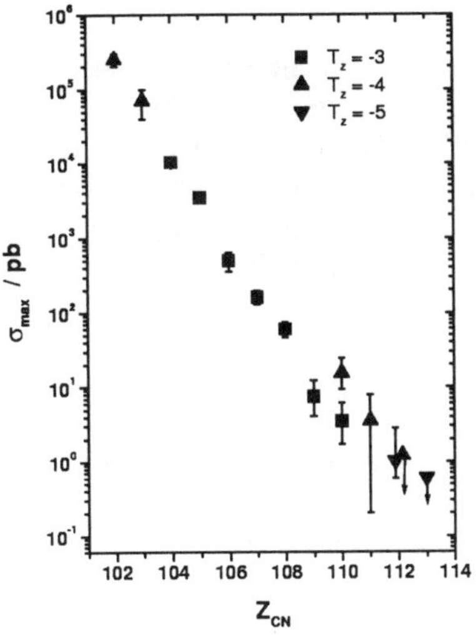

Figure 2. Systematics of 1n- cross-sections for heavy element production in cold fusion reactions. T_z denotes the isospin of the projectile.

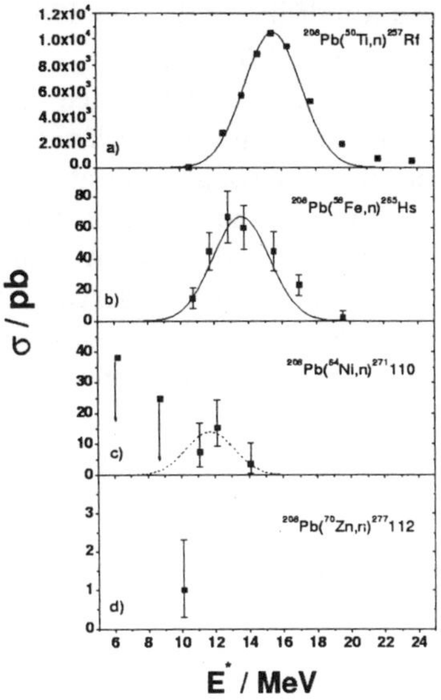

Figure 3. Excitation functions for ^{50}Ti + ^{208}Pb (a), ^{58}Fe + ^{208}Pb (b), ^{64}Ni + ^{208}Pb (c), ^{70}Zn + ^{208}Pb (d).

the maxima of the 1n de-excitation channel with increasing atomic number. While for ^{50}Ti + ^{208}Pb the maximum for the 1n - cross section was obtained at $E^* = (15.6 \pm 0.1)$ MeV (Fig. 3a) it decreased to $E^* = (13.6 \pm 0.2)$ MeV for ^{58}Fe + ^{208}Pb (Fig. 3b). The widths of the 1n - de-excitation functions are narrow, values of (4-5) MeV (FWHM) were obtained from a Gaussian fit to the data.

2.2 Element 110

On the basis of these results an optimum value of $E^* = 12.3$ MeV for the production of $^{269}110$ by $^{208}Pb(^{62}Ni,1n)^{269}110$ was extrapolated. In an experimental run of 12 days (total projectile dose: 2.2 x 10^{18}) we observed four α- decay chains starting with a decay of $^{269}110$. A mean energy $E_\alpha = 11.112 \pm 0.02$

MeV and a half-life of $T_{1/2} = 170^{+160}_{-70}\mu s$ were obtained. The formation cross section was $\sigma = 3.5^{+2.7}_{-1.8}$ pb.
Since for the more neutron rich projectile ^{64}Ni a higher cross section could be expected, allowing to check the validity of our extrapolation, in a succeeding experiment a rough excitation function was measured (Fig. 3c). From fitting a Gaussian to the data the maximum of the 1n-channel was settled at $E^* = (12.4 \pm 0.5)$ MeV. Altogether nine α - decay chains that were attributed to ^{271}110 were observed. They were subdivided into three groups: five events decay with a mean energy $E_\alpha = 10.738$ MeV, two events with $E_\alpha = 10.681$ MeV. The time differences between the implantation of the evaporation residues and the succeeding α-decays $\Delta t(ER - \alpha)$ lay between 0.2 ms and 4.4 ms, resulting in a half-life $T_{1/2} = 1.1^{+0.6}_{-0.3}$ ms. The third group existed of only one decay chain starting with an energy of $E_\alpha = 10.709$ MeV, having $\Delta t(ER - \alpha) = 81$ ms. Because of this significantly longer lifetime, it seems justified to assign this transition to a second level of ^{271}110.

2.3 Element 111

The positive results on the synthesis of element 110 encouraged us to immediately start an attempt to produce an isotope of element 111 by the reaction ^{209}Bi(^{64}Ni,1n)^{272}111. Since the extrapolation of the optimum projectile energy, derived from the doubly even nucleus ^{208}Pb, seemed less certain for the odd-even nucleus ^{209}Bi at that time, we irradiated at three different projectile energies $E^* = 10.0$, 11.6 and 13.0 MeV, starting from the lowest value that seemed meaningful. While no event that could be attributed to ^{272}111 was observed at $E^* = 10.0$ MeV ($\sigma < 2.9$ pb (68 % confidence limit)), one event was registered at $E^* = 11.6$ MeV ($\sigma = 1.7^{+3.3}_{-1.4}$ pb), and two more at $E^* = 13.0$ MeV ($\sigma = 3.5^{+4.6}_{-2.3}$ pb). α- energy and half-life are $E_\alpha = 10.82 \pm 0.02$ MeV and $T_{1/2} = 1.5^{+2.0}_{-0.5}$ ms.

2.4 Element 112

On the basis of the previous results an optimum value $E^* = 10.1$ MeV was chosen for production of ^{277}112 by the reaction ^{208}Pb(^{70}Zn,1n)^{277}112. In an experimental run of about six weeks, where a total projectile dose of 3.4×10^{18} was collected, two α-decay - chains that were attributed ^{277}112 were observed. They started with decays of $E_\alpha = 11.65$ MeV and $E_\alpha = 11.45$ MeV; the half-life obtained is $T_{1/2} = 240^{+430}_{-90}\mu s$. The resulting production cross section is $\sigma = 1.0^{+1.0}_{-0.4}$ pb (Fig. 3d).

2.5 Attempt to produce element 113

A search for the next heavier element with $Z = 113$ was performed in spring 1998. On the basis of the data for lighter elements a cross section between 0.3 pb and 1 pb was expected. During an irradiation period of 46 days, using the $^{70}Zn + ^{209}Bi$ reaction at two slightly different beam energies of $4.97 \cdot A$ MeV (4.5×10^{18} projectiles) and $5.00 \cdot A$ MeV (3.0×10^{18} projectiles) no event was observed that could be assigned to a decay chain starting at element 113. The obtained cross section limits are $\sigma < 0.9$ pb and $\sigma < 1.4$ pb, respectively, resulting in an averaged value of $\sigma < 0.6$ pb at a probability level of 68 %.

3 Enhanced nuclear stability close to $N = 162$

As a consequence of the local maximum of ground state shell correction energies at $N = 162$ and $Z = 108$ (see Fig. 1) an enhanced stability of nuclei against spontaneous fission, and to lower extend also against α- decay is expected in this region [7]. In Fig. 4 we have displayed calculated [7] and experimental fission halflives for even - even isotopes $Z \geq 104$. Qualitatively the experimental data indicate a strong increase in T_{sf} approaching $N = 162$, thus supporting the enhanced stability predicted by theory. However no data are available for $N \geq 162$ so the maximum in T_{sf} is not localized. From this point we may conclude that the experimental data express evidence for a neutron shell but do not allow to position it.
Experimental evidence for a neutron shell also can be obtained from Q_α-systematics for an isotopic chain. A shell crossing is indicated by a local minimum in the Q_α-values. This is demonstrated in Fig. 5. The open squares in the upper part represent calculated Q_α-values from liquid drop masses, which results in a straight line, decreasing with increasing neutron number. Introducing shell effects, a structure with a local minimum at the predicted shell at $N = 162$ (open triangles) becomes evident. It is effected by the differences of the ground state shell correction energies of the mother and daughter nuclei, as shown in the lower part of Fig. 5.

In the case of α - decay of $^{277}112$ the predicted shell at $N = 162$ is crossed by $^{273}110_{163} \xrightarrow{\alpha} {}^{269}Hs_{161}$. In the two observed chains for $^{277}112$ [8] striking different α- energies and 'lifetimes' (i.e. $\Delta t(\alpha(^{277}112) - \alpha(^{273}110)))$ of $E_\alpha = 9.73$ MeV, $\Delta t(\alpha(^{277}112) - \alpha(^{273}110)) = 170$ ms and $E_\alpha = 11.08$ MeV, $\Delta t(\alpha(^{277}112) - \alpha(^{273}110)) = 110$ μs were measured. Because of the large difference in the time distances the two transitions must be assigned to different levels in $^{273}110$. Calculated life-times result in values of 175 ms and 67 μs, respectively, both for $\Delta I = 0$ transitions, which compare well with the

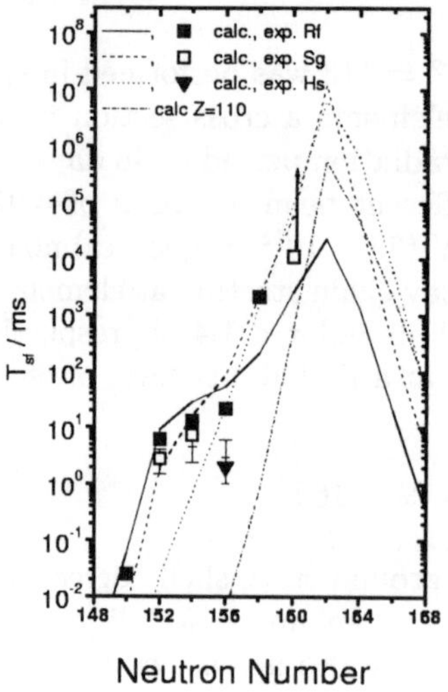

Figure 4. Comparison of calculated [7] and experimental fission halflives for even-even nuclei $Z = 104$ to $Z = 110$.

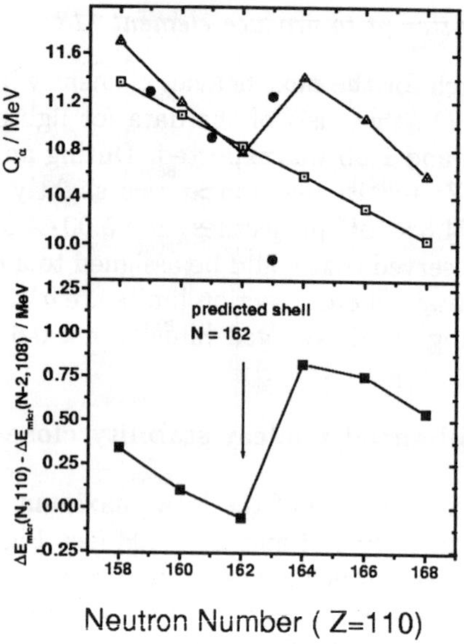

Figure 5. Experimental and calculated Q_α-values for $Z=110 isotopes.$• $Q_{\alpha,exp}$, $Q_{\alpha,LD}$[6], △ $Q_{\alpha,LD+shellcorr.}$[2].

measured time distances. We conclude that both transitions are unhindered and connect analogous states in mother and daughter nuclei. It is evident in Fig. 5 (full dots) that the higher energy causes a local minimum, indicating a shell crossing, while the lower energy indicates a continuation of the trend of decreasing Q_α- values with increasing neutron numbers. Since it is not clear presently which energy has to be attributed to the ground state decay and which to the decay of an isomeric state, we cannot regard our results as a proof (or a disproof) for a neutron shell at $N = 162$. More data are necessary. It certainly would be helpful to produce the even-even isotopes $^{268,270,272}110$ by irradiation of ^{207}Pb with $^{62,64}Ni$ and ^{70}Zn, since ground state to ground state transitions are expected to be the most intense ones in even - even nuclei.

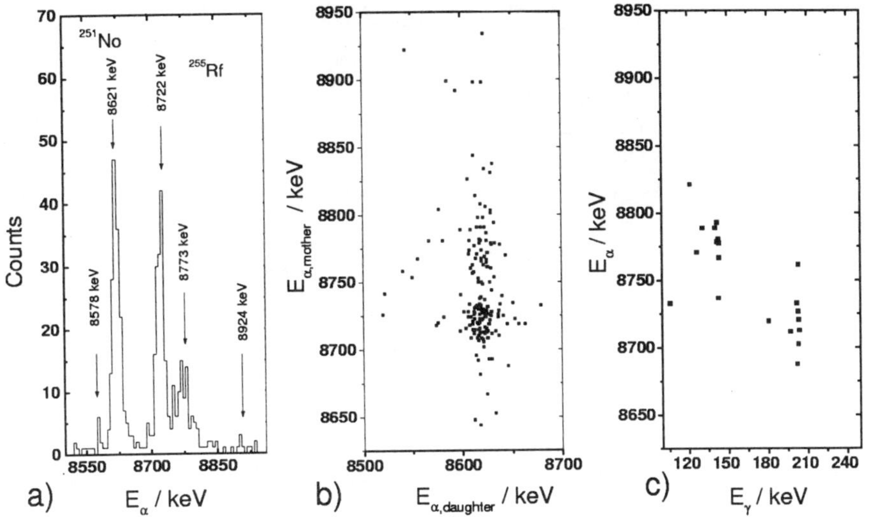

Figure 6. Experimental results for ^{255}Rf; a) α- spectrum of correlated events; b) α - α correlation plot for ^{255}Rf - ^{251}No; c)observed α - γ coincidences for ^{255}Rf.

4 Decay properties of the neutron deficient nuclei $Z = 103 - 105$

The improved sensitivity of our set-up does not only allow us to search for new elements on the pb - level within reasonable irradiation times (in the order of some weeks), but also allows for the production of a 'large' (>100) number of nuclei for isotopes having production cross sections of typically (0.1-10) nb. Thus a more detailed investigation of the α- and sf properties and also α-γ - coincidence measurements become possible.

First detailed results on the α - fine structure had been obtained for ^{257}Rf, [9]. Here we will report on more recent results on ^{255}Rf, ^{257,256}Db, ^{253,252}Lr.

4.1 ^{255}Rf

In a previous study of evaporation residues from ^{50}Ti + ^{206,208}Pb, an indication for the existence of a low lying isomeric state in ^{255}Rf decaying by α-emission with $T_{1/2} \approx 0.9$ s seemed evident, since an abnormally low number of α- decays in the interval $E_\alpha = (8720 - 8730)$ keV was followed by daughter

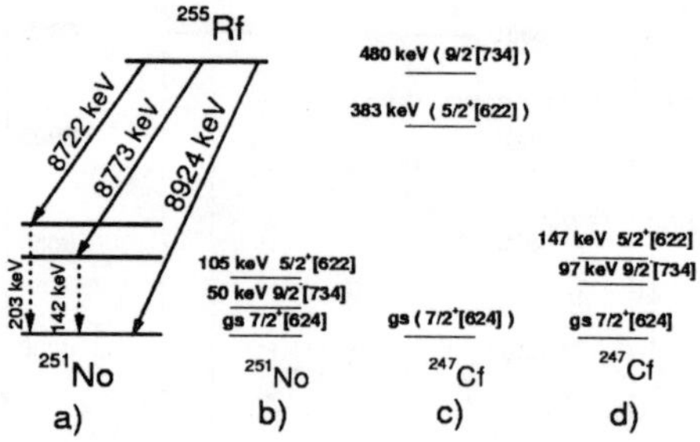

Figure 7. a) tentative decay scheme for ^{255}Rf; b), d) calculated level schemes for ^{251}No, ^{247}Cf [10], c) level scheme attributed to ^{247}Cf (simplified) [11].

decays (^{251}No) [9]. However, neither from theoretical predictions [10] nor from the known levels of lighter $N = 151$ nuclei with even Z numbers (^{251}Fm, ^{249}Cf, ^{247}Cm, ^{245}Pu) [11] having a similar structure, a level sequence allowing the existence of an isomeric state of $T_{1/2} \approx 1$ s could be expected. To clarify this problem we used the ^{207}Pb(^{50}Ti,2n)^{255}Rf reaction to produce an order of magnitude more events of ^{255}Rf. The result is shown in Fig. 6a,b. On the basis of a considerably higher number of observed decays we were not able to confirm an 'abnormal low' correlation rate for $E_\alpha = (8720 - 8730)$ keV within the statistical uncertainty.

In addition we observed a few γ - events in coincidence to the α- decays of ^{255}Rf. Two different groups are indicated (Fig. 6c), a) γ-events with a mean energy of $E_\gamma = 203 \pm 3$ keV essentially in coincidence to α- decays of $E_\alpha = 8722 \pm 10$ keV, and b) $E_\gamma = 142 \pm 3$ keV coincident to $E_\alpha = 8773 \pm 10$ keV. The sums $E_\alpha + E_\gamma = 8925$ keV and $E_\alpha + E_\gamma = 8915$ keV are equal within 10 keV. They are also close to the mean energy $E_\alpha = (8924 \pm 10)$ keV of the group of α-decays in the interval $E_\alpha = (8908 - 8936)$ keV being the α-decays with the highest energies followed by decays of ^{251}No. So we conclude that both γ- transitions lead to the ground state of ^{251}No (Fig. 7a); a calculated level scheme [10] is shown in Fig. 7b. Due to the lowest hindrance factor of HF≈4 the $E_\alpha = 8722$ keV transition is assigned to the 'favored' transition, leading to the analogous Nilsson state. Striking, however, is the low value HF≈13 for the $E_\alpha = 8773$ keV transition. In the lighter $N = 149$ isotones the first excited Nilsson level is settled as $5/2^+[622]$ [11]. The relative intensities of decays into this level are <0.05 in these nuclei. Assigning the $E_\alpha = 8773$ keV - line to the transition $9/2^-[734] \rightarrow 5/2^+[622]$ would result in a striking

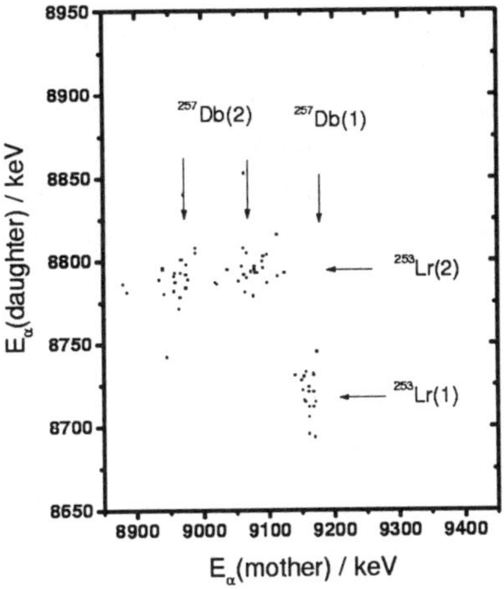

Figure 8. α - α correlation plot for ^{257}Db - ^{253}Lr.

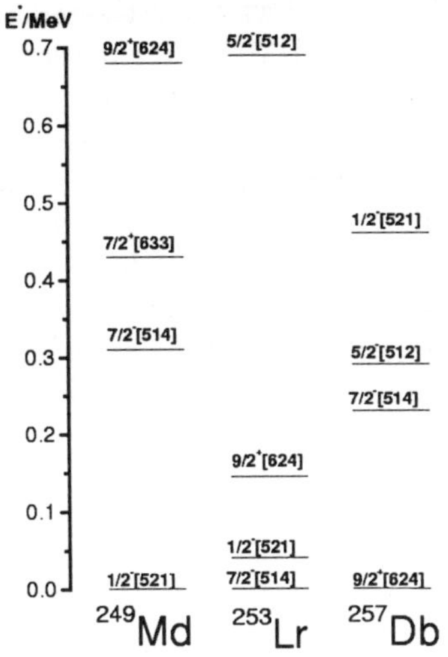

Figure 9. Theoretical level schemes ^{257}Db, ^{253}Lr, ^{249}Md [10].

high relative intensity $i_{\alpha,rel} \approx 0.25$ and hence an abnormal low hindrance factor. In addition, if the γ - decay indeed leads to the ground state in ^{251}No, the $9/2^-[734]$ Nilsson level would lie considerably lower than in the lighter $N = 149$ nuclei.
Thus both observations may indicate a change in the nuclear structure for $N = 149$ isotones between $Z = 98$ and $Z = 102$. On the basis of the present data any definite conclusions would be too speculative. More experimental data are necessary.

4.2 ^{257}*Db*, ^{253}*Lr*

In our first irraditions of ^{209}Bi with ^{50}Ti [12] ^{257}Db and ^{253}Lr were identified by delayed $\alpha - \alpha$ - coincidences to their decay products ^{249}Md and ^{245}Es. Three α-lines of $E_\alpha = 9160, 9071, 8970$ keV were attributed to ^{257}Db, and two lines of $E_\alpha = 8800$, 8722 keV were assigned to ^{253}Lr. A total of 12 α- decays of ^{257}Db correlated to any daughter product ^{253}Lr, ^{249}Md or ^{245}Es had been observed then.

In a recent experiment, the evaporation residue production for the reaction ^{50}Ti + ^{209}Bi was investigated in a bombarding energy range of $E = 4.59 \cdot A$ MeV to $5.08 \cdot A$ MeV. We observed about 200 α-decays of ^{257}Db correlated to any of the daughter products. The previously reported α-energies of ^{257}Db and ^{253}Lr could be reproduced within the error bars (see Table I). With respect to the correlations, however, a behavior was found, that was not realized in the previous experiment due to the small number of observed events: α-decays of ^{257}Db of $E_\alpha = 9163$ keV were only followed by α-decays of ^{253}Lr of $E_\alpha = 8723$ keV, while those ^{257}Db decays of $E_\alpha = 9074$, 8967 keV were only followed by ^{253}Lr decays of $E_\alpha = 8794$ keV (see Fig. 8). No statistically significant difference in the α-decay energies of the granddaughter ^{249}Md was found.

Calculated level schemes [10] at low excitation energies are shown for ^{257}Db, ^{253}Lr, ^{249}Md. A possible candidate for an isomeric state is the $1/2^-[521]$ level. Weisskopf estimates indicate halflives in the order of some tenth of a second for $1/2^- \rightarrow 7/2^-$ transitions for energy differences $\Delta E \approx 100$ keV or lower. In the case of ^{257}Db this means that the $5/2^-$ level should lie above the $1/2^-$. However, definite conclusions on the level structure cannot be drawn on the basis of the present data, thorough α-γ- and α - conversion electrons coincidence measurements are necessary.

Estimation of an sf-branch of ^{257}Db was rendered more difficult by the fact, that as well for the 1n - channel as for the 3n - channel sf is observed via the decay path $^{258}\mathrm{Db} \overset{EC}{\rightarrow} {}^{258}\mathrm{Rf} \overset{sf}{\rightarrow}$ (1n) or $^{256}\mathrm{Db} \overset{EC}{\rightarrow} {}^{256}\mathrm{Rf} \overset{sf}{\rightarrow}$ (3n). The search for sf of ^{257}Db was therefore restricted to bombarding energies where contributions σ(1n) and σ(3n) were less than ten per cent of that of the 2n - channel. An upper limit of three sf events that could be attributed to ^{257}Db(1) or ^{257}Db(2) were registered, resulting in upper limits of $b_{sf} < 0.13$ for ^{257}Db(1) and $b_{sf} < 0.06$ for ^{257}Db(2). Furthermore three α-sf correlations were observed. According to E_α two sf events were attributed to ^{253}Lr(1) and one to ^{253}Lr(2). Resulting branchings are: $b_{sf} = 0.05 \pm 0.04$ for ^{253}Lr(1) and $b_{sf} = 0.01^{+0.01}_{-0.005}$ for ^{253}Lr(2).

4.3 ^{256}Db, ^{252}Lr

Previously unknown α - activities were observed in the ^{50}Ti + ^{209}Bi reactions at the highest irradiation energies $E = 4.97 \cdot A$ MeV and $E = 5.08 \cdot A$ MeV. By delayed α-α - coincidences to known decay products ^{248}Md, ^{248}Fm and ^{244}Cf they were attributed to ^{256}Db and ^{252}Lr. Decay properties are listed in table 1.

In addition to the α- decays also nine sf events were registered at these irradiation energies and attributed to the 3n de-excitation channel. An es-

Table 1. Summary of decay properties of ^{251}No, ^{252,253}Lr, ^{255}Rf, ^{256,257}Db,
* = only one count observed

Isotope	$E_{\alpha 1}$ (keV)	i_{rel}	$T_{1/2}$ (s)	
^{251}No	8621±10	$0.96^{+0.02}_{-0.06}$	0.76±0.03	$b_{sf} < 0.0026$
	8578±15	0.04±0.015		
^{252}Lr	9018±20	0.73±0.26	$0.36^{+0.11}_{-0.07}$	
	8974±20	0.27±0.07		
^{253}Lr(1)	8723±10		$1.49^{+0.30}_{-0.21}$	$b_{sf} = 0.05 \pm 0.04$
^{253}Lr(2)	8794±10		$0.57^{+0.07}_{-0.06}$	$b_{sf} = 0.01^{+0.01}_{-0.05}$
^{255}Rf	8722±10	0.55±0.05	1.64±0.11	$b_{sf} = 0.45 \pm 0.06$
	8773±10	0.25±0.03		
	8797±15	0.07±0.02		
	8692±15	0.035±0.010		
	8831±15	0.035±0.010		
	8887±15	0.025±0.011		
	8924±15	0.025±0.0113		
^{256}Db	9018±20	≈0.64	$1.5^{+0.5}_{-0.4}$	$b_{EC} = 0.35 \pm 0.12$
	8891±20*	≈0.12		
	9075±20*	≈0.12		
	9120±20*	≈0.12		
^{257}Db(1)	9163±10		$0.76^{+0.15}_{-0.11}$	$b_{sf} < 0.13$
^{257}Db(2)	9074±10	0.38±0.08	$1.50^{+0.19}_{-0.15}$	$b_{sf} < 0.06$
	8967±10	0.33±0.08		
	8937±15	0.09±0.04		
	9021±15	0.09±0.04		
	8892±15	0.055±0.031		
	9117±15	0.055±0.031		

timation of T_{sf} for the odd-odd nucleus ^{256}Db on the basis of measured or calculated sf half-lives for the neighbouring even - even nuclei ^{256,258}Sg, ^{254,256}Rf and the sf hindrance factors for ^{255}Rf and ^{257}Db results in a value of $T_{sf} \approx 4000$ s or $b_{sf} \approx 0.0005$ for ^{256}Db. Therefore these sf events are assigned to ^{256}Rf, produced by $^{256}\text{Db} \overset{EC}{\rightarrow} {}^{256}\text{Rf} \overset{sf}{\rightarrow}$. Thus we obtain a value $b_{EC} = 0.35 \pm 0.12$ for ^{256}Db.

Acknowledgments

The experiments were performed in collaboration with S. Hofmann, V. Ninov, P. Armbruster, H. Folger, G. Münzenberg, Ch. Stodel (GSI), M.E. Leino (University Jyväskylä, Finland), A.Yu. Lavrentev, A.G. Popeko, A.V. Yeremin

(JINR Dubna, Russia) and Š. Šáro (Comenius University, Bratislava, Slovakia). The author expresses his gratitude to the UNILAC staff for delivering the high intensity beams and to the staff of GSI target la-boratory, B. Lommel, W. Thalheimer, W. Hartmann and J. Klemm for the skillful preparation of the large area lead and bismuth targets.

References

1. H. Meldner *Arkiv för fysik* **36** 593 (1967)
2. R. Smolanczuk, A. Sobiczewski: *in Proc. EPS Conf. Low Energy Nuclear Dynamics, St. Petersburg (Russia) 1995.* World Scientific, New Jersey, London, Hong Kong 1995, p. 313; *and private communication (1995)*
3. G. Münzenberg et al. *Nucl. Instr. Meth.* **161** 65 (1979)
4. S. Hofmann et al. *Z. Phys. A* **350** 277 (1995)
5. G. Audi, H. Wapstra *Nucl. Phys. A* **595** 409 (1995)
6. W.D. Myers, W.J. Swiatecki *Nucl. Phys. A* **601** 141 (1996)
7. R. Smolanczuk et al. *Phys. Rev. C* **52** 1871 (1995)
8. S. Hofmann et al. *Z. Phys. A* **354** 229 (1996)
9. F.P. Heßberger et al. *Z. Phys. A* **359** 415 (1997)
10. S. Cwiok, S. Hofmann, W. Nazarewicz *Nucl. Phys. A* **575** 356 (1994)
11. R. Firestone, V.S. Shirley (eds.): *Table of Isotopes.* J. Wiley & Sons, New York, Chichester, Brisbane, Toronto, Singapore 1996
12. F.P. Heßberger et al. *Z. Phys. A* **322** 557 (1985)

EXPERIMENTS ON SYNTHESIS OF SUPERHEAVY ELEMENTS WITH Z=110,112 VIA REACTIONS ^{232}Th,^{238}U + ^{48}Ca.

YU.TS. OGANESSIAN, A.V. YEREMIN, V.I. CHEPIGIN, M.G. ITKIS, A.P. KABACHENKO, O. KONSTANTINESKY, M. KONSTANTINESKY, O.N. MALYSHEV, A.G. POPEKO, J. ROHÁČ, R.N. SAGAIDAK, I.V. SHIROKOVSKY, V.K. UTYONKOV

Flerov Laboratory of Nuclear Reactions, JINR, Dubna, Russia

S. HOFMANN, G. MÜNZENBERG

Gesellschaft für Schwerionenforschung, Darmstadt, Germany

M. VESELSKY

Institute of Physics, Bratislava, Slovakia

S. SARO

Department of Physics, Comenius University, Bratislava, Slovakia

N. IWASA, K. MORITA, Y. KOBAYASHI, A. YONEDA

Institute of Physical and Chemical Research (RIKEN), Wako-shi, Saitama, Japan

M. HUSSONNOIS

Institute of Nuclear Physics, Orsay, France

The investigations of the decay properties and formation cross sections of the heaviest isotopes of the 110 and 112 elements with atomic numbers 276, 277 and 282, 283 respectively were performed at the FLNR JINR with the use of the internal and extracted high intensity ^{48}Ca beams. ^{232}Th and ^{238}U targets were used at the experiments. The experiments with internal beam probe include the use of fast chemistry for extraction of the fraction containing 106 element nuclei from ^{232}Th target material and the off line measurements of the possible correlation chains from α decay of ^{268,269}106 and SF of ^{264,265}104. The reaction ^{48}Ca + ^{238}U was investigated at the recoil separator VASSILISSA. The experiments were performed at two beam energies resulting in excitation energies of 33 and 39 MeV. The collected beam doses were 3.5×10^{18} and 2.2×10^{18}, respectively. Two spontaneous fission events were observed at the lower beam energy, which tentatively were assigned to the new neutron rich isotope ^{283}112 produced by the reaction ^{238}U(^{48}Ca,3n)^{283}112. The measured cross-section is $(5.0\ ^{+6.3}_{-3.2})$ pb and the half-life is $(81\ ^{+147}_{-32})$ s. No event was observed at the higher beam energy resulting in an upper cross-section limit of 7.3 pb for the 4n evaporation channel.

1 Introduction

The stability of heavy nuclei depends strongly on shell structure effects. These stabilizing effects increase significantly at closed proton and neutron shells. Beyond uranium the stability of nuclei diminishes rapidly with increasing element number Z. This trend is altered at Z = 100 and N = 152 due to the existence of shell gaps in the single–particle level spectra, which appear there for deformed shapes and which are the reason for the unusual high stability of ^{252}Fm against spontaneous fission (SF).

According to the macroscopic–microscopic theory, the next spherical shell closure for the neutrons beyond N = 126 is predicted at N = 184. The stability of the superheavy nuclei increases sharply, when their neutron number approaches this spherical shell closure. Due to the spherical ground-state and strong ground-state shell-correction energy, the fission barrier is wider and higher than for deformed nuclei, which is the reason for the expected increased stability against SF. For the synthesis of spherical superheavy nuclides it is of advantage, to select reaction partners with the highest possible number of neutrons in order to approach the shell N = 184 as close as possible [1,2].

During the last two decades the so called "cold" fusion reactions using lead and bismuth targets were successful in synthesizing the heaviest transfermium elements with Z = 107–112 [3]. The advantage of this reactions is that only slightly excited (E* ≈ 10–20 MeV) compound nuclei are produced at bombarding energies near the fusion barrier. This is a consequence of the double magic structure of the ^{208}Pb target and the strongly bound projectiles from ^{54}Cr to ^{70}Zn. Low–excitation energies of fragile heavy compound nuclei are considered to be the reason for surviving the prompt fission.

The "hot" fusion reactions using targets of Th to Cf were also applied to synthesize isotopes of the elements Z = 102 to 110 [4]. Due to the higher temperature of the compound nucleus, a number of 4–5 neutrons must be evaporated to reach the ground-state. These reactions allow for the study of the decay properties of more neutron rich isotopes of transfermium elements, which are impossible to synthesize directly by "cold" fusion reactions, e.g. long lived isotopes with neutron numbers close to N = 162 and Z $\geq$ 106. The very asymmetric systems fuse with higher probability at the Coulomb barrier, however, simultaneously the excitation energies of the compound nuclei change from typically 10–20 MeV in the case of "cold" fusion reactions, to 40–50 MeV. This increase of the excitation energy reduces the survival probability of heavy compound nuclei.

The minimum excitation energy in the case of the "hot" fusion reactions can be obtained with the use of a ^{48}Ca beam [1]. Therefore, the compound nucleus ^{286}112 formed in the reaction ^{48}Ca + ^{238}U is less strongly excited. This is a result of the significant mass deficit of the double magic ^{48}Ca. The excitation energy at the Coulomb barrier is only $E^* \approx 35$ MeV, which increases the survival probability of the compound nucleus considerably. The high mass asymmetry in the entrance channel ($A_1/A_2 = 0.2$ and $Z_1Z_2 = 1840$) is also reason for a decrease of the dynamic hindrance of fusion, an effect, which was observed earlier for the more symmetric "cold" fusion reactions [5].

Using a modified version of the statistical code ALICE [6] we could rather well reproduce the measured cross-sections of reactions from ^{12}C+^{248}Cm to ^{34}S+^{244}Pu. For these systems the effective fissility parameter x_{eff} varies from 0.68 to 0.81. Not so far from this range is the effective fissility parameter of the reaction ^{48}Ca + ^{238}U, $x_{eff} = 0.85$, resulting in the compound nucleus ^{286}112. Therefore, we used the same parameter set also for an estimate of the cross-section for production of element 112. The result was 5 to 20 pb for the 3n–evaporation channel and 1 to 5 pb for the 4n channel. These cross-sections would be high enough to be reached in a reasonable irradiation period of about 1 – 1.5 months. However, some improvements of the experimental set-up were necessary, concerning the preparation of the ^{48}Ca beam from the U–400 cyclotron and the detection system at the separator VASSILISSA.

2 Experiment

The production of an intense ion beam from the rare and extremely expensive isotope ^{48}Ca was the key problem of our aim to synthesize isotopes of elements 110 - 112 by "hot" fusion reactions.

Neutral atoms of ^{48}Ca were injected into the plasma of the ECR–4M ion source [7] by controlled heating of a 50 mg sample of metallic calcium. The ^{48}Ca enrichment was 70 %. The amount of calcium reaching the ECR resonance cavity as well as the recovery of the material was controlled and optimized by (γ,n)–activation of ^{48}Ca and measuring the γ rays emitted from ^{47}Ca ($T_{1/2}$ = 4.5 d).

The U–400 cyclotron was modified for an axial injection of the beam from the ECR source. The high charge state of the ions (5^+) extracted from the source allowed for a continuous operation of the cyclotron. This resulted in an increase of the beam intensity by a factor of 2–3 compared with the previously used pulsed mode. As a result of these improvements we reached an internal beam intensity of up to 2 pμA of ^{48}Ca at a material consumption rate of about 0.3 mg/h.

Two sets of experiments with ^{48}Ca (^{232}Th and ^{238}U targets) and a number of test experiments (^{174}Yb, ^{206,208}Pb targets) were performed during period September 1997 – October 1998.

The first search experiment was with the use of ^{232}Th target. It was performed in three runs at November 1997, May and October 1998. In the experiment we used internal beam of Ca ions, because the intensity of the internal beam is factor of 3 higher that external one. This experiment included the use of fast chemistry for extraction of the fraction containing 106 element nuclei from ^{232}Th target material and the off line measurements of the correlation chains from α decay of ^{268,269}106 and SF of ^{264,265}104. Theory [8,9] predicts for the ^{268}106, the granddaughter of ^{276}110 - result of the 4n evaporation channel after fusion of ^{48}Ca and ^{232}Th, the half-life against α and SF decay more than 2 hours. The isotope ^{269}106, the granddaughter of ^{277}110 - result of the 3n evaporation channel after fusion of ^{48}Ca and ^{232}Th could have the half-life against α and SF decay even longer. In this case it could be possible to use fast chemistry and to perform after that off line measurements at the geometry close to 4π. The fraction extracted after chemistry was deposited onto 30 μg/cm^2 thick carbon foil and performed 10 μg/cm^2 thick and 20 mm in diameter source. Carbon foils with that sources were placed in between two silicon detectors (28.5 mm in diameter each) at special vacuum chambers (see Fig. 1). Total registration efficiency for α particles and SF fragments were close to 90 %. Altogether 8 vacuum chambers with pairs of silicon detectors were used in this experiment. In each chamber the measuring time for one source was 3 - 10 days. The suppression factors for long lived transfer reactions products were more than 10^6. Thickness of ^{232}Th targets were about 8 mg/cm^2, more than range of evaporation residues of 110 element but less than ranges of transfer reaction products. Selectivity of the chemical methods allowed to suppress the fraction of U isotopes, probable candidates for long lived background products, by factor of more than 10^5. Typical α spectra recordered during off line measurements are presented in Fig. 2.

The second search experiment was with the use of ^{238}U and ^{232}Th targets, external beam of Ca ions and recoil separator VASSILISSA [10]. It was performed in two runs at March - April (^{238}U target) and June 1998 (^{232}Th target).

The beam was extracted from the cyclotron by the stripping method. The mean beam intensity of ^{48}Ca on the target was 2.2×10^{12} s^{-1}. The beam energy was controlled by measuring the energy of the ions scattered under an angle of 30° from a thin gold foil (200 $\mu g/cm^2$). The two beam energies used in the experiment were (255 $\pm$ 3) and (262 $\pm$ 3) MeV after extraction from the cyclotron.

As target we used ^{238}U with an enrichment of 99.999 % and a thickness

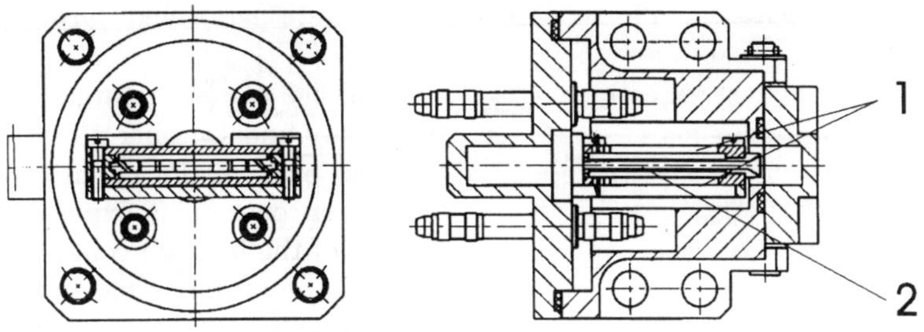

Figure 1. The schematic view of the chamber for off–line measurements of the sources with fractions of 106 element. 1 - semiconductor detectors, 28.5 mm in diameter, distamce between detectors 2 mm; 2 - supporting frame with source onto carbon foil (40 μg/cm^2 thick).

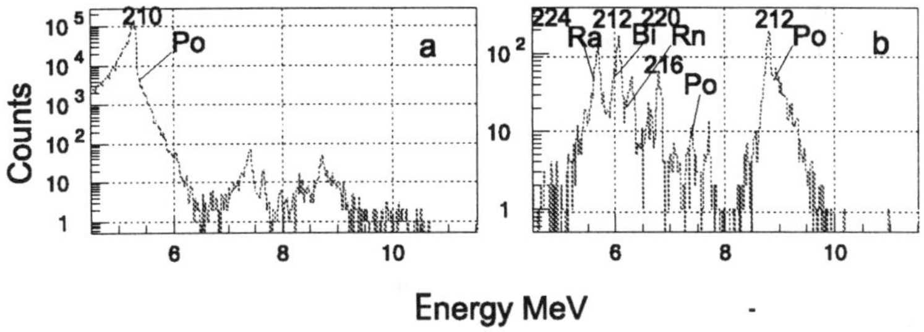

Figure 2. An example of α spectra recordered during off line measurements in October 1998. The beam dose collected – 4.5×10^{17}. Measuring time - 97 hours. a) Spectrum for the source with 106 element fraction, measurement starts 2 hours after stop of irradiation; b) Spectrum for 104 element fraction, measurement starts 12 hours after stop of irradiation.

of 0.3 mg/cm^2. The material was deposited uniformly by electroplating on a thin, 1.6 mg/cm^2, aluminium backing foil of 125 mm diameter, which was fixed between two supporting disks of 50 mm diameter. The disk was rotating with a frequency $\omega = 2000$ rpm. The beam energies were reduced to (231 ± 3) and (238 ± 3) MeV, respectively, up to the middle of the target due to energy loss mainly in the aluminium backing foil. The same target thickness and preparation method were used in the case of ^{232}Th target at the experiment in June 1998. The target design was tested in several experiments using Tb and enriched isotopes of Yb and Pb before the main experiment.

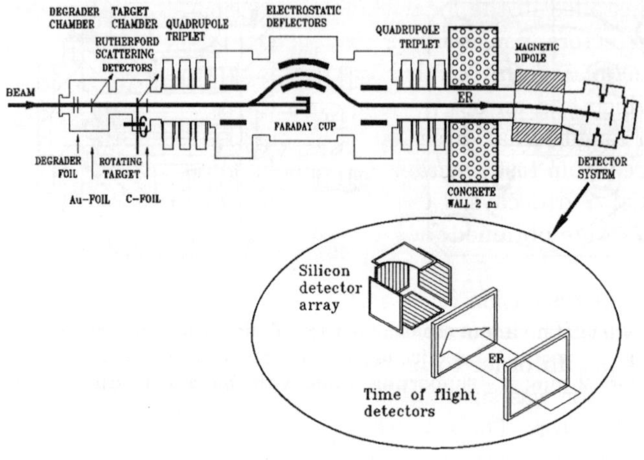

Figure 3. The schematic view of the separator VASSILISSA.

The evaporation residues (ERs) were separated in–flight from the beam particles and other reaction products by the electrostatic recoil separator VASSILISSA [10] (see fig. 3). Scattered low energy projectiles are the main contribution to the background at the focal plane. In order to reduce this background, we installed an additional 8° bending dipol magnet after the second quadrupole triplet of the separator. This magnet allowed for mounting the detectors out of the primary beam direction.

A 3 μm thick degrader foil (mylar) was placed in front of the silicon detectors for reduction of the number of low energetic projectiles reaching the focal plane of the separator.

The field settings were calculated according to a mean charge state of $(19\pm1)^+$ of the ERs and the two mean energies of 38 and 40 MeV, respectively.

The separation efficiency of ERs from the reaction ^{48}Ca + ^{238}U was estimated from preparatory experiments. Cross sections of various xn–evaporation channels were measured for the reactions of ^{159}Tb, ^{174}Yb and ^{206,208}Pb targets irradiated by ^{48}Ca projectiles of different energies. The measurements showed that about 25 % of the $^{286-x}$112 ERs, produced with a ^{238}U target, would be implanted into the detector.

For the registration of the ERs and their radioactive decay a system of time–of–flight (TOF) detectors and a silicon position–sensitive strip–detector array was installed in the focal plane of the separator. After registration by the TOF detectors, consisting of two secondary–electron foil detectors with microchannel plates, the recoiled ions were implanted into the 16–strip silicon detector, which had an active area of 60×60 mm^2. Each strip was position sensitive in longitudinal direction. The position resolution along each strip was measured from test reactions. A value of 0.6 mm (FWHM) was obtained for sequential α–α decays, 1.0 mm for ER–α and 1.5 mm for ER–SF events. These values were obtained for energies of the ERs in the range from 4 to 15 MeV.

The energy resolution for α particles was 20 keV within the energy range from 6 to 9 MeV. The accuracy of the time registration for recording the events was about 1 μs. In order to increase the detection efficiency for α-particles, the strip detector was surrounded by four silicon detectors of the same size as the stop detector. The entire array had the shape of a cube with 60 mm edge length. In the present series of experiments four neighboring strips of the backward detectors were connected galvanicly so that a total of 16 energy sensitive segments was formed, each with a resolution within the range from 120 to 150 keV. The efficiency of the silicon array for detection of α particles with full energy was 85 % of 4π, however, the small energy–loss from escaping α's in the stop detector could be measured only for 50 % of events, because of the low implantation energy of the ERs.

The signals from the TOF detectors were used both for measurement of the velocity of the recoils and for distinguishing the radioactive decays of the previously implanted nuclei. The high efficiency of the TOF detectors allowed for obtaining very clean decay spectra. The time window for measuring decay chains could be significantly widened (up to several hours). The latter was particularly important for the correlation of decays with a long half-life at the presence of a continuously running beam.

The counting rate of all background events at the focal plane detector was only 25–30 Hz. The counting rate on one strip of the detector of α–like events (background without TOF signal) with energies more than 7.5 MeV was 20–30 per hour, and of recoil–like events (background with TOF signal) with energies more than 4 MeV it was 80–120 per hour. The background rate depended on the horizontal position of the strip. From the results of the test reactions we estimated an energy range from 4 to 15 MeV for the signals originating from implanted Z = 112 nuclei.

3 Results

At the experiment with ^{232}Th target and internal beam probe the primary beam energy was $E_L \approx 260$ MeV. The energy loses of the ^{48}Ca beam at the ^{232}Th tagret were about 40 MeV, it covers the width of the excitation functions for 3 and 4 neutron evaporation channels. The experiment was peformed as a number of irradiations with following chemistry procedure (≈ 1.5 - 2 hours) for the extraction of the 106 element fraction after the stop of irradiation. The duration of one irradiation was approximately 8 – 9 hours, beam dose collected was $\approx 5\times10^{16}$. In November 1997 total beam dose collected was 2.4×10^{17} (4 irradiations), 4 samples after the chemistry procedure were prepared and measured. 1 S.F. event (40 MeV + 70 MeV) was detected, the registered time of the event after stop of irradiation was 40 hours. In May 1998 total beam dose collected was 2.5×10^{17} (5 irradiations), 5 samples were prepared and measured. One S.F. event (82 MeV + 54 MeV) with registered time after stop of irradiation 92 hours and one S.F. event (50 MeV + 7 MeV) with registered time after stop of irradiation 126 hours were detected. The background conditions were good enough, during two monthes of the measurements without samples we do not detect any high energy events, which could be explained as SF events. In October 1998 5 irradiations were performed. To accept higher beam intensity internal beam probe with rotating wheel were used (at the previous experiments self supporting standing Th foil was used). Thorium foils were fixed on the edge of the wheel. This construction allows to accept beam intensities up tp 2 pμA. Altogether 5 irradiations were performed, the collected beam doses were 2.6×10^{17}, 9.6×10^{17}, 4.5×10^{17}, 4.0×10^{17} and 3.5×10^{17} respectively. After first and second irradiations more precise chemistry was used. The measurements started after 30 hours for the first irradiation and 17 hours after second one. After second irradiation the source with fraction of 104 element was also prepared. Last three irradiations were followed by fast chemistry ($\approx$ 2 hours) procedure for the extraction of 106 element fraction, after the third irradiation fraction of 104 element was also extracted. No SF events were detected.

During the measurements of the samples counting rate of the α- particles at the energy region from 7.5 to 9 MeV was aproximately 1 event per 1 – 3 hours for each semiconductor detector facing the sample. No short α- SF correlations in the time interval up to 500 seconds were detected.

On the basis of the three detected events we could make the preliminary conclusion, that SF events could not be attributed to the spontaneous fission of the isotope ^{264}104, the product of the α- decay chain of ^{276}110 (the result of the 4 neutron evaporation channel). According to the theoretical predictions [8] the half-life of this isotope against spontaneous fission could be few seconds. The most reasonable explanation of the obtained results is that we detected the spontaneous fission events of isotopes $^{265}104$ or $^{269}106$, the descendants of $^{281}110$, the product of the 3 neutron evaporation channel. The cross-section evaluated for the production of the three fission events is approximately 3 - 5 pb.

Second experimental run was performed with the use of separator VASSILISSA. In a first part of the experimental program we investigated the reactions ^{208}Pb(^{48}Ca,xn)$^{256-x}$No and ^{206}Pb(^{48}Ca,xn) $^{254-x}$No, both at the recoil separator VASSILISSA and at the in-beam time–of–flight fission fragment spectrometer CORSET [11]. At CORSET fusion cross-sections of the compound nuclei were determined by observing the prompt fission events and by measurement of the fission–fragment mass–distribution. For the irradiation of the lead targets the beam energy was varied over a wide range covering the evaporation channels 1n to 3n so that a set of excitation functions was obtained from both experiments. These data allowed for a comparison of the ER cross-sections measured at VASSILISSA and the total fusion cross-sections measured in-beam.

Then, the measurement was repeated at CORSET using a ^{238}U target. Compound-nucleus formation was observed and from the yield as function of the beam energy we concluded that the residue cross-section must be highest for the 3n and 4n channels. Detailed results of these experiments will be published elsewhere, see e.g. [12]. Using the modified ALICE code [6], we calculated excitation energies of $E^* = 33$ and 39 MeV for the maxima of the 3n and 4n channels, respectively. The corresponding beam energies for reactions in the middle of the target thickness are 231 and 238 MeV.

Using these two energies we investigated the reaction ^{48}Ca(^{238}U,xn)$^{286-x}$112 at VASSILISSA. The irradiation started in March 1998. During a period of 25 days a beam dose of 3.5×10^{18} projectiles was collected at the lower energy and during a succeeding period of 15 days a dose of 2.2×10^{18} at the higher energy.

Whereas a number of registered α-particle-like signals was relatively high, only two signals from spontaneous fission were detected at the first irradiation at the lower beam energy.

Both SF events were characterized by a coincident event of a higher energy signal (161.8 and 192.9 MeV) in the stop detector and a lower energy signal (28.0 and 20.7 MeV) in the backward detectors. The first event was measured in the strip number 12 and the second in the strip number 15. The higher strip numbers are in the direction to the lower magnetic rigidity of ions after the deflection magnet. The shared energy results from the implantation close to the surface of the stop detector, in which one fission fragment is stopped and the stopping of the escaping fragment in the backward detectors after some energy loss in the inactive surface layers. The absence of signals from the TOF detectors identifies the coincidence events as radioactive decays and the high energy as fission of an implanted nucleus.

The analysis of events collected in the experiment has been provided to find generic decay links of the implants over a wide range of half–lives. For that purpose the data on the position and energy resolution of the focal plane detector for incoming particles and various decay modes were used.

Possible decay chains of the type ER $\stackrel{E_{ER},t_{ER}}{\longrightarrow}$ α_1 $\stackrel{E_{\alpha_1},t_1}{\longrightarrow}$ α_2 $\stackrel{E_{\alpha_2},t_2}{\longrightarrow}$ $\cdots$ were searched for within time intervals 5 μs $\leq$ t_i $\leq$ 1000 s and 8 MeV $\leq$ E_α $\leq$ 13 MeV. Signals from an α decay could be missed only in the case if an α particle escapes through the open front side of the detector array (15 % probability) or if the lifetime is smaller than $\approx$5 μs and the decay occurs in the dead time of the data acquisition system. The windows for the relative positions were twice the resolution at FWHM, as given before. No such decay chains were found for the two irradiation.

For decay chains of the type ER $\stackrel{E_{ER},t_{ER}}{\longrightarrow}$ α_1 $\stackrel{E_{\alpha_1},t_1}{\longrightarrow}$ $\cdots$ SF $\stackrel{E_{f_1},f_2,t_{SF}}{\longrightarrow}$ the upper limit of the time intervals t_i was extended up to 10,000 s. Such decay chains were also not found for the two irradiation.

The most carefull analysis needs the spontaneous fission of the implanted nucleus. It would result in a delayed coincidence event ER $\stackrel{E_{ER},t_{ER}}{\longrightarrow}$ SF $\stackrel{E_{f_1},f_2,t_{SF}}{\longrightarrow}$. ER-like signals were randomly distributed in the time. The mean rate in the strip number 12 (first event) was θ_1 $\approx$0.001 Hz, in the strip number 15 (second event) it was θ_2 $\approx$0.002 Hz. No bursts of signals were found during the analysis. Both rates were determined for a position window of $\pm$0.8 mm.

To search for the most probable candidate for the parent-recoil we have analyzed the distribution of time intervals between signals. The closest recoil-like signal was detected τ_1=182 sec before the first SF-event. The probability to find 1 random correlated signal during the interval τ_1 at the mean frequency of signals θ_1 is described by the Poisson law. The probability that this correlation is a chance event is 0.15 and that it is the true one is 0.85. For the second SF-event corresponding probabilities are respectively 0.09 and 0.91. Thus we can conclude that the most probable candidates for signals from generic coupled recoils preceding the SF-events are the events coming 182 and 52 sec before.

The position difference determined from the ER and SF signals was 0.36 mm for the first event and 0.75 mm for the second one. For the energies of the implantation signals we measured 3.84 and 4.69 MeV, respectively. These two values are within the range expected for ERs.

The mean value of the two measured time intervals results in a lifetime of $\tau = (117^{+213}_{-\ 46})$ s or a half-life of $T_{1/2} = (81^{+147}_{-\ 32})$ s. However, a longer lifetime cannot be excluded completely. This possibility must be taken into account for the case that the first ER like event preceeding the SF event is not the true parent.

The total kinetic energy (TKE) of the two fission events was obtained by using the known TKE distribution of ^{252}No [13] for calibration. The TKE distribution of ^{252}No was measured during the preparatory experiments using the reaction ^{206}Pb(^{48}Ca,2n)^{252}No. Fig. 4 shows a plot of the ^{252}No TKE distribution together with the energies of two events observed during the ^{48}Ca + ^{238}U irradiation. The TKE values for the two events are 190 and 212 MeV, respectively. A higher energy deficit for the signals of these two events due to a possible higher atomic number as nobelium was not taken into account.

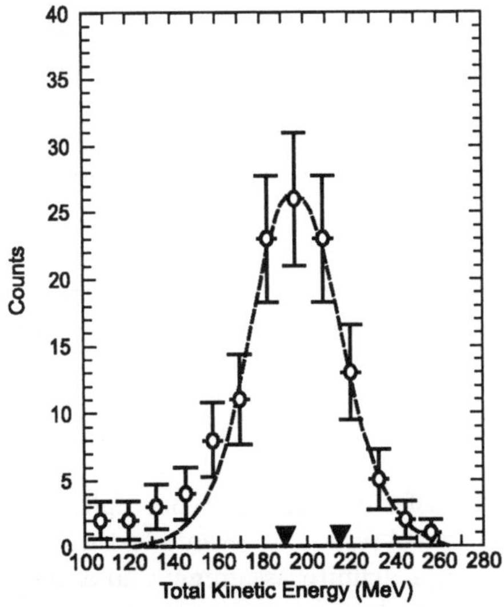

Figure 4. The TKE spectrum for the SF of ^{252}No measured at VASSILISSA using the reaction ^{206}Pb(^{48}Ca,2n)^{252}No. The dashed line is a Gaussian curve fitted to the data. The error bars represent the statistical uncertainties only. The energy calibration was made using the results of Ref. [13]. The energies of the two fission events observed during the ^{48}Ca + ^{238}U irradiation are marked by the filled triangles.

The cross-section evaluated for the production of two fission events at a beam energy of 231 MeV in the middle of the target thickness is $(5.0^{+6.3}_{-3.2})$ pb. The upper cross-section limit obtained at 238 MeV energy is 7.3 pb, calculated at a probability level of 68 %. The error bars represent only the statistical uncertainties, the absolute values have an uncertainty of a factor of 2 due to uncertainties of the determination of the projectile dose and the separator efficiency.

4 Discussion

The assignment of SF events to a certain isotope is not a straight forward task as it is in the most cases of the observation of α–decay chains. Additional information is needed, like, e.g, the measurement of excitation functions or cross bombardment. However, these data are difficult to obtain, if the cross-sections are on a level of few picobarns. Nevertheless, we will present some arguments, which allow for a most plausible assignment of the two SF events observed in irradiation of ^{238}U target by ^{48}Ca.

Fissioning nuclei in the region of ^{238}U, which could be produced by transfer reactions, have either too long half–lives (ground-state SF) or too short half–lives (fission isomers) in order to be candidates for an assignment of the two SF events observed in the reaction ^{48}Ca + ^{238}U. In addition, as measured in the case of the reaction ^{48}Ca + ^{208}Pb, the transfer channels are strongly suppressed by VASSILISSA.

SF half-lives of the order of 100 s or longer, as measured here, are known from the region of neutron rich isotopes of elements from californium to dubnium. These known isotopes or possibly still heavier unknown isotopes could be reached by a ^{48}Ca breake-up reaction. However, in this case the ERs produced by fusion of a ^{48}Ca fragment with ^{238}U have a much smaller velocity as the ERs from a complete fusion and are, therefore, suppressed by VASSILISSA.

The exclusion of transfer products and incomplete fusion products makes the assignment of the two observed SF events to the complete fusion of ^{48}Ca + ^{238}U likely. However, a definite assignment to a specific isotope is made difficult by the lack of any data from previous experiments leading into the region of these relatively neutron rich nuclei. The only one known isotope of element 112 so far is ^{277}112 synthesized in cold fusion reaction [14]. This isotope is an α emitter with a half-life of $T_{1/2} = 0.24$ ms.

The nonobservation of α decay, the unprobable evaporation of protons or α particles from the compound nucleus and the relatively low excitation energy of 33 MeV let us tentatively assign the fission events measured in

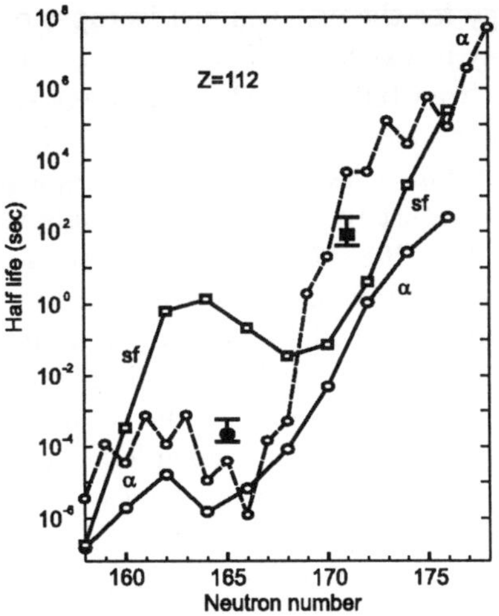

Figure 5. Theoretical predictions for partial α and SF half-lives for isotopes of element Z = 112. Open circles and open squares connected by solid lines: $T_{1/2,\alpha}$ and $T_{1/2,SF}$ from Ref. [8]. Open circles connected by a dashed line: $T_{1/2,\alpha}$ from Ref. [9]. Filled circle: $T_{1/2,\alpha,exp}$ for $^{277}112$ from Ref. [14]. Filled square: $T_{1/2,SF,exp}$ from this work and tentatively assigned to $^{283}112$

this work to the isotope $^{283}112$ produced via a 3n evaporation channel in the reaction $^{48}Ca + ^{238}U \rightarrow ^{286}112^*$.

Results of calculations from Smolańczuk [8] and Möller et al. [9] describing the decay properties of isotopes of element 112 are shown in Fig. 5. The theoretical data were obtained from using the macroscopic-microscopic method. Both calculations were able to reproduce the decay properties of the so far known nuclei of the heavy elements rather well within an accuracy of about 1 to 2 orders of magnitude. This can be seen by comparing the measured α half-life of $^{277}112$ (neutron number N = 165) with the calculations (Fig. 5). For this isotope SF is expected to have a much longer half-life than α decay.

The calculations predict a steep rise of the half-lives with increasing neutron number. This is a result from the decreasing negative shell–correction energy and a transition of the nuclear shape from well deformed quadrupole to spherical, which both increase the stability and the fission barrier and thus the partial α and SF half-lives. Compared to the SF half-life calculations performed for the even–even nuclei as shown in Fig. 5, hindrance factors due to a specialization energy for the odd neutron isotopes of up to about three orders of magnitude may be expected. Therefore, the calculated SF half–life for $^{283}112$ is in the range from 100 to 1,000 s, which agrees with the experimental half-life of (81^{+147}_{-32}) s.

The calculated partial α half–lives also increase steeply with increasing neutron number. They start to become comparable with [8] or even exceed [9] The SF half–lives at N = 168–170. For almost all isotopes shown in Fig. 2 the calculations of Ref. [9] result in the longer half-lives, and still, in the case of the known α emitter $^{277}112$, the prediction is a factor of ten smaller than the experimental value. Concerning $^{283}112$ only a lower limit of $T_{1/2,\alpha} \geq 190$ s can be deduced from the experiment. This value is in agreement with the data from Möller et al. and does not completely contradict the data from Smolańczuk, taking into account that his values were calculated only for the even isotopes.

The cross–section for the production of element 112 is higher (5.0 pb) using the hot fusion reaction and a ^{48}Ca beam as the value obtained by using the cold fusion reaction and a ^{70}Zn beam (1.0 pb) [14]. The relatively high cross–section is probably a result of the relatively low excitation energy of the compound nucleus that can be achieved at the fusion barrier using ^{48}Ca. The negative result of the irradiation at the higher beam energy and the obtained upper limit of 7.3 pb shows that the 4n channel does not have a significantly higher cross-section as the 3n channel.

The irradiation of ^{238}U by ^{48}Ca was a first experiment at our Laboratory using an improved experimental set–up at VASSILISSA, which allows for investigation of the hot fusion reaction on a cross–section level of 1 pb. The investigated reaction ^{48}Ca + ^{238}U results in ERs, which are far from the region of known isotopes of the heavy elements. Moreover, no α decay chains were observed, whose properties would provide additional spectroscopic information in order to establish a final assignment of the observed reaction product. Therefore, we continue our experiments with an investigation of the reaction ^{48}Ca + ^{232}Th $\rightarrow$ $^{280}110^{*}$. Using this reaction, ERs should be produced halfway the known isotopes in the region of deformed heavy elements and the isotope $^{283}112$. The results could hopefully provide more data in order to establish a trend of the decay properties and cross–sections for the synthesis of isotopes by hot fusion on the way to the region of spherical superheavy elements. Using the energy $E_{targ.} = 238 \pm 2$ MeV ($E^{*} = 39$ MeV) we investigated the reaction ^{48}Ca(^{232}Th,4n)^{276}110 at VASSILISSA. The irradiation started in June 1998. During a period of 15 days a beam dose of 1.8×10^{18} projectiles was collected. No spontaneous fission events were detected and no $\alpha - \alpha$ correlations were observed in the whole $\alpha-$ particle energy range from

8 to 12 MeV in the time interval of up to 1000 sec. The upper cross-section limit obtained at 238 MeV energy is 9.1 pb, calculated at a probability level of 68 %.

5 Conclusion

The expected decay properties of the isotopes with Z = 112 are quite specific. According to the calculations provided by Smolańczuk [8], the even – even isotopes $^{282}112$ and $^{284}112$ have partial half–lives for α– decay respectively: T_α = 0.005 s. and 1.0 s. Their spontaneous fission half–life ($T_{s.f.}$) is slightly higher than T_α. At the same time, the isotope $^{280}110$, which is the α–decay daughter of the $^{284}112$ nucleus, will mainly decay by spontaneous fission. As for the chain of sequential decays of the other isotope $^{282}112$, all products with $Z < 106$ will have $T_{s.f.} \ll T_\alpha$ too. For the isotope $^{283}112$ the predictions are less clear, because the odd number of neutrons may lead to significant limitations for both α– decay and spontaneous fission. Here, a strong competition between the two modes of decay is also highly expected.

Note, that calculations performed by Möller et. al. [9], predict for the same nuclei the half-life versus α– decay to be hundreds and thousands times longer than T_α from [8]. However, it seems less significant, since it does not change the main decay properties of the isotopes with Z = 112, produced in the reaction $^{48}Ca + {}^{238}U$. Either they will decay by spontaneous fission, or by the chains of their α– decay (one or several) will end by the spontaneous fission anyway. On the other hand, modern separation techniques allow synthesizing and identifying the ER, if their half lives are more than one microsecond. It covers the range of all the predictions and thus the experimental results obtained using these techniques can be used as a direct test of the predictive capabilities of the existing theoretical models.

We report a new attempt to synthesize super heavy spherical nuclei in the reactions $^{48}Ca + {}^{232}Th$ and $^{48}Ca + {}^{238}U$. As compared with all previous experiments with ^{48}Ca ions, the sensitivity of the present experiments is more than 100 times higher. Three spontaneous fission events were detected in the reaction $^{48}Ca + {}^{232}Th$ and two spontaneous fission events were detected in the reaction $^{48}Ca + {}^{238}U$ at the beam energy below the Coulomb barrier. As it follows from the data analysis, the events were most probably triggered by the decay of the even – odd isotopes $^{277}110$ and $^{283}112$, respectively, formed in the 3n – evaporation channel. The half-life of the new isotope $^{283}112$ is about 100 sec. It could also be that it undergoes α– decay with a branching ratio $b_\alpha \leq 50$ %.

The present experiments are the first ones in our long term research program with the ^{48}Ca beam, dedicated to the synthesis and study of the properties of super heavy elements. We strongly intend to increase the intensity of the beam, which will allow us to continue these experiments in order to collect more statistics and to try to synthesize other isotopes with Z = 110 - 114. Also is is planned to improve chemical methods in order to obtain more clean and detailed information, when chemical extraction of the long–lived fractions of 104 and 106 elements and off–line α and SF spectrometry are used in experiments.

Acknowledgments

This work was performed partially under the financial support of the Russian Foundation for Basic Research, contract N 96-02-17209 and INTAS, contract N 96-662.

References

1. Yu.Ts. Oganessian, in: *Proc. of Int. Conf. Nuclear Physics at the Turn of the Millenium: "Structure of the Vacuum and Elementary Matter"*, Wilderness, South Africa, (World Scientific, Singapore), (1996) 11.
2. A.V. Yeremin, V.K. Utyonkov and Yu.Ts. Oganessian, in: *Proc. of Tours Symposium on Nuclear Physics III,* September 2 - 5, 1997, Tours, France, p. 16–28.
3. S. Hofmann, Rep. Prog. Phys. **61**, (1998) 639.
4. Yu. A. Lazarev, et. al., Phys. Rev. C **54**, (1996) 620.
5. J.P. Blocki, et. al., Nucl. Phys. A **459**, (1986) 145.
6. B.I. Pustylnik, in: *Proc. of the Int. Conf. on Dynamical Aspects of Nuclear Fission*, Častă–Papiernička, Slovak Republic, Aug. 30–Sept. 4, (1996) 121.
7. V.B. Kutner et. al., in: *Proc. of 15th Int. Conf. on Cyclotrons and their Applications*, 14–19 June, 1998, Caen, France.
8. R. Smolańczuk, Phys. Rev. C **56**, (1997) 812.
9. P. Möller et. al., Atomic Data and Nuclear Data Tables **59**, (1995) 185 and P. Möller et. al., "Nuclear properties for astrophysics", http://t2.lanl.gov/publications/astro/normal.html
10. A.V. Yeremin et. al., Nucl. Instr. and Meth. B **126**, (1997) 329.
11. M.G. Itkis et. al., Il Nuovo Cim. (to be published).
12. A.V. Yeremin et. al., JINR Rapid Communications, (1998) (in press).
13. E.K. Hulet Yad. Fiz. **57**, (1994) 1165.
14. S. Hofmann et. al., Z. Phys. A **354**, (1996) 229.

SEARCH FOR THE MULTIMODAL FISSION OF ^{256}No, ^{270}Sg AND ^{271}Hs IN REACTIONS OF SUBBARRIER FUSION

M. G. Itkis[a], N. A. Kondratiev[a], E. M. Kozulin[a], L. Krupa[a], Yu. Ts. Oganessian[a], I. V. Pokrovsky[a], E. V. Prokhorova[a] and A. Ya. Rusanov[b]

[a]Flerov Laboratory of Nuclear Reactions, JINR, Dubna, Russia,
[b]INP, Alma-Ata, Kazakhstan

1. Introduction

One of the most interesting aspects of modern nuclear physics is synthesis of heavy and superheavy elements and study of their properties. In this connection experimental investigation of the fusion-fission process of compound nuclei with Z> 100 at low excitation energies is of special importance. With this purpose at the U-400 accelerator of FLNR the fission cross sections, mass and energy distributions of fission fragments were measured in the reactions ^{48}Ca + ^{208}Pb $\rightarrow$ ^{256}No, ^{22}Ne + ^{248}Cm $\rightarrow$ ^{270}Sg and ^{22}Ne + ^{249}Cf $\rightarrow$ ^{271}Hs in the compound nucleus excitation energy range from 12 to 60 MeV. The interest in the reactions ^{22}Ne + ^{248}Cm and ^{22}Ne + ^{249}Cf is connected first of all with the fact that the number of neutrons in the compound nuclei ^{270}Sg and ^{271}Hs is 164 and 163 respectively. In this case one can expect that at low excitation energies the fission fragment mass distribution demonstrates properties of bimodal fission determined by structural peculiarities of the deformation potential energy surface, mush as it takes place in spontaneous fission of ^{258}Fm and ^{201}Md ($Z \geq 50$, $N \approx 160$).

The interest in reactions with ^{48}Ca ions is explained by their importance for the presently realized at the FLRN program of the superheavy ion synthesis in the region of nuclei with Z=114 - 116 and N=182 - 184, for which a considerable increase in their stability to spontaneous fission and α-decay has been predicted. That's why it is very important to measure the compound nucleus cross section for reactions of the ^{48}Ca + U, Pb, Cm - type, on the basis of which one is able to predict the fission cross section and survivability of superheavy nuclei.

2. Experiments

The experimental study of the fission of superheavy elements was performed on the U-400 accelerator at the Flerov Laboratory of Nuclear Reactions, Dubna. The targets were 100-200 $\mu g/cm^2$ thick layers of ^{208}Pb, ^{248}Cm and ^{249}Cf produced by deposition on carbon backings 50-60 $\mu g/cm^2$ thick. Time-coincident fission fragments were registered by the CORSET double-arm time-of-flight spectrometer [1]. Each arm of the spectrometer included a start detector built on the basis of microchannel plates with an electrostatic mirror and two stop position-sensitive (x-, y-sensitivity) detectors also built on the basis of microchannel plates 6x4 cm each. The start detectors were placed at a distance of 4 cm from the target. The minimal start-stop flight distance was 10 cm. So, the spectrometer consisted of 2 start detectors and 4 stop detectors, the solid angle being 360 msr. The position resolution of the stop detectors was ±0.1°. The estimated mass resolution was 3-5 a.m.u.

In the experiment carried out on the beam of ^{22}Ne and ^{48}Ca the arms of the spectrometer were set at such correlation angles as for it to register the energies and masses of fragments up to the ration 5. The data were processed by regular techniques [2,3] on the basis that the fission was a two-body process. Special attention was paid to the angular folding correlations both in the plane and out of the plane of the reaction. Only those registered events were selected and then analysed that corresponded to a two-body process of total momentum transfer. Fission events associated with incomplete fusion reactions, sequential fission, were rejected. The date processing took account of the energy losses of fragments in the target itself, its backing and in the electron emitters of the start detectors. In addition, because the geometrical dimensions of the spectrometer did not allow registration of the whole set of the energy-mass ratios of fragments with equal probability, we made corrections for its geometrical efficiency.

3. Reaction $^{48}Ca + ^{208}Pb$

The products of the reaction $^{48}Ca + ^{208}Pb$ were measured at six energies of ^{48}Ca projectiles: E_{lab}=211, 218, 220, 225, 231 and 242 MeV. To give a typical example, shown in Fig.1 are the MED of these products for the energies E_{lab}=231 and 211 MeV. These energies of projectiles correspond to the excitation energies of the compound nucleus of ^{256}No E^*=34 and 18 MeV. Shown from top to bottom are: the two-dimensional matrices (TKE-Mass), mass distribution, dependence of TKE on the mass of fragments. For energy of 231 MeV, the MED of fragments corresponding to a rather wide vicinity around of A/2 are typical for heated nuclei: the mass yields are close to a gaussian, TKE(M) is close to a parabola. For high asymmetry, in the region of the masses of light fragments equal to 65-80 and the masses additional to them, "shoulders" are seen, that is, increased yield as compared to that expected for the fission of the compound nucleus. In addition, TKE has an increased value, which suggests that the scission configuration for these masses is a compact one. It is possible if at least one of the fragments formed is close to a magic one. This condition is met for the light fragment with mass of 70-85, whose nucleon composition can be one of three combination of nucleons close to magic numbers of neutrons and protons: a) at Z=28, then, from the standard hypothesis that the charge is proportional to the mass, N=42 and,

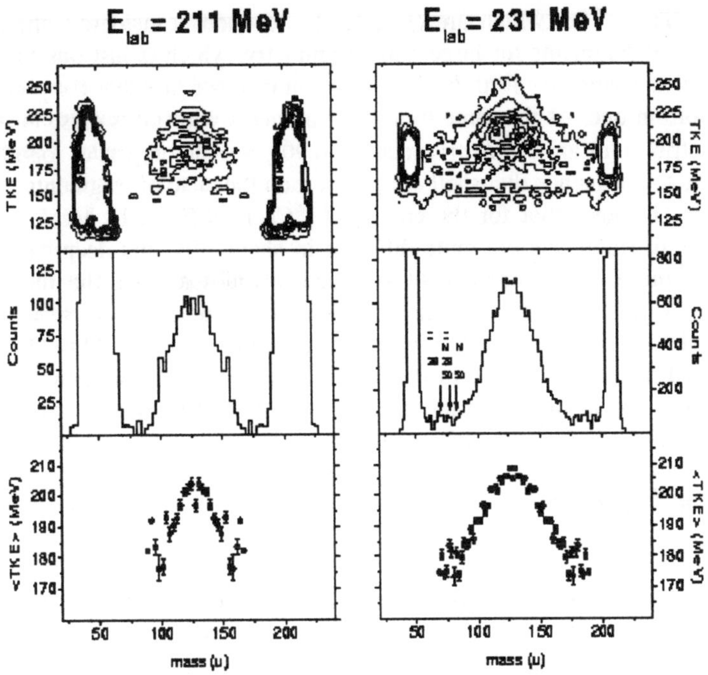

Figure 1. Fission fragment mass-energy distributions for ^{256}No, formed in the reaction ^{48}Ca+^{208}Pb. From top to bottom: two-dimensional matrices TKE-m, mass distributions and dependencies <TKE>(m) for E_{lab}=211 and 231MeV.

correspondingly, M=70; b) at N=50, on the basis of the same hypothesis, Z=33 and, correspondingly, M=83; c) at Z=28 and N=50, M=78, if the masses of the fragments do not follow the above rule. The heavy fragments additional to the light fragments are not formed near any spherical shell. In Fig. 4 the corresponding light masses are indicated with arrows. It is clearly seen that it is in the region of the masses of fragments indicated with arrows that the "shoulder" experimentally observed appears.

Recall that the initial excitation of the nucleus E^*=34 MeV, therefore the structure of the fission fragments of the compound nucleus cannot demonstrate itself so clearly. Therefore we think this effect to be connected with quasi-fission, which does not result in the classical compound nucleus. This assumption is indirectly corroborated by works [4, 5], according to which in the reactions ^{40}Ar +

^{208}Pb and $^{50}Ti + ^{208}Pb$, which are close to the reactions considered, the angular distributions of fragments for large mass asymmetry, which is just our case, show clearly defined asymmetry forth-back, which points to the fact that the process is a non-equilibrium one. At the same time, for fragments with mass close to A/2, the angular distributions are symmetric relative to 90°. So, the properties found can be attributed to the influence that shell effects exert on the yield of fragments.

Fig. 1 shows that for the energy of ^{48}Ca ions E_{lab}=211 MeV, the mass distributions of fragments are quite different in shape. For this deep subcoulomb energy, the quasi-fission component is practically absent, but the mass yields themselves, which are characteristic of low-excited nuclei, show structure. An increased yield of heavy fragments is observed in the region of M=132-134 (the first standard fission mode) and in the region of M=140-150 (the second standard fission mode). So, at low excitation energy, the MED of the fission fragments of ^{256}No have composite, multimodal structure.

4. Fission of ^{270}Sg and ^{271}Hs

We investigated the fusion-fission reactions $^{22}Ne + ^{248}Cm$ and $^{22}Ne + ^{240}Cf$ at the projectile energies equal to 102 and 127 MeV. These energies result in the initial excitation energy of the compound nucleus of ^{270}Sg E^*=28 and 50 MeV. As for the fission of ^{271}Hs, only the preliminary results will be presented, and we will mainly discuss only the first reaction. For that reaction, the lab frame Coulomb barrier calculated according to the Bass model [6] is B_{bass}=117 MeV, therefore the lower energy of the bombarding ions in the experiments, which was 102 MeV, was below this barrier by 15 MeV, which resulted in the fusion cross section and consequently the fission cross section being decreased by a factor of $\sim 1.5 \times 10^3$ as compared to those corresponding to the energy of projectiles equal to 127 MeV.

There were three reasons why such reactions were chosen to study fission modes.

First, the potentialities of the conventional technique of investigating the properties of the fragments of superheavy elements as applied to studying of the spontaneous fission of the nuclei with Z=105, 106 and so on are most likely to have been exhausted because it seems to be an intractable experimental problem to collect a sufficiently large number of nuclei of those elements and then to investigate their fission. Therefore, in our view, the most reasonable and promising technique is to investigate the MED of the fragments of superheavy elements in subbarier fusion-fission reactions, which allow one to have low excitation energies.

Secondly, the nucleus of ^{270}Sg chosen to be investigated has 164 neutrons, and its fission into equal parts results in fragments with the magic neutron number 82. Thirdly, the fact that the target and ion are deformed in the ground state must result in higher subcoulomb fusion cross sections, which allows one to have lower excitation energies, at which shell effects responsible for the multimodal nature of fission begin playing their role.

Fig. 2 presents the main results of the measuring of the MED of the fragments of the compound nucleus of ^{270}Sg for two values of the energies of ^{22}Ne projectiles. For energy of 127 MeV, 9.2×10^5 fission events were registered; for energy of 102 MeV, 1.2×10^4. Shown in Figs. 2a and 2b are the two-dimensional TKE-mass matrices of the fragments. For high energy (Fig. 2b), we see a triangular distribution with rounded edge slopes typical for rather strongly heated nuclei. Its properties are close to those predicted by the liquid-drop model [7,8]. For the neon energy of 102 MeV (Fig. 2a) $E^*=28$ MeV, the two-dimensional matrix is different in shape from the above-described one. It has a sharp top and is definitely elongated towards the higher TKE. Fig. 2c presents a comparison of the mass distribution of the fragments for both energies. It is clearly seen that for the lower energy unlike for the high energy, the mass distribution is definitely nongaussian in shape and can be considered as manifestation of different fission modes. One mode, rather narrow, is associated with masses near $A_{CN}/2$, the other mode, rather wide, with asymmetric masses, the latter being so wide as for the mass yields at its edge slopes to be higher than those for high excitation energy. It is only possible if the cooling of the nucleus is accompanied by manifestation of shell effects in the region of the masses of the fragments from 200 to 210, that is, near magic lead. This effect can be interpreted in two ways. Recent theoretical calculations of the potential energy surface for ^{252}Cf, which is lighter in mass, predict that there is a "lead" fission valley [9] lying in the part of very high energy and therefore unobservable experimentally in the spontaneous fission of ^{252}Cf. The superheavy nuclei with Z=114 and 120 are also predicted to have such a valley [10]. ^{270}Sg, investigated by us, is very likely to have such a valley, because the light fragment also proves to be close to the magic one with Z=28.

The increased mass yield of fragments in the region of A=200 at low excitation energy may also be associated with the quasi-fission component, which, in our view, is most likely. The properties of the quasi-fission of superheavy nuclei was studied in works [11] and discussed in [12]. It was shown [10] that in the reaction ^{238}U + ^{27}Al, which is close to the reaction discussed, at the energy of the bombarding uranium near the Coulomb barrier, on the slopes of the prevailing symmetric fission there appear "shoulders" due to the increase in the yield of reaction products just right in the mass region 200-210, which is the magic lead region. They are connected with the non-equilibrium quasi-fission component and stand out sharply in the mass-angle correlations. Because the angular distributions of the fragments were not the subject of this work and the angular range of the spectrometer used was not too wide, it is difficult to put any accurate interpretation on the fact that the yield of the fragments increases in this mass region.

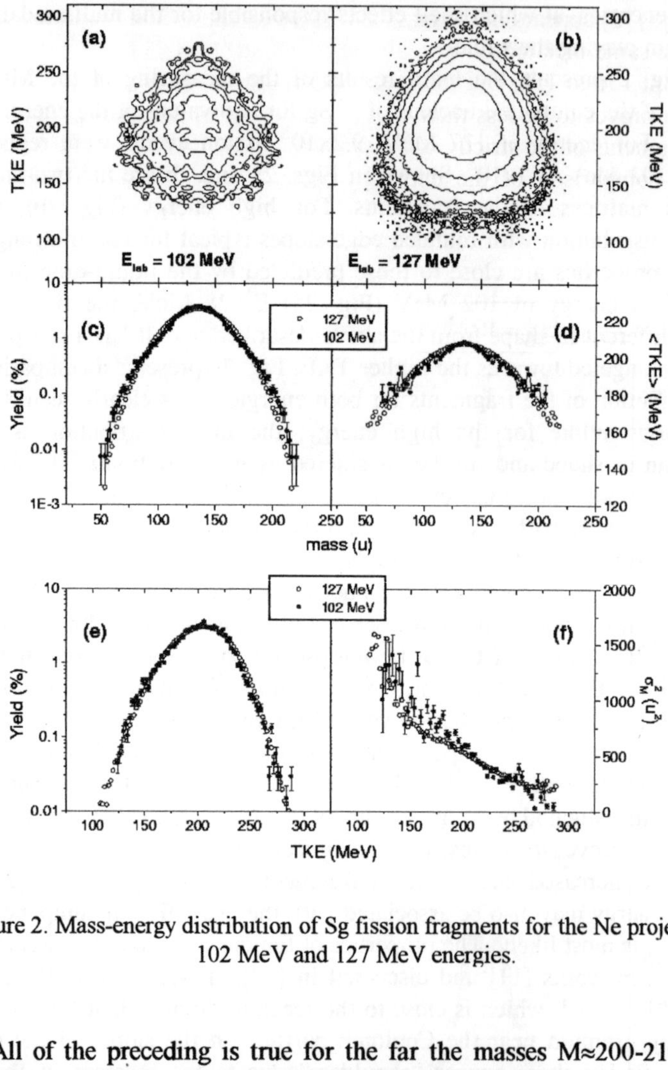

Figure 2. Mass-energy distribution of Sg fission fragments for the Ne projectiles of 102 MeV and 127 MeV energies.

All of the preceding is true for the far the masses M≈200-210 and the masses additional to them. However, the bend of the mass distribution in Fig. 2c, which is due to the fact that the narrow distribution becomes wide near the fragment mass M≈160, cannot be accounted for neither by a "lead valley" nor by quasi-fission. Here other modes come into play. The narrow symmetric component, in our opinion, is associated with the fact that both fragments have the spherical neutron shell with N=28. Spontaneously fissioning heavy nuclei never have 164 neutrons. Even as this number is approached and the number changes from 157 to 158, nuclei of fermium show metamorphoses: the narrow energetic super short fission mode [13] appears in the mass and energy yields of the fragments. Nuclei of Rf show the same metamorphoses, although not so clearly, as the neutron number changes from 156 to 158. But the contribution of the super short mode to the total distribution proves to be less than that observed for ^{258}Fm. This is due to the fact that the fragments of the strictly symmetric fission of ^{262}Rf are non-magic.

A comparison of the energy characteristics of the fragments of ^{270}Sg

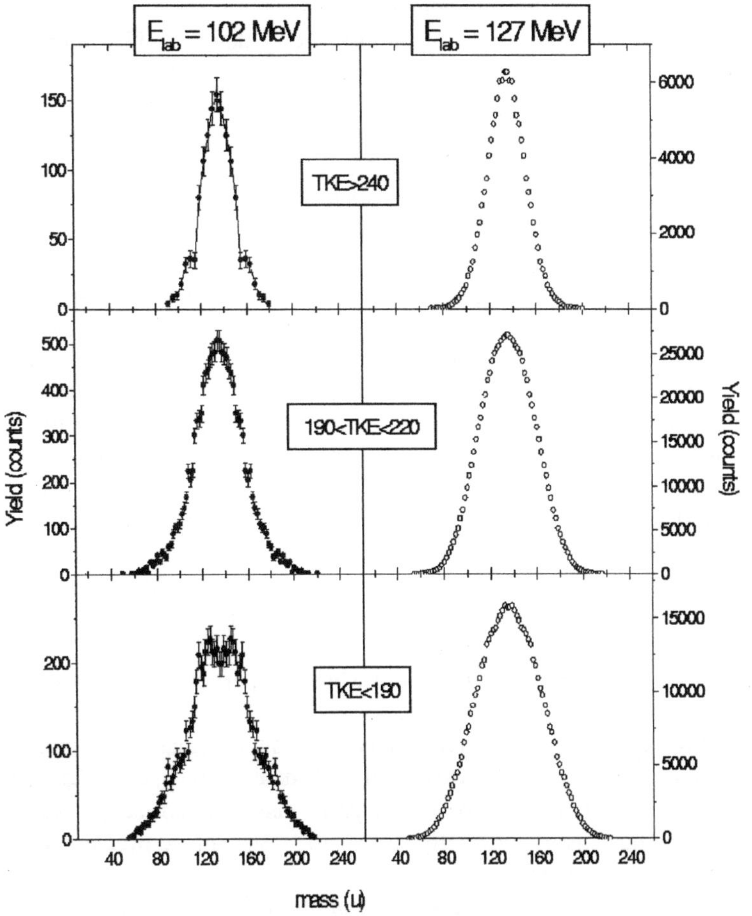

Figure 3. ^{270}Sg fission fragment mass distribution for specific TKE ranges

shows that the average TKE of both characteristics is practically the same, but the characteristics differ in shape. As shown in Fig. 2d, the parabola is steeper in the case of symmetric masses. It is clearly visible in Fig. 2e, which presents a comparison of Y(TKE), for lower excitation, in the region of TKE=220-240 MeV, the yields are lower, the distribution is narrower for the low excitation of the nucleus than for the high excitation. In the region of TKE>250 MeV, the closed dots are practically coincident with open ones and form a high-energy "arm" on the yield slope. Our approximate estimate of the average TKE for the high-energy mode is ~240 MeV, which is in good agreement with other data [13, 14].

Fig. 2f shows the fragment mass distribution σ^2_M as a function of TKE. For lower energy, this characteristic slopes steeply down, as TKE increases, and it approaches zero at the maximal TKE, which suggests that the mass distributions of fragments for this region of TKE are very narrow.

Fig. 3 shows the mass distributions of fragments for the selected ranges of TKE. For E_{lab}=102 MeV and high TKE>240 MeV, the sum of two fission modes, one of which is sharp, narrow and the other is wide, is clearly visible. The nature of the high-energy component has already been discussed, now we will dwell on the wide one. Its properties appear to be close to those of the standard II mode, but there are some peculiar features in the case being considered. As discussed above, this mode is believed to be connected with the manifestation of the deformed shell with Z=56. Then, the light fission fragment of Sg will have Z=50, that is, a magic characteristic of the heated nucleus and whose properties are close to those of the liquid-drop component proton number. Assuming N and Z of fragments to be proportional to their masses, the light fragment with Z=50 will have N=77 and the mass M=127. As shown at the bottom of the Fig. 3, at E_{lab}=102 and TKE<190 the top of the distribution is two-humped in shape, and the peak corresponding to the light fragment is at M=126. Although the statistics is not high and the error is great, it is possible that we observe manifestation of this mode.

In addition, the same figure shows that at TKE<190 MeV there is a wide low-energy pediment that is most likely to be connected with the fission mode characteristic of the heated nucleus and whose properties are close to those of the liquid-drop component.

Fig. 4 presents the preliminary data on the mass distributions of the fragments of ^{271}Hs at E_{lab}=102 MeV. Only 365 pair fission events were collected in the experiment, but a sharp peak is clearly visible at M≈A/2, because, as for neutrons, this nucleus is practically similar to ^{270}Sg. So, we probably also observe manifestation of the super short mode.

Thus, we have observed manifestation of the multimodal structure of the MED of the fission fragments of ^{270}Sg and ^{271}Hs at middle excitation energies. At $E_{lab}(^{22}Ne)$=102 Mev, narrow symmetric structure was found to be present in the total mass distribution of the fragments. This structure is connected with

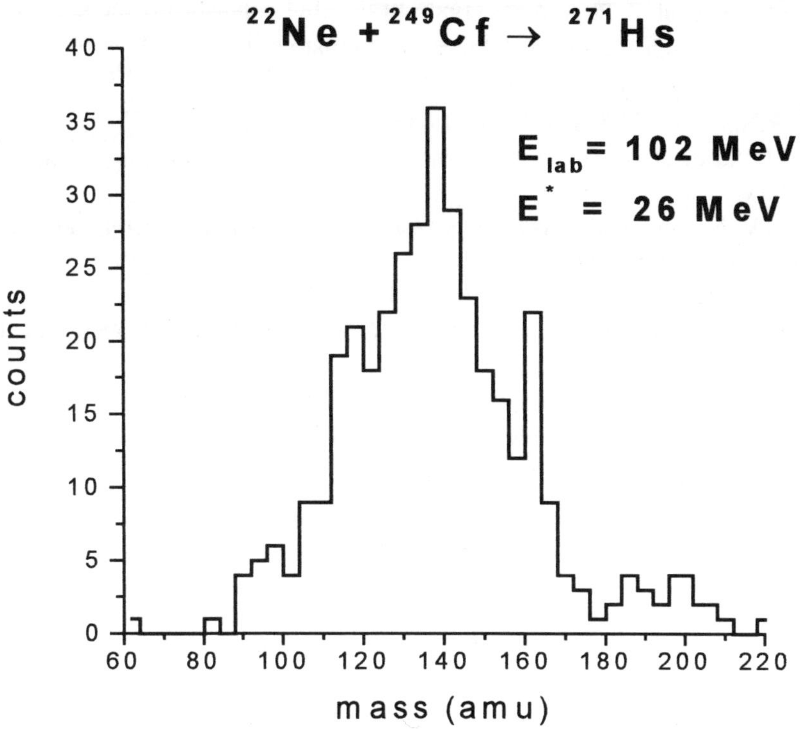

Figure 4. ^{271}Hs fission fragment mass distribution for the ^{22}Ne energy of 102 MeV.

manifestation in both fragments of the magic neutron number N=82. The wide structure that forms the pediment of the distribution is likely to consist of two modes. So, it can be stated that the investigation of the deep subcoulomb fission of superheavy nuclei at rather low energies gives abundant information about the properties of fission.

References

1. E.M.Kozulin et al., in *Heavy Ion Physics*, scientific Report 1995-1996, JINR, FLNR, Dubna (1997) p. 215; N.A.Kondratiev et al., Fourth Int. Conf. on *Dynamical Aspect of Nuclear Fission (DANF'98)*, Casta-Papiernicka, Slovak Republic, October 1998. (in press).
2. Chubaryan G.G. et al. Phys. At Nucl. **56** (1993) 286.
3. Toke J. et al. Nucl. Phys. **A440** (1985) 327.

4. H.Keller et al. Z. Phys. **A326** (1987) 313.
5. K.Lutzenkirchen et al. Nucl. Phys. **A452** (1986) 351.
6. R.Bass Nucl. Phys. **A231** (1974) 45.
7. J.R.Nix, and W.J.Swiatecki, Nucl. Phys. **71** (1965) 1.
8. G.D.Adeev et al., Sov. J. Part. Nucl., **19** (1988) 529; Nucl. Phys **A502** (1989) 405 c.
9. Yu.V.Pyatkov et al. Nucl. Phys. **A624** (1997) 140; Izv. RAN, ser. Fiz. **60** (1996) 30, (in Russian)
10. R.A.Cherghescu et al. J. Phys. **G23** (1997) 1715.
11. W.Q.Shen et al. Phys. Rev. C **36** (1987) 115; B.B.Back et al. Phys. Rev. C **41** (1990) 1495
12. M.G.Itkis and A.Ya.Rusanov, Phys. Part. Nucl. **29** (1998) 160.
13. Hullet E.K. et al. Phys. rev. Lett, **56** (1986) 313; Phys. Rev. C **40** (1989) 770.
14. Lane M.R. et al. Phys. Rev. C **53** (1996) 2893.

SYNTHESIS OF SUPERHEAVY ELEMENTS
A FLUCTUATION-DISSIPATION DYNAMICAL APPROACH

T. WADA, T. TOKUDA, K. OKAZAKI, M. OHTA
Department of Physics, Konan University, 8-9-1 Okamoto, Kobe 658-8501, Japan
E-mail: wada@konan-u.ac.jp

Y. ARITOMO
Flerov Laboratory of Nuclear Reactions, JINR, Dubna, Moscow region 141980, Russia

Y. ABE
Yukawa Institute for Theoretical Physics, Kyoto University, Kyoto 606-8501, Japan

The dynamical process of synthesizing superheavy elements is studied on the basis of the fluctuation-dissipation dynamics. For the study of the symmetric incident channel, a two-dimensional Langevin equation is used for the first stage of the process, from the contact of two nuclei to the time when the relative kinetic energy completely dissipates. For the description of the following stage, a two-dimensional Smoluchowski equation is adopted taking into account the temperature dependent shell correction energy and the competition between fission and neutron evaporations are described with the statistical model. Owing to the stronger diffusion, the fusion probability is larger for higher incident energy. For lower incident energy, owing to the restoration of the shell correction energy, fission is suppressed and the survival provability is larger. The competition between these two factors brings about the optimum energy around 20-30 MeV. ¿From the study of the isotope dependence of the production cross section, the survival probability is found to be very sensitive to the separation energy of neutron. A three-dimensional model is proposed to study the mass-asymmetric incident channels. The results show the importance of the use of the neutron-rich beams and targets.

1 Introduction

The synthesis of the superheavy nucleus (Z=114, N=184) which is predicted to be a new doubly magic nucleus is a long-standing important subject in nuclear physics [1]. The macroscopic fission barrier vanishes at this atomic number and the nucleus is stabilized against fission due to the strong shell correction energy. This special feature makes the synthesis of this element challenging. Recent findings of the new elements were accomplished mainly with the so-called *cold fusion* reactions using Pb or Bi targets [2]. In this approach, by using the nuclei which have large shell correction stability as a target, we can produce the compound nuclei in very low excitation energy, and thus we can expect large survival probability against fission. On the other hand, number of neutrons in the compound system can be increased by using the actinide targets with the combination of the neutron-rich projectiles like ^{48}Ca. In this case, the excitation energy becomes higher than the former case, hence called *warm* or *hot fusion* reactions.

To understand the mechanism of synthesizing superheavy elements, we have to treat the whole fusion-fission process, from the touching of two colliding nuclei to the formation of spherical compound nucleus, and further to the reseparation. For this purpose, we propose to use the framework of the fluctuation-dissipation dynamics, which is expressed by the Kramers (Fokker-Planck) equation or by the Langevin equation. The framework is necessary and appropriate because the formation of spherical compound nucleus can be accomplished only by fluctuation of collective motion in superheavy region.

After the formation of the spherical compound nucleus, it needs to survive against fission decay and to be cooled down by particle evaporations. In estimating the competition between fission decay and particle evaporations in superheavy region, it is necessary to take into account the stabilization due to the shell correction energy which is predicted to be as large as 10 MeV. The important point is that the shell correction energy depends on the excitation energy; at high excitation energy, the shell correction energy disappears.

In the present paper, we first study the symmetric incident channels. We divide the whole dynamical process into two stages; the first stage describes the process from the touching of the two colliding nuclei to the complete dissipation of the initial kinetic energy of relative motion, and in the second stage, the formation of spherical compound nucleus and the competition between fission and neutron evaporation take place.

In the preliminary analysis[3], to study the dynamics in the second stage, we employ the one-dimensional Smoluchowski equation in the elongation degree of freedom taking into account the temperature dependent shell correction energy. We pointed out a new possibility of synthesizing the superheavy element with relatively hot fusion reaction. However, for the first stage, the treatment with the Smoluchowski equation is not appropriate because it is not capable of treating the effect of the initial momentum. Moreover, the difference between the fusion path and the fission path is neglected in the one-dimensional model.

In the new model, we employ the two-dimensional Langevin equation including the elongation and necking degrees of freedom, in the first stage. Then in the second stage, we calculate the evaporation residue cross section of superheavy elements using the two-dimensional Smoluchowski equation.

The next task is to extend our framework to incorporate with the mass-asymmetric incident channels. By changing the incident mass-asymmetry, we can control the problem of the fusion hindrance[4]. For this purpose, a three-dimensional model is proposed. The formation probability of the spherical compound nuclei is calculated with a three-dimensional Langevin equation including the elongation, necking, and mass-asymmetry degrees of freedom. After the formation of the compound nuclei, the survival probability is estimated by solving a one-dimensional Smoluchowski equation taking into account the particle evaporations and the temperature dependent shell correction energy.

In Section 2, the fusion mechanism of two massive nuclei is discussed and the thermal fluctuation around the mean trajectory is introduced. We briefly explain our framework in Section 3. Results for the excitation function of the evaporation residue cross section are presented in Section 4. In Section 5, the isotope dependence is discussed in connection with the neutron separation energies. Mass-asymmetric incident channels are treated in Section 6 with a three-dimensional model. Summary is given in Section 7.

2 Fusion of massive nuclei

The hindrance of fusion is observed in the symmetric-like fusion reactions forming $Z > 80$ nuclei [5], where Z denotes the atomic number of the fused system. Several interpretations have been given for this phenomenon [6]. Our interpretation is based on the work by Swiatecki and his coworkers [4]. We consider the dynamical motion in nuclear deformation space under a strong friction.

For light systems ($Z < 80$), fusion occurs when the two nuclei come into contact overcoming the incident Coulomb barrier. Because the Coulomb repulsion between two nuclei is rather small, the attractive nuclear force makes the system to form a spherical compound nucleus and fusion occurs with large probability. This situation is described successfully by the Bass model [7].

For heavy systems ($Z > 80$), the Coulomb repulsion becomes strong and starts to compete with nuclear attraction even after the contact of two nuclei. Therefore, to go to the spherical region, the system needs extra energy after the contact. This situation is described by the extra push model [4].

For superheavy mass systems ($Z > 100$), the situation drastically changes. The Coulomb repulsion is extremely strong so that the nuclear force cannot keep the system to form a spherical compound nucleus. Only the quantum effect, the shell correction energy, can stabilize the system from fissioning. In this case, even when we put large extra energy, almost all the colliding nuclei go down the potential slope and decay by fission immediately. Only by the thermal fluctuation around the mean trajectory, the system can come close to the spherical shape. The fluctuation reflects the fact that the energy flow is not necessarily from the collective degree of freedom to nucleonic degrees of freedom but the flow in the reverse direction is also important. The fluctuating force can be expressed as a random force and the trajectory can be traced by means of the Brownian motion. From this viewpoint, the fusion probability is a smooth function of the incident energy and there is a finite probability to fuse with a relatively small extra energy. For the synthesis of superheavy elements, this is the main mechanism to form spherical compound nucleus.

3 Two-dimensional model for symmetric channels

For the symmetric incident channels where the phenomenon of fusion-hindrance is prominent, we divide the whole fusion-fission process into two stages. The first stage is from the contact of the two colliding nuclei to the complete damping of the initial kinetic energy. In the second stage, the system may form a spherical compound nucleus or may reseparate into two fragments immediately; when the spherical nucleus is formed, the competition between neutron evaporation and fission takes place.

In the first stage, the fusion process is described with the multi-dimensional Langevin equation to treat the effect of the initial relative momentum properly. We adopt the two-dimensional nuclear deformation space with the two-center parametrization [8]. The neck parameter ϵ is fixed to 1.0 in the present calculation, deformation parameters δ_1 and δ_2 of the colliding nuclei are taken to be equal, *i.e.*, $\delta_1 = \delta_2 = \delta$, and we set the asymmetry parameter α=0. We treat z_0 (distance between two potential centers) and δ as the two collective parameters to be described by the Langevin equation. The multi-dimensional Langevin equation is given in the following form,

$$\begin{aligned}
\frac{dq_i}{dt} &= \left(m^{-1}\right)_{ij} p_j, \\
\frac{dp_i}{dt} &= -\frac{\partial V}{\partial q_i} - \frac{1}{2}\frac{\partial}{\partial q_i}\left(m^{-1}\right)_{jk} p_j p_k - \gamma_{ij}\left(m^{-1}\right)_{jk} p_k + g_{ij} R_j(t),
\end{aligned} \qquad (1)$$

where V is the potential energy, m_{ij} and γ_{ij} are the shape-dependent collective inertia and dissipation tensors, respectively. The normalized random force $R_i(t)$, is assumed to be a white noise, *i.e.*, $\langle R_i(t)\rangle = 0$ and $\langle R_i(t_1)R_j(t_2)\rangle = 2\delta_{ij}\delta(t_1 - t_2)$. The strength of the random force g_{ij} is given by $\gamma_{ij}T = g_{ik}g_{jk}$, where T is the temperature of the compound nucleus calculated from the excitation energy as $E_x = aT^2$ with a denoting the level density parameter. The potential is calculated as the sum of a generalized surface energy [9], Coulomb energy, and the centrifugal energy with the moment of inertia of the rigid body. Hydrodynamical inertia tensor is adopted with the Werner-Wheeler approximation for the velocity field, and the wall-and-window one-body dissipation [10] is adopted for the dissipation tensor. At t=0, each trajectory starts from z_0=1.6R_0 and δ=0 corresponding to the contact configuration with the initial velocity in the z_0 direction, where R_0 denotes the radius of the spherical compound nucleus. Excitation energy of the composite system E_x is calculated for each trajectory as,

$$E_x = E_0 - \frac{1}{2}\left(m^{-1}\right)_{ij} p_i p_j - V(q), \tag{2}$$

where E_0 is given as $E_0 = E_{\rm cm} - Q$ with Q denoting the Q-value of this reaction and $E_{\rm cm}$ the incident energy in the center-of-mass frame.

As the probability of forming spherical compound nucleus is extremely small in the case of superheavy element synthesis, the treatment with the Langevin equation is not practical. Therefore in the second stage after the complete dissipation of the initial kinetic energy, the evolution of the probability distribution $P(q,l;t)$ in the collective coordinate space is described with the Smoluchowski equation,

$$\frac{\partial}{\partial t}P(q,l;t) = \left(\gamma^{-1}\right)_{ij}\left[\frac{\partial}{\partial q_i}\left\{\frac{\partial V(q,l;t)}{\partial q_j}P(q,l;t)\right\} + T\frac{\partial^2}{\partial q_i \partial q_j}P(q,l;t)\right]. \tag{3}$$

The angular momentum of the system is expressed by l. The temperature dependent shell correction energy is added to the macroscopic potential energy,

$$V(q,l;t) = V_{\rm macro}(q,l) + V_{\rm shell}(q)\Phi(t), \tag{4}$$

where $V_{\rm macro}$ is the same as that used in the previous stage and $V_{\rm shell}$ is the shell correction energy at $T = 0$.

The temperature dependence of the shell correction energy is extracted from the free energy calculated with single particle energy[11]. The temperature dependent factor $\Phi(t)$ in Eq. (4) is parametrized as $\Phi(t) = \exp\left(-aT^2(t)/E_d\right)$, following the work by Ignatyuk *et al.*[12]. The shell-damping energy E_d is chosen as 20 MeV. The compound nucleus is cooled down through particle evaporations, which is expressed in terms of the cooling curve $T(t)$. The particle evaporations are simulated by the statistical model code SIMDEC[11].

We calculate the evaporation residue probability $d(E_0,l;t)$ as a function of time, which is the probability of being inside the fission barrier,

$$d(E_0,l;t) = \int_{\text{inside saddle}} P(q,l;t)dq. \tag{5}$$

The evaporation residue cross section $\sigma_{\rm EV}$ is calculated from the evaporation residue probability in the final stage of the cooling process as,

$$\sigma_{\rm EV} = \frac{\pi\hbar^2}{2\mu_0 E_{\rm cm}}\sum_l (2l+1)d(E_0,l;t=\infty), \tag{6}$$

where μ_0 denotes the reduced mass in the entrance channel.

4 Excitation function of the evaporation residue cross section

In this section, we present the excitation function calculated with the present dynamical model, considering the reaction ${}^{149}_{57}\mathrm{La} + {}^{149}_{57}\mathrm{La} \rightarrow {}^{298}114$. According to the calculation using the Langevin equation, the initial relative kinetic energy is found to dissipate completely around $t_{\mathrm{diss}} = 0.5 \times 10^{-21}$s. Therefore, it is valid to switch from the Langevin equation to the Smoluchowski equation at this time. The distribution of the Langevin trajectories at $t = t_{\mathrm{diss}}$ is adopted as the initial distribution of the Smoluchowski equation.

The time dependent feature of the probability $d(E_0, l = 10; t)$ is shown in Fig. 1 for five different incident energies E_0 = 15, 20, 30, 40, and 50 MeV. Up to the time of $(10 - 20) \times 10^{-21}$s, the probability density in the region of the compact configuration is supplied by diffusion from the contact region and its yield increases rapidly. But during that time, the main part of the probability descends the potential slope and the supply stops. For larger incident energy like E_0 = 50 MeV, the initial distribution comes to more compact shape and also the diffusion coefficient $D = T/\gamma$ is larger. Therefore, the peak value increases monotonically with E_0. However, due to the low fission barrier, the probability accumulated in the compact configuration area diffuses back over the fission barrier. The probability d decreases drastically, and the final probability remains this region is small. On the other hand, for low incident energy such as E_0 = 15 MeV, an extremely small probability can come to the compact configuration area. As the fission barrier, however, is high, most of the probability survives against fission.

The height of the peak is essentially determined by the diffusibility into the compact configuration area, while the decrease from the peak value to the final yield at $t_\infty = 2000 \times 10^{-21}$s is determined by how fast the fission barrier glows enough by the restoration of the shell correction energy. Thus, the final yield surviving in the compact configuration area is determined by the two factors; the diffusibility depending on the temperature and the restoration of the shell correction energy.

In terms of the obtained values of $d(E_0, l; t_\infty)$, we calculate the evaporation residue cross section σ_{EV} by summing up the partial wave contributions according to Eq. (6). The excitation function of σ_{EV} for the ${}^{149}\mathrm{La} + {}^{149}\mathrm{La} \rightarrow {}^{298}114$ reaction is shown in Fig. 2 by solid line. For this reaction system, the Bass potential barrier height [7] is 320 MeV in center-of-mass frame and corresponds to $E_0 = 9$ MeV in the compound nucleus as shown in Fig. 2. The excitation function shows a bell shape having a maximum around $E_0 \sim 25$ MeV which is produced by the new mechanism discussed above, *i.e.*, the competition between high fusion probability at high incident energy due to large

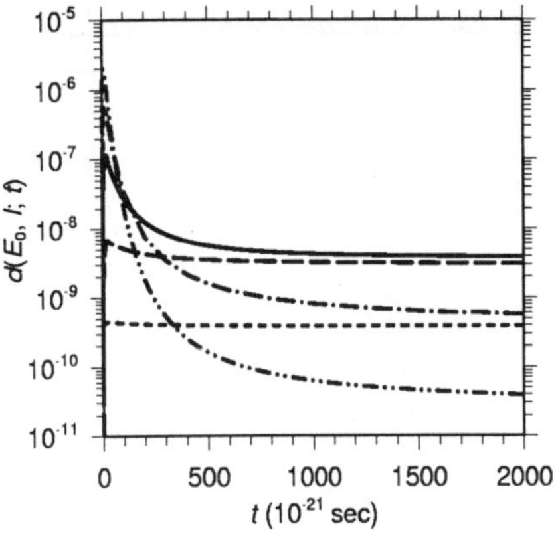

Figure 1: Time evolution of the probability density in the compact configuration region $d(E_0, l = 10; t)$ for five excitation energies; E_0 = 15 (short-dashed), 20 (long-dashed), 30 (solid), 40 (dot-dashed), and 50 MeV (dot-dot-dashed).

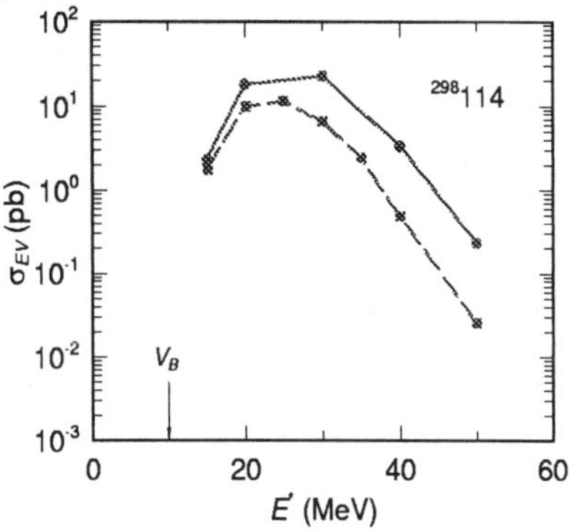

Figure 2: Excitation function of the evaporation residue cross section for ${}^{149}_{57}\mathrm{La}$ + ${}^{149}_{57}\mathrm{La} \rightarrow {}^{298}114$ reaction. The Bass barrier energy is indicated by the arrow. The solid line denotes the result with the present model, while the dashed line denotes that with the one-dimensional model.

diffusibility and the high survival probability at low excitation energy due to the restoration of the fission barrier.

In Fig. 2, the result with the one-dimensional model is also shown in dashed line. The essential feature is common in the present model and in the one-dimensional model. The difference is essentially due to the neglect of the initial kinetic energy in the one-dimensional model.

5 Isotope dependence of the cross section

In this section, we investigate the dependence of the evaporation residue cross section on neutron numbers with the one-dimensional model. Large evaporation residue cross section is expected for neutron-rich isotopes, because they generally have large neutron decay widths and thereby favored in the survival probability.

We calculated the evaporation residue cross section for a series of Z=114 isotopes from N=176 to 184 with the one-dimensional model [3]. We used different cooling curves for each isotope while we neglected the isotope dependence of the energy surface. The energy surface for N=184 was used for all cases. Figure 3 shows the calculated evaporation residue cross section for ${}^{145,146,147,148,149}_{57}\mathrm{La}$ + ${}^{145,146,147,148,149}_{57}\mathrm{La} \rightarrow {}^{290,292,294,296,298}114$ reactions as functions of initial excitation energy E_0. The isotope dependence of evaporation residue cross section is found to be very strong. Larger evaporation

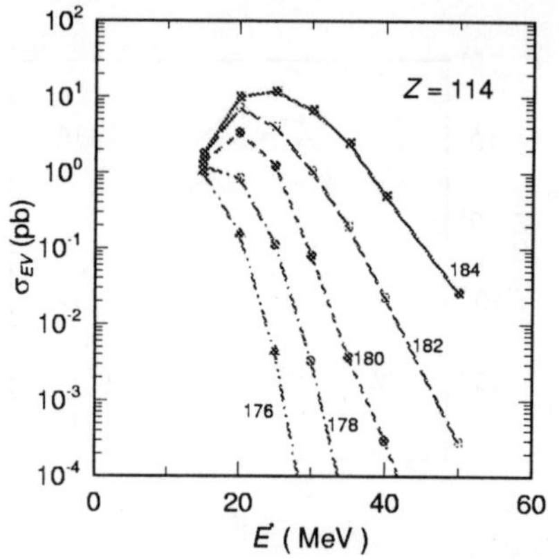

Figure 3: Isotope dependence of the excitation function of the evaporation residue cross section for Z=114. Figures denote neutron numbers.

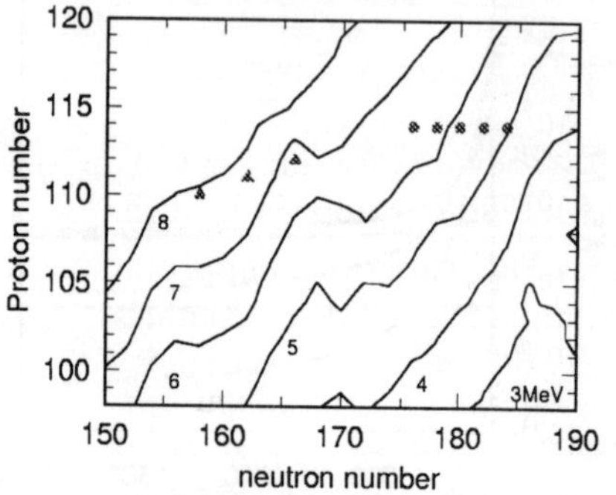

Figure 4: Contour map of the neutron separation energy $\langle B_n \rangle$. Three triangles indicate the isotopes produced in GSI. The dots indicate those which are investigated in the present work.

residue cross section is seen in neutron rich nucleus as is expected. The excitation function changes from the monotonically decreasing one to the bell shape one with the increase of the neutron number.

In Fig. 4, a contour map of the neutron separation energy averaged over four successive neutron emissions $\langle B_n \rangle$ is displayed [13]. The figures associated with lines give the values of $\langle B_n \rangle$. Naturally, neutron-rich isotopes have smaller B_n's than neutron-deficient ones. When the neutron separation energy is small, the cooling due to the neutron evaporation is fast. In this case, owing to the restoration of the shell correction energy, fission barrier arises rapidly to keep the compound nucleus from fissioning. Therefore, we can observe the competition between large diffusibility at high temperature and high survival probability at low temperature. On the other hand, when the neutron separation energy is large, the cooling is slow. In this case, the fission barrier recovers slowly and we obtain monotonically decreasing excitation function. The cooling speed is very important in this dynamical mechanism.

Three triangles in Fig. 4 indicate isotopes which are synthesized in GSI by the *cold fusion*. The values of $\langle B_n \rangle$ are $7-8$ MeV, thus we cannot expect quick cooling which gives rise to the enhancement at higher energies. On the other hand, the dots indicate those which are investigated in the present work with $\langle B_n \rangle$ equal to $5-6$ MeV. It is expected that even more neutron rich isotopes are more favorable for the enhancement of the residue cross sections. Thus, an exploration of experimental feasibility to reach the neutron rich side of the superheavy elements is an extremely interesting and urgent subject.

6 Mass-asymmetric incident channels

So far, we only have treated the symmetric incident channels. But in both cold fusion and hot fusion reactions, experiments have been performed using mass-asymmetric incident channels. A three-dimensional model is developed to extend our approach to the mass-asymmetric case.

There are three important factors in considering the effectivity of the mass-asymmetric incident channels for the synthesis of superheavy elements. The first one is the magnitude of the hindrance for fusion. As it is essentially determined by the magnitude of the Coulomb repulsion, it is expected to be weaker in mass-asymmetric incident channels, and therefore we can expect larger probability of the formation of spherical compound nuclei for more mass-asymmetric case. The second one is the minimum excitation every $E^*_{\rm min}$ which can be attained at the incident energy corresponding to the Bass barrier height. Smaller value of $E^*_{\rm min}$ is preferable for larger survival probability. $E^*_{\rm min}$ is calculated as $E^*_{\rm min} = E_{\rm Bass} - Q$, where $E_{\rm Bass}$ is the center-of-mass energy corresponding to the Bass barrier [7] and Q is the Q-value of this reaction. Both of them decrease in their magnitude as the mass-symmetry becomes larger. And the resulting $E^*_{\rm min}$ is found to increase with the increase of the mass-asymmetry. The last one is the number of neutrons in the compound system. As we learned from the study of the isotope dependence of the evaporation residue cross section, the separation energy of neutrons strongly affects the speed of the cooling. Faster cooling means the faster restoration of the shell correction energy, hence results in the larger survival probability. On the other hand, in our framework, the formation probability is not very sensitive to the neutron numbers. Thus, if we neglect the problem of the intensify of the beam, use of the neutron-rich beam is favored.

In our three-dimensional model, as the magnitude of the fusion hindrance is weaker for mass-asymmetric channels, we calculate the probability of the formation of spherical compound nuclei above the Bass barrier energy by the Langevin approach without approximating it by the Smoluchowski equation. The method of calculating the formation cross section with the Langevin equation is found in Ref. 14. Below the Bass barrier energy, the formation probability is estimated by treating the barrier penetration problem in WKB approximation. After the formation of the spherical compound nuclei, a one-dimensional Smoluchowski equation is used along the symmetric fission path to estimate the survival probability including the temperature dependent shell correction energy. The effect of the particle evaporation is treated as the same way as the previous models. We adopt again the two-center parametrization for the model space: the collective coordinates are Z_0, δ, and the mass-asymmetry

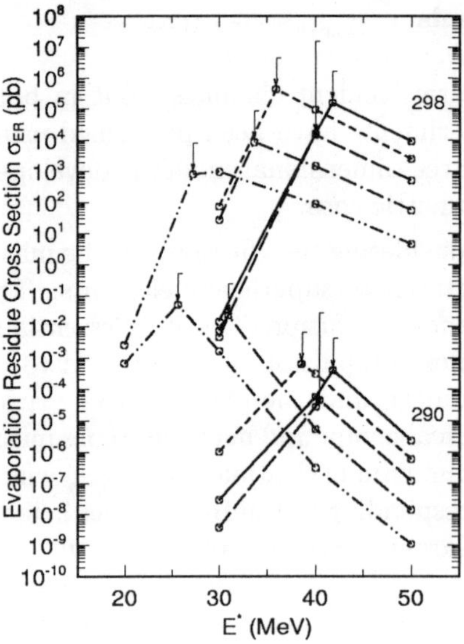

Figure 5: Excitation function of σ_{EV} for mass-asymmetric incident channels with various values of α: $\alpha = 0.70$ (solid), 0.65 (short dashed), 0.60 (long dashed), 0.55 (dot-dashed), and 0.50 (dot-dot-dashed). Two groups of lines correspond to different mass number of the compound systems, $A =$ 298 and 290. Arrows denote the corresponding Bass barrier energies.

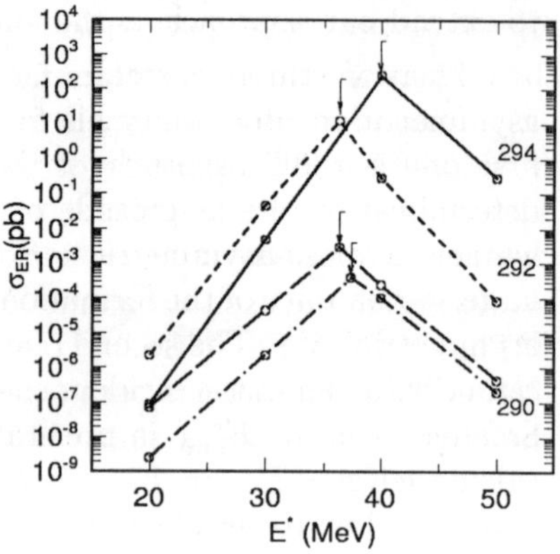

Figure 6: Excitation function of σ_{EV} for realistic incident channels. Figures attached to the lines denote the mass number of the compound systems. Solid line is for ^{258}Fm+^{32}Si, short dashed line for ^{244}Pu+^{48}Ca, long dashed line for ^{238}U+^{52}Ti, and dot-dashed line for ^{234}Th+^{56}Cr. Arrows denote the corresponding Bass barrier energies.

parameter α; $\alpha = (A_1 - A_2)/(A_1 + A_2)$, where A_1 and A_2 denote the mass numbers of the target and the projectile, respectively.

In this model space, there are two paths for the fusion reaction. One is essentially the reverse process of fission; two nuclei form a deformed mononuclear system and it changes its shape into a sphere along Z_0 coordinate. The other is the path through the Bussinaro-Gallone saddle point along α coordinate. For very asymmetric case, the latter process is not negligible. But usually, the main process is found to be the former one.

The formation probability becomes larger for larger mass-asymmetric case. However, at the same time, $E^*_{\rm min}$ becomes higher. The survival probability against fission strongly depends on the excitation energy of the compound system. When the neutron separation energy is not small enough, the survival probability decreases very rapidly.

Figure 5 shows the calculated evaporation residue cross section with various incident channels forming the element $Z = 114$ with N=184 and $N = 176$. Five values of the mass-asymmetry parameter α are taken; $\alpha = 0.70, 0.65, 0.60,$

0.55, and 0.50. In the figure, arrows denote the corresponding Bass barrier energies. The formation probability doesn't depend on the neutron number very much in the present model. In the case of $\alpha = 0.70$, the formation probability is about 20% and it is of the order of 10^{-4} for $\alpha = 0.50$. It is seen that the lines are divided into two groups according to their neutron numbers. For $N = 184$ ($A = 298$), since the decrease of the survival probability with the excitation energy is small, strongly mass-asymmetric channels which have large formation probability have larger cross section. On the other hand, for $N = 176$ ($A = 290$), since the survival probability decreases rapidly, the resulting evaporation residue cross section is very small and mildly mass-asymmetric channels have larger cross section.

In Fig. 6, the excitation function of σ_{EV} is shown for four *realistic* channels, ^{258}Fm+^{32}Si, ^{244}Pu+^{48}Ca, ^{238}U+^{52}Ti, and ^{234}Th+^{56}Cr. Among them, the ^{244}Pu+^{48}Ca channel has been studied in FLNR, JINR, Dubna. It is seen that the cross section has a sharp peak around the Bass barrier energy and the maximum value becomes larger for more neutron-rich system. The absolute value of the cross section depends on the values of the parameters, which have not been thoroughly examined yet.

7 Summary

The fluctuation-dissipation dynamics in the deformation parameter space is shown to be a necessary and an appropriate way to describe the fusion-fission process in heavy mass region. We use the multi-dimensional Langevin equation for the first stage and the Smoluchowski equation for the following stage taking into account the particle evaporations and the temperature dependent shell correction energy. We apply this approach for the synthesis of superheavy elements and estimated the evaporation residue cross sections starting from the contact of two colliding nuclei.

With the two-dimensional model for the symmetric incident channel, it is shown that there exists the optimum temperature or the excitation energy of compound system for the synthesis of superheavy elements due to the competition between the diffusibility for fusion and the restoration of the shell correction energy against fission. For the symmetric case, the maximum cross section is obtained around $E_0 = 20{\sim}30$ MeV. The characteristic features are the same as those obtained with the previous one-dimensional analysis.

It is found that the position of the optimum energy strongly depends on the time scale of the cooling by neutron evaporation. Large cross section is obtained for neutron-rich system because of the fast cooling. The excitation function of the evaporation residue cross section changes from a monotonically decreasing shape to a bell shape as the number of neutron increases.

A three-dimensional model is used to study the mass-asymmetric incident channels. It is shown that the strongly asymmetric channel results in large formation probability, but it results in high excitation energy at the Bass barrier energy. Since the survival probability strongly depends on the neutron separation energy of the compound system, the use of the neutron-rich beam and target is highly desirable. The evaporation residue cross section is estimated for the mass-asymmetric channels which can be studied experimentally. The maximum cross section is expected around the Bass barrier energy.

Acknowledgements

This work is partly supported by the Grant-in-Aid for Scientific Research (No. 10640287) of the Japan Ministry of Education, Science, Sports and Culture.

References

1. P. Armbruster, Annu. Rev. Nucl. Sci., **35** (1985) 135; Y.T. Oganessian and Y.A. Lazarev, *Treatise on Heavy-Ion Science* (Plenum, 1985) pp.3-251; G. Münzenberg, Rep. Prog. Phys., **51** (1988) 57.
2. S. Hofmann *et al.*, Z. Phys., **A350** (1995) 277; S. Hofmann *et al.*, ibid. 281; S. Hofmann *et al.*, Z. Phys. **A354** (1996) 229.
3. Y. Aritomo, T. Wada, M. Ohta, and Y. Abe, Phys. Rev., **C55** (1997) R1011; Y. Aritomo, T. Wada, M. Ohta, and Y. Abe, Phys. Rev., **C59** (1999) 796.
4. W.J. Swiatecki, Nucl. Phys., **A376** (1982) 275.
5. K.-H. Schmidt and W. Morawek, Rep. Prog. Phys., **54** (1994) 949.
6. W. Reisdorf, *Lecture Notes in Physics*, **317** (Springer, 1988) p.26; N.V. Antonenko *et al.*, Phys. Lett., **B319** (1993) 425; C.E. Aguiar *et al.*, Nucl. Phys., **A517** (1990) 205.
7. R. Bass, *Nuclear Reactions with Heavy Ions* (Springer, 1980).
8. J. Maruhn and W. Greiner, Z. Phys., **251** (1972) 431; K. Sato, A. Iwamoto, *et al.*, Z. Phys., **A288** (1978) 383.
9. H.J. Krappe, J.R. Nix, and A.J. Sierk, Phys. Rev., **C20** (1979) 992.
10. J.R. Nix and A.J. Sierk, Nucl. Phys., **A428** (1984) 161c.
11. M. Ohta, Y. Aritomo, T. Tokuda and Y. Abe, Proc. of Tours Symp. on Nuclear Physics II (World Scientific, Singapore, 1995) p.480.
12. A.V. Ignatyuk *et al.*, Sov. J. Nucl. Phys., **21** (1975) 255.
13. P. Möller *et al.*, Atomic Data and Nuclear Data Tables, **59** (1995) 185.
14. T. Tokuda, T. Wada, and M. Ohta, Prog. Theor. Phys., **101** (1999) in press.

FUSION-FISSION DYNAMICS IN THE REACTIONS FOR THE SUPERHEAVY NUCLEUS PRODUCTION

G. FAZIO, G. GIARDINA

Istituto Nazionale di Fisica Nucleare, Sezione di Catania, and Dipartimento di Fisica dell'Universitá di Messina, 98166 Messina, Italy

F. HANAPPE

Universitè Libre de Bruxelles, 1050 Bruxelles, Belgium

A. NASIROV

Bogoliubov Laboratory of Theoretical Physics, JINR, 141980 Dubna, Russia

L. STUTTGÈ

Institut de Recherches Subatomiques, F-67037 Strasbourg Cedex2, France

The investigation on the synthesis of superheavy elements is an interesting and present research field [1,2,3,4,5,6] in order to study the fusion dynamics and the decay properties of nuclei close to the stability island at Z=114. By analyzing the evaporation residues (ER), we study the heavy-ion induced reactions (^{48}Ca, ^{64}Ni, ^{68,70}Zn) on the ^{238}U, ^{208}Pb and ^{209}Bi targets that lead to the Z=110, 112 and 113 compound nuclei.

We investigate the capture cross section of the two interacting nuclei that lead to a dinuclear system (DNS), the quasifission-fusion competition in the entrance channel and the fission–evaporation competition along the de-excitation cascade of the compound nucleus. The evaporation residue cross section is determined from the fusion cross section taking into account the surviving of the compound nucleus against fission, and the complete de-excitation cascade is calculated by an advanced statistical model.

The study of the entrance channel dynamics allows us to determine the beam energy window which is favorable to the fusion (for the Ni + Pb, Zn + Pb, Ca + U and Zn + Bi reactions), while the shell correction to the fission barrier and the dissipative effects influence the compound nucleus evolution. This effect also determines the fission–evaporation competition which leads to the residual nucleus in the superheavy nucleus formation. We also calculate the Γ_n/Γ_{tot} ratio at each step of the de-excitation cascade of the compound nucleus.

1 Introduction

It is known that the cross sections for the production of heavy elements by fusion reaction and using Pb and Bi targets decrease down to a value of about 10 pb for the 109 element formation. It is expected that for the Z=110-113 element formation, when using Pb, U, Bi targets, the cross section production decreases down to about 1 pb. In the last four years new experiments [1,2,3,4] were carried out (by increasing both the efficiencies of the experimental set-up and the beam current intensity) in order to produce the Z=110-113 heavy elements in the cold fusion reactions at GSI, by observing the evaporation residues (ER) after 1, 2 neutron emission, or in hot fusion reactions at Dubna, after 3n, 4n, 5n from the compound nucleus. In addition to the experimental difficulties for the low cross section production (about few pb), another problem arises about the fusion calculation and the de-excitation cascade in order to analyze the experimental data of the evaporation residues. It is impossible to reach an agreement with the experimental data if one does not include the quasifission contribution in competition with complete fusion in order to describe the entrance channel dynamics. According to the dinuclear system (DNS)- concept the fusion of nuclei is considered as a two stage process: the first stage is capture (formation of the DNS after dissipation of part of the relative kinetic energy and orbital angular momentum); the second stage is fusion (transformation of the dinuclear system in a compound nucleus by a continuous multinucleon transfer from a light nucleus to the heavy one, retaining the individuality of nuclei having a shell structure). In this second stage, fusion is in competition with quasifission, that is, the decay of the dinuclear system in two fragments before formation of the compound nucleus. Thus, fusion is treated as a motion in the charge (mass) asymmetry variable, and capture and quasifission are a consequence of the relative motion of nuclei in DNS. Use of the optical model does not allow one to study the dynamics of capture. The microscopic model [7,8,9] on the other hand includes the peculiarities of the realistic single-particle states of the projectile and target nuclei and can be applied to analyse the capture stage before fusion - which is the main subject of this paper. The radial friction coefficients, the dynamic change of the nucleus-nucleus potential and the reduced mass of the relative motion are determined from the evolution of the coupling term between the relative and intrinsic motions. The kinetic energy dissipation in deep-inelastic heavy ion collisions is a gradual process, and the internuclear nucleon exchange and particle-hole excitation are the main reasons for the dissipated energy.

The energy window for the capture was revealed. If there is no capture, nuclei cannot be fused due to the smallness of the interaction time and excitation energy generated from the relative kinetic energy. Therefore, the present model is required to analyse the fusion of light nuclei at high energies and the fusion of massive nuclei at all collision energies where quasifission plays a decisive role in decreasing the fusion cross section [10,11]. As a result of the calculation we find that: *i*) a number of partial waves with angular momentum lead to capture (*i.e.*, those partial waves which have been trapped in a potential minimum of the entrance channel; the critical value of angular momentum l_{cr} for a given collision energy can be calculated dynamically);

ii) some amount of the excitation energy is available before complete fusion at a given collision energy and impact parameter or orbital angular momentum;

iii) a beam energy window is favorable to the fusion: the low limit (E_{min}) of this window is defined by the dynamical barrier in the entrance channel before the capture, and the upper limit (E_{max}) appears because friction forces cannot provide the intense initial kinetic energy loss necessary to trap the dinuclear system in a pocket.

After fusion, the de-excitation cascade of the compound nucleus was analyzed in detail in the framework of the advanced statistical model (ASM) approximation [12]. With this model it is possible to trace the Γ_n/Γ_{tot} values at each step of the de-excitation cascade and to have information on the consequences for the fission process dynamics. The model allows us to consider the influence of the nuclear viscosity and the dynamical dependence of the shell correction to the fission barrier on the calculated ER cross sections. The angular-momentum dependence of the shell correction to the fission barrier, and the effects of the nuclear viscosity and dynamical deformation for these fissile excited nuclei are considered. Particular attention is devoted to the determination of the level densities. These are caculated in the non-adiabatic approach allowing for the rotational and vibrational enhancements. These collective effects are gradually removed above a certain energy. In the case of the rotational enhancement this energy is related to the Coriolis force wich couples intrinsic and collective motions. Our level densities acquire a dynamic aspects through the dipendence of the Coriolis force and of the rotational enhancement of the nuclear shape, which is, in turn, obtained form the classical model of the rotating liquid drop. Intrinsic level densities are calculated using the Ignatyuk approach [13], which takes into account shell-structure effets and pairing correlations. Use of the correct level densities is of fundamental importance for this analysis as they determine the phase space available for each channel, it is essentially this that governs any statistical decay. In the case of the ER production one should also carefully consider the low-energy level densities since this is the energy interval in which most of the ERs are formed.

That is why we use the superfluid model of the nucleus [14] in our calculations with the standard value of pairing correction $\Delta = 12/\sqrt{A}$. The yrast lines are automatically included in our calculations by the requirement that the total excitation energy should be higher than the rotational one, otherwise the level density is set to zero.

2 General remarks on the formalism

The role of the entrance channel in the fusion-fission reactions leading to nearly the same compound nucleus was studied in several experiments [15,16,17] but the fusion mechanism has not been fully understood yet. This fact is very significant for the fusion of massive nuclei where there are large and qualitative differences between capture and fusion processes. The entrance channel effects were also discussed in [18,19] where authors affirmed that in order to fully understand the various data obtained in heavy ion fusion-fission reactions one must include the precompound nucleus dynamics in the analysis of these reactions. We study the precompound nucleus dynamics through the evolution of DNS (after capture) to complete fusion (that is in competition with the quasifission process). The precompound emission of the gamma-rays, neutrons, protons and alpha particles is not considered in this paper; this effect will be the subject of a future work. The excited compound nucleus can undergo fission or emit particles before the reaction products (evaporation residues, or fission fragments, or light particles) are detected.

In the suggested model, the relevant variables are energies and occupation numbers of single-particle states of nucleons in nuclei (the intrinsic degrees of freedom), the relative motion of nuclei at the first (capture stage), and the mass (charge) asymmetry degrees of freedom at the second (fusion) stage. The competition between complete fusion and quasifission of the dinuclear system formed after capture, under further evolution, is taken into account by the comparison between the phase spaces of the initial and final states. In contrast to [20], the capture probability $T_l^{capture}(E,l)$ is estimated in a dynamic approach, considering the evolution of the coupling between the relative motion of nuclei and intrinsic motion of their nucleons. In the capture stage the system is trapped in a potential wall due to the dissipation of the initial kinetic energy. It was realized by the self-consistent solution of the equations of the intrinsic degrees of freedom and relative distance [7,8]. The formed DNS lives for long enough (several 10^{-21} s) to evolute from the projectile- and target-like configuration to become mononucleus (compound nucleus). This evolution corresponds to the motion on the mass (charge) asymmetry axis.

Note that for massive nuclei, a barrier could arise on the mass asymmetry

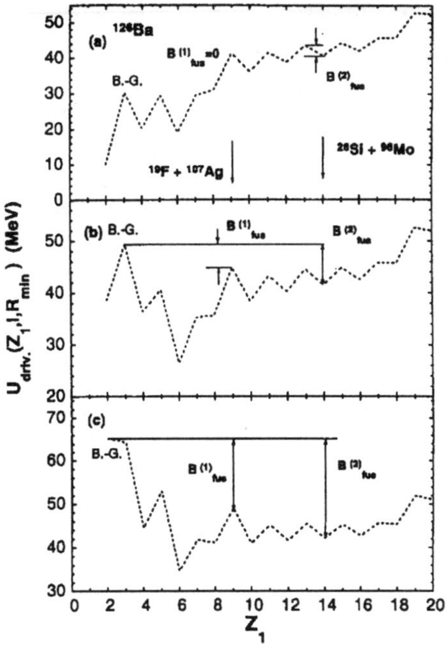

Figure 1. The driving potential (dashed curves) of compound nucleus ^{126}Ba calculated for different values of angular momentum: $l = 0$ (a), $l = 20\hbar$ (b), and $l = 30\hbar$ (c). The different values of the entrance mass asymmetry are shown by arrows ($Z_1 = 9$ and 14). $B^{(1)}_{fus}$ and $B^{(2)}_{fus}$ are intrinsic barriers in the way to fusion for the ^{19}F + ^{107}Ag and ^{28}Si + ^{98}Mo reactions, respectively.

axis on the way to fusion. This barrier was called an intrinsic fusion barrier in [20]. It is connected with the mass asymmetry degree of freedom. It is equal to the difference between the bottom of the potential surface corresponding to the initial charge asymmetry and its value on the highest saddle point (similar to the Businaro–Gallone point; see Fig. 1b,c) which is at smaller values of mass (charge) asymmetry.

The height of this barrier and the wall depth of the nucleus-nucleus potential (see Fig. 2) determine the competition between quasifission and complete fusion at a given excitation energy of the dinuclear system.

The nucleon flow between nuclei assumed to be quasistationary and motion on the mass asymmetry axis can be considered in a statistical way. The

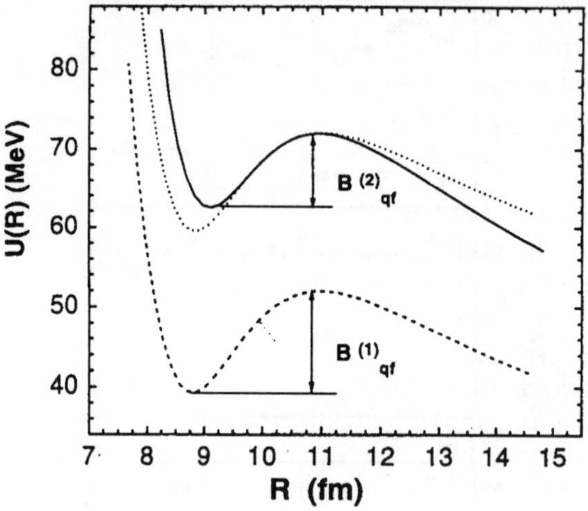

Figure 2. The nucleus-nucleus potential $U(R)$ calculated for the $^{28}Si + ^{98}Mo$ (solid curve) and $^{19}F + ^{107}Ag$ (dashed curve) reactions as a function of internuclear distance. For an easier comparison of two pockets (in width and depth) the dashed curve is shifted (dotted curve) up to overlapping with the solid curve. $B^{(1)}_{qf}$ and $B^{(2)}_{qf}$ are quasifission barriers for the corresponding reactions.

nucleus-nucleus potential was calculated by taking into account the nuclear shape deformation which was considered as the low energy collective first 2^+ state of nuclei. The deformation parameter β_2 was obtained from the experimental $B(E2)$ values [21]. It was assumed that these states are easily populated at the near Coulomb barrier energies and are not damped during the considered charge asymmetry of the dinuclear system. That is because the decay time of these degrees of freedom is of the order of 10^{-12} s [21], much larger than the relaxation time of the relative motion and evolution of the mass asymmetry. For the driving potential calculation (Fig. 1) the shape parameter of nuclei in DNS was changed as a function of Z, using the table given in [21].

The evaporation residue cross section is related to the partial fusion cross section $\sigma_l^{fus}(E,l)$, as well as to the probability $W_{sur}(E,l)$ that the compound nucleus survives fission during the de-excitation cascade at the bombarding energy E:

$$\sigma_{er}(E) = \sum_{l=0}^{\infty} \sigma_l^{cap}(E) P_{CN}(E,l) W_{sur}(E,l) \tag{1}$$

where the partial fusion cross section is $\sigma_l^{fus} = \sigma_l^{cap}(E)P_{CN}(E,l)$ and the partial capture cross section is $\sigma_l^{cap}(E) = \frac{\lambda^2}{4\pi}(2l+1)T_l^{capture}(E,l)$. The factor $P_{CN}(E,l)$ is used to take into account the decrease of the fusion probability due to dinuclear system break-up competing with fusion. At first this factor was determined in [20]. If DNS has been formed and lives long enough we can calculate the P_{CN} probability of compound nucleus formation using the driving potential of this system. The latter contains information how to reach fusion and on the intrinsic barrier B^*_{fus}.

Knowing the intrinsic fusion barrier B^*_{fus} versus the mass asymmetry axis (see Fig.1) and the depth of the potential pocket B_{qf} versus the radial distance axis R (see Fig.2) from the driving potential we can calculate the competition between fusion and quasifission. This barrier is determined by the difference between the driving potential maximum in reaching fusion and the value corresponding to the entrance mass asymmetry in the driving potential (Fig.1). The depth of the potential pocket B_{qf} is considered like a quasifission barrier for the given dinuclear system.

The calculation of competition between fusion and quasifission processes can be made according to principles of the statistical theory by comparing the level density at the top point (after which the dinuclear system goes to the fusion through smaller mass asymmetry values) of the $U(Z,A;l,R_m)$ driving potential (see for example Fig. 1 b,c) with the level density at the bottom of the exit barrier pocket in the $V(Z,A;l,\mathbf{R})$ internuclear potential (see Fig.2). The driving potential, playing the main role in fusion dynamics, was calculated as follows:

$$\begin{aligned} U(Z,A;l,R_m) &= B_1(Z;A) + B_2(Z_P+Z_T-Z;A_P+A_T-A) + \\ &+ V(Z,A;l,R_m) - B_0, \end{aligned} \tag{2}$$

where B_1 and B_2 are the binding energies [22,23] of the nuclei in a dinuclear system, $V(R_m) = V(Z,A,R_m)$ is the minimum value of the nucleus–nucleus interaction potential at the bottom of the pocket; B_0 is the binding energy of the compound nucleus. For the given total charge and mass numbers, $Z_{tot} = Z_1 + Z_2$ and $A_{tot} = A_1 + A_2$, the A/Z ratio of considered fragment was determined from the minimum value of $U(Z,A;l,R_m)$.

The available difference in calculation of the capture and fusion cross sections was required to introduce $P_{CN}(E,l)$. Analytically this appears as follows:

$$P_{CN}(E,l) = \frac{\psi_{fus}(E^* - B^*_{fus}(R_m,l))}{\psi_{fus}(E^* - B^*_{fus}(R_m,l) + \psi_{qf}(E^* - B_{qf}(R_m,l))}, \tag{3}$$

where

$$\psi_K(E^* - B_K(R_m, l)) = \frac{g}{2\sqrt{g_1 g_2}} \cdot$$

$$\cdot \frac{\exp\left[2\pi\sqrt{g(E^* - B_K(R_m, l))/6}\right]}{\left[\frac{3}{2}g(E^* - B_K(R_m, l))\right]^{\frac{1}{4}} (E^* - B_K(R_m, l))\sqrt{48}} \quad (4)$$

Here $E^* = E - V(R_m, l)$ is the excitation energy of the dinuclear system, at the relative distance R_m where an adiabatic nucleus-nucleus potential $V(R_m, l)$ has a minimum; $g = (g_1 + g_2)/2$, and g_1 and g_2 are the level densities of single particle states near the Fermi surface; B^*_{fus} and B_{qf} are the intrinsic fusion and quasifission barriers, respectively. $W_{sur}(E)$ is the survival probability of the compound nucleus against fission.

The $T_l^{capture}(E, l)$ factor is connected with the relative kinetic energy and angular momentum losses which were determined from the equations of motion

$$\mu(\mathbf{R}(t))\ddot{\mathbf{R}} + \gamma_R[\mathbf{R}(t)]\dot{\mathbf{R}}(t) = -\frac{\partial V[\mathbf{R}(t)]}{\partial \mathbf{R}} \quad (5)$$

$$\frac{dL}{dt} = \gamma_\theta[\mathbf{R}(t)]\left(\dot{\theta}R^2_{eff} - \dot{\theta}_1 R^2_{1eff} - \dot{\theta}_2 R^2_{2eff}\right) \quad (6)$$

where $\mathbf{R}(t)$ is the relative motion coordinate, $\dot{\mathbf{R}}(t)$ is the corresponding velocity; $\dot{\theta}$, $\dot{\theta}_1$ and $\dot{\theta}_2$ are angular velocities of the dinuclear system and its two fragments, respectively, while R_1 and R_2 are the fragment radii. We assume

$$R_{eff} = \frac{R + R_1 + R_2}{2} \quad , \quad R_{1(2)eff} = \frac{R_{1(2)}}{R_1 + R_2} R \quad .$$

The γ_R and γ_θ friction coefficients, a change of the nucleus-nucleus potential during interaction of nuclei,

$$V[\mathbf{R}(t)] = V_0[\mathbf{R}(t)] + \delta\hat{V}(\mathbf{R}), \quad (7)$$

and the dynamic contribution $\delta\mu(\mathbf{R})$ to the reduced mass

$$\mu(\mathbf{R}) = mA_T A_P/(A_T + A_P) + \delta\mu(\mathbf{R}) \quad (8)$$

are calculated from the estimation of the coupling term between the relative motion of nuclei and intrinsic excitation of nucleons in them [8].
The dynamics of the capture is sensitive to the nucleus-nucleus potential

$$V_0(\mathbf{R}) = V_C(\mathbf{R}) + V_{nucl}(\mathbf{R}) + V_{rot}(\mathbf{R}), \quad (9)$$

where $V_C(\mathbf{R})$, $V_{nucl}(\mathbf{R})$, and $V_{rot}(\mathbf{R})$ are the Coulomb, nuclear and rotational potentials, respectively. The nuclear shape is important to calculate the

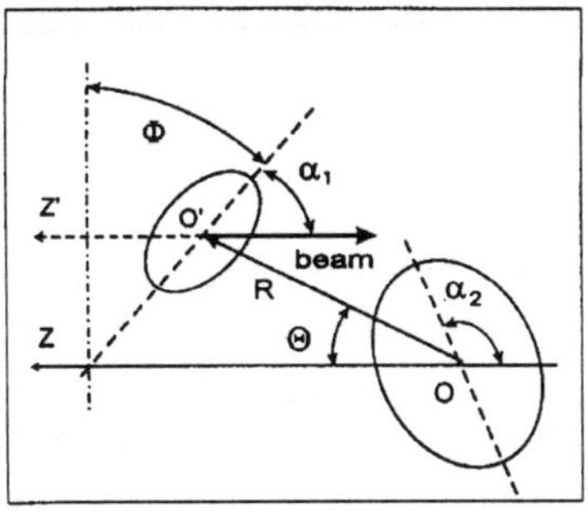

Figure 3. A sketch of the interaction between the deformed projectile-nucleus (the axial symmetry axis forms an angle α_1 with the beam) and the deformed target-nucleus (the axial symmetry axis forms an angle α_2 with the beam), with the impact parameter $b = R\cos\Theta$. The planes, in which the axial symmetry axes of nuclei lie, cross the Oz line and form an angle Φ.

Coulomb and nuclear interaction between colliding nuclei. Thus, the Coulomb interaction of deformed nuclei (with a quadrupole deformation) was calculated according to the expressions from [24]:

$$V_C(R) = \frac{Z_1 Z_2}{R} e^2 + \frac{Z_1 Z_2}{R^3} e^2 \cdot$$
$$\cdot \left\{ \left(\frac{9}{20\pi} \right)^{1/2} \sum_{i=1}^{2} R_{0i}^2 \beta_2^{(i)} \mathcal{P}_2(\cos\alpha_i) + \frac{3}{7\pi} \sum_{i=1}^{2} R_{0i}^2 \left[\beta_2^{(i)} \mathcal{P}_2(\cos\alpha_i) \right]^2 \right\} \quad (10)$$

where Z_i, $\beta_2^{(i)}$ and α_i are (for each fragment) the atomic number, the quadrupole deformation parameter and the angle (see Fig. 3) between the beam and the axial symmetry axis of the fragment $i (i = 1, 2)$, respectively; $R_{0i} = r_0 A_i^{1/3}$, with $r_0 = 1.15$ fm. $\mathcal{P}_2(\cos\alpha)$ is the second term of the Legendre second type polynom.

The nuclear part of the nucleus-nucleus potential is calculated using the folding procedure between the effective nucleon-nucleon forces $f_{eff}[\rho(x)]$ suggested by Migdal [25] and the nucleon density of projectile- and target-nucleus

$$V_{nucl}(\mathbf{R}) = \int \rho_1^{(0)}(\mathbf{r} - \mathbf{R}_1) f_{eff}[\rho(r)] \rho_2^{(0)}(\mathbf{r} - \mathbf{R}_2) d^3\mathbf{r} \quad (11)$$

where $\rho = \rho_1^{(0)} + \rho_2^{(0)}$; $\mathbf{R}_i$ $(i = 1, 2)$ is the position of the center of mass of each fragment i. The nucleon densities are assumed as a Fermi distribution

$$\rho_i^{(0)}(\mathbf{r},\mathbf{R}_i(t),\alpha_i,\beta_2^{(i)})=\left\{1+\exp\left[\frac{|\mathbf{r}-\mathbf{R}_i(t)|-R_{0i}(1+\beta_2^{(i)}Y_{20}(\alpha_i))}{a}\right]\right\}^{-1}. \tag{12}$$

The shape of nuclei of the dinuclear system changes by the evolution of the mass asymmetry degrees of freedom: $\beta_2 = \beta_2(Z,A)$. In order to calculate the potential energy surface as a function of charge number, we use the $\beta_2^{(2^+)}$ deformation parameter from the data in [21]. In [26,27,28,29] it was shown that the strong coupling to excited states leads to a set of Coulomb barriers of different heights and "weights". In our approach, relative motion was treated by including excitation of collective first 2^+ state of nuclei. Taking into account the deformation and orientation of nuclei allows us to consider nuclear collisions at so-called subbarrier energies.

The orientation influences the geometrical cross section of the collision not so much as the value of l_{cr} determining the reaction cross section. In collisions with "tip-tip" orientation of projectile- and target-nucleus the size of the wall in the nucleus-nucleus potential is larger (deeper and wider) than in the case of "side-side" orientation and allows one to trap the trajectories with larger orbital angular momentum. Consequently the capture cross section for the collisions near the "tip-tip" orientation of nuclei is much larger than for the ones near "side-side" orientation.

The atomic number for every intermediate charge asymmetry of the DNS nuclei was fixed from the minimum of the potential energy surface, the latter being considered as a function of $A_1 = A_{tot} - A_2$.

The rotational potential is calculated in the usual way:

$$V_{rot}(\mathbf{R}) = \hbar^2 \frac{l(l+1)}{2\mu R^2}. \tag{13}$$

Taking the deformation of colliding nuclei into account means using the nucleus-nucleus potential depending on the mutual orientations of the symmetry axes of the colliding nuclei. As a result the value and position of the entrance barrier and the pocket size in potential are dependent on the relative orientation of the symmetry axes of colliding nuclei. The final fusion cross section is obtained after averaging $\sigma_l(E)$ over all orientations. These effects are very important to interpret the results obtained at subbarrier energies for all nuclei.

By our AMS was possible to take into account the dynamical aspect of the compound nucleus evolution on the fission-evaporation competition along the de-excitation cascade. As far as the fission barriers are concerned, we use the rotating droplet model predictions (angular momentum dependent) as parameterized by Sierk [30] and allow for angular momentum and temperature fade-out of the shell corrections [31]. It is expressed by the formula for the actual fission barrier used in our calculations:

$$B_f(J,T) = c\ B_f^m(J) - f(T)\ g(J)\ \delta W, \tag{14}$$

with

$$f(T) = \begin{cases} 1 & T \leq 1.65 \text{ MeV} \\ k\exp(-mT) & T > 1.65 \text{ MeV}, \end{cases}$$

and

$$g(J) = \{1 + \exp[(J - J_{1/2})/\Delta J]\}^{-1},$$

where $B_f^m(J)$ is the parametrized macroscopic fission barrier [30] depending on J angular momentum, $\delta W = \delta W_{sad} - \delta W_{gs} \simeq -\delta W_{gs}$ is the microscopic (shell) correction to the fission barrier taken from the tables [32] and the constants for the macroscopic fission barrier scaling, temperature and angular momentum dependencies of the microscopic correction are chosen to be as follows: $c = 1.0$, $k = 5.809$, $m = 1.066$ MeV^{-1}, $J_{1/2} = 24\hbar$ (for nuclei with an A mass number of about 210) and $\Delta J = 3\hbar$. These constants were obtained as the result of analysis of a large set of the data on the fissility of nuclei in the Lu–At range obtained in the reactions induced by light (p, t and α) and heavy (^{12}C and ^{18}O) projectiles [31]. This makes our shell corrections become dynamical quantities too. The $J_{1/2}$ parameter slowly decreases increasing the A mass number. For nuclei with $A > 250$ we use $J_{1/2} = 20\hbar$.

Dissipation effects, which delay fission, are treated according to [33,34]. These include Kramer's stationary limit and an exponential factor applied to Kramer's fission width to account for the transient time, after which the statistical regime is reached. The systematics obtained by Bhattacharya et al. [35] gives the possibility of taking into account the incident energy per nucleon ϵ and compound nucleus mass A_{cn} dependencies of the reduced dissipation coefficient β (the ratio between the γ friction coefficient describing the coupling to the fission degree of freedom and the m reduced mass of the system; this ratio characterizes the dissipative and diffusion motion equation) in the calculations making use of the simple form:

$$\beta(\epsilon, A_{cn}) = a\epsilon + bA_{cn}. \tag{15}$$

We use the same set of parameters for all superheavy nuclei and there are not free parameters in our calculations. By this model it is possible to trace the Γ_n/Γ_{tot} values at each step of the de-excitation cascade.

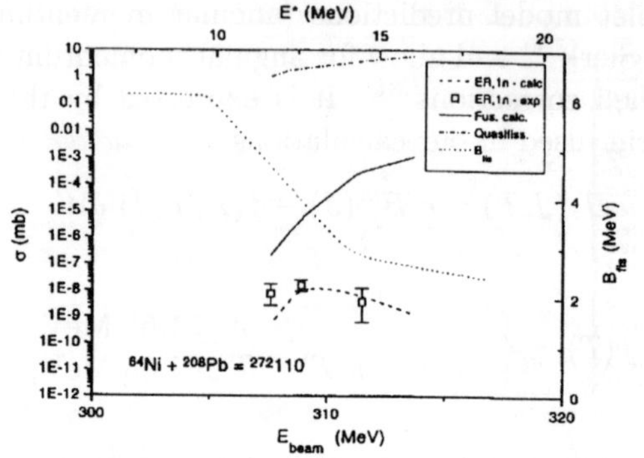

Figure 4. The results for the ^{64}Ni $+^{208}$Pb reaction that leads to the ^{272}110 compound nucleus. The lines represent the fusion cross section (full line), the quasifission contribution (dotted-dashed line), the ER production (dashed line) after 1 neutron emission from the compound nucleus. Squares are the experimental points [1]. Dotted line represents the B_{fis} fission barrier.

3 Result and discussion

We present the results of the investigation on massive systems that lead to the formation of the Z=110, 112 and 113 superheavy elements, and we compare our calculation with the available experimental data.

Figure 4 shows the fusion cross section (full line) for the ^{64}Ni + ^{208}Pb reaction that forms the ^{272}110 excited compound nucleus by a *cold fusion reaction*. In the same figure we report the quasifission contribution (dotted-dashed line), the evaporation residue cross-section (dashed line) after 1 neutron emission from the excited compound nucleus, and the fission barrier (dotted line).

In this figure we also report the experimental data [1] of the evaporation residues. We can observe the full agreement between experimental data and calculation. It is possible to analyze the ratio (10^{-7} — 2×10^{-4}) between the complete fusion and quasifission, and the ratio (2×10^{-3} — 10^{-5}) between the evaporation residue production and fusion. At low beam energies (lower than 307 MeV) the evaporation residue (ER) cross section is limited by the low value of the fusion formation. At high beam energies (higher than 313 MeV) the ER production decreases because the fission process strongly increases. This last effect is related to the fade-out of the fission barrier, due to the damping of the shell correction.

Figure 5 shows the results obtained for the ^{48}Ca + ^{238}U reaction that lead to the ^{286}112 excited compound nucleus. From this figure one can observe

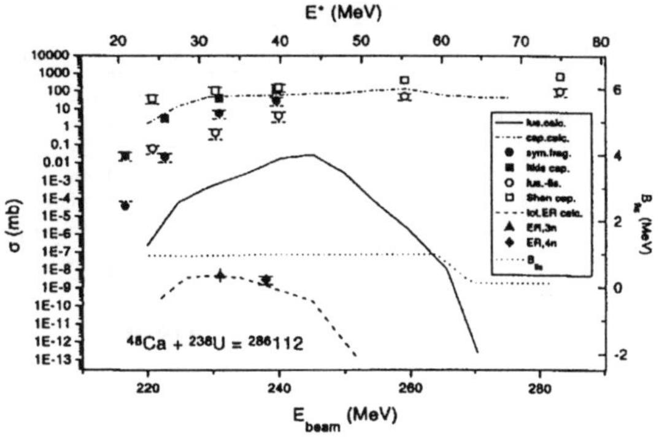

Figure 5. As Fig. 4, but for the ^{48}Ca $+^{238}$U reaction that leads to the ^{286}112 compound nucleus. The lines represent our calculation: dashed-dotted line is the capture cross section, full line is the fusion cross section, dashed line is the total ER production after 2, 3 and 4 neutron emission, dotted line is the fission barrier for the compound nucleus. The symbols are the experimental data: open squares and open circles represent [37] the capture cross section and fusion-fission process respectively; full squares and full circles represent [36] the capture cross section and the symmetric fragment production respectively. Triangle and diamond are experimental data [36] of the ER cross section after 3n and 4n emission respectively.

the beam energy window ($E_{beam} \simeq 218 - 262$ MeV, corresponding to about $E^* = 21 - 58$ MeV) that allows for the complete fusion formation (> 1 nb). In this energy interval and for the above-mentioned *hot fusion reaction*, the ratio between fusion and quasifission is about $10^{-8} - 2\times10^{-4}$ and the ratio between the evaporation residue cross section and complete fusion is about 10^{-4} — 5×10^{-8}. At lower beam energies the ER production is limited by the fusion formation, at higher beam energies the ER cross section is limited by both the low fusion formation and the fade-out of the fission barrier ($B_{fis} < 0.12$ MeV) that strongly increases the fission process.

In Fig. 6 we report the excitation functions for the ^{48}Ca (^{238}U, xn) $^{286-xn}$112 reaction after the x (x=2,3,4) neutron emission from the excited compound nucleus. In this figure we also report the experimental ER cross section [36] obtained by the $^{286-3n}$112 spontaneous fission events. For the $^{286-4n}$112 residual nucleus the experimental estimation of the upper limit is reported. The agreement between the calculation and experimental data is very good, even if until now only partial experimental mearurements are known.

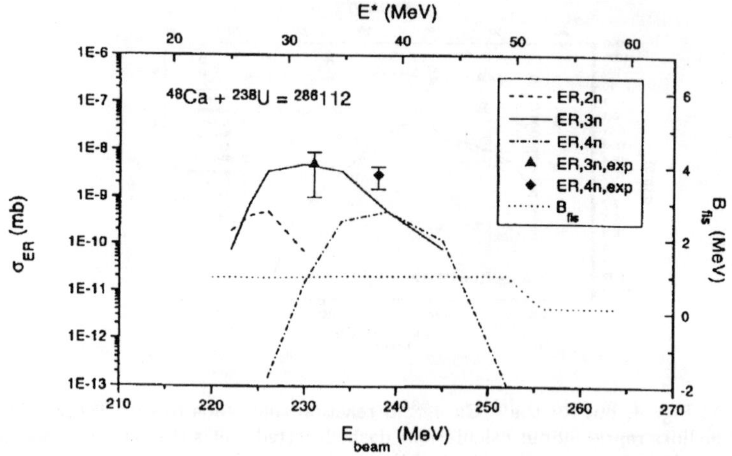

Figure 6. The excitation functions for the ^{48}Ca (^{238}U, xn) $^{286-xn}$112 reaction. Dashed, full and dashed-dotted lines represent the individual ER calculations after x=2, x=3 and x=4 neutron emission respectively, from the ^{286}112 compound nucleus. Triangle and diamond are experimental data [36] of the ER cross section after 3n and 4n emission respectively.

In Fig. 7 we report the U(Z, A; l, R_m) driving potential against the mass asymmetry parameter, for the Z=112 composite system. If we use an asymmetric entrance channel (for example the ^{48}Ca + ^{238}U system) we observe a B_f^* intrinsic fusion barrier of about 13.5 MeV in order to overcome (by changing mass (charge) between the two nuclei of DNS, as a motion in the mass (charge) asymmetry variable) the top point of the driving potential (corresponding to smaller values of mass (charge) asymmetry), and the dinuclear system to go to the complete fusion. If we use a more symmetric entrance channel (for example ^{130}Sn + ^{158}Sm) there is a B_f^* intrinsic fusion barrier of about 48 MeV to reach the fusion.

Therefore, in the ^{48}Ca + ^{238}U case and an excitation energy of DNS lower than 13.5 MeV the composite system evolves (moving the asymmetry mass parameter along the direction where the driving potential is lower) through a quasifission process producing symmetric fragments, but at such an excitation energy the yield is very low. For higher DNS excitation energies the process can lead to the complete fusion, in competition with the quasifission process. In this last case the quasifission is due to both symmetric and asymmetric contributions, but with a ratio between the two contributions depending on the DNS excitation energy.

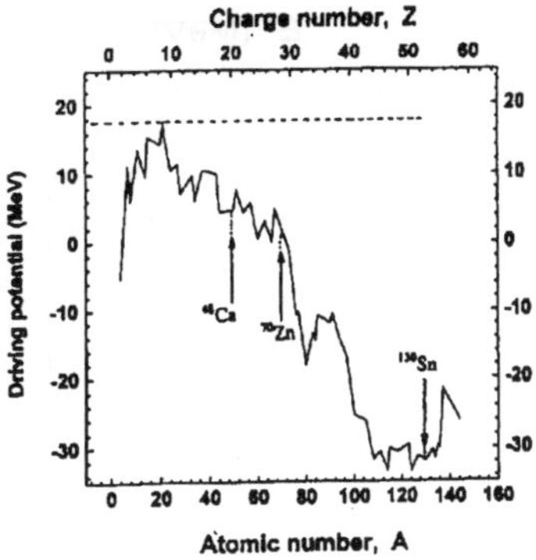

Figure 7. The $U(Z,A;l,R_m)$ driving potential against the mass asymmetry parameter, for the Z=112 and A=286 composite system. The arrows indicate the entrance channel with the ^{48}Ca, ^{70}Zn and ^{130}Sn beams. In the first case (asymmetric entrance channel) the intrinsic fusion barrier is $B_f = 13.5$ MeV because DNS overcomes the Businaro–Gallone point and fuses, in the last case (close symmetric entrance channel) is $B_f = 48$ MeV. For ^{70}Zn beam, B_f is about 16 MeV.

In addition, in Fig. 5 the experimental data of capture cross section [36,37], the symmetric fragment cross section [36], the ER production [36] after 3n and 4n emission and the fusion-fission production [37] for the reaction induced by ^{48}Ca (where an intrinsic fusion barrier $B_f^* \simeq 13.5$ MeV exists) are also reported. In the same figure we report, for comparison, our capture cross section calculation (dotted-dashed line), fusion cross section (full line) and the total ER production (dashed line) after 2, 3 and 4 neutron emission from the excited ^{286}112 compound nucleus. As one can see, the capture cross section calculation and the total ER calculation are in agreement with the experimental data of [36,37], whereas the fusion cross section calculation is lower than the symmetric fragment production [36] and the fusion cross section [37] extracted from the experimental fragment production. Therefore, the fragment production should be mainly contributed from the quasifission process, whereas the fission products after fusion at the various beam energies can reach the fusion cross section value. In fact, for the beam energy range correspondent to the experimental data of ER, the ratio between our ER calculation and the fusion cross section calculation is about $10^{-5} - 10^{-6}$, whereas the ratio between the experimental ER production and the experimental symmetric fragment production is about $5 \times 10^{-8} - 5 \times 10^{-9}$ (corresponding to a very small ER value, considering that the B_{fis} fission barrier is about 1 MeV).

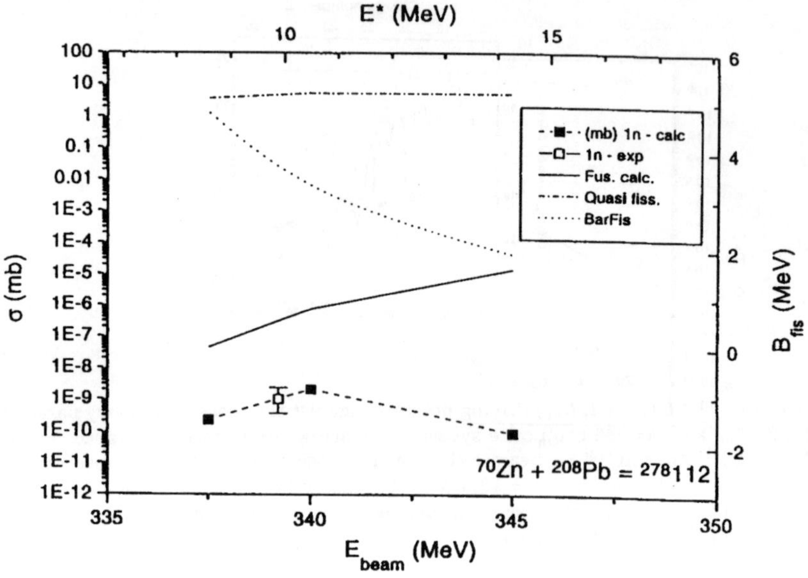

Figure 8. As Fig. 4, but for the $^{70}Zn + ^{208}Pb$ reaction that leads to the $^{278}112$ compound nucleus. The square is the ER experimental point [1] after 1 neutron emission from the compound nucleus.

Figure 8 shows the results for the $^{70}Zn + ^{208}Pb$ reactions that leads to the $^{278}112$ compound nucleus. For the above-mentioned reaction induced by ^{70}Zn, the intrinsic fusion barrier is $B_f^* \simeq 16$ MeV (see Fig. 7). In this figure we report the calculation for an energy range around the experimental point (only one at $E^* = 10$ MeV) obtained at GSI [1]. In the explored energy range and for the above-mentioned *cold fusion reaction* the ratio between the fusion and capture cross section calculation is about 1.5×10^{-7}. This value is lower than 10^{-4} — 10^{-5} (in correspondence of the experimental ER data) of the reaction induced by ^{48}Ca. This result is due to the fact that for the *cold fusion reaction* (induced by ^{70}Zn) the excitation energy is close to (or lower than) the intrinsic fusion barrier, whereas for the *hot fusion reaction* (induced by ^{48}Ca) the excitation energy is higher than the intrinsic fusion barrier. The ratio between the ER cross section and fusion calculation is about 3×10^{-2} (for the reaction induced by ^{70}Zn). This value is much higher than 10^{-6} for the reaction induced by ^{48}Ca. This is because the fission barrier is $B_{fis} \simeq 3.6$ MeV (for the compound nucleus obtained by the ^{70}Zn beam), while the fission

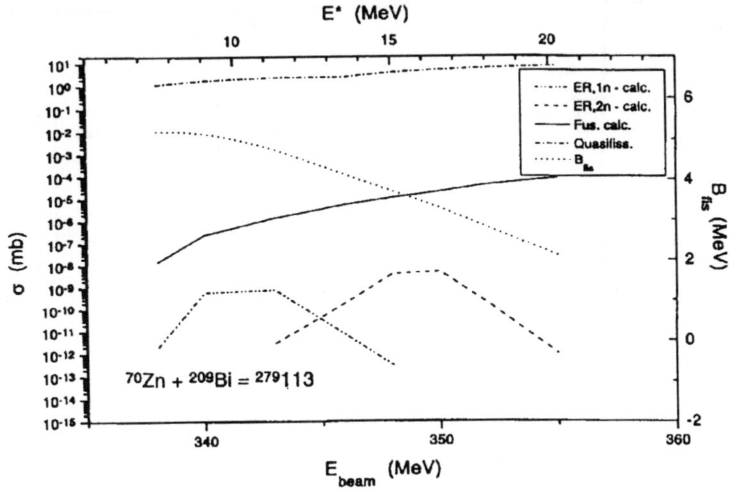

Figure 9.

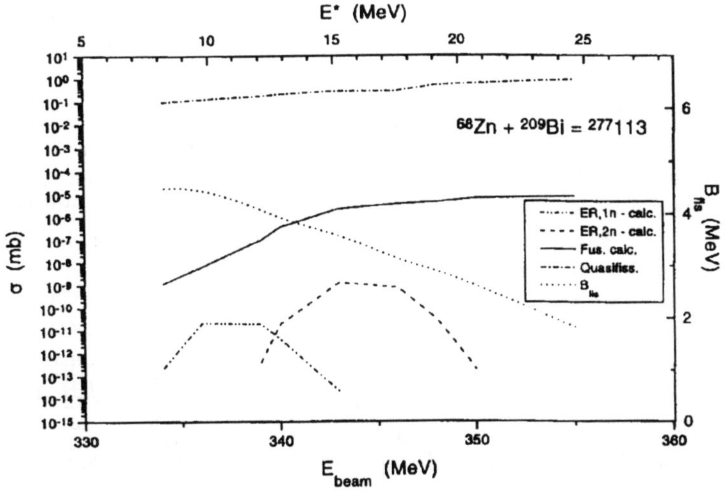

Figure 10.

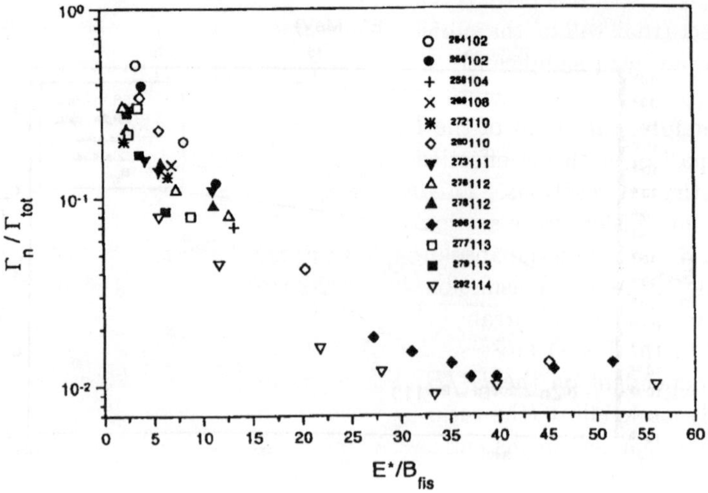

Figure 11.

barrier is $B_{fis} \simeq 1$ MeV for the reaction induced by ^{48}Ca. Our ER calculation agrees well with the experimental ER production [1].

From these considerations it is possible to affirm that by *cold fusion reactions* it is very difficult to synthesize superheavy elements (and to observe the evaporation residue products) because the excitation energy of the dinuclear system is generally lower than (or close to) the intrinsic fusion barrier, while by *hot fusion reactions* the E^*_{DNS} excitation energy is higher than the intrinsic fusion barrier.

In Fig. 9 we report the calculation for the $^{70}\text{Zn} + ^{209}\text{Bi} \rightarrow ^{279}113$ reaction, while in Fig. 10 we report the results for the $^{68}\text{Zn} + ^{209}\text{Bi} \rightarrow ^{277}113$ reaction. By comparing the ER cross sections for the two above-mentioned reactions, it is possible to observe the effect due to the isotopic effects of the beam.

In Fig. 11 we present the systematics of the Γ_n/Γ_{tot} ratio, at first step of the de-excitation cascade of the compound nucleus, for reactions that lead to heavy and superheavy nuclei with Z=102-114. This figure shows the trend Γ_n/Γ_{tot} ratio against E^*/B_{fis} for many reactions at different energies. As one can see the Γ_n/Γ_{tot} ratio is strongly correlated (within a factor 2) to the E^*/B_{fis} ratio. It is to be noted that the emission of the gamma rays, neutrons, protons and alpha particles is not considered in the present model.

This effect (that will be the subject of a future work) can be relevant at higher energies, and can significantly change the Γ_n/Γ_{tot} trend at higher E^*/B_{fis} values.

Therefore, the trend of the ER production versus the excitation energy (in comparison with the effective fission barrier) has the same behaviour for all investigated reactions. This means that the ratio between the ER production and fusion cross section is mainly related (for all reactions) to the E^*/B_{fis} value, while the absolute ER value is related to the competition between the fusion and quasifission during the evolution of the dinuclear system (after capture) in the entrance channel. Therefore, the ER production (normalized to the fusion cross section) for the heavy and superheavy nuclei is mainly dependent on the E^*/B_{fis} ratio, while the fusion process is related to the dynamic effect (the ratio between the intrinsic fusion barrier and the quasifission barrier) in the entrance channel.

References

1. S. Hofmann, *GSI-Preprint-98-16*, Darmstadt, March 1998, and references therein.
2. Yu. Ts. Oganessian, *Proc. of Int. Conf. Nucl. Phys. at the Turn of the Millenium*, Wilderness, South Africa (World Scientific, Singapore, 1996 page 11).
3. A. Ghiorso *et al*, *Nucl. Phys.* A **583**, 861c (1995).
4. Yu. A. Lazarev *et al*, *Phys. Rev.* C **54**, 620 (1996).
5. G. Giardina, *J. Phys.* G **23**, 1285 (1997).
6. Y. Abe *et al*, *J. Phys.* G **23**, 1275 (1997).
7. G. G. Adamian *et al*, *Phys. Rev.* C **53**, 871 (1996).
8. G. G. Adamian *et al*, *Phys. Rev.* C **56**, 373 (1997).
9. G. G. Adamian *et al*, *Z. Phys.* A **347**, 203 (1994).
10. R. V. Jolos *et al*, *JINR Preprint E4-97-302* (Dubna, 1997).
11. G. G. Adamian *et al*, *Nucl. Phys.* A **633**, 409 (1998).
12. A. D'Arrigo *et al*, *Phys. Lett.* B **34**, 1 (1994).
13. V. A. Ignatyuk *et al*, *Sov. J. Nucl. Phys.* **21**, 255 (1975).
14. V. A. Ignatyuk *et al*, *Sov. J. Nucl. Phys.* **29**, 450 (1979).
15. H. Gäggeler *et al*, *Z. Phys.* A **316**, 291 (1984).
16. R. J. Charity *et al*, *Phys. Rev.* C **56**, 873 (1997).
17. J. F. Liang *et al*, *Phys. Rev.* C **56**, 908 (1997).
18. G. V. Ravi Prasad and V. S. Ramamurthy, *Phys. Rev.* C **54**, 815 (1996).
19. G. V. Ravi Prasad *et al*, *Phys. Rev.* C **51**, 2833 (1995).
20. N. V. Antonenko *et al*, *Phys. Lett.* B **319**, 425 (1993); *Phys. Rev.* C **51**,

2635 (1995).
21. S. Raman *et al*, *Atomic Data and Nuclear Data Tables* **36** 1, (1987).
22. A. H. Wapstra and G. Audi, *Nucl. Phys.* A **432**, 1 (1985).
23. P. Möller and J. R. Nix, *Preprint LA-UR-86-3983* (Los Alamos Nation. Labor., 1986).
24. C. Y. Wong, *Phys. Rev. Lett.* **31**, 766 (1973).
25. A. B. Migdal, in *Theory of the Finite Fermi–Systems and Properties of Atomic Nuclei* (Moscow, Nauka, 1983 (in Russian)).
26. N. Rowley *et al*, *Phys. Lett.* B **254**, 25 (1991).
27. J. X. Wei *et al*, *Phys. Rev. Lett.* **67**, 3368 (1991).
28. N. Rowley. *Nucl. Phys.* A **538**, 205c (1992).
29. J. R. Leigh *et al*, *Phys. Rev.* C **47**, R437 (1993).
30. A. J. Sierk, *Phys. Rev.* C **33**, 2039 (1986).
31. A. D'Arrigo *et al*, *J. Phys.* G **20**, 365 (1994).
32. P. Möller and J. R. Nix, in *At. Data Nucl.* Data Tables **26**, 165 (1981).
33. P. Grangé and H. A. Weidenmüller, *Phys. Lett.* B **96**, 26 (1980).
34. E. M. Rastopchin *et al*, *Sov. J. Nucl. Phys.* **53**, 741 (1991).
35. C. Bhattacharya *et al*, *Phys. Rev.* C **53**, 1012 (1996).
36. M. Itkis *et al*, to be published in *Proc. of the Int. Conf. INPC98* (Parigi, 1998).
37. W. Q. Shen *et al*, *Phys. Rev.* C **36**, 115 (1987).

THE EXPERIMENTAL PRESENT AND FUTURE OF SUPERHEAVY NUCLEI SYNTHESIS

Š. ŠARO AND M. MATOŠ

Faculty of Mathematics and Physics, Comenius University, Mlynská Dolina, SK-84215 Bratislava, Slovakia and GSI Darmstadt, Germany

The synthesis of nuclei with Z over 100 always belonged to experiments at laboratory detection limits. The last large scale upgrade of experimental technique enabled to synthesize new nuclei with Z = 110, 111, 112. At these experiments no such a significant shell stabilizing or isospin effect was observed, which could eliminate the continuous decrease of cross sections of both "cold" and "hot" fusion reactions. For the synthesis of nuclei with $Z \geq 113$ cross sections below 1 pb are expected. To synthesize new superheavy nuclei with higher charge, the laboratory technique, to be used at the proposed reactions, should be again considerably upgraded. New target systems are considered, able to accept higher beam intensities. Some of the new superheavy evaporation residues will probably decay by fission. At present no adequate detection methods are used at superheavy element research, enabling an unambiguous identification of single fissionable evaporation residues. Possibilities to develop such detector systems are presented.

1 Introduction

The understanding of the structure and properties of the heaviest atomic nuclei is one of the long term programs of nuclear physics. Until 1940 ^{238}U was the heaviest known nucleus. From that time 20 transuranium elements were synthesized and more than 200 isotopes of these elements were unambiguously identified. Fig. 1 shows the upper end of the chart of nuclei. The figure contains three type of data: Identified isotopes of new elements, nuclear reactions where only cross section limits were determined and nuclear reactions supposed to be carried out in a near future. The comparison of experimental and predicted alpha half-life values shows, that recent theoretical calculations based on the Nilsson-Strutinsky model give for even-even nuclei half-life values, which are in good agreement with the measured values [1,2] shown in Fig. 2. From the experimental point of view is very important, that the predicted half-life values for the available target-projectile combinations are longer than the experimentally critical 1 μs, and are short enough to apply the powerful $\alpha - \alpha$ correlation method of evaporation residue identification [1,2,3,4]. Extrapolation of the known half-life values to atomic numbers of evaporation residues over 112 indicates for the available target - projectile combinations measurable half-life values at least for the next several steps to higher Z.

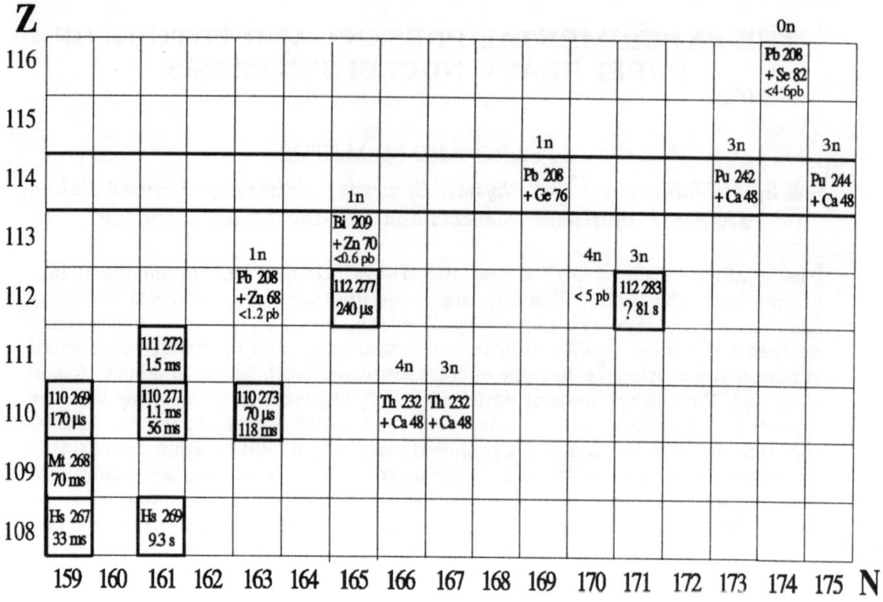

Figure 1. The upper end of the table of nuclei.

The predictive force of the known production cross section calculations is not satisfactory for planning of long beam time experiments. Recent calculations made by R. Smolańczuk [5] reproduce the measured production cross sections for the 1n reaction channel of Pb and Bi targets up to Z = 112. R. Smolańczuk predicts for the synthesis of the hypothetical spherical superheavy nuclei unexpectedly high production cross sections. For the reaction of $^{208}Pb(^{82}Se, 1n)^{289}116$ the predicted production cross section is 32 pb, and for the reaction of $^{208}Pb(^{86}Kr, 1n)^{293}118$ 1.8 nb. These values are in contradiction with the trend of continuous decrease of experimental production cross sections with Z as it is shown in Fig. 3. The stabilizing effect of the supposed spherical shape of these nuclei may be overestimated and the role of quasifission is probably underestimated. The predictive force of this calculations can not be tested with the experimental limit, achieved at the velocity separator SHIP (GSI) for the reaction of $^{82}Se +^{208} Pb$, because this experiment was running at excitation energies calculated for the 0n de-excitation channel.

We are still too far from good understanding of all basic physical properties of the heaviest nuclei and the mechanism of their creation. To improve

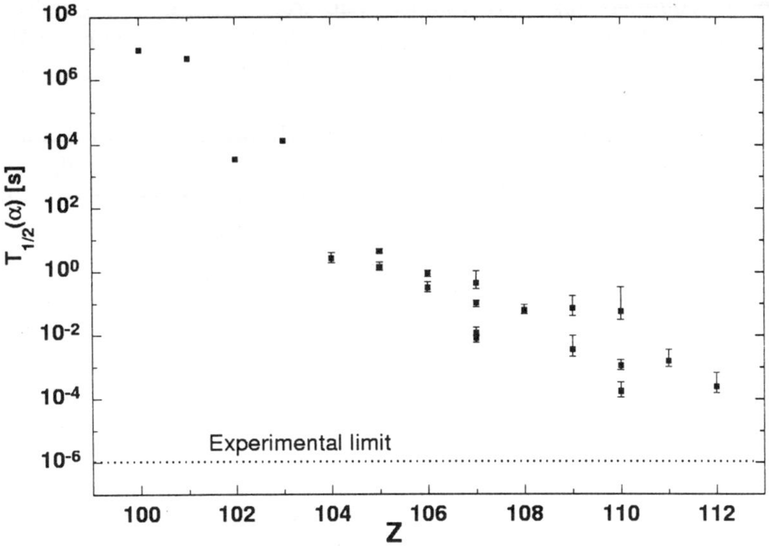

Figure 2. Dependence of experimental α-decay half-life values on the atomic number Z.

the model images of this area of nuclei more experimental points at the upper and of the chart of nuclei are needed. Especially around Z = 114, where model calculations, based on the Nilsson-Strutinsky approach [6] predict a closed proton shell with strong shell correction energies and spherical shapes of nuclei [2,4]. Our possibilities for experimental testing of this area of nuclei are limited. First of all with available combinations of stable or long living target and projectile nuclei. We need more neutron rich combinations of target and projectile nuclei. This can be achieved only with radioactive beams. In spite of the great progress in secondary beam intensities we are today still several orders of magnitude below the necessary level. In the case of Pb and Bi based complete fusion reactions even the known most neutron rich radioactive nuclei of Zn, Ga and Ge do not allow to get close to the supposed double magic nucleus of $^{298}114_{184}$. The situation is similar also for the possible target - projectile combinations for the hot fusion 3n, 4n, 5n de-excitation channels.

2 Cross-section limitations

The primary source most of the experimental problems, concerned with the synthesis of superheavy nuclei beyond Z = 110, is the low production cross section of complete fusion reactions. The cross sections of the cold fusion reactions, based on Pb and Bi targets and 1n de-excitation channel is steadily decreasing. For Z = 112 it decreased to the level of 1 pb as it is shown in Fig. 3. Linear extrapolation to Z = 114 gives for the 1n channel complete fusion reaction a cross-section about 0.1 pb. Higher cross-section values can be expected taking into consideration the higher isospin value for the heaviest available projectile nuclei of Germanium - ^{76}Ge. But the stabilizing effect of the isospin is not large enough to manifest itself over other reaction parameters as can be deduced from the data, given in Tab. 1.

The present sensitivity of the UNILAC-SHIP complex (GSI Darmstadt) for the detection of one heavy element α-decay chain is about one decay chain per 10 days at a cross-section of 1 pb. The second facility, designed for heavy element research, the U400-VASSILISSA complex (JINR Dubna) has a similar sensitivity.

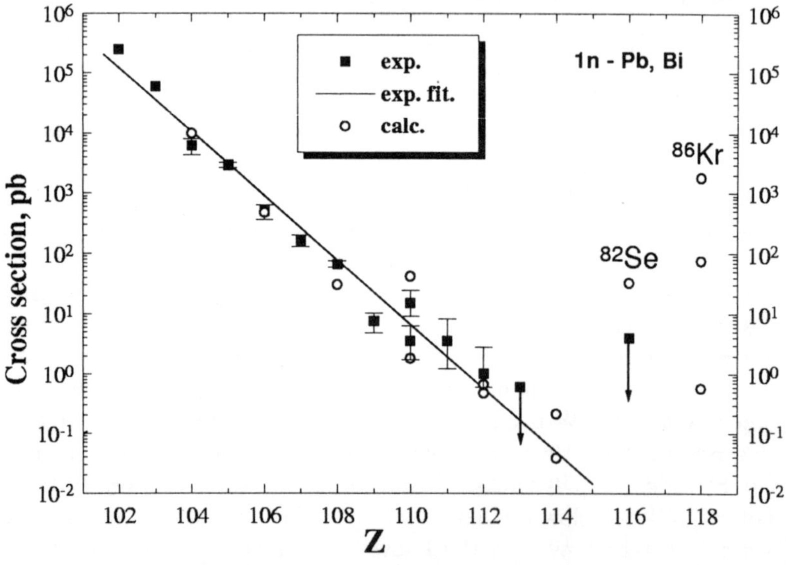

Figure 3. The highest measured (full squares) production cross sections of isotopes of transfermium elements for the 1n de-excitation channel and Pb and Bi targets and values (open circles) calculated by R. Smolańczuk [5].

To go further and to have a realistic probability to synthesize several nuclei of elements around Z = 114 there is a need to improve the sensitivity to the level of one identified decay chain within 10 - 15 days at a cross section of about 0.1 pb. There are two parameters which can substantially contribute to the sensitivity: the beam intensity on the target and the target ability to accept high beam intensity.

Table 1. Experimental cross sections for the heaviest nuclei.

Reaction	Reaction type	Evaporation residue	Exp. cross section	Ion isospin	Ref.
$^{50}Ti+^{204}Pb$	1n	$^{253}104$	0.19±0.05 nb	3	7
$^{50}Ti+^{206}Pb$	2n	$^{254}104$	2.4±0.2 nb	3	7
$^{50}Ti+^{208}Pb$	1n	$^{257}104$	6.1±1.8 nb	3	8
$^{50}Ti+^{209}Bi$	2n	$^{257}105$	2.1±0.8 nb	3	9
$^{50}Ti+^{209}Bi$	1n	$^{258}105$	2.9±0.3 nb	3	9
$^{51}V+^{209}Bi$	2n	$^{258}106$	38±13 pb	2.5	7
$^{54}Cr+^{209}Bi$	2n	$^{261}107$	36^{+22}_{-14} pb	3	10
$^{54}Cr+^{209}Bi$	1n	$^{262}107$	163±34 pb	3	10
$^{58}Fe+^{207}Pb$	1n	$^{264}108$	$3.2^{+6.1}_{-2.6}$ pb	3	11
$^{58}Fe+^{208}Pb$	1n	$^{265}108$	66±8 pb	3	12
$^{58}Fe+^{209}Bi$	1n	$^{266}109$	7.5±2.7 pb 6	3	12
$^{62}Ni+^{208}Pb$	1n	$^{269}110$	$3.5^{+2.7}_{-1.8}$ pb	3	13
$^{64}Ni+^{208}Pb$	1n	$^{271}110$	15^{+9}_{-6} pb	4	13
$^{48}Ca+^{232}Th$	3n	$^{277}110$		4	
$^{64}Ni+^{209}Bi$	1n	$^{272}111$	$3.5^{+4.6}_{-2.3}$ pb	4	14
$^{68}Zn+^{208}Pb$	1n	$^{275}112$	< 0.6 pb	4	13
$^{70}Zn+^{208}Pb$	1n	$^{277}112$	$1.0^{+1.8}_{-0.4}$ pb	5	15
			$1^{+1.3}_{-0.7}$ pb		16
$^{48}Ca+^{238}U$	3n	$^{283}112$	5.0± 2 pb	4	17
$^{70}Zn+^{209}Bi$	1n	$^{278}113$	< 0.6 pb	7	18
$^{48}Ca+^{244}Pu$	3n	$^{289}114$		4	
$^{76}Ge+^{208}Pb$	1n	$^{283}114$		6	
$^{82}Se+^{208}Pb$	1n	$^{289}116$		7	
$^{82}Se+^{208}Pb$	0n	$^{290}116$	< 4 − 6 pb	7	13
$^{86}Kr+^{208}Pb$	1n	$^{293}118$		7	

The ECR heavy ion sources and injectors are gradually increasing the delivered beam intensities above 10^{13} pps and there are projects to achieve beyond several years intensities above 10^{14} pps. Present heavy ion beam intensities, like Ni or Zn, on the rotating target in front of the velocity filter

SHIP can reach about 8×10^{12} pps. This is more than two times higher than the rotating Pb target foil can accept without a significant loss of quality. On the other side in 1999 a new heavy ion source and injector at GSI will be able to inject into UNILAC as many as 6×10^{13} pps of heavy ions and in a future 3×10^{14} pps for heavy element experiments at SHIP [19]. Similar intensities are expected also in other nuclear research centers, namely at RIKEN.

2.1 Target

There are several possibilities how to create conditions for the irradiation of the rotating Pb or Bi target foils at SHIP with higher beam intensities. The possibilities considered in a previous work [20] are not able to increase the beam intensity more than 2 - 3 times. The only type of Pb or Bi targets, able to accept beam intensities considerably over 10^{13} pps are vapor or cluster jets. The optimum Pb target thickness is about 450 $\mu g/cm^2$. The equivalents Pb vapor or cluster amount is $1.3 \times 10^{18}/cm^3$. The vapor jet can have an effective length of 5 cm and the corresponding vapor density will be $2.6 \times 10^{17}/cm^3$. This vapor density of lead is achieved at 1430 °C as it is shown in Fig. 4.

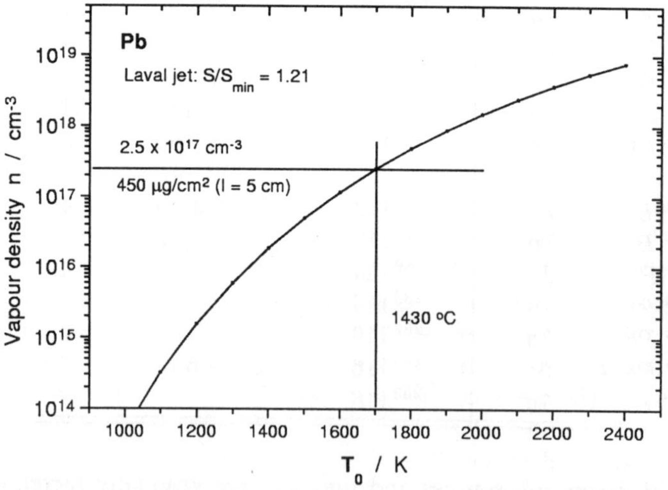

Figure 4. Saturated Pb vapor density as the function of temperature. Conditions for vapor density, equivalent to 450 $\mu g/cm^2$ Pb target foil are indicated.

At this temperature the atoms or molecules of lead have a velocity of 400 m/s, therefore the consumption of lead will be about 65 kg/hour. To have a reasonable consumption the target system is proposed to be closed as much as possible. A schematic view of the possible design of such a Pb or Bi vapor-cluster jet target is shown in Fig. 5. From the oven with a Laval nozzle in top the vapor stream will create a volume of 5 cm^3 with the desired density of $2.6 \times 10^{17}/cm^3$. The expanding material will condense on the surface of a system of condensers, having a temperature slightly over the melting point of lead. The condensed melted lead is collected at the bottom of the vacuum tank and pumped back into the oven. A system of collimators in the condensers allow the evaporation residues and the beam to get out from the target vessel. According to our estimations the amount of lead, escaping from the target vessel, can be in the case of pulsed regime of operation well below 1 g/day.

3 Identification of reaction products

3.1 The $\alpha - \alpha$ correlation method

The main decay mode of all known nuclei beyond Z = 106 is α-decay. This allows the safe identification of both - mass number A and atomic number Z of an evaporation residue. The method is based on the correlation between the known α-decay characteristics of a chain of daughter α-emitters following the α-decay of the unknown mother nucleus. The reliability of the method is increased when the detector system is position sensitive and the pulses caused by α-particles are detected from the same position as the initial pulse from the incoming ER. An additional increase of the reliability of the method is achieved with time and position correlation between the pulse created with the ER in the analyzing detector and pulses from time-of-flight (TOF) detectors in front of the analyzing detector. The $\alpha - \alpha$ correlation method was first time successfully used at synthesis of a new element by G. Münzenberg et al. [21] in 1981 at the identification of the first nuclei of element 107. At all the later discoveries of heavier atomic nuclei this method of identification was used , last time in 1996 at the identification of the first two nuclei of element 112 [15]. In Fig. 6 the $\alpha - \alpha$ correlation chain of $^{277}112$ is shown.

There are two principal limitations for the application of this method. First, the investigated heavy ER have to be an α-emitter and at least some members of the α-decay chain should have known α-particle energies and half-life values. The second limitation are the background conditions. Without any background the method could be used, in principle, for $\alpha - \alpha$ correlation chains of nuclei with half-lives as long as hours, days or even longer. At real

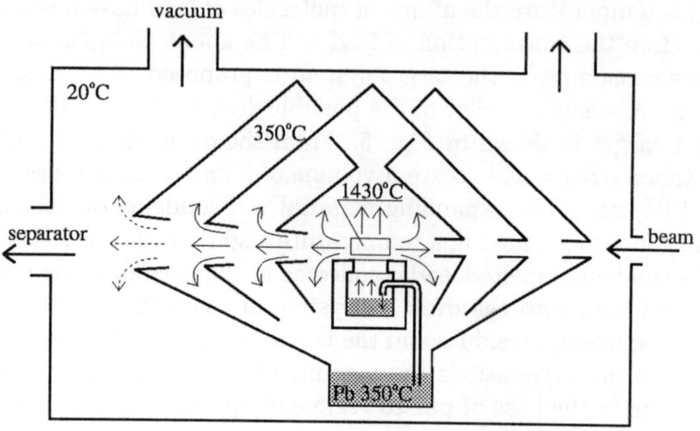

Figure 5. Schematic view and operation principle of the proposed Pb vapor jet target.

background conditions, an unambiguous assignment of detected energy pulses to particular α emitters is limited to minutes. Better background conditions can be achieved for the longer living members of the α-decay chains, when the pulses created by the first fast decaying members of the decay-chain are triggering the switch off the beam for some minutes, or if necessary even for a longer period. At this condition the prolongation of the correlation time could help to "bridge" unknown decays of long lifetimes, which finally will end in known decays. It will enable an unambiguous identification of alpha-decay chains with several unknown members.

3.2 Fission fragment identification

There is no physical reason to suppose, that all not yet synthesized nuclei above Z = 106 will have a significant α-decay branch. In the last experiment, carried out at the VASSILISSA kinematic separator in Dubna, the reaction $^{238}U(^{48}Ca, 3n, 4n)^{283,282}112$ was investigated [22] . At the beam energy of 231 MeV, corresponding to an excitation energy of 33 MeV, two fission events were detected with a position sensitive semiconductor detector array. After a careful analysis of all experimental parameters, related to these two events, the authors came to the conclusion that the observed fission fragments are, with very high probability, the decay products of $^{283}112$. But the used method of identification can not give an unambiguous proof that such rare single fission events have their origin in the decay of the supposed nucleus.

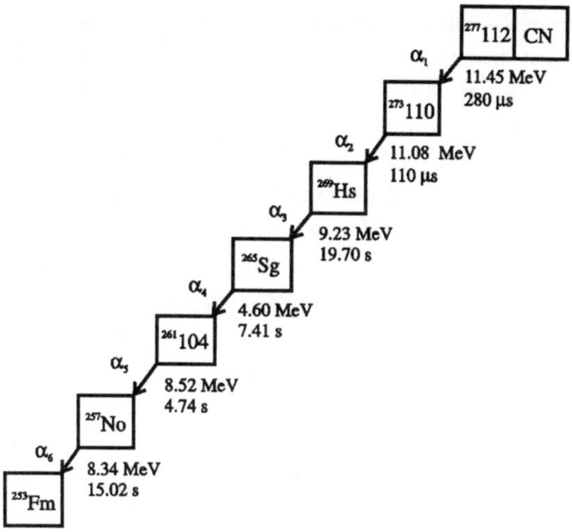

Figure 6. One of the two alpha decay chains obtained using the $\alpha - \alpha$ correlation method at the identification of the new element Z = 112 [15].

Let us consider, what possibilities we have for an unambiguous identification of a fissionable superheavy nucleus. A nucleus is identified when his atomic number Z and mass number A are identified using a direct or an indirect method of identification. Position sensitive silicon detectors, often used to identify α-decay chains, can not give any information about Z of the impact nucleus. The output energy signal from a silicon detector has a nonlinear deformation for heavy ions caused by the plasma effect. For this reason the correlated TOF and energy measurements will give mass values of heavy ERs, which do not allow to make a definite mass assignment.

Bragg spectrometry

From the point of view of possible methods for A and Z identification is useful to distinguish between short- and long- living ERs. ERs with a half-life longer than about 20 ms can be slowed down to a thermal velocity and deposited on the surface of a very thin foil. A double Bragg spectrometric ionization chamber could have a resolving power to determine the atomic numbers Z_1 and Z_2 of the independently detected fission fragments. Because of the emission of prompt neutrons, the mass numbers can not be determined in a Bragg spectrometer. Ten years ago a double ionization chamber was developed (Fig. 7) and connected behind the kinematic separator VAS-

SILISSA [23]. The ERs reached the common cathode as drifting positive ions after the process of slowing down in the gas filling. The ERs were collected with an efficiency close to 100%. The fission fragments were detected independently in the ionization chambers. The drift time of the ERs to the common cathode was about 40 ms with the possibility to reduce this time to about 10 ms. Taking into consideration this time limit, this device is proofing the possibility to design a double Bragg ionization chamber for atomic number identification of fission fragments. A possible limitation is connected with the fact, that fission fragments of neighboring very heavy nuclei have their Bragg peaks very close each to other.

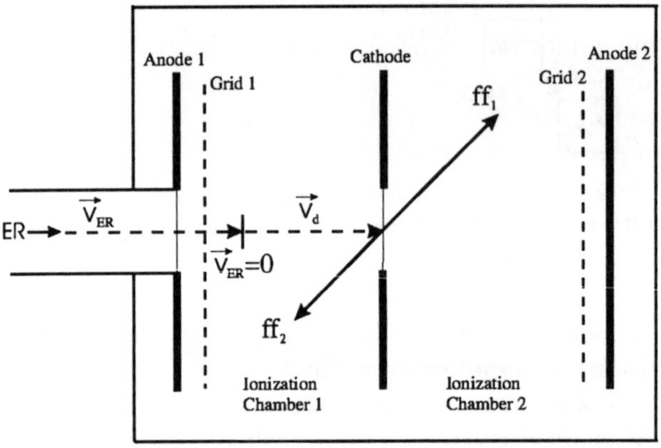

Figure 7. Schematic view of the double ionization chamber, developed for alpha and fission fragment detection at the kinematic separator VASSILISSA [23].

Penning ion traps

There is a possibility to use a Penning ion trap to identify very heavy reaction products. In a Penning ion trap the mass separated radioactive or stable ions are stored and their mass is determined by a measurement of the cyclotron frequency in a magnetic field. In experiments with ions up to ^{230}Ra the Penning trap mass spectrometer at ISOLDE separator allowed the determination of mass values with an accurateness of $\Delta M/M = 10^{-7}$ for all isotopes [24]. There are planes to install a Penning trap together with a gas-filled laser ion source and a Paul filter behind the velocity separator SHIP. This combination can perform first times atomic, nuclear and mass spectroscopy of transuranium nuclei. The use of a Penning trap behind SHIP could solve the problem of identification of non-alpha decaying heavy nuclei.

In a Penning trap there is a possibility, in principle, to investigate the properties of single ions. This unique property belongs only to the trap itself. The reaction products, emerging from the kinematic separator, have to be first decelerated and after that captured by the trap. The efficiency of this process is rather low, less than one percent, and the stopping process on a in a two stage trap takes about 1 second. For the use of Penning ion traps for identification of reaction products, at picobarn cross section levels, the efficiency of the process of cooling and consequent capturing should be increased to a value close to 100%.

Coincidence γ-spectrometry and roentgen-spectrometry of single fission events

The atomic number Z of a heavy fissionable nucleus could be determined, in principle, by γ-spectrometric methods. Coincidence γ-spectra of both fission fragments (Z1,A1) + (Z2,A2) can be measured with a γ-sphere. Such γ-spheres are operating in many laboratories. The basic problem of the possible use of γ- spectrometry for the identification of a single fissionable nucleus is the background spectrum of such a spectrometer. Also the level schemes of the fission fragments should be known. At the moment it is not clear, if the background spectrum of a coincidence γ- spectrometer can be suppressed to the desired extremely low level. And probably also the level schemes most of the possible fission fragments are not known.

Better conditions can be expected for the identification of both fission fragments if their coincidence roentgen spectra will be measured and analyzed. Both possibilities, coincidence γ-spectroscopy and coincidence roentgen-spectroscopy of fission fragment should be carefully analyzed.

4 Conclusions

The predictive force of the last half-life calculations is strong enough to expect measurable half-lives for nuclei of elements 113 and 114. Recent theoretical calculations, based on the Nilsson-Strutinsky model predict for the available target-projectile combinations half-lives which are longer than the experimentally critical 1 μs and short enough to use the $\alpha - \alpha$ correlation method of identification.

In spite of new attempts to present more reliable calculated cross section values for the synthesis of superheavy nuclei, created in Pb(HI,1n) reactions, the presented cross section values for Z above 112 remain very uncertain to use them as a guide for future experiments.

New ECR heavy ion sources and injectors under development will deliver in the near future, at least at GSI Darmstadt and at RIKEN, heavy ion beam intensities several times 10^{13} pps and later even more. This gives the possibility to synthesize new superheavy nuclei at a cross-section level of 0.1 pb at a reasonable beam time.

The conditions for high beam acceptance of Pb and Bi targets can be achieved by using vapor- or cluster-jet targets. There are no principal physical or technical limitations to design such a target system.

There is a need to investigate and determine which possibilities we have to develop detector systems, able to determine Z and A of heavy evaporation residues undergoing fission. If the efficiency of the process of slowing down of evaporation residues will be solved, ion traps can give an excellent mass resolution in the case of any heavy nucleus. There are several principal possibilities for Z and A determination, which should be the subject of further investigation.

The authors thanks S. Hofmann, P.F. Heßberger, G. Münzenberg, A.N. Popeko and A.V. Yeremin for many valuable discussions. The Slovak VEGA Grant No. 1/5186/98 is gratefully acknowledged.

References

1. R. Smolańczuk, Preprint SINS/Th/4/1997.
2. A. Sobiczewski, Physics of Particles and Nuclei **25** (1994) 295.
3. P. Möller and R. Nix, J. Phys. G: Nucl. Part. Phys. **20** (1994) 1681.
4. R. Smolańczuk et al., Phys. Rev. **C 52** (1995) 1871.
5. R. Smolańczuk, Preprint SINS/Th/2/1998, August 1998.
6. V.M. Strutinsky, Nucl. Phys. **A 95** (1967) 420.
7. F.P. Heßberger et al., GSI Sci. Report 1995 9.
8. F.P. Heßberger et al., Z. Phys. **A 321** (1985) 317.
9. F.P. Heßberger et al., Z. Phys. **A 322** (1985) 557.
10. G. Münzenberg et al., Z. Phys. **A 333** (1989) 163.
11. G. Münzenberg et al., Z. Phys. **A 324** (1988) 489.
12. S. Hofmann et al., Z. Phys. **A 358** (1997) 377.
13. F.P. Heßberger et al., Contibution to 'Tours Symposium on Nuclear Physics III', Tours, France, Sept. 2-5, 1997
14. S. Hofmann et al., Z. Phys. **A 350** (1995) 281.
15. S. Hofmann et al., Z. Phys. **A 354** (1996) 229.
16. S. Hofmann, GSI Preprint-98-16.
17. Yu. Ts. Oganessian, Proceedings of the 2nd International Conference on Exotic Nuclei and Atomic Masses, 23-27 June, 1998 Shanty Creek Resort, USA.

18. F.P. Heßberger et al., Int.Conf., DANF'98, Oct.19-23, 1998, Častá – Papiernička, Slovakia, to be published in Acta Physica Slovaca.
19. D. Habs, GSI Report 97-02, February 1997.
20. Š. Šaro et al., Heavy Ion Physics, Proc. VI. Int. School-Seminar, Dubna, Russia, 22-27 Sept. 1997, p.445., World Sci. Publ.
21. G. Münzenberg et al., Z. Phys. **A 300** (1981) 107.
22. A.V. Yeremin et al., Int.Conf., DANF'98, Oct.19-23, 1998, Častá – Papiernička, Slovakia, to be published in Acta Physica Slovaca.
23. Š. Šaro et al., Nucl. Instr. Methods **A 330** (1993) 125.
24. H.-J. Kluge, G. Bollen, Nucl. Instr. Methods **B 70** (1992) 473.

COLD SPONTANEOUS FISSION PROCESSES OF ^{252}Cf AND THE STRUCTURE OF NEUTRON RICH Ba AND La NUCLEI

J.H. HAMILTON[1,2], A.V. RAMAYYA[1], J.K. HWANG[1], W. GREINER[1,2,3], S.J. ZHU[1,3,4], A. SANDULESCU[1,2,5], A. FLORESSCU[1,2,3], J. KORMICKI[1], G. TER-AKOPIAN[1,3,5], Y. OGANESSIAN[5], A.V. DANIEL[1,3,5], G.S. POPEKO[5], J. KLIMAN[5,6], M. MORHÁČ[6], J.D. COLE[7], R. ARYAEINEJAD[7], M.W. DRIGERT[7], W.E. COLLINS[8], W.C. MA[9], E.F. JONES[1], L.K. PEKER[1], P.M. GORE[1], G. DRAFTA[1], B.R.S. BABU[1], G. WANG[4], J.K. DENG[4]

[1]*Physics Department, Vanderbilt University,Nashville, TN 37235, USA*
[2]*Institut für Theoretishe Physik der J.W. Goethe Universität, Frankfurt, Germany*
[3]*Joint Institute for Heavy Ion Research, Oak Ridge, TN 37831, USA*
[4]*Physics Department, Tsinghua University, Beijing, P.R. China*
[5]*Joint Institute for Nuclear Research, 141980 Dubna, Russia*
[6]*Institute of Physics SASc, 84228 Bratislava, Slovak Republic*
[7]*Idaho National Engineering and Environmental Lab., Idaho Falls, ID 83415, USA*
[8]*Physics Department, Fisk University, Nashville, TN 37208, USA*
[9]*Physics Department, Mississippi State University, MS 39762, USA*
E-mail: hamiltj1@ctrvax.vanderbilt.edu

Rare cold (zero neutron emission and low excitation energy) binary fission, cold α and ^{10}Be-accompanied ternary fission and cold fission with two neutrons emitted at scission and new nuclear structure effects have been observed in the spontaneous fission of ^{252}Cf. Evidence for stable octupole deformation in ^{143,144,146}Ba and ^{145,147}La are presented. The roles of rotation in both enhancing and quenching stable octupole deformation are observed. In ^{145}La one has a new class of shape coexistence with symmetric and asymmetric rotational bands.

1 Introduction

Studies of the prompt gamma-rays emitted in spontaneous fission (SF) of ^{252}Cf and ^{248}Cm are providing a wealth of new insights into the fission process and the broad range of new structures of neutron rich nuclei (for example see [1–4]). Cold zero neutron emission processes can be considered as the ultimate examples of cluster radioactivity first predicted by Sandulescu and Greiner [5]. Our work on cold binary fission [6,7] is summarized including the first odd Z,A- odd ZA cold SF. Ternary hot fission, with the third fragment being an α particle has been known for a long time [8]. We have made the first measurements of the yields of correlated pairs in cold α ternary fission [9] and carried out theoretical calculations [10]. The total yields of cold ternary fission also have been reported recently [11]. Very recently we made the first yield measurement of the correlated pairs in cold ^{10}Be SF [12]. Finally we present evidence for the first example of cold fission (here meaning very low energy neutrons) with two neutron emission at scission [13] and compare the yield with the known hot 2n SF yields [14]. These rare processes give new insights into the fission process and its relationship to the theory of the multifragmentation of nuclei.

Table 1. Average cold binary fission yields

	A_L / A_H	Y_{exp}	Y_{the}	$Y_{the}^{(ren)}$
Zr / Ce	100/152	0.010(2)	0.38	0.004
	102/150	0.020(4)	2.82	0.033
	103/149	0.030(6)	4.21	0.049
	104/148	0.010(2)	1.03	0.012
Mo / Ba	104/148	0.010(2)	0.47	0.005
	105/147	0.040(8)	5.39	0.063
	106/146	0.040(8)	0.61	0.007
	107/145	0.070(14)	3.07	0.036
	108/144	0.030(6)	7.45	0.087
Tc / Cs	109/143	0.090(18)	11.03	0.128
Ru / Xe	110/142	0.060(12)	3.78	0.044
	111/141	0.10(2)	7.12	0.083
	112/140	0.020(4)	0.59	0.007
	114/138	0.020(4)	1.17	0.014
Pd / Te	116/136	0.050(20)	2.35	0.027

Likewise, many new insights into the structure of neutron rich nuclei have been gained from SF studies [1-4]. In this paper we consider only the evidence for stable octupole deformation in ^{143,144,146}Ba and ^{145,147}La. The experimental procedures are described elsewhere [1,3].

2 Cold binary processes

Since the neutronless binary yields are much smaller than those with neutrons emitted, double gating techniques are necessary to extract the yields either for cold binary or ternary fission. Earlier we reported the first results for the correlated pairs in cold binary fission in ^{252}Cf [6,7] and ^{242}Pu [15]. Subsequently we have extracted more detailed yields of cold binary fission as reported here. Two methods were used to extract yields.

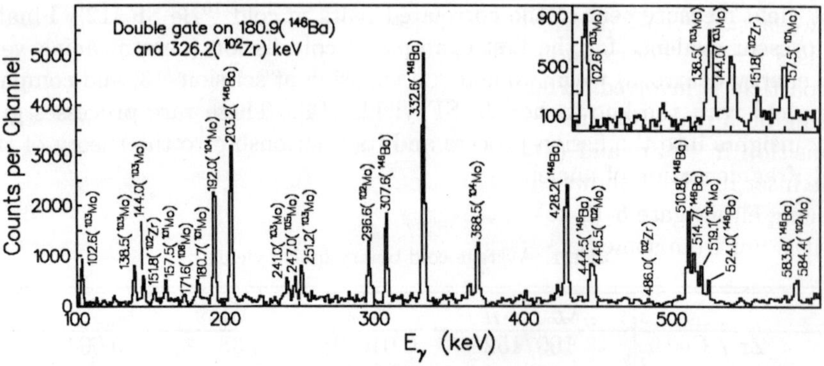

Figure 1. The spectrum is obtained with a double gate set on the $2^+ \rightarrow 0^+$ transition at 180.9 keV in ^{146}Ba and on the $4^+ \rightarrow 2^+$ transition at 326.2 keV in ^{102}Zr. The clear peak at 151.8 MeV corresponds to the $2^+ \rightarrow 0^+$ transition in ^{102}Zr. A number of transitions in ^{146}Ba are seen because of the 324.2 keV ^{146}Ba transition in the gate.

In method 1, the cold (neutronless) binary fragmentations and 1 to 6 neutron evaporation channels were studied by setting double gates on the $2^+ \rightarrow 0^+$ and $4^+ \rightarrow 2^+$ transitions in an even-even fragment (e.g. heavy or light) and then measuring the intensities of the $2^+ \rightarrow 0^+$ transitions in the correlated fragments (e.g. light or heavy). In method 2 the yields are determined, for example, by setting double gates on the $2^+ \rightarrow 0^+$ transition in ^{146}Ba and $2^+ \rightarrow 0^+$ transitions in the Mo isotopes and using the $4^+ \rightarrow 2^+$ transition in ^{146}Ba as a measure of the relative yields. The cold binary fission yields are shown in Table I along with our theoretical values based on a cluster model similar to the one-body model used for the description of cluster radioactivity [7]. The trends, e.g. variations with A and Z, of the theoretical yields are in good agreement with the experimental yields. All even Z, odd A-odd A pairs are predicted to be larger than the even A - even A pairs and this is observed in every case. Moreover the largest theoretical yield is predicted for the odd Z - odd Z ^{109}Tc-^{143}Cs and this is true within experimental errors.

3 Cold ternary fission

While hot ternary fission has been known for a long time, with the third particle generally being an α particle [8], cold ternary fission has not been directly observed in terms of the correlated pairs until our work [9]. Meutter and coworkers [11] have extracted total yields for ternary SF. In the case of cold (neutronless) alpha ternary fission of ^{252}Cf, one looks at the correlation between two even-Z fragments with the sum of charges Z=96 and sum of masses A=248. For example, ${}^{146}_{56}$Ba and ${}^{102}_{40}$Zr are the partners when a ternary α -particle is involved. The ^{248}Cm was negligible in our source. In Fig. 1, gating on a transition in ^{146}Ba and ^{102}Zr, one sees the 151.8 keV ($2^+ \rightarrow 0^+$) transition in ^{102}Zr, and (4^+ - 2^+) in ^{146}Ba to establish cold α-ternary SF. The inset in Fig. 1 shows the region around 151.8 keV (^{102}Zr) transition. By moving either gate 5-10 keV either way, the 151.8 keV transition disappears so it is not from some background effect.

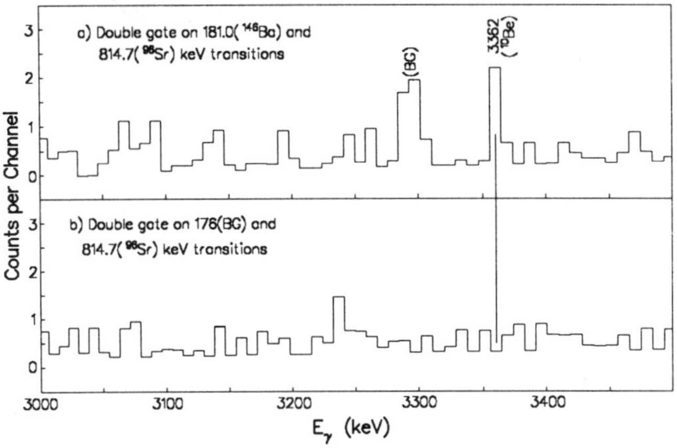

Figure 2. a) Coincidence γ-ray spectrum obtained by double gatting on the 814.7, $2^+ \rightarrow 0^+$ transition in ^{96}Sr and 181.0 keV, $2^+ \rightarrow 0^+$ transition in ^{146}Ba. b) Coincidence γ -ray spectrum obtained by double gatting on the 814.7, $2^+ \rightarrow 0^+$ transition in ^{96}Sr and on 176.0 keV background peak, which is shifted from 181.0 keV $2^+ \rightarrow 0^+$ transition in ^{146}Ba. BG : Background.

These and similar spectra clearly demonstrate that one has observed for the first time correlated pairs in cold α -ternary SF.

In Table II, the first cold neutronless alpha ternary fission yields for correlated pairs are given. The theoretical yields [10] normalized to the total experimental yields shown in columns 3 and 6 of Table II in general, are in good agreement with the experimental yields.

The experimental odd-A - odd-A SF yields are higher than the yields for the even-even neighbors, as found theoretically. The differences may be related to the differences in level densities near the ground states in odd-odd and even-even nuclei. The enhanced experimental yields for the cold α -ternary fission in which the heavier partners are Ba isotopes could be attributed to

Table 2. The alpha ternary isotopic yields Y_{exp} obtained per 100 fission events [9].

α Partner nuclei	Y_{exp} (%)	Y^{ren}_{the}	α Partner nuclei	Y_{exp} (%)	Y^{ren}_{the}
$^{92}_{36}$Kr–$^{156}_{60}$Nd	0.002(1)	0.002	$^{96}_{38}$Sr–$^{152}_{58}$Ce	0.008(3)	0.010
$^{98}_{38}$Sr–$^{150}_{58}$Ce	0.014(6)	0.017	$^{99}_{38}$Sr–$^{149}_{58}$Ce	0.018(9)	0.016
$^{100}_{38}$Sr–$^{148}_{58}$Ce	0.021(10)	0.027	$^{101}_{38}$Sr–$^{147}_{58}$Ce	0.014(11)	0.010
$^{100}_{40}$Zr–$^{148}_{56}$Ba	0.038(12)	0.017	$^{101}_{40}$Zr–$^{147}_{56}$Ba	0.082(10)	0.058
$^{102}_{40}$Zr–$^{146}_{56}$Ba	0.009(4)	0.017	$^{103}_{40}$Zr–$^{145}_{56}$Ba	0.084(29)	0.050
$^{104}_{40}$Zr–$^{144}_{56}$Ba	0.017(8)	0.016	$^{106}_{42}$Mo–$^{142}_{54}$Xe	0.018(7)	0.031
$^{107}_{42}$Mo–$^{141}_{54}$Xe	0.030(14)	0.017	$^{108}_{42}$Mo–$^{140}_{54}$Xe	0.007(3)	0.014
$^{112}_{44}$Ru–$^{136}_{52}$Te	0.011(6)	0.028	$^{116}_{46}$Pd–$^{132}_{50}$Sn	0.006(3)	0.048

the static octupole deformation reported in this region, since octupole shapes at the scission configuration could significantly lower the Coulomb barrier and increase the penetrability between the final fragments.

Next we searched for cold neutronless ternary fission with the third particle being a heavier cluster such as ^{10}Be. Contrary to α-ternary SF with α excited state energies >20 MeV so γ emission would not be seen, ^{10}Be has an excited 2^+ state 3.368 MeV. When setting a double gate on the $2^+ \to 0^+$ 814.7 and 181.0 keV transition in ^{146}Ba and ^{96}Sr, respectively, we see the 977 keV transition in ^{96}Sr (Fig. 2) and a peak at $\approx$3368 keV (observed at 3362 keV). Assuming that the peak at 3362 keV is the $2^+ \to 0^+$ transition in ^{10}Be, we set one gate on it and the other gate on the $2^+ \to 0^+$ transition in ^{96}Sr. The coincidence spectra shows the 977.5 keV, $4^+ \to 2^+$ transition in ^{96}Sr and the 181.0 and 332.6 keV peaks in ^{146}Ba. With a double gate on the 3362(^{10}Be) and 181.0(^{146}Ba) keV peaks, the 814.7 keV ($2^+ \to 0^+$) transition in ^{96}Sr is seen. The yield for the ^{10}Be ternary process is obtained from the double gate on the 181.0 (^{146}Ba) and 814.7 keV (^{96}Sr) transitions. The yield to the first excited state of ^{10}Be is $\approx 4(2) \times 10^{-4}$ per 100 SF and to the ground state ≈ 0.

The major question is why is the 3362 keV line not Doppler broadened since the 2^+ state lifetime is 125 fs and the stopping time of ^{10}Be is 1250 fs. A non-Doppler shifted peak is also observed by Mutterer et al. [11] with NaI detectors in ^{10}Be ternary fission. One possible explanation is that after ^{10}Be formation ^{96}Sr - ^{10}Be - ^{146}Ba form a nuclear molecule with ^{10}Be held in the potential well for a time $\approx 10^{-12}$ s during which it γ decays before breaking up into ternary SF. If this is true, many exciting new possibilities for molecular configurations and nuclear structures are opened up. We will gain further insight into this question, in our experiment with light charged

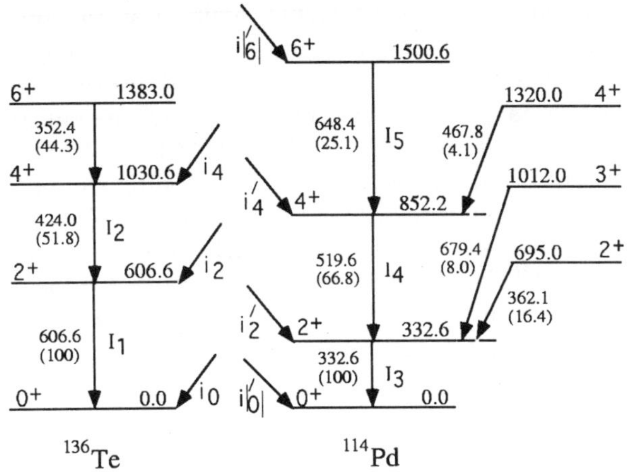

Figure 3. Partial level schemes of ^{136}Te and ^{114}Pd.

particle detectors inside Gammasphere so we can gate on the LCP and triple fold γ rays.

Recently we have discovered evidence for a correlated pair in cold fission in which two neutrons are emitted from the neck presumably with low kinetic energy. Most neutrons emitted in fission process are evaporated isotropically from the accelerated primary hot fragments with an average kinetic energy of 2.34 (5) MeV [16]. Bowman et al. [16] reported that $\approx$10 % of the neutrons are emitted at 90° to the scission axis of the two fragments with an average kinetic energy of 0.39(6) MeV. Later Budtz-Jorgensen and Knitter [17] observed a much lower yield of 1.1(3) % for cold neutron emission. Continuing our work on rare ternary SF, we looked for cold fission with two neutrons emitted at scission.

The partial lower spin level schemes of ^{136}Te and ^{114}Pd are shown in Fig. 3. The transition intensities as observed in normal SF of ^{252}Cf are shown in parenthesis. In general, the 2^+ state in either nucleus is populated by discrete γ transitions (I_d) and by statistical γ rays (I_s) after normal fission where neutrons are evaporated so $I(2^+ \rightarrow 0^+) = I_s + I_d$. When cold 2n emission SF occurs, this equation will be $I(2-0) = I_s + I_d + i_c$ where i_c is the population of a particular state via cold fission with 2n emission at scission.

How can one extract i_c from the experimental data? For a typical level scheme as shown in Fig. 3, consider what happens for two partner nuclei,

^{136}Te and ^{114}Pd, with two cold neutron(s) emitted at scission in addition to normal hot fission. In Fig. 3, I_1, ... I_6 are the intensities of the transitions as shown in Fig. 3. The labels i_0, ... i_6 and i_0' ... i_6' are the population of the states from cold ternary fission when two neutron(s) are emitted. For a gate on the transition I_1, the ratio $R_1 = I(4^+ \rightarrow 2^+)/(2^+ \rightarrow 0^+) = I_4/I_3$. In this ratio, all the discrete and statistical transitions feeding and i_2 and i_4 (cold 2n fission) direct population of the 2^+ and 4^+ levels in ^{136}Te are included.

If only hot fission occurs, then any single gate or double gate on any two transitions in ^{136}Te will give the same relative γ-transition intensities in ^{114}Pd. Only if there is a second process like cold fission that populates only one or two low-lying states in these two partners will a double gate and single gate in ^{136}Te give different relative intensities in ^{114}Pd. So, if we see a difference in relative intensities in low spin states, there must be a second process that populates only low spin states. Remember the neutrons come out first, leaving excited ^{136}Te-^{114}Pd which decay by γ-emission. So in any one process, every γ-ray in ^{136}Te is in coincidence with every γ-ray in ^{114}Pd and there is no difference in single and double gates. To show that normal side feeding in ^{136}Te from hot fission is not responsible for the difference, we looked at the $8 \rightarrow 6$ - $6 \rightarrow 4$ and $6 \rightarrow 4$ - $4 \rightarrow 2$ double gates (where the 8^+, 6^+, and 4^+ states have quite different side feedings compared to the 2^+ state but presumably no cold neutron emission feedings). In both these double gates, the ^{114}Pd intensities are the same as in the $4 \rightarrow 2$ - $2 \rightarrow 0$ double gate, to show the difference is not related to side feeding.

For a double gate on the transitions I_1 and I_2, the ratio R_1 will be modified to: $R_2 = (I_4 - \Delta I_{24} - \Delta I_{26})/(I_3 - \Delta I_{22} - \Delta I_{24} - \Delta I_{26})$. Terms like ΔI_{24} represent the coincidence intensity to the 4^+ state in ^{114}Pd because of the population of the 2^+ state in ^{136}Te in the cold 2n ternary SF process. Since i_2 is excluded in the double gate, the intensity of I_4 will be reduced by the corresponding coincidence intensities like ΔI_{24} and ΔI_{26}. Similarly, I_3 also will be reduced. In the single and double gates, one also has: $R_1' = I(6^+ \rightarrow 4^+)/I(2^+ \rightarrow 0^+) = I_5/I_3$ and $R_2' = (I_5 - \Delta I_{26})/(I_3 - \Delta I_{22} - \Delta I_{24} - \Delta I_{26})$. Without i_c (i_2) feeding the 2^+ state directly, the ratios R_1 and R_2 are equal. If there is cold fission with two neutrons emitted at scission, these two ratios will be different, and similarly for R_1' and R_2'.

The relative γ -ray intensities of the transitions in ^{114}Pd and ^{112}Pd were determined for the above gates. The observed ratios R_1, R_2, R_1' and R_2' for ^{136}Te and ^{114}Pd are 0.67(5), 0.55(4), 0.25 (2) and 0.43(3) respectively. These yield $R_2 < R_1$ and $R_2' > R_1'$ which indicate that the dominant correlation is from ΔI_{24}. From these data and the above equations, we obtain an average value of $\Delta I_{24}/I_3 = 0.34$ [11]. From this value and the total yield for the 2n channel (normal + cold) 0.18 per 100 SF events the yield for the cold fission with 2n emission is 0.06 per 100 SF events. From the coincident intensities of the 606.6 and 424.0 keV transitions in ^{136}Te when the gates were set on the 2^+ - 0^+ and 2^+ - 0^+ plus 4^+ - 2^+ transitions in ^{114}Pd, there is negligible correlation between the 2^+ state in ^{114}Pd and the 2^+, 4^+ or 6^+ states in ^{136}Te with 2 neutrons emitted. Moreover, we detected no spin correlation between ^{136}Te and ^{112}Pd with 4 neutrons emitted so there is no cold 4n SF. Since ^{114,116}Pd are deformed and ^{136}Te with N=84 should be spherical and ^{138}Te with N=86 deformed, the cold binary fission more easily proceeds via deformed ^{138}Te and deformed ^{114}Pd as two fragments rather than through deformed ^{116}Pd and spherical ^{136}Te. These several new cold SF processes provide significant insight into the cold rearrangement of clusters of nucleons and multifragmentation.

4 Stable octupole deformation in ^{143,144,146}Ba and ^{145,147}La

Stable octupole deformation may be induced in nuclei where the Fermi level lies between single particle orbitals of the type $< N, L, j >$ and $< N+1, L+3, j+3 >$ (e.g., $p_{3/2}$ -$g_{9/2}$, $d_{5/2}$ -$h_{11/2}$, $f_{7/2}$ -$i_{13/2}$, etc.), because such orbitals couple strongly through the octupole potential term Y_{30} [18,19]. Evidence for stable octupole deformation is found in Xe, Ba, and Ce nuclei centered around Z=56, N=88 [1,3,20-24] as theoretically expected. We first consider the evidence in ^{143,144,145}Ba and then look at new results for ^{145,147}La. Of particular importance will be the role of nuclear rotation in enhancing stable octupole deformation as the nuclear rotation increases in some cases and quenching ($\beta_3 \to 0$) of stable octupole deformation as the nuclear rotation increases in other cases as theoretically predicted [25]. In ^{145}La bands built on both symmetric and asymmetric shapes are observed for the first time [21,26,27].

In our first report, new high spin states in $^{142-146}$Ba, and the first evidence for octupole deformation in an odd-A nucleus in this region in N=87, ^{143}Ba were found [20]. From our 1995 Gammasphere data, we extracted new insights into the importance of stable octupole deformation in these nuclei as well as ^{140,148}Ba and also Xe and Ce nuclei [3,21]. Our new level schemes for ^{143,144,146}Ba are shown in Fig. 4. Similar level schemes were simultaneously extracted from Eurogam II data in SF of ^{248}Cm [23,24]. Recently we have added new insights in ^{143}Ba. Important contributions of Urban et al. [23] and Jones et al. [24] were the establishment of the spins and parities of many levels assigned earlier on systematics in the even- and odd-A Ba isotopes from directional correlation and polarization measurements.

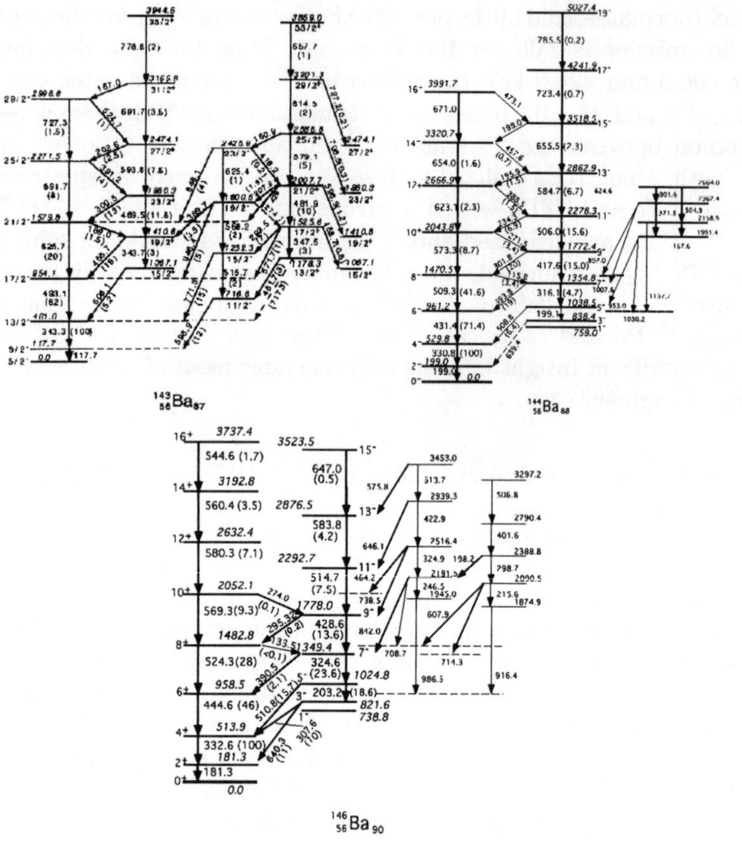

Figure 4. Level schemes for ^{143,144,146}Ba.

A striking feature seen in Fig. 4 is the intertwined $E1$ transitions between the positive and negative parity bands to quite high spin in ^{144}Ba but not continuing above 10^+ in ^{146}Ba. These $E1$ transitions are strongly enhanced in ^{142,144}Ba with $B(E1) \sim 1 \times 10^{-3}$ W.u. to give evidence for stable octupole deformation. However $B(E1) \sim 5 \times 10^{-6}$ W.u. in ^{146}Ba even when seen. This drop in $E1$ strength in ^{146}Ba was first noted by Phillips et al. [22]. In the moments of inertia for $^{142-148}$Ba, one sees in ^{146}Ba a backbend in J_1 and a sharp rise and drop in J_2 above 10^+ that is not seen in ^{142,144,148}Ba where only smooth upbends are seen. Cranked shell model calculations for the Ba-

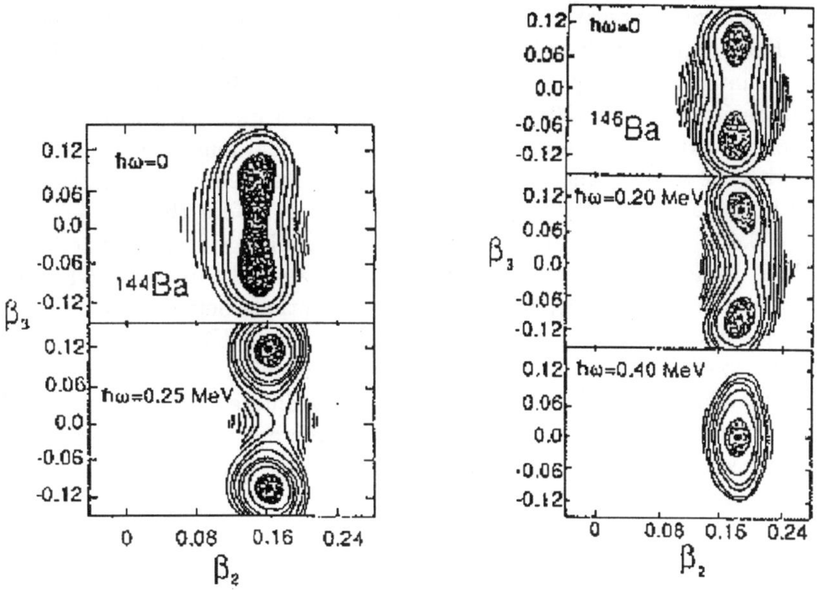

Figure 5. Cranked shell model calculations for Ba.

Ce nuclei indicate that around N=88 (^{144}Ba) at intermediate spins stable octupole deformation increases and the octupole minima are much better separated (see Fig. 5) [25]. In ^{144}Ba, where there is no backbend in J_1, the new data to 16^+ indicate no reduction in the $E1$ strength for these higher spins compared to the lower spin states. In fact, the average D_0 values are significantly larger for the higher spin states than for the lower spin states, to indicate rotation enhances the stable octupole deformation in ^{144}Ba in agreement with theory. A shape transition towards $\beta_3 = 0$ is predicted in ^{146}Ba at frequencies above 0.3 MeV (Fig. 5) after the alignment of $\nu i_{13/2}$ and $\pi h_{11/2}$ pairs [25]. The new data for ^{146}Ba show the ground band is crossed above 10^+ (Fig. 4), and no intertwined connecting transitions are seen above 10^+. These data [3,20,21] provided the first evidence for the theoretical prediction of the vanishing of β_3 with increasing rotation above $\hbar\omega = 0.3$ MeV.

When reflection asymmetric stable octupole deformation occurs in odd-A nuclei, the level patterns are similar to rotational bands in reflection-asymmetric molecules including two pairs of parity doublets with the same spins but opposite parities in each doublet, the simplex $s = i$ and $-i$ doublets [28]. Each doublet pair is intertwined by enhanced $E1$ transitions. The first two, $E1$ intertwined opposite parity bands were observed in an odd-A nucleus in this region in ^{143}Ba [20] (Fig. 4) and were interpreted as the $s = -i$ members of the expected doublet pairs. Jones et al. [24] found the $s = i$ doublet pair with the same spins but opposite parities. Our newer data [3,21] likewise show both parity doublets. $B(E1) \sim 1 \times 10^{-3}$ W.u. for the $s = -i$ doublet and $\sim 0.4 \times 10^{-3}$ W.u. for the new $s = i$ doublet, values similar to those in^{144}Ba. In addition, in our most recent work we find three new $E1$ crossing transitions and five new presumably $M1$ transitions between the $s = i$ and $-i$ bands which were not reported in ref. 23. These $E1$ and $M1$ transitions between the two doublets measure the degree of mixing of the $s = i$ and $-i$ doublets. The levels of ^{145,147}Ba are strikingly different from those of ^{143}Ba [1,3] with their ground bands having quite different structures with no evidence for stable octupole deformation. This change between ^{143}Ba and ^{145}Ba is surprising since the ^{145}Ba core, ^{144}Ba has the strongest D_0's, most enhanced $E1$'s and largest $\beta_3 \sim 0.10$ in this region.

Neutron rich $^{145-147}$La with Z=57 and N=88,90 lie between the Ba(Z=56) and Ce(Z=58) nuclei where stable octupole deformation is reported. Thus, they are expected to be good candidates for octupole deformation. As our extensive study of these nuclei from ^{252}Cf SF was being completed [27], evidence for octupole correlations in $^{145-147}$La up to intermediate spins were reported from SF studies of ^{248}Cm [26]. However, evidence for stable octupole deformation was not conclusive up to the intermediate spins observed in that work [26], where levels to $31/2^-$, $17/2^+$, $23/2^+$, and $25/2^-$ were seen in bands labeled 1,3,4,5, in our Fig. 6. In ^{145}La Fig. 6, we established eleven new levels and twenty-two new γ transitions mostly at high spin and the parity change between bands -4 and -5 [27] beyond that of ref. 26. Particularly important is the discovery of a new high-spin band (2) with intertwined crossing transitions to band (1). In ^{147}La, Fig. 6 six new levels and nine new transitions primarily at high spins were identified.

Bands -3 and -4 in ^{145}La, from the ground state to I^π $(19/2^+)$ or $(21/2^+)$, form a strong- coupled band with well-deformed symmetric rotor shape. This strong-coupled ground band in ^{145}La is assigned a $\pi g_{7/2}$ configuration coupled to the ^{144}Ba core. This band is not seen in ^{147}La. Band -4 shows a sharp upbend in J_1 at the $23/2^+ \rightarrow 19/2^+$ transition that is not seen in band -5 to suggest the $19/2^+$ (or $23/2^+$) to $31/2^+$ levels are not a continuation of band -4. The large and remarkably constant eight $B(E1)/B(E2)$ ratios for bands -4 and -5 from $19/2^+$ to $33/2^-$ indicate a shift from a symmetric shape to an asymmetric octupole (or strong octupole correlation) shape around spin

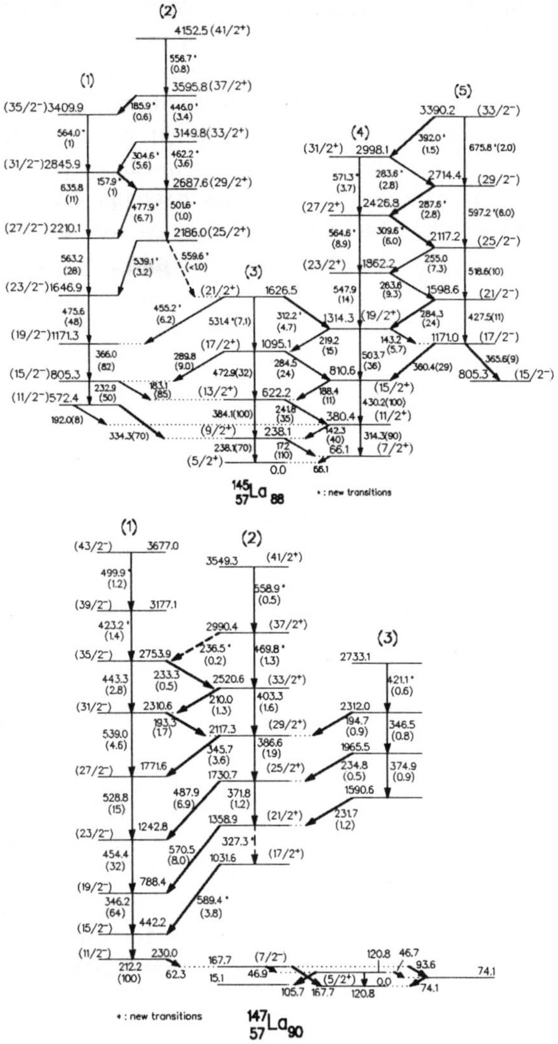

Figure 6. Level schemes of ^{145,147}La.

($19/2^+$) in ^{145}La. The average of these ratios, 1.3 x 10^{-6} fm^{-2}, is nearly twice those found in its ^{144}Ba core and in the parity doublets in ^{143}Ba. This first evidence for bands built on a stable octupole asymmetric deformed shape (or with strong octupole correlations) coexisting with bands built on a symmetric deformed shape confirm the expectations of Nazarewicz [29] and Chasman [30] that octupole-deformed and reflection-symmetric shapes may occur in the same odd-A nucleus, depending on the properties of the single particle orbitals. For bands -1 and -3 at low spin and for bands -1 and -2 at high spin in ^{145}La, one sees quite different behavior. The low $B(E1)/B(E2)$ ratios for bands -1 and -3 indicate weaker octupole correlations, as noted earlier [26]. Our new band-2 seen at the higher spins and our extenstion of band-1 now give strong additional evidence for strong octupole correlation [21,27] for bands -1 and -2 at higher spin in ^{145}La. Similar behavior is seen in ^{147}La for bands 1 and 2. The bands 1 and 2 in ^{145,147}La assigned $s = +i$ are very similar in structure. A comparison of the $s = +i$ band levels in ^{145}La and ^{147}La with those of the N=88 and 90 isotones ^{144}Ba and ^{146}Ba shows that the energies of both the positive and negative bands between the isotones are remarkably similar to each other with backbends in ^{144}Ba, ^{145}La, and ^{146}Ba, ^{147}La at the same $\hbar\omega$. Thus, the $s = +i$ bands in ^{145}La and ^{147}La most probably originate from the single-particle $\pi h_{11/2}$ orbital coupling with the ^{144,146}Ba cores, respectively. It is quite interesting that at high spins, the $B(E1)/B(E2)$ ratios in the odd-Z ^{145}La and ^{147}La nuclei are larger than in the even-even cores, ^{144,146}Ba. In summary, in ^{145}La, a quadrupole ground state collective band with well-deformed symmetric shape coexists and competes with octupole deformed bands with asymmetric shape. Above spin 19/2 in ^{145,147}La, the asymmetric shape is rotation enhanced. This is the first observation for these structural characteristics in odd-Z nuclei in this region.

Acknowledgments

Vanderbilt U., Fisk U. and Mississippi State U. work is supported by the Dept. of Energy Grants No. DE-FG05-88ER40407, DE-FGO5-95ER40939 and DE-FGO5-97EER41056, the Joint Institute for Heavy Ion Research by U. Tennessee, Vanderbilt U., Oak Ridge National Laboratory, and DOE Grant DE-FG05-87ER40311 with the U. Tennessee, and Tsinghua U. by the National Natural Science Foundation and Sc. Found. for Nuclear Industry, China. JHH would like to thank the Alexander von Humbolt Foundation for support.

References

1. J.H. Hamilton et al. *Prog. Part. Nucl. Phys.* **35**, 635 (1995)
2. I. Ahmed, W.R. Phillips *Rep. Prog. Phys.* **58**, 1415 (1995)
3. J.H. Hamilton et al. *Part. Nucl. Phys.* **38**, 273 (1997)
4. *See papers in Fission and Properties of Neutron Rich Nuclei (Eds. J.H. Hamilton, A.V. Ramayya).* World Sci., Singapore 1998
5. A. Sandulescu, W. Greiner *Rep. Prog. Phys.* **55**, 1423 (1992)

6. J.H. Hamilton et al. *J. Phys. G* **20**, L85 (1994)
7. A. Sandulescu et al. *Phys. Rev.* **54**, 258 (1996)
8. C. Wagemans: *The Nuclear Fission Process.* CRC Press, London 1990
9. A.V. Ramayya et al. *Phys. Rev. C* **57**, 2370 (1998)
10. A. Sandulescu et al.: *Int. J. Mod.* Phys. *E* (1998)
11. M. Mutterer et al.: *DANF proceedings (to be published)*
12. A.V. Ramayya et al. *Phys. Rev. Lett.* **81**, 947 (1998)
13. J.K. Hwang et. al.: *to be published*
14. G.M. Ter-Akopian et al. *Phys. Rev. Lett.* **77**, 32 (1996)
15. Y.X. Dardenne et al. *Phys. Rev. C* **54**, 206 (1996)
16. H. Bowman et al. *Phys. Rev.* **126**, 2120 (1962)
17. C. Budtz-Jorgensen, H. Knitte: *in Proc. Semin. Fission (Ed. C. Waagemans).* Report BLG586, SCK/CEN Mol, Belgium, 1986, p. 91
18. G.A. Leander et al. *Nucl. Phys. A* **388**, 452 (1982)
19. W. Nazarewicz et al. *Nucl. Phys. A* **429**, 269 (1984)
20. S.J. Zhu et al. *Phys. Lett. B* **357**, 273 (1995)
21. S.J. Zhu et al.: *in Fission and Properties of Neutron Rich Nuclei (Eds. J.H. Hamilton, A.V. Ramayya).* World Sci., Singapore 1998, p. 234; J.H. Hamilton et al.: *in Highlights of Modern Nuclear Structures (Ed. A. Corvello).* World Sci., Singapore 1998, *(in press)*
22. W. R. Phillips et al. *Phys. Rev. Lett.* **57**, 3257 (1986)
23. W. Urban et al. *Nucl. Phys. A* **613**, 107 (1997)
24. M.A. Jones et al. *Nucl. Phys. A* **605**, 133 (1996)
25. W. Nazarewicz, S. Tabor *Phys. Rev. C* **45**, 2226 (1992)
26. W. Urban et al. *Phys. Rev. C* **54**, 945 (1997)
27. S.J. Zhu et al.: Phys. Rev. C (Feb. 1999) *(in press)*
28. W. Nazarewicz, P. Olanders *Nucl. Phys. A* **441**, 420 (1985)
29. W. Nazarewicz *Nucl. Phys. A* **520**, 333c (1990)
30. R.R. Chasman *Phys. Lett. B* **6B**, 7 (1980)

EMISSION OF ^{5}He AND ^{7}He IN SPONTANEOUS TERNARY FISSION OF ^{252}Cf

M. Mutterer [1], Yu.N. Kopach [1,2], P. Singer [1], M. Klemens [1], A. Hotzel [3], D. Schwalm [3], P. Thirolf [3], M. Hesse [4], and F. Gönnenwein [4]

[1] *Institut für Kernphysik, Technische Universität, 64289 Darmstadt, Germany*

[2] *Frank Laboratory of Neutron Physics, JINR, Dubna, Russia*

[3] *Max-Planck-Institut für Kernphysik, 69115 Heidelberg, Germany*

[4] *Physikalisches Institut der Universität, 72076 Tübingen, Germany*

Abstract

Formation of the neutron-unstable odd-N helium isotopes ^5He and ^7He in light-charged-particle (LCP) accompanied spontaneous fission of ^{252}Cf was investigated by an angular correlation measurement of neutrons, fission fragments and LCPs. Neutrons were measured with the Darmstadt-Heidelberg 4π $NaI(Tl)$ Crystal Ball (CB) spectrometer, the fission fragments and LCPs with the kinematic spectrometer CODIS. The ternary ^5He and ^7He particles (life-times $\leq 4 \times 10^{-21}\,s$) were identified by the angular distribution of their decay neutrons centered at the directions of motion of ^4He and ^6He particles. The probabilities for ^5He (^7He) accompanied fission were determined to be 0.23 (0.21) relative to ^4He (^6He) accompanied fission. We have, furthermore, deduced the energy and angular distributions of the residual ^4He and ^6He nuclei, and the correlated neutron time-of-flight spectra. Neutrons coming along with Li particles were also observed, and attributed to the neutron-breakup of excited ^8Li.

PACS numbers: 25.85.Ca, 24.75.+i, 23.20.En, 29.30.Ep.

1 Introduction

In the rare ternary fission process ($\simeq 1/260$ relative to binary fission, for ^{252}Cf)[1] emission of a large variety of predominantly neutron-rich isotopes from all the elements up to about silicon has been observed [2]. The most abundant light charged particle (LCP) species are ^3H (6.5%), ^4He (87%), ^6He (3.5%), $^{7,8,9}Li$ (0.5%), and $^{9,10,11}Be$ (1.6%), the yields of heavier LCPs dropping down drastically with increasing mass and/or nuclear charge number.

The LCPs can be created with some probability also in excited states [3], some of them decaying by the emission of neutrons (e.g., the neutron-rich helium isotopes ^6He and ^8He, and the odd-odd lithium isotope ^8Li). Also, particles which are unstable against neutron decay already in their ground states can be emitted, such as the odd-N isotopes ^5He, ^7He and ^{10}Li. The very short life-times ($\leq 10^{-20}$ s) of these nuclei are comparable with the period of LCP acceleration in the Coulomb field of the emerging fission fragments. Neutrons from the decay of such "exotic" LCPs are emitted preferentially in the direction of LCP motion, the width of their angular distributions reflecting the LCP velocity distributions at the time of decay, and the Q-value in the neutron breakup channels involved. Hence, precise data on the kinematical parameters of these decays can provide information about both the dynamics of fission and the structure of light exotic nuclei. Since LCPs are focused nearly orthogonal to the direction of fission an angle sensitive detection of neutrons, LCPs, and fragments allows one to disentangle the neutrons from the LCP decay and the more frequent prompt fission neutrons which, being emitted from the moving fragments, have angular distributions peaked at the fission axis.

An enhancement of neutron emission into the direction of motion of ternary α-particles compared to the opposite direction was observed first by Nefedov et al. [4]. In turn, Cheifetz et al. [5], and Graevskii et al. [6], measured the correlated neutron and α-particle energy spectra, and interpreted their results by the formation of ^5He. In the present experiment we have measured, for the first time, the complete angular correlation of neutrons, LCPs, and fission fragments. In addition to the decay of ^5He we were able to investigate also the similar emission of neutrons coming along with the much rarer ^6He and Li particles which was never observed before.

2 Experimental setup

In the experiment[3], the Darmstadt-Heidelberg Crystal Ball (CB) [7] serving as a homogeneous and highly granulated 4π neutron detector was combined with an efficient detection system for fission fragments and ternary particles. The CB is a dense spherical package of 162 large (20 cm long) $NaI(Tl)$ crystals of high γ-ray detection efficiency ($\geq 90\%$). Simultaneously, the CB registers neutrons, mainly by the (n,n'γ) reaction, with a rather high efficiency of $\simeq$ 60%. The γ-rays and neutrons are separated by time-of-flight. The ^{252}Cf sample (4×10^3 fissions/s, thin backing) and the assembly of detectors for measuring fragments and LCPs were mounted inside a CH_4-filled aluminum vessel positioned at the center of the CB. This kinematic spectrometer "CODIS" consists of a Frisch-gridded 4π twin ionization chamber (IC) for measuring fission-fragment energies and emission angles, and of a ring of 12 ΔE-E telescopes (made from ΔE IC's and silicon PIN diodes, with a solid angle of $\pi/4$) surrounding one half of the fragment IC for measuring the LCPs.[3] Angular resolutions (fwhm) were $\leq 5\,^{o}$both for the fragments and the LCPs. In four weeks of measurements we have registered, in coincidence with neutrons (and γ-rays), 1.2×10^6 ternary fission events accompanied by α-particles, 8×10^4 events with ^{3}H, 2.5×10^4 events with ^{6}He, and 2.5×10^3 events with the emission of Li and Be each. Simultaneously, 7×10^7 binary fission events were recorded, the data acquisition rate for this event type being reduced by a factor of 1/64. All crystals in the CB subtend equal solid angles ($\pi/40$, each), but relative neutron (and γ-ray) efficiencies varied up to $\simeq$ 20% due to the absorption in the detector assembly mounted inside. The corresponding corrections were derived from the ^{252}Cf measurement itself. The relative efficiency matrix for neutron detection was determined with isotropically distributed fission neutrons as measured in coincidence with fragments from binary fission emitted over the full solid angle from the source.

3 Data analysis and results

For analyzing the neutron angular correlations we sorted the measured emission angles of registered triple coincidence events between neutrons, LCPs, and fragments into a properly chosen coordinate system xyz (see Fig 1). The z axis is arbitrarily chosen as the direction of motion of the fission fragments from the light mass group (LF). The x axis is in the plane made up by the three charged reaction products (the fragment pair and the LCP), and close to the direction of LCP motion (e.g., ternary α-particles are known to be emitted at a mean angle of 83° with respect to LF, and with an angular distribution of 20^0 width fwhm). The y axis is accordingly perpendicular to that plane making a right-handed coordinate system xyz.

Transforming the measured spatial distribution of neutron intensity into this system has allowed us to study details of the n-LF-LCP-angular correlation, e.g. by projecting the neutron yields per unit solid angle on the planes $(i, j) = (z, x), (z, y), (x, y)$, with ϕ_{ij} being the respective polar angles in the planes (i, j). The left-hand side of Fig. 2 displays, for the cases of ^3H, ^4He and ^6He emission, the resulting projections w_{zx} on the plane zx (in which the momenta of the LCP and the fragments are coplanar), as a function of the polar angle ϕ_{zx}. The patterns show the well known emission of prompt fission neutrons preferably into the directions of fragment motion, with higher intensity along fragments from the light mass group, at $\phi_{zx} = 0$. In the cases of ^4He

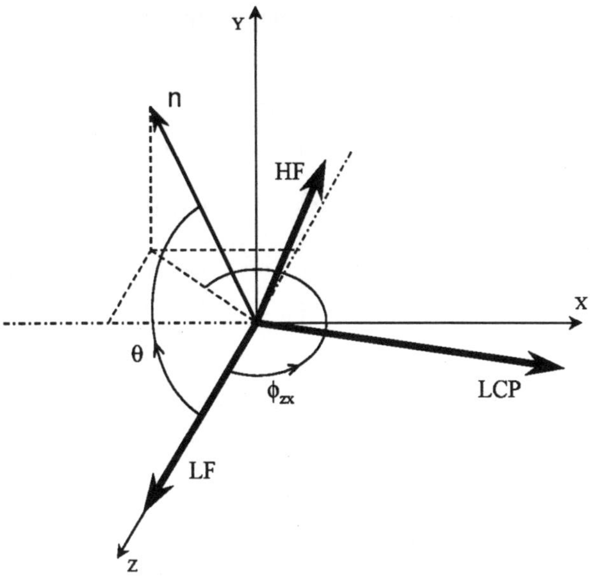

Figure 1: Coordinate system xyz for analysing the measured angular distribution of neutron intensity with respect to the motion of LCPs and fission fragments (see text).

and ^6He emission, neutron intensity is observed to be slightly higher around the angle of LCP emission ($\phi_{zx} \simeq 83°$, marked by an arrow) compared to the opposite direction. In contrast, there is no such difference in the neutron yield between forward and backward angles with respect to the direction of motion of ternary tritons. The enhancement in neutron yield along the direction of ^4He and ^6He motion is more clearly manifested in the projections w_{xy} (i.e. perpendicular to the fragment motion), and, even more, when only events are selected for which the θ angles lie within angular limits $60° \leq \theta \leq 120°$, i.e. about orthogonally to the fission axis. Such conditioned projections w_{xy} of the measured neutron intensity are displayed at the right-hand side of Fig. 2, as a function of ϕ_{xy}. We observe a rather wide peak in the neutron yield along the direction of ^4He particles, being imposed on the "background" from the more frequent prompt fission neutrons. The corresponding peak in the case of ^6He is narrower, and no such surplus neutrons along the LCP direction are observed when ternary fission is accompanied by tritons.

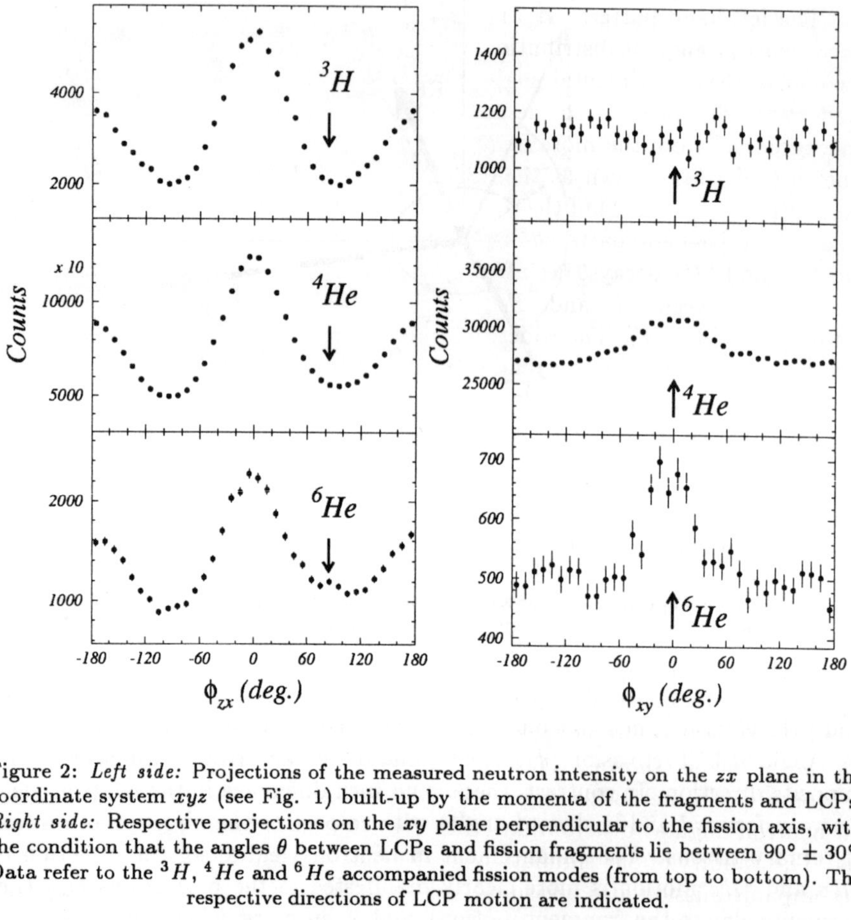

Figure 2: *Left side:* Projections of the measured neutron intensity on the zx plane in the coordinate system xyz (see Fig. 1) built-up by the momenta of the fragments and LCPs. *Right side:* Respective projections on the xy plane perpendicular to the fission axis, with the condition that the angles θ between LCPs and fission fragments lie between $90° \pm 30°$. Data refer to the ^3H, ^4He and ^6He accompanied fission modes (from top to bottom). The respective directions of LCP motion are indicated.

For an analysis of the present data in terms of the neutron decays of ^5He and ^7He we have calculated neutron angular distributions resulting from the break-up of these LCPs during motion, simulating the time evolution of LCP velocity during acceleration with trajectory computations that are able to reproduce measured LCP energy spectra reasonably well[8]. Neutron emission was assumed to proceed isotropically in the LCP rest frames, and from the break-up of the ground states in ^5He and ^7He (spectroscopical data from Ref. [9]):

$$^{5}He(3/2^-) \rightarrow {}^{4}He + n \quad \tau = 1.1 \times 10^{-21}\text{s} \quad Q = 0.85 \text{ MeV}$$
$$^{7}He(3/2^-) \rightarrow {}^{6}He + n \quad \tau = 4.1 \times 10^{-21}\text{s} \quad Q = 0.44 \text{ MeV}$$

The left-hand patterns in Fig. 3 depict the projections w_{xy} of the calculated neutron angular distributions, and the corresponding distributions with account of the experimental angular resolution. The latter distributions were used as the fit functions to the data, together with a function describing the "background" from the more frequent neutrons evaporated by the fragments. The resulting fits, shown as the central patterns in Fig. 3, represent a good description ($\chi^2/N \simeq 1$) of the data, with no constraint on the θ angles in this case. The right-hand patterns display the separated neutron components from the ^{5}He and ^{7}He decays, i.e. after background subtraction. The difference in the width between ^{5}He and ^{7}He is well reproduced by the calculated curves (shown as solid lines). The wider distribution in case of ^{5}He compared to ^{7}He is due to the shorter lifetime, and the higher neutron energy in the cm system. The ^{5}He nuclei decay closer to the fissioning nucleus (average distance $\simeq$ 12 fm) and, related to that, have gained lower velocity, on average.

Special care was taken in evaluating the shape of the back-ground curves which may leave some ambiguity in the procedure of extracting the angular distributions and yields of the extra neutrons in question: In ternary fission, LCP emission imposes a noticeable recoil momentum upon the fission fragments, causing their linear momenta to deviate from collinearity. In case of the He isotopes the average "recoil angle" is $\simeq 4.5°$, (see e.g. Ref. [10]). Accordingly, the intensity of prompt fission neutrons is slightly supressed towards the LCP direction, and slightly enhanced in the opposite hemisphere. We have simulated the respective projections w_{xy} of the intensity of prompt fission neutrons on the xy plane (shown as dashed lines in the central patterns of Fig. 3) by a Monte Carlo method, using precisely known ^{252}Cf binary fission data [11] on neutron angular distributions with respect to fragment motion as the basis. The validity of our simulation was tested by comparison with data on Be accompanied fission also measured in the present work. In this case the recoil angle is larger ($\simeq 10°$) and no neutron decay from LCPs occurs.

The probability for emission of the short-lived ^{5}He and ^{7}He isotopes in the ternary fission of ^{252}Cf was finally deduced from the measured ratios of LCP neutrons to prompt fission neutrons, taking for the average multiplicity of fission neutrons, $\bar{\nu_p}$, values of 3.1 (2.8) for the ^{4}He (^{6}He) accompanied fission modes, respectively, as deduced also from the present experiment (see also Ref. [1]). For ^{4}He and ^{6}He energies above our registration thresholds of 8 MeV, and 10 MeV, respectively, we have obtained for the relative yields $^{5}He/^{4}He$,

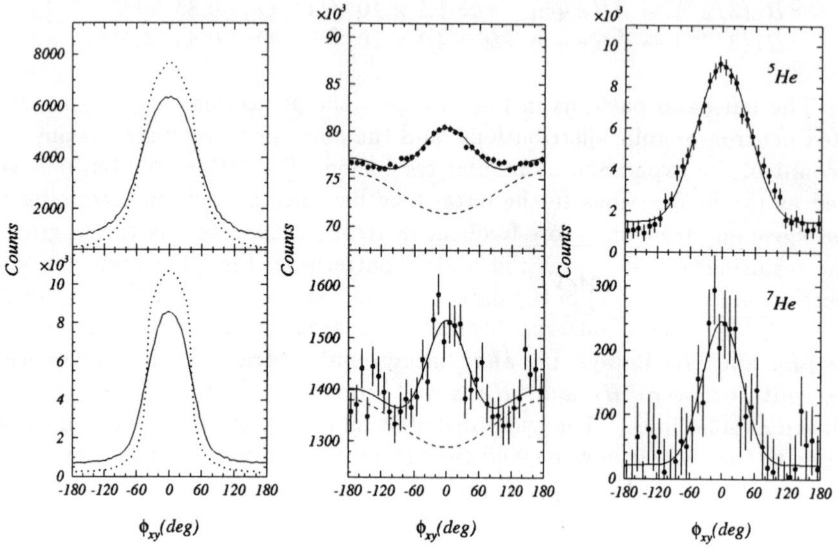

Figure 3: Different steps in the analysis of the neutron angular correlations in ^5He and ^7He accompanied fission. The following neutron angular distributions w_{xy} are shown: *left:* calculated ones, without (dotted line) and with account of experimental resolution (solid line), *central:* measured ones, with the background from prompt fission neutrons (dashed line) indicated, and *right:* measured ones, after back-ground subtraction. For details, see text.

and $^7He/^6He$, a value of 0.19 each. A correction for the sub-threshold events (see below) gives our final results:

$$^5He/^4He = 0.23(5), \quad \text{and} \quad ^7He/^6He = 0.21(5)$$

The errors given in brackets are dominated by the estimated uncertainty on the background correction; statistical errors being below the 5% level. We also note that we assumed neutron registration by the CB with constant efficiency over neutron energy, and did not consider possible reabsorption of neutrons that are emitted backwards towards the nearby fissioning nucleus (i.e. a shadowing effect). Our value for the $^5He/^4He$ ratio compares reasonably with 0.10 ± 0.02 given by Cheifetz et al. [5], considering that only a fraction of the neutrons from ^5He were registered in that work since neutron detectors were placed along the direction of the α-particles only.

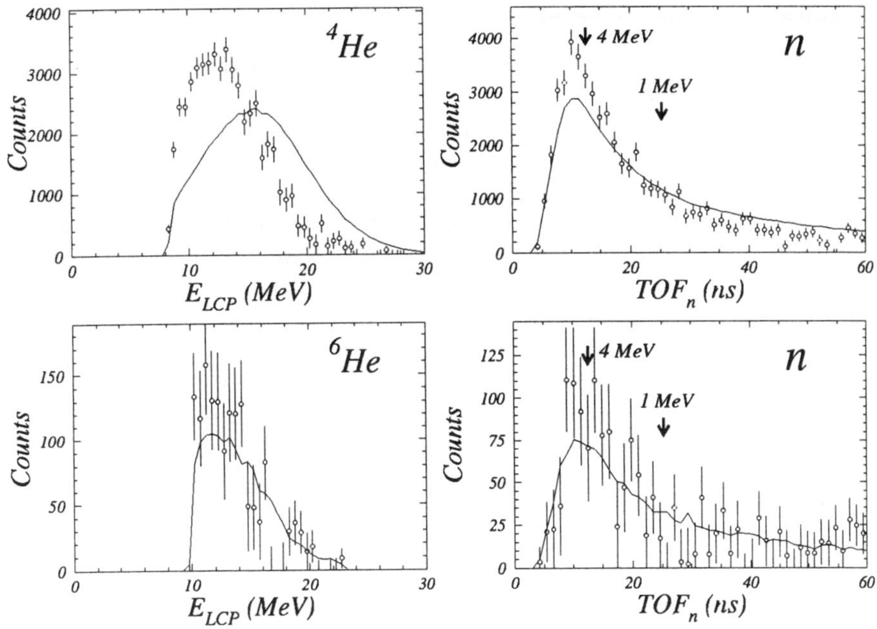

Figure 4: ^4He and ^6He energy spectra (left) and neutron time-of-flight spectra (right), for neutron emission angles directed towards the direction of motion of these particles. Solid lines show the corresponding ungated distributions, normalized to the same areas. The cutoff in the LCP energy measurement is 8 MeV (10 MeV) for ^4He (^6He). The neutron energies marked by arrows were calculated from the mean distance of 35 cm between source and NaI crystals

We have, furthermore, analysed the energy spectra of the neutrons and of the ^4He and ^6He particles which are associated with the ^5He and ^7He accompanied fission modes. This was performed approximately by setting constraints on the measured neutron angular distributions (Fig. 3, central part). We divided the projections w_{xy} into two regions: region **I** at angles $-90^0 < \Phi_{xy} < +90^0$, with neutrons directed towards the LCP motion, and region **II** at angles $\phi_{xy} < -90^0$ and $\phi_{xy} > +90^0$, with neutrons going in the opposite hemisphere. For several measured parameters (LCP energy, neutron TOF, and LCP emission angle relative to the fission axis.) we evaluated two distinct coincidence spectra with the condition that there were neutrons registered either in the region **I** or **II**. Each spectrum **II** was then multiplied with the background ratio (0.97) in the above regions, and the difference between the two spectra was evaluated.

We show in Fig. 4 the resulting energy spectra of the "secondary" ^4He and ^6He particles, and the corresponding neutron TOF spectra, in comparison with the normal (ungated) E_{LCP} and TOF_n spectra for ^4He and ^6He accompanied fission, respectively. The α-particles from ^5He have about 3 MeV lower average energy than the ternary α-spectrum, while the difference in the ^6He spectra is below 1 MeV. This behaviour could be well reproduced by the trajectory calculations mentioned above which were used to correct our ^5He and ^7He yields for below-threshold events (see above). The neutron TOF spectra from Fig. 4 indicate that neutron energies from the decays of both helium isotopes are, on average, higher than those of the prompt fission neutrons of ^{252}Cf. We note that the present results on the energies of both decay products from ^5He are in full agreement with the previous n-α-correlation experiments of Refs. [5,6]. We have analysed analogously the LCP angular distributions with respect to the direction of fragment motion. We find no significant difference in the angular distributions of the α's and the ^6He particles between the gated and ungated spectra. This is readily understood by the rather low recoil energies in the rest frame of the ^5He (^7He) particles, and by the fact that the residual ^4He (^6He) nuclei are, unlike the neutrons, further affected by the fragment Coulomb fields. It is, thus, evident that it is only by a neutron correlation measurement (as in Refs. [5,6] and in the present work) that the odd-N He-accompanied fission modes can be identified.

4 Summary and discussion

We have investigated the formation of the neutron-unstable ^5He and ^7He nuclei in the ternary fission of ^{252}Cf(sf) by an angular correlation measurement of neutrons, fission fragments and LCPs. The use of high-efficiency angle-sensitive detectors has yielded precise data on emission probabilities and kinematical parameters.

The present data on the emission probabilities allow us to complete the yield systematics of the helium isotope chain within mass numbers from 4 to 8, using known data on the ratios $^6He/^4He$ and $^8He/^4He$ (Refs. [12,13]). This is done in Fig. 5. Apparently, there is no neutron even-odd effect on the helium particle yields. Such effect is well known to appear on the yields of LCPs from other elements, the emission of isotopes with even neutron numbers being generally favoured compared to their odd-N neighbours [1]. The rather high $^5He/^4He$ and $^7He/^6He$ ratios of $\simeq$ 20% are also in conflict with what

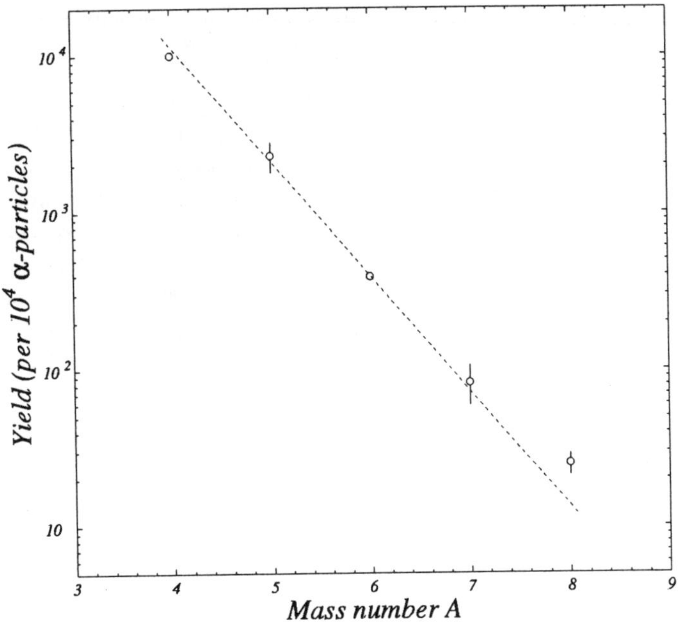

Figure 5: Yields of helium isotopes ($4 \leq A \leq 8$) in ternary fission of ^{252}Cf, normalized to 10^4 α-particles. The ^6He and ^8He yields are from Refs.[12,13].(dashed line, to guide the eye.)

is postulated by available theories [14,15,16]. As an example, Halpern's picture [14] of describing LCP yields in terms of the "average energy costs" for their production estimates only a 5% ^5He rate relative to ^4He. At present, we consider two alternative explanations which could possibly help solving this discrepancy:

1. There is reason to assume that the LCP yields are not only governed by the energetics of the emission process but also by some preformation probability in the fissioning nucleus[17]. Indeed, taking for the relative He-isotope yields versus mass number A just the corresponding ratios of the "bulk spectroscopical factors", $S_A = S_\alpha^{(A-1)/3}$ ($S_\alpha = 0.0068$), introduced in the theory of Cluster radioactivity [18] for describing preformation, results in a plot rather close to the data in Fig. 5. This interpretation, however, would fail in describing others than the above helium isotope

yields, and also the very rare emission of ^3He (measured so far only in thermal-neutron induced fission of ^{235}U and ^{239}Pu [1,19]).

2. A less serious problem for a generally valid theoretical description of the ternary fission process is to introduce nuclear structure effects into present theoretical calculations. In a systematic statistical approach theoretical yields should be multiplied with the statistical weight, $(2I_i + 1)$, with I_i being the spin of the LCPs in states i. Such a factor was considered earlier by Valskii [20] who established an interpolation formula for the description of measured relative LCP yield, rather successfully for atomic numbers $Z \leq 8$. Taking this assumption for granted, theoretical yield ratios $^5He/^4He$, and $^7He/^6He$ will increase by a factor of four, due to the $3/2^-$ spin of the $^{5,7}He$ ground states, and the 0^+ spin of $^{4,6}He$. Thus, the consideration of the microscopical LCP final state density would bring calculated yields into close accordance with the present data.

The neutron angular distributions, measured for the first time in the present experiment, could be well reproduced by a trajectory calculation for the LCPs, with assuming the formation of the ^5He and ^7He nuclei only in the neutron-unstable ground states. It is, however, reasonable to assume that other LCP decay channels might give minor contributions to the measured neutron yields as well [6,20]. Possible candidates are the two-neutron break-up reactions of excited ^6He and ^8He nuclei, and the break-up of ^5He in the about 4 MeV excited state with spin $1/2^-$. We note that ternary fission with the emission of neutron-rich helium isotopes provides a source of prompt fission neutrons that are emitted about orthogonally to the fission axis. These neutrons contribute, albeit little, to the ^{252}Cf neutron spectrum used as a reference standard in nuclear measurements. The dominant part of surplus neutrons comes from ^5He, with about one neutron in every 1500 binary fission events.

We have also deduced some other characteristics of the ^5He- and ^7He-accompanied fission modes, such as the energy and angular distributions of the residual ^4He and ^6He nuclei and the correlated neutron time-of-flight spectra. As shown in Fig. 4, the ^5He decay contributes to a ternary α- spectrum mainly at low energies. The enhanced low-energy α-particle yield compared to a Gaussian shape, which was frequently discussed in the literature (see, e.g. Refs. [21,22]), can thus qualitatively be explained by ternary ^5He emission. Our observation that about 20% of registered α-particles are actually due to ^5He emission implies that LCP yield systematics have to be corrected accordingly which, for practical reasons, often use the yield of the most frequent ternary

α-particles for normalization.

In the present experiment, neutrons coming along with Li particles were also observed. These neutrons have a quite narrow angular distribution w_{xy} peaked along the Li-particle motion and are, according to a trajectory calculation, attributed to the neutron-breakup of the second excited state ($E^* = 2.26$ MeV) in ternary ^8Li. Here, the poor counting statistics did not permit a detailed analysis of the kinematic parameters. In this context, it is interesting to note that another decay channel from ternary Li particles was discussed by Feather[23] who interpreted the measured coincident emission of tritons and α-particles[24] to originate from the decay of ^7Li when created in its excited state at 4.63 MeV.

This work forms part of the PhD Theses of P. Singer (TH Darmstadt, 1997) and Yu.N. Kopach (JINR, Dubna, to be finished in 1999), and was supported in parts by the BMBF, Bonn (06DA461 and 06TU669). One of us (Yu.K) would like to thank the DAAD, Bonn for a research grant. We are indepted to N.V. Kornilov for providing the ^{252}Cf neutron data, and to A. Göpfert and S.R. Neumaier for making available computer codes for performing the trajectory calculations.

References

1. M. Mutterer and J.P. Theobald, in *Nuclear Decay Modes*, ed. D. Poenaru, (IOP Publ., Bristol, England, 1996) Chapt. 12.
2. M. Hesse *et al.*, in *Proc. Int. Conf. on Dynamical Aspects of Nuclear Fission, DANF96, Častá Papiernička, Slovakia* , ed. J. Kliman and B.I. Pustylnik, (JINR Dubna, 1996) p. 238.
3. M. Mutterer *et al.*, in *Proc. Int. Conf. on Fission and Properties of Neutron Rich Nuclei, Sanibel Island, Fl.*, ed. J.H. Hamilton and A.V. Ramayya, (World Scientific, Singapore, 1998) p. 119.
4. V.N. Nefedov *et al.*, *Sov. J. Nucl. Phys.* **3**, 337 (1966).
5. E. Cheifetz *et al.*, *Phys. Rev. Lett.* **29**, 805 (1972).
6. A.P. Graevskii, and G.E. Solyakin, *Sov. J. Nucl. Phys.* **18**, 369 (1974), and references therein.
7. V. Metag *et al.*, *Lecture Notes in Physics* **178**, 163 (1983).
8. W. Baum, Ph.D. Thesis, TU Darmstadt (1992), unpublished.
9. F. Ajzenberg-Selove, *Nucl. Phys.* A **490**, 1 (1988).
10. P. Heeg *et al.*, *Nucl. Instr. Meth.* A**278**, 52 (1989).
11. N.V. Kornilov, private communication.
12. Z. Dlouhy *et al.*, in *Proc. Int. Conf. on Fifty Years Research in Nuclear*

Fission, Berlin, ed. D. Hilscher *et al.*, *Report HMI-B 464*, 43 (1989).
13. V. Grachev *et al. Sov. J. Nucl. Phys.* **47**, 622 (1988).
14. I. Halpern, *Annu. Rev. Nucl. Sci.* **21**, 245 (1971).
15. V.A. Rubchenya, and S.G. Yavshits, *Z. Phys.* A **329**, 217 (1988).
16. A. Pik-Pichak, *Phys. Atom. Nuclei* **57**, 906 (1994).
17. O. Serot *et al.*, *Nuclear Fission and Fission-Product Spectroscopy: 2nd International Workshop, Seyssins, France*, ed. G. Fioni *et al.*, (American Institute of Physics, Woodbury, N.Y., 1998) p. 306.
18. R. Blendowske *et al.*, in *Nuclear Decay Modes*, ed. D. Poenaru, (IOP Publ., Bristol, England, 1996) Chapt. 7.
19. A.A. Vorobyov *et al.*, *Phys. Lett.* B **40**, 102 (1972).
20. G. Valskii, *Sov. J. Nucl. Phys.* **24**, 140 (1976).
21. C. Wagemans, in in *Particle Emission from Nuclei, Vol. III*, ed. D.N. Poenaru and M.S. Ivascu, (CRC Press, Boca Raton, FL., USA, 1988) Chapt. 3.
22. C. Wagemans, in in *The Nuclear Fission Process*, ed. C. Wagemans, (CRC Press, Boca Raton, FL., USA, 1991) Chapt. 12.
23. N. Feather, *Proc. R. S. E.* **71**, 21 (1974).
24. S.K. Kataria *et al.*, Preprint IAEA-SM-174/63 (1973).

ANGULAR CORRELATIONS IN TERNARY FISSION INDUCED BY POLARIZED NEUTRONS*

F. GÖNNENWEIN, P. JESINGER, A. KÖTZLE, K. SCHMIDT
Physikalisches Institut, Universität Tübingen, Auf der Morgenstelle 14
D 72076 Tübingen, Germany

A.M. GAGARSKI, G.A. PETROV, V.I. PETROVA
PNPI of RAS, Gatchina, Leningrad district, 188350, Russia

G. DANILYAN, V.S. PAVLOV, V.B. CHVATCHKIN
Institute for Theoretical and Experimental Physics, Moscow 117259, Russia

M. MUTTERER, S.R. NEUMAIER
Institut für Kernphysik, Technische Universität,
D 64289 Darmstadt, Germany

V. NESVIZHEVSKY, O. ZIMMER, P. GELTENBORT
Institut Laue-Langevin, F 38042 Grenoble, France

K. KOROBKINA
Kurchatov Institute, Moscow, 123182, Russia

*This contribution is devoted to the memory of Boris I. Pustylnik, to the scientist for his seriousness and honesty, to the colleague for his modesty, to the friend for his reliable friendship.

Ternary fission in the reaction ^{233}U(n_{th},f) has been studied with polarized neutrons inducing the process. The experiment was performed at the Institut Laue-Langevin, Grenoble. In one set of experiments parity non-conservation in ternary fission was investigated. The main emphasis was, however, put on the search for a triple correlation between the helicity of the longitudinally polarized neutrons and both, the emission directions of fragments and ternary particles. The angular correlation function being [5 analyzed is $B = \hat{\mathbf{s}} \cdot [\hat{\mathbf{p}}_{LF} \times \hat{\mathbf{p}}_{TP}]$ with $\hat{\mathbf{s}}$ the neutron spin, $\hat{\mathbf{p}}_{LF}$ and $\hat{\mathbf{p}}_{TP}$ the momenta of the light fission fragment and the ternary particle, respectively, all vectors being normalized to unit length. The choice of this specific correlation function was motivated by a similar correlation explored in the free decay of polarized neutrons. There the correlation signals a violation of time reversal invariance. The deviation from a uniform distribution is parametrized by a factor $(1 + DB)$ in the correlation with D a constant called the correlation coefficient. In the present experiment the triple correlation coefficient D for ternary fission of ^{233}U with cold neutrons was measured to be $D = -(2.35 \pm .05) \cdot 10^{-3}$. The result quoted is a raw value not yet corrected for experimental resolution. The large size of this novel effect almost certainly precludes a violation of time reversal invariance as the mechanism at work.

1 Introduction

The discovery of both, ternary and quaternary fission was reported in a short note in 1947 by Tsien San-Tsiang et al[1]. In nuclear emulsions soaked with U-salts and irradiated with neutrons these authors observed events with three and even four prongs, all starting from a common vertex. Decay processes of heavy nuclei with three charged particles in the final state are called ternary fission and, correspondingly, those with four charged particles are called quaternary fission. Compared to binary fission with only two fission fragments, ternary fission is a rare process with the ratio of ternary/binary yields being typically some 2×10^{-3} for thermal neutron induced fission and 4×10^{-3} for spontaneous fission. Quaternary fission is even rarer, the yield ratio quaternary/binary fission being typically at the 10^{-6} level for nuclear decays at excitation energy not exceeding a few MeV.

While quaternary fission has not received much attention in the past, ternary fission has been extensively studied. Our knowledge on this process has been reviewed in two recent well documented compilations[2,3]. Generally, besides the two main fragments emitted in ternary fission, the third particle is a light charged particle. In 90% of all cases ternary fission is accompanied by an alpha-particle. In a stopping medium the ranges of these alphas are long compared to the ranges of alphas from natural radioactivity. This observation was already made at the very discovery of ternary fission. It points to the rather large average alpha energy of almost E = 16 MeV. In a further 7% of all cases the ternary particles are tritons. Elements beyond helium as ternary particles contribute less than .5% to the total yield, with the heaviest ternary particles identified so far being n-rich Si-isotopes[4]. Further distinct characteristics of light particles from ternary fission are their kinetic energy and their angular distributions. In the energy distributions both, the average energy $\langle E \rangle$ and the width FWHM increase with the nuclear charge Z of the ternary particle. Surprisingly, however, the ratio FWHM/$\langle E \rangle$ stays pretty constant and close to FWHM/$\langle E \rangle = 1$. It means that the distributions extend down to zero energy. This peculiar feature may be understood in case the ternary particles are born near the constriction zone of a deformed and fission prone nucleus. At scission the ternary particles are then emitted from a space region in between the two main fragments. Hence the Coulomb forces between the fragments and the light particle may partly compensate each other which explains the occurence of very low kinetic energies. Angular distributions confirm the above conjecture as to the birthplace of ternary particles. The distributions are usually given as the probability of emission as a function of the angle between the ternary particle and the light fission fragment. The overwhelming majority of ternary particles are observed roughly at right angles to the direction of flight of fragments. For light particles being ejected from a point in the space between the two main fragments, the focusing effect of the combined Coulomb fields should indeed lead to a sharply peaked angular distribution at right angles to the line of flight of the fragments. In the discussion of experiments to be described here it will therefore be taken for granted that in a dumb-bell model for the scission configurations the ternary particles are emitted from the neck region joining the two eventual fission fragments.

In recent years very complex multiparameter experiments in spontaneous fission have been carried out. In one series of studies ternary fission is traced by identifying in a 4π γ-spectrometer of high resolution the accompanying main fragments[5]. In another type of experiments the charged particles are measured with ionization chambers and PIN diodes while both the neutrons and gammas are registered in a 4π crystal ball[6,7].

Ternary fission with the characteristics described is typical for low energy fission. However, even at excitation energies of nuclei where multifragmentation with the appearance of intermediate mass fragments is a dominating mode in heavy ion reactions, the decay of neck-like structures has been reported[8,9]. The increase in probability for this process with excitation energy E^* is surprisingly slow and reaches only a factor of about two at $E^* = 200$ MeV compared to spontaneous fission.

Another branch of ternary fission research makes use of polarized neutrons to induce fission. Ternary fission being a rare process the choice of reactions suited to the case at hand is limited. Since there are efficient methods to polarize neutrons at or close to thermal energies, and since intense thermal neutron beams are available from reactors, these studies have so far all been performed for low energy neutrons. Only two experiments have been reported in the past. In one experiment a correlation between the polarization vector of the neutron beam and the alpha particle emission direction was analyzed[10]. In a second experiment the correlation between neutron spin direction and fragment momentum was measured for events with a ternary particle in coincidence[11]. The non-zero outcome for the correlation in this study points to parity-non-conservation in ternary fission.

The experimental work to be reviewed here is closely related to these last two studies, but nevertheless introduces a new facet of the fission process. The reaction $^{233}U(n_{th},f)$ was analyzed with cold neutrons inducing fission, the neutrons being almost 100% polarized. The experiment was performed at the Institut Laue-Langevin in Grenoble. The focus was placed on the study of ternary fission. An experimental setup was chosen which resembles the pattern of decay studies of polarized neutrons devised to search for violation of time reversal invariance. The setup allows to study both, parity non-conservation and a triple correlation between neutron spin and the momenta of fission fragments and ternary particles, which at least formally appears to violate time reversal invariance.

2 Experimental setup

As recalled in the above, ternary fission is a rare decay mode of fission prone nuclei. Investigating in ternary fission angular correlations between the momenta of the fission fragments and the ternary particles for polarized neutrons inducing the reaction requires, therefore, a setup which is optimally adapted to the process. Otherwise there is no chance to observe with good statistics correlation coefficients anticipated to be small. For example, the parity-non-conservation effect (PNC effect) in fission is typically at the 10^{-4} level as known since the discovery of the effect[12].

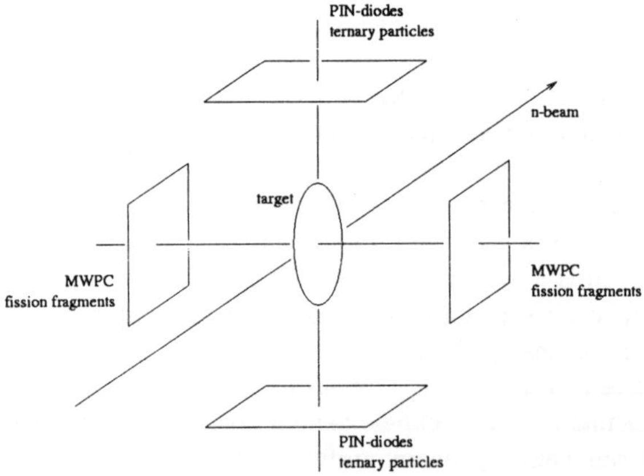

Figure 1. Experimental layout.

The experimental layout chosen is sketched in the schematic drawing of Fig. 1. A cold neutron beam (wavelength $\lambda = 4$ Å) from a neutron guide of the ILL highflux reactor was polarized by a magnetized supermirror(polarization $P = 96\%$) and traversed a detector chamber. The beam was running horizontally and the flux of the polarized beam was $\Phi = 6 \times 10^8$ n/(cm$^2\cdot$ s). A target with a total amount of 5 mg of enriched ^{233}U homogeneously evaporated onto a Ti substrate (diameter 8 cm, thickness 45 μg/cm^2) was placed in the center of the chamber. The neutron beam hit the target at grazing angles. Fission fragments (FF) were intercepted by a pair of multi-wire-proportional-counters (MWPC) mounted on a common horizontal axis through the target. The axis was oriented perpendicularly to the beam. The MWPCs were operated with ultra-pure CF_4 at a pressure of about 11 mb as the counting gas. Each of the MWPCs subtended a solid angle of 27.8% of 4π at the site of the target. Finally, the ternary particles (TP) were detected with two sets of 19 PIN diodes each. The diodes were installed on a common vertical axis through the target. Each PIN diode had an active surface of 3×3 cm^2 and each of the two sets subtended a solid angle of 18.8% of 4π at the target. In order to avoid pile-up the diodes had to be shielded from fission fragments by a thin Al foil. Since TPs are emitted roughly perpendicularly to the fission axis, the present setup with the detectors for FFs and TPs being oriented at right angles relative to each other ensures maximum efficiency.

For the fragments only a timing signal is derived from the MWPCs. Coincident events in the two MWPCs facing each other are considered to be valid fission events. The time-of-flight difference for these events allows to infer quite reliably whether a fragment from the light or heavy group has travelled to the left or right MWPC. Of course, the measured time-of-flight difference carries also some information on the velocity ratio and hence mass ratio of the two fragments. But the mass resolution is not very high, firstly, because both the target and the MWPCs have large sizes without any position sensitivity being available and, secondly, because even for a given geometry of the event and at a fixed mass ratio the velocity difference and likewise the time-of-flight difference observed is not unique. This is due to the finite width of the energy distributions for any fixed mass ratio. By contrast, the information collected for the ternary particles by the PIN diodes is of higher quality. Exploiting the granularity of the diode arrays, the number of each diode recording a charged particle in coincidence with a valid fission event is kept track of. In addition, for one of the two arrays the energy spectra of the TPs were registered diode by diode. Following standard procedures both, the time-of-flight spectra of the MWPCs and the energy spectra of the PIN diodes were digitized and stored on magnetic tape for off-line analysis. A separate data acquisition system allowed to monitor the data influx on-line and to find quickly results which served as a guide in the different options for running the experiment.

The spin orientation of the neutron beam could be turned around at will with the help of magnetic guiding fields. For the present study two spin orientations were especially important. In one part of experiments, said to be of the "PNC-type", the neutron spin was pointing either to the left or right fragment detector, i.e. parallel or anti-parallel to the fragment momentum. In a novel type of experiments in ternary fission, the main effort was, however, put into the analysis of events for the neutron polarization being longitudinal and pointing either in or opposite to the neutron direction of flight. Both the fragment and the ternary particle momenta are in this case orthogonal to neutron spin. This type of experiment is called a "triple correlation" experiment because correlations between neutron spin, the fragment and ternary particle momenta are coming into play.

Since the asymmetry and correlation effects being investigated are expected to be small, the well established spin flip technique was employed. The technique allows to compensate asymmetries in the geometry and efficiency of detectors. The spin was flipped every second and the spin state was registered together with the data from the detectors. Some further tests on false asymmetries were made by turning around the magnetic guiding fields.

The count rate measured was about 10^6 binary fission events/s, while for ternary fission it was close to 10^3 ternary events/s. Only ternary events were recorded on magnetic tape. With a measuring time of roughly 50 days the present study very probably has accumulated the largest amount of ternary data ever observed in a single experiment.

3 Experiment on Parity Non-Conservation in ternary fission

The search for Parity Non-Conservation (PNC) in ternary fission was not the main incentive for starting the present series of experiments. Nevertheless, the measurement of PNC effects in the ternary decay process $^{233}U(n_{th},f)$ provided a welcome means to check the proper functioning of the detectors and the electronics. Since more details on the physical background of PNC in fission are given in a companion contribution to this conference reporting on the investigation of PNC in binary fission[13], the presentation here will be deliberately brief.

PNC in nuclear reactions is conventionally traced to the weak interaction. Well-named, this interaction is by a factor of 10^{-7} weaker than the strong interaction responsible for nuclear reactions like fission. However, it is pointed out by theory[14,15] that several enhancement mechanisms could be at work. Especially important for heavy nuclei is the so-called dynamical enhancement where it is argued that the high level density of nuclei excited by neutron capture will favor the mixing of levels with opposite parity at the compound stage. With enhancement factors of 10^3 in fissile nuclei the predicted PNC effects are 10^{-4} in order of magnitude. For binary fission this prediction has been confirmed in many studies of PNC[12,13].

Characteristic for the weak interaction is a pseudoscalar interaction

Hamiltonian. Hence, the signature for PNC in experiment is the appearance of a non-vanishing pseudoscalar observable. A pseudoscalar quantity is readily found in case fission is induced by polarized neutrons. The scalar product between the unit vector of spin polarization $\hat{\mathbf{s}}$ and the unit vector of fragment momentum $\hat{\mathbf{p}}_{\mathrm{LF}}$, i.e.

$$A = \hat{\mathbf{s}} \cdot \hat{\mathbf{p}}_{\mathrm{LF}} = \cos\Theta \tag{1}$$

is in fact a pseudoscalar. Evidently, in the equation Θ is the angle between the two unit vectors in question. In eq. (1) care has to be taken that the momentum of a specific fragment, conventionally a fragment from the light mass group, is considered. The index L in the fragment momentum $\hat{\mathbf{p}}_{\mathrm{LF}}$ is a reminder of this specification. In fact, if no distinction is made between the fragments from the light and heavy group flying apart collinearly, the angles θ and $(\pi - \theta)$ are mixed up and the average $\langle\cos\Theta\rangle$ will vanish on trivial grounds.

The PNC effect, if at all present, then becomes manifest as a non-vanishing expectation value $\langle A\rangle$ of the observable A averaged over the full solid angle. In experiment, however, usually the asymmetry in the angular distribution $W(\Theta,\phi)d\Omega$ relative to neutron spin is measured as

$$W(\Theta,\phi)d\Omega \sim (1 + \alpha\cos\Theta)d\Omega\ . \tag{2}$$

The relation between the PNC asymmetry coefficient α and the average $\langle A\rangle$ of the observable A is

$$\langle A\rangle = \langle\cos\Theta\rangle = (1/3)\alpha\ . \tag{3}$$

With detectors subtending only relatively small solid angles as seen from the target, the standard technique in PNC studies has been to place these detectors parallel or antiparallel to the beam polarization, i.e. at $\Theta = 0$ and $\Theta = \pi$ for maximum sensitivity in eq. (2). The spin flip technique is perfectly adapted to this scheme. With $N_\uparrow$ and $N_\downarrow$ the number of counts for a small (in principle infinitely small) detector placed at $\Theta = 0$ and spin parallel and antiparallel to the line of flight of fragments, the PNC asymmetry is found from

$$\alpha = (N_\uparrow - N_\downarrow)/(N_\uparrow + N_\downarrow) \tag{4}$$

In actual practice data are taken with detectors of finite size and neutron polarizations smaller than 100%. The raw data have then to be corrected for these shortcomings.

For studying PNC asymmetries with the detector arrangement shown in Fig. 1 the neutron polarization has to be transversal to the beam direction and pointing towards the fragment detectors, i.e. either to the left or to the right MWPC. The spin flip will just switch from one spin situation to the other. With the setup in Fig. 1 it is possible to analyze in parallel PNC effects for both, binary and ternary fission. Since compared to binary fission the ternary process is so rare, it is sufficient for evaluating the binary PNC asymmetry to just make sure by a coincidence between the two fragment MWPCs that a valid fission event has been recorded and to discard any information on the ternary particles. By contrast, PNC effects in ternary fission are found by insisting that in coincidence with a valid fission event a third charged particle has been detected. It has to be stressed that in the present PNC experiment no information on the ternary particle beyond its mere existence is required to find the PNC effect in ternary fission.

The PNC asymmetry for binary fission of $^{233}U(n_{th},f)$ was measured with great precision in an experiment run in parallel to the present one. The results are quoted in a paper submitted to this conference[13]. The binary PNC asymmetry is there found to be

$$\alpha^{bin} = (+3.96 \pm 0.17) \cdot 10^{-4} \ . \tag{5}$$

From the present experiment the following PNC effect is obtained in ternary fission of the same reaction $^{233}U(n_{th},f)$:

$$\alpha^{tern} = (+3.5 \pm 0.8) \cdot 10^{-4} \ . \tag{6}$$

This value for α^{tern} takes into account the finite sizes of detectors and of neutron polarization but has not yet been corrected for background due to chance coincidences between valid fission events and scattered ternary particles. This correction will tend to further increase the α^{tern}-value. However, the size of this correction is anticipated to be small and will not exceed 10%.

Comparing the sizes of the PNC asymmetries in binary and ternary fission in eqs. (5) and (6), respectively, it is most interesting to realize that they are equal within experimental uncertainty. This result is in full agreement with the conclusions reached in a former study of PNC effects in ternary fission[11]. It should be pointed out that, except for the order of magnitude, the absolute value of α is of not much significance simply because it seems that theory will not be able to predict a precise value for a given reaction any time soon. By contrast, the mere equality of PNC effects in binary and ternary fission of the same compound nucleus carries highly valuable information on the fission process. This is expounded in detail in a former paper[11]. Very briefly, the arguments put forward are that in low energy fission the angular distributions of fragments are governed by the sequence of levels at the saddle point. This statement is taken to hold true also for angular distributions violating parity. The equality of PNC effects in the two decay variants of binary and ternary fission is, hence, considered as an experimental proof that at least up to the saddle point the paths followed in the potential energy surface from the compound stage to scission are identical. It is only close to the scission point that binary and ternary fission become different processes. This should be true for the overwhelming majority of ternary charged particles. Evidently this finding is fully compatible with the conclusions drawn from the energy and angular distributions of ternary particles. As discussed in the introduction these indicate that, for very elongated scission shapes, the birthplace of ternary particles is in the space between the two main fragments right at scission.

4 Experiment on triple correlations in ternary fission

The main purpose of the present experiment was to explore a triple correlation in ternary fission between the spin of the neutron inducing fission, the momentum of a fission fragment (e.g. a fragment from the light mass group), and the momentum of the ternary particle. More precisely, an angular correlation is sought between the three above vectors, all normalized to unit length.

The observable for this triple correlation is chosen in analogy to the observable B in the decay of a polarized neutron:

$$B = \hat{\mathbf{s}} \cdot [\hat{\mathbf{p}}_e \times \hat{\mathbf{p}}_\nu] \tag{7}$$

where $\hat{\mathbf{s}}$ is the unit vector of neutron spin, and $\hat{\mathbf{p}}_e$ and $\hat{\mathbf{p}}_\nu$ are the unit vectors of the momenta of the electron and the neutrino. This observable is odd under time reversal, but even under the parity operation. A non-vanishing average value for B might, hence, signal a violation of Time Reversal Invariance (TRI). Already several years ago it was remarked that an observable very similar to B in eq. (7) is accessible in ternary fission[16,17]. In fact, it was proposed to measure in ternary fission the quantity

$$B = \hat{\mathbf{s}} \cdot [\hat{\mathbf{p}}_{\mathrm{LF}} \times \hat{\mathbf{p}}_{\mathrm{TP}}] \tag{8}$$

where $\hat{\mathbf{s}}$ is now a unit vector in the spin direction of the neutron inducing fission, and $\hat{\mathbf{p}}_{\mathrm{LF}}$ and $\hat{\mathbf{p}}_{\mathrm{TP}}$ are the unit vectors of the momenta of a light fragment (LF) and the ternary particle (TP). The reason why a specific fragment (here the light one) has to be selected for the calculation of B is similar to the one already given with regard to the observable A in eq. (1): keeping in mind that the light and heavy fragment have opposite momenta, the indiscriminate use of the light or heavy fragment's momentum in eq. (8) would lead to a trivial zero expectation value $\langle B \rangle = 0$ upon averaging.

Recalling the experimental setup in Fig. 1 it is seen that, for a neutron beam which is longitudinally polarized in or against the direction of flight, the fragment and the TP detectors are mounted in such a way as to maximize the triple correlation function of eq. (8). The setup should, therefore, be most sensitive for any correlation of the type proposed.

The correlation function B is running between the extreme values $B = -1$ and $B = +1$. It can readily be shown that, for fixed spin direction, isotropic emission of fission fragments into 4π and emission of TPs at right angles to the fragments but rotationally symmetric with respect to fragment momentum, the B values are distributed uniformly between -1 and +1. Though the emission pattern of the TPs is idealized, it comes pretty close to reality. In fact, Monte Carlo simulations of the B distribution taking into account the experimentally observed angular distribution of TPs relative to the fission axis show that basically the above uniform distribution of B is maintained also for this more realistic approach. Both for the idealized and the realistic B distribution the average $\langle B \rangle$ will vanish under the above assumptions: $\langle B \rangle = 0$. To observe a non-vanishing expectation value $\langle B \rangle$ therefore requires to deviate from these assumptions. Again in analogy to the β-decay of polarized neutrons the following form of the triple correlation is suggested:

$$W(\hat{\mathbf{p}}_{\mathrm{LF}}, \hat{\mathbf{p}}_{\mathrm{TP}}) d\Omega_{\mathrm{LF}} d\Omega_{\mathrm{TP}} \sim (1 + DB) P(\hat{\mathbf{p}}_{\mathrm{LF}} \cdot \hat{\mathbf{p}}_{\mathrm{TP}}) d\Omega_{\mathrm{LF}} d\Omega_{\mathrm{TP}} \tag{9}$$

being understood that the polarization of the neutron inducing fission is kept fixed. Reversing the polarization — as in the spin flip technique — will reverse the sign of B. The quantity D in eq. (9) is taken to be a constant called the "correlation coefficient" in the following. The factor $P(\hat{\mathbf{p}}_{\mathrm{LF}} \cdot \hat{\mathbf{p}}_{\mathrm{TP}})$ describes the angular dependence in the emission of the ternary particle TP relative to the light fragment LF.

In eq. (9) terms accounting for some further subtle effects like PNC, left-right or fore-aft asymmetries in the emission of fragments relative to the neutron have been omitted. For the detector geometry and the beam polarization chosen this is well justified. All information on the triple correlation is now contained in the correlation coefficient D. For the idealized emission pattern of TPs at right angles to the LF it is easily shown that the following relation holds between the expectation value $\langle B \rangle$ of the observable determined in an ideal detector system covering 4π of solid angle and the correlation coefficient D:

$$\langle B \rangle = (1/3) D \, . \tag{10}$$

For the distribution $P(\hat{\mathbf{p}}_{\mathrm{LF}} \cdot \hat{\mathbf{p}}_{\mathrm{TP}})$ of TPs relative to the fission axis as observed in experiment the relation (10) represents an excellent approximation.

Eq. (10) for the triple correlation coefficient should be compared to the corresponding eq. (3) for the PNC effect. Incidentally the numerical factors are almost identical.

In actual practice, for detectors as arranged in Fig. 1 and making use of the spin flip technique the correlation coefficient is found from

$$DB_{\uparrow} = (N_{\uparrow} - N_{\downarrow})/(N_{\uparrow} + N_{\downarrow}) \ . \qquad (11)$$

Here $N_{\uparrow}$ and $N_{\downarrow}$ are the count rates of fragment-ternary particle coincidences for the neutron polarization pointing either in or against the neutron's direction of flight, respectively, i.e. positive or negative neutron helicity. $B_{\uparrow}$ is the value of B from eq. (8) for positive neutron helicity and the detector combination chosen. For the setup in Fig. 1 $B_{\uparrow} = +1$ or $B_{\uparrow} = -1$, depending on the MWPC-PIN diode pair. Strictly speaking this relation is valid only for infinitely small detectors placed at right angles to each other and both detectors at right angles relative to the neutron polarization. The raw data taken with the apparatus of Fig. 1 have, therefore, to be corrected for the finite size of detectors, finite neutron polarization, the efficiency in discriminating light from heavy fragments and background coincidences. These corrections have not yet been fully evaluated and, hence, a preliminary raw value for D will be given below. All corrections will tend to increase the final value for D. It should also be noted that in the experimental setup of Fig. 1 there are 4 pairs of MWPCs for the fragments and Pin diodes for the TPs, so the coefficient D in eq. (11) can be determined independently 4 times in one experiment. Besides the usual tests of measuring D for the magnetic guide fields reversed and for a depolarized neutron beam, the 4 independent measurements of D provide a convenient means to check the apparatus for hidden instrumental asymmetries. In experiment all tests were passed and the value for D given represents an average over all combinations of fields and detectors. More detailed information can be found in a report on the comparisons of all tests having been performed[18].

From the raw data the following value for the triple correlation coefficient D in ternary fission is deduced:

$$D = -(2.35 \pm .05) \cdot 10^{-3} \qquad (12)$$

The triple correlation experiment evidently yields a non-zero result. The surprisingly large size of the effect was totally unexpected. Referring once more to Fig. 1, the negative sign of D in eq. (12) means e.g. that, for a neutron with positive helicity (spin oriented parallel to neutron momentum) and a light fission fragment flying to the left, the probability for the ternary particle to be emitted downwards is slightly larger than being emitted upwards. Formally the non-vanishing of the coefficient D or, with eq. (10), of the expectation value $\langle B \rangle$ of the correlation function B of eq. (8) indicates that TRI could be violated. However, there are two serious objections to such a reasoning. First, the mere size of the effect is much too large to be credibly traced to a violation of TRI. Currently it is estimated that TRI violating effects should be smaller than PNC effects by a factor of about 10^{-3}. Upon comparing α^{tern} in eq. (6) with D in eq. (12) from the present experiments the opposite appears to be true, i.e. the PNC effect is smaller than a presumed TRI effect. Besides this worrying feature there is a second objection of moment: final state interactions in the decay of a nucleus can mimic a violation of TRI. This fact has been repeatedly emphasized by theory (for a recent review see ref. [19]). The challenge to further experimental investigations is then to find the clue to the mechanism of final state interactions at work.

5 Further analysis of triple correlations

The data collected in the present experiment on triple correlations in ternary fission contain much more detailed information than is summarized and therefore buried in the coefficient D of eq. (12).

Probably first of all one should ask whether the triple correlation is depending on the mass and/or energy of the fission fragments or the ternary particles. As already stated when describing the experimental setup, the mass and/or energy resolution for the fragments in the present experiment is not outstanding. Nevertheless a rather clearcut result appears to emerge as discussed in the following.

In the two panels of Fig. 2 the distribution of time-of-flight (TOF) differences between the two fragments is given on top, while on bottom the coefficient D times the effective B for a given detector pair and neutron helicity is shown as a function of the flight-time difference. The left peak in the TOF difference spectrum corresponds to events where the light (faster) fragment is flying towards the left detector and, correspondingly, the heavy (slower) fragment to the right detector of Fig. 1. For the right peak the situation is just reversed. It is seen in Fig. 2 that light and heavy fragments can be distinguished though the separation is not perfect. Since from mass and momentum conservation the velocity ratio of fragments is just the inverse of their mass ratio, events farer away from symmetry in the TOF difference spectrum correspond to more asymmetric mass splits. The strict one-to-one correspondence is, however, spoiled by the fact that even for a given mass division a quite broad kinetic energy and thus also velocity distribution is observed. The mass assignment is therefore not unique. Nevertheless, the

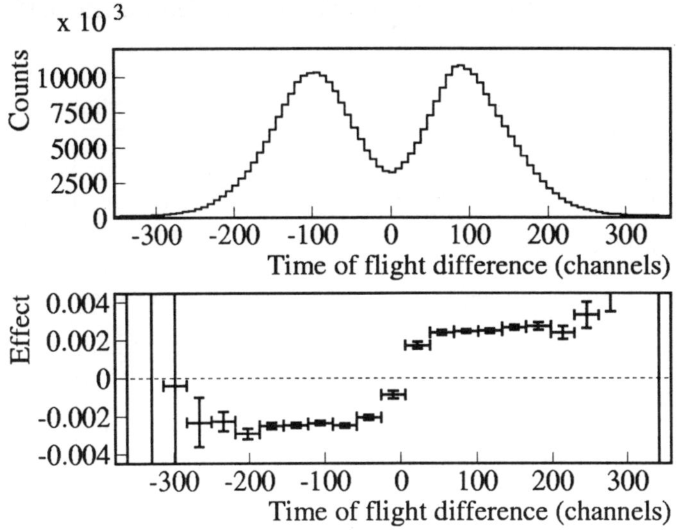

Figure 2. TOF-dependence of coefficient D.

average mass ratio is a function of the TOF difference and the TOF spectrum in Fig. 2 is considered to be a more or less fair representation of the fragment mass spectrum.

The most spectacular feature of the quantity DB depicted on bottom of Fig. 2 is the change in sign between the two peaks of the TOF or mass spectrum. This is, in fact, precisely the behaviour to be expected from the ansatz $(1 + DB)$ in eq. (9) for the angular correlation, with the definition of the correlation function B as given in eq. (8): since in the two peaks the direction of flight of the light fragment is inverted, also the sign of B is reversed and the experimental result in Fig. 2 may, therefore, be taken as a justification for the ansatz made. It should be noted again that the B in question here is the effective B of the setup of Fig. 1 with B being not exactly $B = \pm 1$ but very nearly so. A further remarkable fact to be read from Fig. 2 is the constancy of DB in each of the two peaks or, stated otherwise, the constancy of D throughout. This statement is only impaired close to symmetry and for very asymmetric fission. Near symmetry it is the finite mass resolution which is held responsible for making the distinction between light and heavy fragment ambiguous and thus leading to smaller DB values. For very asymmetric fission, scattered events contribute significantly and as expected the DB values are dropping. In summary, at the present level of resolution, the analysis does not disclose any dependence of the triple correlation coefficient D on fragment mass and/or energy.

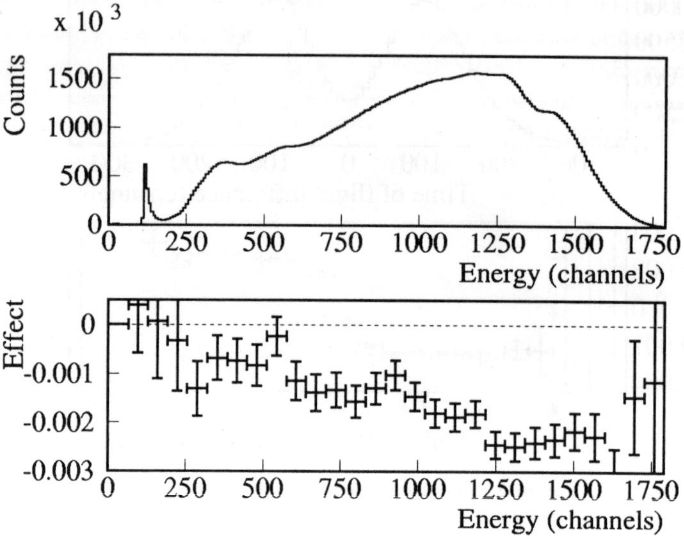

Figure 3. TP-energy-dependence of coefficient D.

In sharp contrast to the fission fragments, the dependence of the correlation coefficient D on the properties of the ternary particles is very pronounced. Though not all of the 38 PIN diodes in operation had depletion layers of adequate thickness to stop all of the more energetic hydrogen or helium isotopes, sufficient statistics was accumulated in one of the two arrays of diodes where it made sense to register energy spectra. Results are provided in Fig. 3. In the top panel, the sum of the energy spectra recorded for ternary particles is depicted. The fine structure in the shape of the spectrum is due to the fact that for not identical depths of depletion layers the response varies from diode to diode. This gives the sum spectrum shown a bumpy shape. The peak to the left, i.e. at the lowest energies, is background due to noise which deliberately has not been suppressed completely. The dependence of the coefficient D on the energy of the TPs is given in the bottom panel of Fig. 3. A very clear and spectacular correlation between D and the TP energy comes into view. At very small energies close to noise the triple coefficient D starts out with $D = 0$, as it should, and reaches its largest values for the highest kinetic energies of the TPs. The question then is where these energetic ternary particles are coming from. While conventionally TP emission is associated with elongated scission shapes of the fissioning nucleus, among those the more compact ones will generate TPs of high kinetic energies due to the larger Coulomb repulsion. Alternatively, however, large energies of TPs could also result from large momenta the TPs already have while being ejected at scission. To decide which of the two alternatives is responsible for large TP energies is not obvious. Nevertheless, any theory of the triple correlation in ternary fission will have to account for the experimental observation that the correlation is enhanced for the more energetic ternary particles.

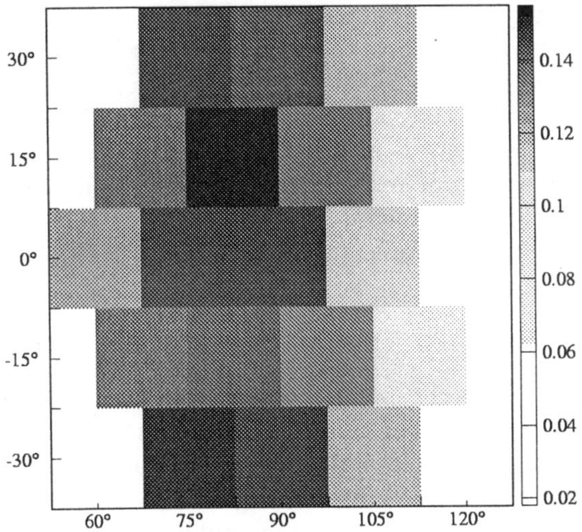

Figure 4. Angular distribution observed in the present experiment.

Finally, the granularity of the detectors for the ternary particles should be exploited. With the 19 PIN diodes per detector being read out individually, it becomes possible to trace the angular distributions of the TPs described in eq. (9) by the factor $P(\hat{\mathbf{p}}_{\rm LF} \cdot \hat{\mathbf{p}}_{\rm TP})$. This distribution is known to be centered at an angle of $\Theta_{\rm TP} = 82^\circ$ with respect to the light fission fragment and to have a width of 20° (FWHM). The angular distribution observed in the present experiment is reproduced in Fig. 4. In spite of the coarse graining the distribution is seen to be perfectly compatible with the data from literature[2,3].

From a theoretical point of view an obvious mechanism for a final state interaction, easily to be confused with a violation of TRI, is a spin-orbit interaction of the outgoing particles[20]. Either electromagnetic or strong in character, this interaction could depend on the spin orientation of the neutron inducing fission and thereby simulate a TRI violation. In experiment this should become visible as a shift in the angular distribution of the TPs, increasing or decreasing the average angle Θ_{TP} introduced above. In the angular distributions themselves the shift should be difficult to disclose in case the shift were very small. But a very sensitive test for a shift is to determine the correlation coefficient D for the individual PIN diodes of the array. If this type of final state interaction is responsible for the observed non-vanishing of the triple correlation, then the correlation D should exhibit a characteristic variation when moving through the array. The experimental result is clearcut: with D staying constant all over the array within statistical uncertainty, experiment does not give the slightest indication for a final state interaction of the type discussed. Stated otherwise, the factor $P(\hat{\mathbf{p}}_{LF} \cdot \hat{\mathbf{p}}_{TP})$ in eq. (9) does not depend on the spin helicity of the neutron inducing fission. All the dependence of the triple correlation on spin polarization is contained in the factor $(1 + DB)$ of eq. (9). It means that the azimuthal distribution of TP emission angles is no longer uniform around the fission axis defined by the momentum of the light fragment. This is considered to be a notable result which should steer further studies.

6 Summary

In the present work a novel type of experiment in ternary fission of ^{233}U has been performed. For polarized cold neutrons inducing fission, angular correlations between neutron spin and the momenta of outgoing particles were analyzed. Pairs of detectors for fission fragments and ternary particles were mounted in a plane perpendicular to the neutron beam and placed at right angles relative to each other. Two different types of experiments could be performed with the detector setup staying in place and only turning around the neutron polarization from transversal to longitudinal. In PNC-type experiments the neutron spin was transversal and pointed towards the fragment detectors. Measuring asymmetries in the angular distribution of fission fragments probes the pseudoscalar quantity $A = \hat{\mathbf{s}} \cdot \hat{\mathbf{p}}_{LF}$. A non-vanishing expectation value $\langle A \rangle$ for A is a fingerprint for PNC in ternary fission. Confirming former studies it was observed that indeed there is a non-zero asymmetry and that the size of the PNC effect is the same for binary and ternary fission. This is interpreted as telling that ternary fission is only showing up very late close to scission. The result lends further support to the view that the majority of ternary particles are ejected from the space between the fragments forming the neck of a basically binary scission configuration.

Triple correlations in ternary fission were not investigated in the past. In these experiments the neutron polarization was longitudinal, and for the detector arrangement chosen, maximum sensitivity is achieved for the triple correlation function $B = \hat{\mathbf{s}} \cdot [\hat{\mathbf{p}}_{LF} \times \hat{\mathbf{p}}_{TP}]$. The prototype for the correlation function was introduced in the study of β-decay of polarized neutrons. The intriguing feature of B is that it changes sign upon time reversal. A non-zero expectation value $\langle B \rangle$ of the observable B in the β-decay of neutrons would indeed imply a violation of TRI. In the presence of nuclear forces like in fission this conclusion is, however, no longer valid. The present investigation is facing therefore an ambiguity. On one hand, a non-vanishing expectation value for B was established, the order of magnitude being $\langle B \rangle = 10^{-3}$. On the other hand, the size of $\langle B \rangle$ is considered to be much too large for violation of TRI to play the decisive role. It should perhaps not be lost out of sight, though, that ternary fission itself is a rare process having only a probability of some 10^{-3} compared to binary fission.

In a further analysis of the accumulated data it was tried to find parameters of the fission process controlling the outcome of the correlation. Apparently the correlation $\langle B \rangle$ does not depend on fragment masses and/or energies. In sharp contrast, a clear dependence of the correlation $\langle B \rangle$ on the kinetic energies of the TPs was observed. The correlation $\langle B \rangle$ is most pronounced for the highest energies of the TPs. This is an interesting effect but, unfortunately, there is an ambiguity as to the origin of these energetic TPs. They either stem from rather compact scission configurations or they are already ejected with higher than average energy at scission. At least tentatively the second alternative seems to be more relevant with respect to the correlation $\langle B \rangle$ in question since compact scission shapes should also lead to larger than average fragment energies while no dependence of $\langle B \rangle$ on these energies could be disclosed in experiment.

Also the angular distributions of TPs relative to the light fragment (LF) came under scrutiny. A surprising result emerged. It appears that the distribution of polar angles Θ_{TP} between the momenta of LF and TP is neither shifted nor widened or narrowed down upon flipping the neutron spin. This observation is in conflict with expectation for theories invoking a spin-orbit coupling for the outgoing ternary particles. Instead, the non-zero triple correlation comes about by a non-uniform and neutron spin dependent distribution of azimuthal angles of TPs around the fission axis identified here with the direction of flight of the LF. These findings are still awaiting a theoretical understanding. Possible couplings between the spin of incoming neutrons and the spin of the fragments or the angular momentum of the ejected TPs give ample space for speculations. But before venturing into speculations it will be safer to continue with experiments since there are too many questions still open. Among them are especially important ones like the dependence of the correlation on the type and/or the spin of both, the compound nucleus and the ternary particles.

Helpful discussions with K. Dietrich, W. Furman and especially A. Barabanov are gratefully acknowledged. Part of this work was supported by grants from the BMBF under the contract number 06TU669 and by the program "Fundamental'naya Yadernaya Fizika".

References

1. Tsien San-Tsiang et al., Phys. Rev. **71** (1947) 382
2. C. Wagemans in *The nuclear fission process*, ed. C. Wagemans (CRC Press, Boca Raton, USA 1991) p 545
3. M. Mutterer and J.P. Theobald in *Nuclear Decay Modes*, ed. D. Poenaru (IOP Publ. Bristol, UK 1996) chapter 12
4. F. Gönnenwein et al. in *Nuclei far from stability*, ed. R. Neugart et al. (IOP Conf. Series 132, 1993) p 453
5. A.V. Ramayya et al., Phys. Rev. **C57** (1998) 2370
6. M. Mutterer et al. in *Fission and properties of neutron rich nuclei*, eds. J.H. Hamilton and A.V. Ramayya, (World Scientific, Singapore 1 998) p 119
7. Yu. N. Kopach et al., Phys. Rev. Lett. **82** (1999) 303
8. M. Sowinski et al., Z. Phys. **A324** (1986) 87
9. C.P. Montoya et al., Phys. Rev. Lett. **73** (1994) 3070
10. A. Ya. Alexandrovich et al., Leningrad Nuclear Physics Institute, preprint 1057, Leningrad 1985
11. F. Gönnenwein et al., Nucl. Phys. **A567** (1994) 303
12. G. Danilyan et al., JETP Lett. **26** (1977) 186
13. A. Kötzle et al. in *Dynamical aspects of nuclear fission*, (World Scientific), to be published
14. V.V. Flambaum and O.P. Sushkov, Phys. Lett. **94B** (1980) 277
15. V.E. Bunakov and V.P. Gudkov, Z. Phys. **A321** (1985) 271
16. K. Schreckenbach, Int. ILL Report 88 SCO9T, Grenoble, 1988
17. K. Schreckenbach et al. in *Time reversal invariance and parity violation in neutron reactions*, eds. C.R. Gould et al., (World Scientific, Singapore 1994) p 187
18. P. Jesinger et al. in *Nuclear fission and fission-product spectroscopy*, eds. G. Fioni et al., (AIP Conf. Proc. 447, Woodbury USA, 1998) p 395
19. V.E. Bunakov and L.B. Pikelner, Progr. Part. Nucl. Phys. **39** (1997) 337
20. A.L. Barabanov, private communication

PARITY NONCONSERVATION IN NUCLEAR FISSION: DOES IT DEPEND ON FRAGMENT MASS/ENERGY?

A. KÖTZLE, P. JESINGER, F. GÖNNENWEIN
*Physikalisches Institut, Auf der Morgenstelle 14, 72076 Tübingen, Germany**

G.A. PETROV, V.I. PETROVA, A.M. GAGARSKI
PNPI of RAS, Gatchina, Leningrad district, 188350, Russia

G. DANILYAN
ITEP, Moscow 117259, Russia†

O. ZIMMER, V. NESVIZHEVSKY
Institut Laue-Langevin, 38042 Grenoble, France

Parity Nonconservation (PNC) in thermal neutron induced fission is observed as an asymmetry of the angular distribution of the light fragment momentum direction $\hat{\boldsymbol{p}}_{\rm LF}$ relative to the neutron spin $\hat{\boldsymbol{\sigma}}$: $W(\theta) \sim (1 + \alpha_{\rm nf} \cdot P_{\rm n} \cdot (\hat{\boldsymbol{\sigma}} \cdot \hat{\boldsymbol{p}}_{\rm LF}))$. The aim of this work was to search for a mass- and total kinetic energy (TKE) dependence of $\alpha_{\rm nf}(m_{\rm LF}, {\rm TKE})$ in the neutron induced fission of ^{233}U and to test the predicted angular distribution of fragments $W(\theta)$. Altogether more than $2 \cdot 10^{10}$ fission events with high mass/energy resolution were collected. This corresponds to an increase in statistics compared to previous experiments [1] of about a factor of 20. The preliminary analysis of the mass/energy dependence shows no significant variation of $\alpha_{\rm nf}$ for different fragment masses and total kinetic energies, whereas the question concerning the angular dependence could be answered definitely with a precision unreached up to now: $W(\theta) \sim \cos\theta$.

1 Introduction

Parity Nonconservation (PNC) in nuclear fission induced by polarized cold neutrons is observed as an asymmetry of the fragment angular distribution relative to the direction of the neutron spin:

$$W(\theta) \sim (1 + \alpha_{\rm nf} \cdot P_{\rm n} \cdot (\hat{\boldsymbol{\sigma}}_{\rm n} \cdot \hat{\boldsymbol{p}}_{\rm LF})) = (1 + \alpha_{\rm nf} \cdot P_{\rm n} \cdot \cos\theta) \tag{1}$$

where: $\hat{\boldsymbol{\sigma}}_{\rm n}$ spin direction of the incoming neutron (unit vector),
$\hat{\boldsymbol{p}}_{\rm LF}$ momentum direction of the light fragment (unit vector),
θ angle between $\hat{\boldsymbol{\sigma}}_{\rm n}$ and $\hat{\boldsymbol{p}}_{\rm LF}$,

*This work is supported by BMBF, Bonn, under contract no. 06Tü669.
†Supported by the RFBR (96-02-17152) and program "Fundamental Nuclear Physics"

P_n degree of neutron polarization,
α_{nf} size of the asymmetry.

Since the scalar-product of the axial vector $\hat{\boldsymbol{\sigma}}_n$ and the polar vector $\hat{\boldsymbol{p}}_{LF}$ transforms like a pseudo-scalar, any non-vanishing α_{nf} means that parity is not conserved in this reaction.

PNC in nuclear fission has been observed for the first time about 20 years ago [2] in the neutron induced fission of ^{235}U and since that time in eight more (n,f)-reactions (see table 1). What was astonishing is the big size of the effect: the asymmetry α_{nf} was found to be typically some 10^{-4}. One has to keep in mind that the relative strength of the weak interaction (being responsible for PNC-effects) compared to the strong interaction (being responsible for nuclear reactions) is estimated to be some 10^{-7}. This fact prompted the suggestion that somewhere in the course of the fission process some enhancement mechanisms must be at work. Mainly Sushkov and Flambaum [3], and Bunakov and Gudkov [4] developed the theory of PNC in fission. Both theories point to the mixing of nuclear states at the compound stage of nuclei undergoing neutron induced fission. The high level density in these nuclei will favour the mixing of states with opposite parity but the same spin, and thus entail PNC interference effects. Calculations of the size of this so-called ***dynamical enhancement*** (taking into account the level density and the strength of the residual interaction) give a factor of about 1000. This theory is supported by the experimental fact that PNC-asymmetry coefficients fluctuate dramatically and may even change sign, when the incident neutron energy is shifted by less than 1 eV, i. e. the typical energy scale for the distance between levels in a heavy compound nucleus after thermal neutron capture [5]. As a consequence, also the sign and the size of $\alpha_{nf}(E_n\text{=thermal})$ should fluctuate randomly for different fissioning systems. According to table 1 this seems to be the case. Yet, a further puzzling feature of PNC in fission is the question, how this mixing of compound nucleus (CN) states being responsible for the asymmetry can survive on the way from the CN over the fission barrier to the scission point. At this point it has to be recalled that, as first emphasized by A. Bohr [6], on top of the barrier the nucleus has to cool down, since all its excitation energy is transformed into deformation, and, therefore, only few and widely spaced transitional states are accessible to the nucleus. For rotationally symmetric deformations of nuclei the collective part in the wave function of a quasi-stationary state over an asymmetric fission barrier is given by the wave function of a pear-shaped top. The Sushkov-Flambaum theory

Table 1. Global asymmetry coefficients α_{nf} observed up to now in neutron induced fission.

target	$\alpha_{nf} \times 10^4$	ref.
^{229}Th	$-5,6 \pm 0,2$	7
^{233}U	$+3,67 \pm 0,06$	7
^{235}U	$+0,84 \pm 0,06$	8
^{237}Np	$+0,86 \pm 0,42$	9
^{239}Pu	$-5,0 \pm 0,1$	10
^{241}Pu	$-0,89 \pm 0,04$	7
^{241}Am	$-1,00 \pm 0,25$	7
^{245}Cm	$-0,26 \pm 0,06$	9
^{249}Cf	$+0,13 \pm 0,13$	9

argues that, therefore, the transitional states are in fact closely spaced doublets differing only in parity, very similar to the well-known doublets in the tunneling vibration of the NH_3-molecule. This narrow splitting of levels with opposite parity at the barrier now gives the mixed states the commensurate chance to preserve their mixing also during the way over the saddle to the scission point. PNC in nuclear fission in principle being understood — it can now be used as a tool to test theories about the role played by the saddle point to steer the mass distributions of fission fragments.

Summarizing, questions partly still open and to be answered by further PNC-experiments in nuclear fission are:

1. Is α_{nf} a function of the fragments mass and total kinetic energy? Or, in other words: Which fragment properties are already set at the saddle point?

2. Can the $\cos\theta$-dependence be confirmed? Up to now most of the experiments have been performed with a fixed hardware collimator and hence an averaged $\overline{\cos\theta}$ for the solid angle of detection. An exception is the work of Vesna et al. [11], where multifilament proportional counters have been used to measure the superposition of the left-right and the parity nonconserving asymmetry.

2 Experimental Setup and Methods

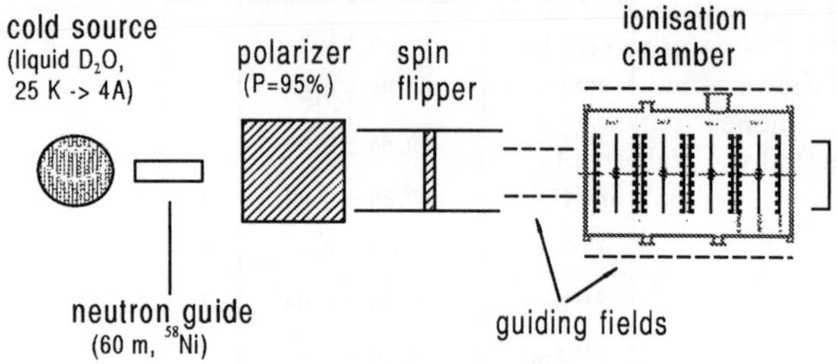

Figure 1. Experimental setup at the cold polarized beam PF1 of ILL.

The experiment was carried out at the cold polarized beam PF1 of the high flux reactor of the Institute Laue-Langevin in Grenoble. The setup is shown in fig. 1. Neutrons coming from the horizontal cold source with a wavelength of about 4 Å were guided by a 60 m long neutron guide to the polarizer. The polarizer was a super-mirror with a cross-section of $6 \times 12\,\mathrm{cm}^2$. Slow neutrons have the advantage to be readily polarized; here the polarization was $P_n = (95 \pm 1)$ % and the polarized neutron flux was $6 \cdot 10^8 \frac{1}{\mathrm{cm}^2 \cdot \mathrm{s}}$. A current sheet spin flipper placed behind the polarizer allowed to reverse the spin direction. Depending on the status of the flipper (current on or off), a static magnetic guide field rotated the spin adiabatically in or opposite to the direction of flight of the neutron and the information about the spin status was transmitted to the data acquisition system. Downstream the polarizer a weak but quite homogeneous guiding field parallel to the beam ensured the neutron polarization to stay constant.

Behind the polarizer a fourfold back-to-back ionisation chamber was placed. In each of the four sections a target of $5\,\mu g$ ^{233}U on a thin carbon backing ($25\,\mu g/\mathrm{cm}^2$) covered a circular aperture in the center of the common cathode between the two Frischgrid-shielded anodes. As material inside the chamber only stainless steal and ceramics was used to reduce gasing out. The chamber was operated with high purity CF_4-gas at a pressure of 400 mbar and a constant flow of about $2\,\ell/\mathrm{h}$. For each single event all three pulseheights (*pha* from anode *a*, *phb* from anode *b* and *phc* from the cathode) were written on magnetic tape (altogether about 100 GByte of data have been taken). Since the total charge of electrons accumulated at the anodes and the corresponding pulseheights delivered by the electronics is to good approximation proportional to the kinetic energies of fission fragments being stopped in the gas volume, the mass ratio of complementary fragments may be inferred from the ratio of pulseheights in the two chambers. This yielded a mass resolution of 0.75 amu in the cold region, rapidly deteriorating as soon as neutron evaporation sets in. A scatterplot of kinetic energies is shown in fig. 2, where the two "humps" indicate the asymmetric mass-distribution of $^{233}U(n,f)$. The structure at the highest total kinetic energy indicates resolved fragment mass ratios, so-called *masslines*, which demonstrate the excellent resolving power of the chamber.

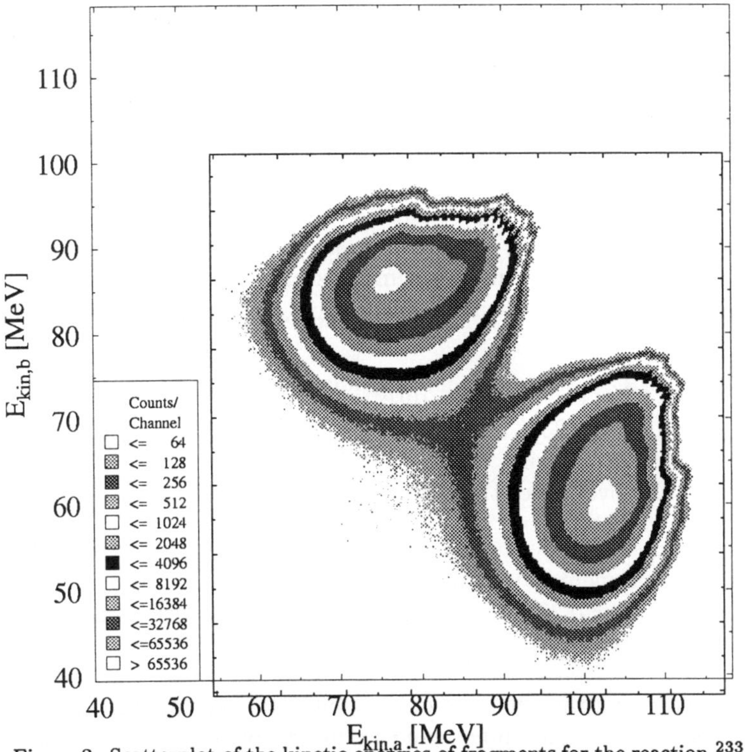

Figure 2. Scatterplot of the kinetic energies of fragments for the reaction $^{233}U(n_{th},f)$ under study (selected solid angle 80°...90° relative to the target plane).

The standard spin-flip technique was used to overcome instrumental asymmetries (like backing effects) and long-term drifts in electronics and neutron beam intensity. Every second the direction of the spin of the incoming neutrons was reversed by a current sheet spin flipper. With this method two scatterplots of kinetic energies like fig. 2 were obtained: one for spin in direction of flight of the incoming neutron and one for the opposite direction. With these two sets of data the experimental asymmetry can be calculated as

$$\alpha_{\mathrm{nf}}^{\mathrm{exp}} = \alpha_{\mathrm{nf}} \cdot P_{\mathrm{n}} \cdot \overline{\cos\theta} = \frac{(N\uparrow - N\downarrow)}{(N\uparrow + N\downarrow)} \tag{2}$$

for each cell of the scatterplot matrix, giving the PNC-value for different mass- and TKE regions ($N\uparrow$ and $N\downarrow$ is the number of counts with spin in direction of flight of the incoming neutron and opposite to direction of flight of the incoming neutron, respectively).

3 Results and Discussion

3.1 Global values

- The global asymmetry value (averaged over all masses and total kinetic energies) was found to be $\alpha_{\mathrm{nf}} = (+3.96 \pm 0.17) \cdot 10^{-4}$ and thus slightly higher than values found in literature. This can be understood, since with the method used in this experiment scattered events, which are decreasing the size of the effect, could be identified and eliminated (see below).

- The values for the two possible directions of the guiding field are: $\alpha_{\mathrm{nf,field}\uparrow} = (+4.12 \pm 0.23) \cdot 10^{-4}$ and $\alpha_{\mathrm{nf,field}\downarrow} = (+3.78 \pm 0.24) \cdot 10^{-4}$, respectively. Since reversing the guiding field also reverses the spin of the neutrons, the overlap of these two values demonstrates the absence of any false asymmetry.

- The test measurement with an unpolarized neutron beam yielded $\alpha_{\mathrm{nf,unpol.}} = (-0.25 \pm 0.23) \cdot 10^{-4}$, which is compatible with zero, as expected.

These figures demonstrate the correct working of the whole setup and the evaluation procedure. A crucial point (and an important hint for the design of further experiments) is the identification and the treatment of scattered events. If the light fragment has lost some part of its energy due to inelastic scattering either in the backing, in the target itself, or with counting-gas nuclei, it may be identified to have the wrong mass. These "wrong" events decrease the size of the experimental asymmetry. Usually only the channels with a sizable yield (some fixed percentage of the maximum of the spectrum) are used for the calculation of α_{nf}, since here the contribution by scattered events is relatively small. Nevertheless, how dramatically the value of α_{nf} changes with the cutoff-value is shown in fig. 3, where the experimental asymmetry coefficient α_{nf}^{exp} from eq. 2 is plotted as a function of the percentage level chosen for the evaluation.

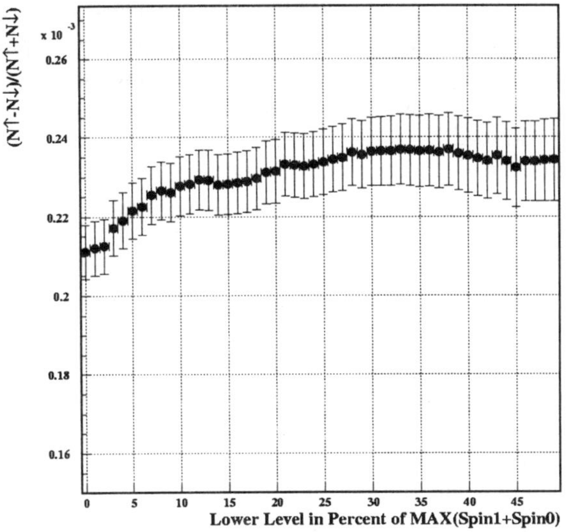

Figure 3. Experimental asymmetry for the full solid angle versus cutoff-value.

Here only those channels of the scatterplots are taken into account for the calculation of α_{nf}, having more counts than the cutoff-value indicated on the abscissa[a]. As can be seen, the value of α_{nf} increases by about 10% when going from no cutoff at all to higher cutoffs. This dependence of the size of the asymmetry on the cutoff-value due to scattered events is getting stronger when going to larger angles and/or to thicker targets. So, to make high precision measurements of the absolute value of α_{nf}, only thin targets and a small solid angle should be used. Unfortunately this is in contradiction to the needs of high statistic. Even worse, this background is not the same for all masses and TKE's. For example, the chance to include scattered events in the calculation of α_{nf} is larger for small TKE than for large TKE, since scattering will always lead to underestimating the TKE.

[a]Of course these values are highly correlated. That explains the smooth shape of the curve.

3.2 Angular dependence

From theory it is predicted that according to eq. 1 the experimental asymmetry α_{nf}^{exp} depends in a characteristic way on the angle θ between spin and momentum direction. For the test of this predicted shape of the angular distribution $W(\theta) \sim (1+\alpha_{nf} \cdot P_n \cdot \cos\theta)$ with $\alpha_{nf} = \text{const}$ it has been made use of the fact, that the normalized cathode pulseheight $pcn = phc/(pha + phb)$ senses the angle between the direction of emission of the fragments and the chamber axis (normal to the cathode and target plane) in the following way: Free electrons produced by a fission fragment which is ionising the atoms of the counting gas are drifting from the location of their formation towards the anode. Since the cathode is not electrically shielded by a Frisch-grid, the signal generated by these drifting charges lasts until they reach the Frisch-grid of the anode. So, the closer to the cathode they are formed, the longer the electrons drift and the higher is the cathode signal. Since the cathode signal also depends on the fragments energies, it has to be divided by the anode signals to find the angle θ. For longitudinally polarized neutrons flying along the chamber axis as in the present experiment θ is also the angle relevant for PNC studies, viz. the angle between fragment momentum and neutron spin. Consequently, when plotting the count-rate versus the normalized cathode pulseheight, one obtains a rectangular distribution ranging from $\cos\theta = 1$ (emission along the chamber axis) to $\cos\theta = 0$ (emission parallel to the cathode plane). In such a plot different angular ranges can, hence, be selected for the calculation of the respective experimental asymmetry as a function of angle θ, viz. $\alpha_{nf}^{exp}(\cos\theta) = \alpha_{nf} \cdot P_n \cdot \cos\theta$ with α_{nf} being the PNC coefficient and P_n the degree of neutron polarization. The angular resolution is about $\Delta\cos\theta \simeq 0.05$ [12].

For 16 angular ranges the measured asymmetry values are plotted in fig. 4 (data points) together with a least-square-fit (line). The raw experimental asymmetry α_{nf}^{exp} for a given angle is calculated according to eq. 2 under the constraint of a fixed normalized cathode pulseheight. A difficulty here is that for angles θ approaching $\theta = 90°$ no data can be taken since the target holder (a thin Al frame) is screening this angular region. For all other angles the data are in excellent agreement with the correlation $\alpha_{nf}^{exp} \sim \cos\theta$ for the asymmetry. This result indicates that the ansatz $W(\theta) \sim (1 + \alpha_{nf} \cdot P_n \cdot \cos\theta)$ for the angular distribution made in eq. 1 with $\alpha_{nf} = \text{const}$ as a measure of PNC violation in fission is well justified.

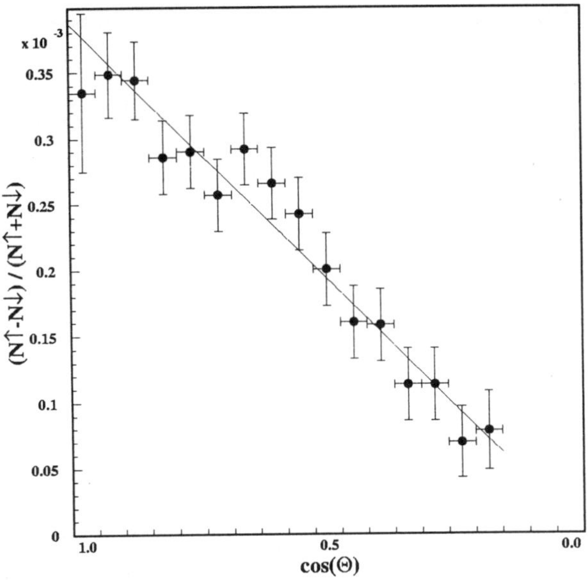

Figure 4. The measured raw asymmetry $\alpha_{nf}^{exp} = \alpha_{nf} \cdot P_n \cdot \cos\theta = (N\uparrow - N\downarrow)/(N\uparrow + N\downarrow)$ of the light fission fragments relative to the spin of the incoming neutron.

3.3 Mass- and TKE-dependence

In fig. 5, the mass- and the TKE-dependence of α_{nf} are plotted. In the regions with a sizable yield, they exhibit no significant deviation from a constant value. The slightly lower values for small and big light fragment masses may be due to scattered events, since (assuming about the same background of scattered heavy fragments over the whole mass range) in regions with a small mass yield the contribution of scattered events leading to lower α_{nf}-values is relatively large. Mathematically the question, whether there is a TKE-dependence of α_{nf} or not, could not uniquely be answered by a χ^2-test, since both assumptions, a constant α_{nf} and a TKE-dependent one, yielded a very good χ^2. However, the value found for the TKE-dependence was very small: $(0.013 \pm 0.011) \cdot 10^{-4}$/MeV. Another argument against any dependence of α_{nf} on TKE can again be found in the scattered events. Since along with a decrease of the size of the effect due to scattered events there is a shift to lower TKE-values, the lower α_{nf}-values for lower TKE may simply reflect the unability to eliminate the scattered events completely. It appears, therefore, that wheter there is a mass/energy dependence of the

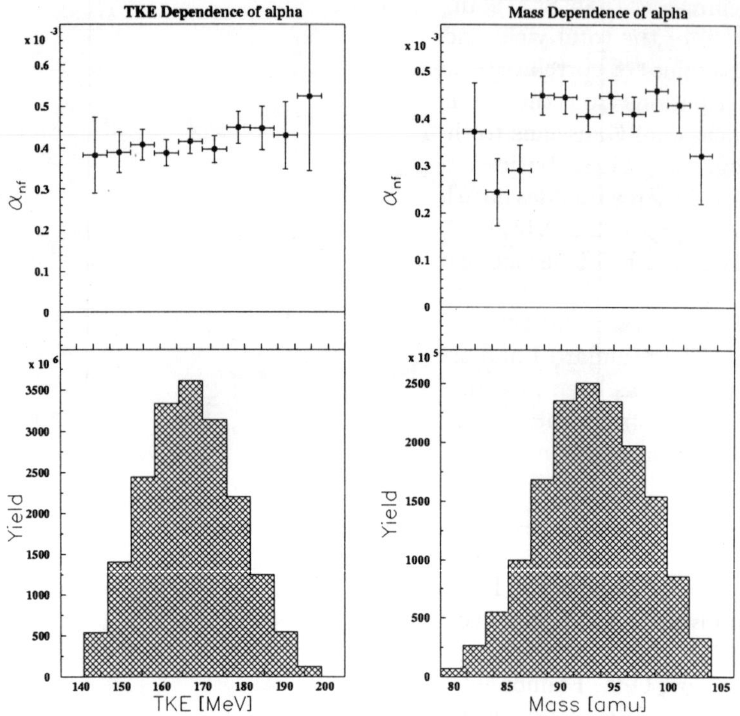

Figure 5. Dependence of α_{nf} on total kinetic energy (left) and on light fragment mass (right).

PNC coefficient α_{nf} is linked as much to the appropriate statistics as to our appreciation of the background.

The experiment was also analyzed in terms of Brosa-modes [13] (Standard I and Standard II). Brosa et al. calculated the potential energy surface of a fissioning nucleus as a function of geometrical shape parameters. In this potential energy landscape the nucleus may choose between various channels on the way to disintegration. Since these channels end up at different values of the shape parameters, fragments following different channels will have different mass- and energy-distributions. In low energy fission of the lighter actinides the so called Standard I- and Standard II-modes are the most important ones. They both exhibit an asymmetric mass distribution (St.I slightly more symmetric than St.II). In the fission of ^{233}U Standard II amounts to about 75% of the total yield and Standard I to about 25%. For the analysis the scatterplots of correlated fragment energies have been deconvoluted into the contributions from the two modes. This was accomplished by fitting two two-dimensional Gaussians to the data. This fit reproduces the experimental mass- and energy-distribution quite well. According to this fit the asymmetry-values for the Brosa-modes have been found to be: $\alpha_{\mathrm{nf,St.I}} = (4.10 \pm 0.27) \cdot 10^{-4}$ and $\alpha_{\mathrm{nf,St.II}} = (3.92 \pm 0.15) \cdot 10^{-4}$, respectively. Hence, within the statistical accuracy of about 6% per mode reached in the present experiment, there is no evidence for the PNC asymmetries to differ in the two standard Brosa modes. This experimental result can be taken as a confirmation of one of Brosa's predictions: The Standard I and the Standard II modes bifurcate only once the fission barrier has been passed. Thus, at the saddle point, where the angular distribution of fragments is fixed [14] for both, parity conserving and parity violating fission reactions, the two modes are not yet separated.

References

1. Graf, U., Gönnenwein, F., Geltenbort, P. et al., *Z.Phys.* **A351**, 281 (1995)
2. Danilyan G.V., Vodennikov, B.D., Dronyaev et al., *JETP.Lett.* **26**, 186 (1977)
3. Sushkov, O.P., Flambaum, V.V., *Sov.J.Nucl.Phys* **33**, 31 (1981)
4. Bunakov, V.E., Gudkov, V.P., *Nucl.Phys.* **A401**,93 (1983)
5. Petrov G.A., Valskii, G.V., Petuhkov, A.K. et al., *Nucl.Phys.* **A502**, 297 (1989)
6. Bohr, A., Proc. Int. Conf. Peaceful Uses Atomic Energy, Geneva 1955, United Nations, New York, 1956, Vol.II, p.151.
7. Alexandrovich, A.Ya., Gagarski, A.M., Petrov, G.A. et al., *Nucl.Phys.* **A576**, 541 (1994)
8. Borovikova, N.V., Vesna, V.A., Egorov, A.I. et al., *JETP.Lett.* **30**, 495 (1979)
9. Gagarski, A.M., Krasnoschokova, I.A., Petrov, G.A. et al., Proc. ISINN-6, Dubna 1998.
10. Petuhkov, A.K., Petrov, G.A., Stepanov, S.I. et al., *JETP.Lett.* **32**, 300 (1980)
11. Vesna, V.A., Knyaz'kov, V.A., Kolomenskii, E.A. et al., *JETP.Lett.* **31**,663 (1980)
12. Kaufmann, J., Untersuchung der $^{238}U(n,f)$-Reaktion, *PhD-Thesis*, University of Tübingen, 1996.
13. Brosa, U., Großmann, S., Müller, A., *Phys.Rep.* **197**, 167 (1990)

14. Wheeler, J.A., in: Marion, J.B. and Fowler, J.L., eds., *Fast Neutron Physics* (Wiley & Sons Interscience, New York, 1963).

YIELDS OF TERNARY PARTICLES FROM THE REACTIONS $^{229}Th(n_{th},f)$, $^{233}U(n_{th},f)$ AND $^{239}Pu(n_{th},f)$

M. WÖSTHEINRICH[A], R. PFISTER[A], F. GÖNNENWEIN[A],
H.O. DENSCHLAG[B], H. FAUST[C], S. OBERSTEDT[C]

[a] *Physikalisches Institut, Universität Tübingen, D-72076 Tübingen**
[b] *Institut für Kernchemie, Universität Mainz, D-55099 Mainz*
[c] *Institut Laue-Langevin, F-38042 Grenoble*
E-mail: marcus.woestheinrich@uni-tuebingen.de

Ternary particles (TP's) from the thermal neutron induced fission of ^{229}Th, ^{233}U and ^{239}Pu have been studied. Their energy distributions and yields have been determined using the mass separator Lohengrin at the high flux reactor of the Institut Laue-Langevin (ILL). Special emphasis was put on the search of the heaviest ternary particles. As a new result it has been found that the mass and/or charge spectra of the heaviest TP's at the limit of detectability of present techniques are strongly correlated with the mass and/or charge of the fissioning compound nucleus. The actual limit of detection reached is about 1 TP in 10^{10} fission reactions. At this limit, in Th-fission the heaviest TP's are Ne-isotopes, while for ^{239}Pu-fission the observable TP's are shifted to Si-isotopes. The energy distributions and yields of several TP's having so far not been reported from the reactions $^{229}Th(n_{th},f)$, $^{233}U(n_{th},f)$ and $^{239}Pu(n_{th},f)$ are given. The dependence of the yields of the TP's on the fissility of the fissioning compound nucleus for actinides ranging from Th to Cf is shown. Experimental yields are compared to predictions in the framework of the Halpern model.

1 Introduction

Ternary fission is a process where at scission of the disintegrating nucleus three particles are formed. First evidence for such a process was found more than fifty years ago [1,2]. In thermal neutron induced fission ternary particles are produced in about two out of thousand fissions. In these processes one obtains two fission fragments and one light particle, the ternary particle (TP). The TP is in about 92% of these cases an α-particle. Particles with a nuclear charge of more than two are very rare. They can be observed in only about 1% of all cases of ternary fission. An overview of ternary fission is given in [3,4]. For the $^{229}Th(n_{th},f)$ reaction only α- and triton-particles have been determined in the past [5]. For the $^{233}U(n_{th},f)$ reaction TP's up to ^{10}Be [6] and for the $^{239}Pu(n_{th},f)$ reaction TP's up to ^{20}O [7] are known from literature.

The aims of the present experiments were the search for heavy ternary particles as well as the determination of their energy distributions and yields.

*Supported by the BMBF, Bonn, under contract number 06Tü669

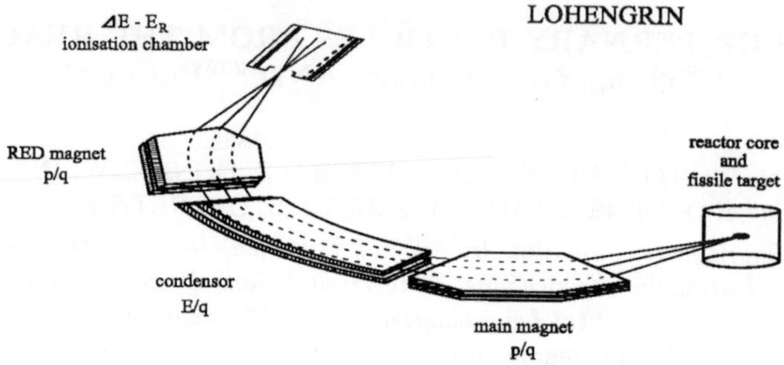

Figure 1. Experimental setup: mass separator Lohengrin with RED-magnet and ΔE-E ionisation chamber.

Of special interest was the comparative study of the heaviest TP's as a function of the fissioning compound nucleus, which are detectable with the present techniques. Since in the meantime also ternary fission of ^{242}Pu*, ^{243}Am*, ^{246}Cm* and ^{250}Cf* [8–10] has been investigated, a broad range of actinides is now available for comparison.

2 Experimental Setup

For the determination of the energy distributions and yields of the TP's the mass separator Lohengrin [11] at the Institut Laue-Langevin (ILL) was used (see Fig. 1). Fission in targets with specific thickness of about 300 μg/cm^2 is induced by thermal neutrons from the reactor core. The flux at the target position is 5,3·10^{14} n/(cm^2·s). Fission products emerging from the target reach the mass separator Lohengrin. In this spectrometer they are separated according to their mass over ionic charge A/q as well as their kinetic energy over ionic charge E_{kin}/q ratio by a magnet and an electrostatic condenser. An additional RED-magnet (reverse energy disperson-magnet) focuses all particles leaving the exit slit of the mass spectrometer on the entrance window of an ionisation chamber. In this $\Delta E - E$ ionisation chamber the fission products are identified as to mass and energy and their nuclear charge is measured.

Determination of Yields

For the determination of the yields it is necessary to know the ionic charge distribution $F(v, q)$ and the energy distribution $Y(E_{\text{kin}})$ of the ternary particles. The separation of the particles depends on their ionic charge q. Due to

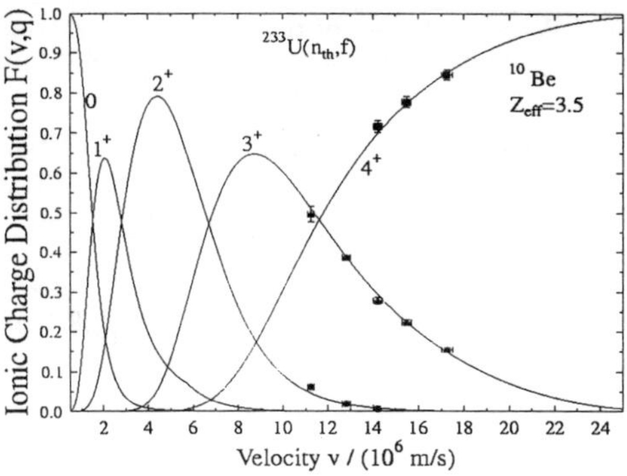

Figure 2. Ionic charge distribution of ^{10}Be. The curves are calculated, each marker represents one measurement. Z_{eff} is a fit parameter.

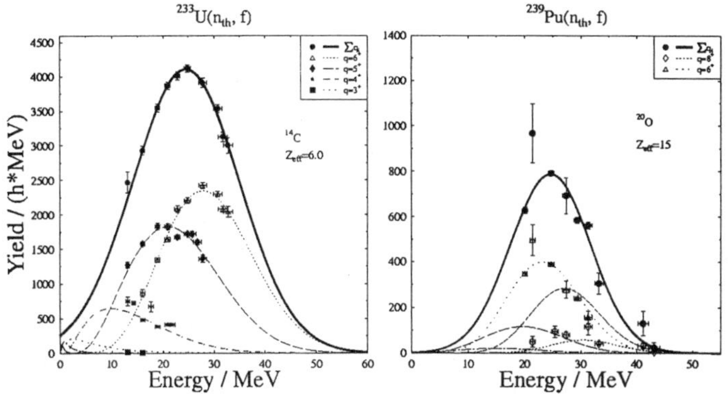

Figure 3. Energy distributions of the ternary ^{14}C and ^{20}O from the reactions ^{233}U(n_{th},f) and ^{239}Pu(n_{th},f).

limited beam time it is not possible to perform a measurement for every ionic charge and energy. For a TP with the velocity v the ionic charge distribution $F(v,q)$ represents the relative contribution of the ionic charge q to all possible ionic charges. With the knowledge of $F(v,q)$ it is sufficient to perform only one measurement at one ionic charge for each energy. In Fig. 2 the $F(v,q)$ distributions of ^{10}Be are shown. Each marker represents one measurement, the curves are calculated. The ionic charge distribution is parameterized by

Table 1. Mean energy $\overline{E}$, FWHM and yield of the TP's of the reaction $^{229}Th(n_{th},f)$. The yield is given relative to the yield of the ternary α-particle which is set to 10^4.

TP	This Measurement $\bar{E}$ / MeV	FWHM / MeV	Yield / $Y(\alpha)\cdot 10^{-4}$	Wagemans [5] $\bar{E}$ / MeV	FWHM / MeV	Yield / $Y(\alpha)\cdot 10^{-4}$
^3H				—	—	392±71
^4He				—	9.4±0.4	10^4
^7Li	—	—	10.4±1.7			
^9Be	—	—	3.8±2.3			
^{10}Be	18.4±0.5	17.5±1.2	75.7±7.1			
^{12}Be	—	—	1.1±0.5			
^{14}C	—	—	20.9±2.4			
^{15}C	23.8±0.9	15.9±1.8	4.3±0.4			
^{18}N	—	—	$(1.2\pm0.9)\cdot10^{-1}$			
^{20}O	22.9±2.8	21.5±3.8	1.8±0.5			
^{21}O	24.8±3.8	12.9±12.1	$(1.4\pm0.6)\cdot10^{-1}$			
^{21}F	—	—	$(2.5\pm2.0)\cdot10^{-2}$			
^{24}Ne	—	—	$(2.9\pm2.4)\cdot10^{-2}$			

a modified reduced χ^2 function. Z_{eff} is a parameter which is used to fit the modified reduced χ^2 function to the experimental data. For further details see [12].

Since the functions $F(v,q)$ fix the relative probability of the ionic charges the global yield of a TP as a function of its energy can be found. In Fig. 3 some energy distributions are shown. Global yields (summed over all ionic charges) are indicated by full points, thick lines represents the global energy distributions. These distributions are found from Gaussian fits to the data. With the help of $F(v,q)$ the global energy distributions are decomposed into the energy distributions at given ionic charges. These energy distributions for individual ionic charges are indicated in Fig. 3 by thin lines.

The yields of the TP's are finally determined by an integration over the global energy distributions.

3 Results

3.1 $^{229}Th(n_{th},f)$

For the $^{229}Th(n_{th},f)$ reaction 17 different TP's were observed, of which the heaviest particle was ^{24}Ne. The count rate for this ternary particle in the different settings of the mass separator Lohengrin was about 1 event/day. In

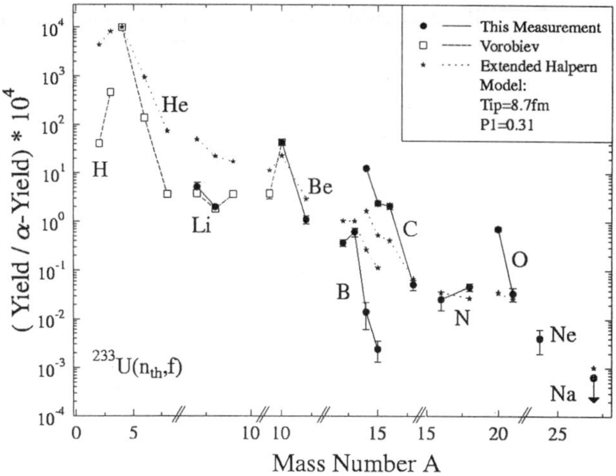

Figure 4. Comparison between the experimental yields of TP's from this measurement [12] and Vorobiev [6] with the calculated yields of the extended Halpern model. In the case of ^{24}Ne the calculated and experimental yield is "by chance" equal. The ^{27}Na data point represents an upper limit.

practice this count rate sets a limit to the detectability of rare isotopes from ternary fission. Due to poor statistics the yields of the TP's ^{6}He, ^{9}Li, ^{12}B, ^{15}B, ^{18}C and ^{20}N could not be evaluated. In Table I mean energies $\overline{E}$ and FWHM of kinetic energy distributions, and yields of the ternary particles from the reaction ^{229}Th(n$_{th}$,f) are shown. In case no energy data are indicated in Table I this means that not sufficient measurements were available to establish the energy distributions from experiment. For the determination of yields, $\overline{E}$ and FWHM had to be guessed by extrapolation from neighboring TP's and /or other compound nuclei. The yield is given relative to the yield of the ternary α-particle which is set to 10^4.

3.2 $^{233}U(n_{th},f)$

For the ^{233}U(n$_{th}$,f) reaction 25 different TP's were observed. It was possible to determine the yields of 17 ternary particles [12]. The heaviest TP searched for was ^{27}Na. However, only an upper limit was obtained with no positive detection.

The dependence of the ternary yield on mass or charge of the particles was analyzed in the framework of the extended Halpern model. In this model [13]

Table 2. Mean energy $\overline{E}$, FWHM and yield of TP's from the reaction ^{239}Pu(n_{th},f). The yield is given relative to the yield of the ternary α-particle which is set to 10^4. For ^{34}Si an upper limit is given, it was not observed.

TP	This Measurement $\overline{E}$ / MeV	FWHM / MeV	Yield / $Y(\alpha) \cdot 10^{-4}$
^{10}Be	16.4±0.1	19.2±0.6	49±1
^{17}N	24.2±0.9	17.6±0.9	$(1.9\pm0.1)\cdot10^{-1}$
^{20}O	24.6±0.2	16.7±0.5	$(6.8\pm0.3)\cdot10^{-1}$
^{21}F	—	—	$(2.5\pm0.4)\cdot10^{-2}$
^{24}Ne	28.8±1.3	20.1±1.7	$(2.5\pm0.3)\cdot10^{-2}$
^{27}Na	—	—	$(2.0\pm0.5)\cdot10^{-3}$
^{30}Mg	—	—	$(2.8\pm1.5)\cdot10^{-3}$
^{32}Al	—	—	$(1.6\pm1.6)\cdot10^{-4}$
^{34}Si	—	—	$\leq 3.5\cdot10^{-4}$

the energy costs which are necessary to switch from binary to ternary fission are calculated. By weighting the energy costs with the binary fragment distribution it is taken into account that for a given TP many different mass and charge splits of the fragments may contribute. The distance between the tips of the fragments directly after scission is called the tip-distance. The tip-distance can be estimated from the total kinetic energy of the fragments. The fragments with the smallest kinetic energy originate from scission configurations with the largest deformations or equivalently with the largest tip distance between undeformed fragments. From such configurations ternary particles are thought to be emitted. High energy costs are expected to be unfavourable for ternary fission. The ternary yield is hence parameterized by an exponential function $Y \sim \exp(-E_{\text{cost}} \cdot P1)$ with only one fit parameter $P1$.

In Fig. 4 experimental yields from this and Vorobiev's [6] measurement are compared with the calculated yields from the extended Halpern model. As seen from Fig. 4 the general trend as a function of mass is quite well reproduced while the odd-even effect appears to be grossly underestimated.

3.3 ^{239}*Pu*(n_{th},*f*)

For the ^{239}Pu(n_{th},f) reaction 36 different TP's were found, with 23 of them having never been reported before from this reaction. The new TP's range from the light ^{6}He up to the heavy ^{33}Si. For this reaction the data are still

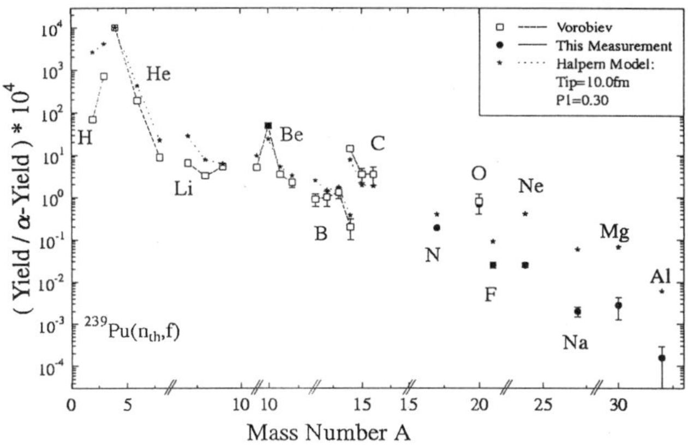

Figure 5. Comparison between the experimental yields of TP's from this measurement and Vorobiev [7] with the calculated yields of the Halpern model. In the case of ^{20}O the calculated and experimental yield is "by chance" equal.

under evaluation. In Table II the mean energies $\overline{E}$, FWHM and yields of the TP's which were evaluated until now are shown.

Figure 5 shows a comparison of experimentally obtained data and calculated yields for the reaction $^{239}Pu(n_{th},f)$. Experimental data are from this measurement and from Vorobiev [7]. The yields are calculated with the Halpern model. In contrast to the yields calculated for the $^{233}U(n_{th},f)$ reaction only one mass and charge split of the fragments is taken into account. The heavy fragment is set to be the doubly magic ^{132}Sn and it is assumed that the ternary particle is emitted by the complementary light fragment. Without weighting of the energy costs with the binary fragment distribution a better agreement is obtained between experimental and calculated yields, at least for ternary particles up to fluorine. For still heavier TP's, however, there is a violent disagreement.

4 Discussion

In Fig. 6 the yields of ternary particles for several thermal neutron induced fission reactions are compared. For this survey only the yields of the most abundant isotopes for each element have been retained. The yields are given relative to α-particles, with the α-particle yield being arbitrarily set to 10^4. The yields are shown as a function of the mass number A of the isotopes

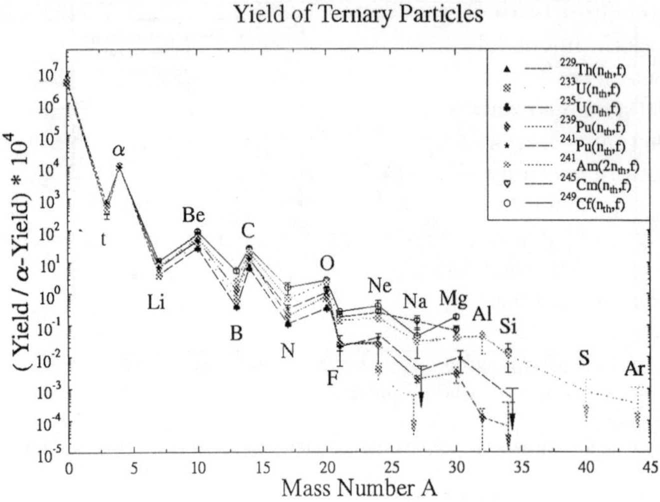

Figure 6. Yields of ternary particles as a function of their mass number A for several actinides. The yields are given relative to the α-particle, whereby this yield is given as 10^4. Arrows indicate upper limits.

in question. For $A = 0$ the binary fission yield normalized to the above α-ternary yield is included. One can observe a dramatic decrease of the TP yields with their mass. Furthermore, systematically a higher yield for even charge numbers of TP's is seen. This odd-even effect is smoothed for the heavier elements. The yields of the lighter TP's from the different reactions lie close together (e.g. t, ^{7}Li, ^{10}Be). For higher masses of the particles, however, a widening spread in the yields for the different reactions is observed. As a rule, the heavier the compound nucleus the larger the ternary yields, especially for the heavy TP's. The mass or charge number of the heaviest TP's at the limit of detectability of present techniques is strongly correlated with the mass and/or charge of the fissioning compound nucleus. In Th- and U-fission the heaviest TP's are Ne-isotopes, in Pu-fission Al-isotopes and in Am- and Cf-fission the heaviest TP's are shifted to Si-isotopes. At a comparable level of sensitivity the lightest fragment from binary Am-fission has a mass of $A = 68$ [14]. This means that in the mass spectrum of particles emitted in the fission process induced by thermal neutrons, there is at a 10^{-9} level in yield still a gap in observation between the heaviest ternary TP's and the lightest fragments from binary reactions.

The systematics in the limits of the heaviest TP's as a function of the compound system might be explained by recent evidence in an experimental study of spontaneous fission of ^{252}Cf [15]. In this experiment it was discovered that the total average excitation energy $\overline{TXE}$ still left in the ternary fissioning system actually decreases rapidly with increasing mass and/or charge of the TP. Also a remarkably clear correlation between the TP yield and $\overline{TXE}$ emerges. Extrapolating this correlation down to the yield levels measured for Si-isotopes e.g. in ^{243}Am-fission, one is led to the conclusion that in these limiting cases the system runs out of energy. Stated otherwise, for the heaviest observed TP's, ternary fission is a cold process. Having identified the limit of TP emission as the cold frontier of phase space, it is finally seen why the limit of TP emission is shifting with the mass and/or charge of the compound nucleus undergoing fission. Since the mean neutron number in fission increases with the mass of the fissile nuclei and since this number is a measure of $\overline{TXE}$, it is concluded that more energy $\overline{TXE}$ is available in the heavy fissioning systems. Hence the limit of TP emission with ever increasing energy costs shifts to heavier fissioning systems providing more energy.

References

1. T. San-Tsiang, H. Zah-Wei, R. Chastel, L. Vigneron: *Phys. Rev.* **71**, 382 (1947)
2. G. Farwell, E. Segrè, C. Wiegand: *Phys. Rev.* **71**, 327 (1947)
3. C. Wagemans: *in The Nuclear Fission Process (Ed. C. Wagemans)*, CRC Press, Boca Raton 1991, p. 545
4. M. Mutterer, J.P. Theobald: *in Nuclear Decay Modes (Ed. D.N. Poenaru)*, IOP Publ., Bristol 1996, p. 487
5. C. Wagemans: *in Proc. Int. Workshop on Dynamical Aspects on Nuclear Fission, Smolenice (Slovakia) 1991 (Eds. J. Kristiak and B.I. Pustylnik)*, Dubna 1992, p. 139
6. A.A. Vorobiev et al. : *Phys. Lett.* **30 B**, 332 (1969)
7. A.A. Vorobiev et al. : *Sov. J. Nucl. Phys.* **20**, 248 (1975)
8. U. Köster et al. : *in Proc. of the 2nd Int. Workshop on Nuclear Fission and Fission-Product Spectroscopy, Seyssins (France) 1998 (Eds. G. Fioni, H. Faust, S. Oberstedt and F.-J. Hambsch)*, AIP Conference Proceedings 447, Woodbury (USA) 1998, p. 315
9. M. Hesse: *PhD thesis*, Univ. Tübingen (1997)
10. M. Davi: *PhD thesis*, Univ. Mainz (1997)
11. E. Moll et al. :*Kerntechnik* **19**, 374 (1977)
12. M. Wöstheinrich et al. : *in Proc. of the 2nd Int. Workshop on Nuclear*

Fission and Fission-Product Spectroscopy, Seyssins (France) 1998 (Eds. G. Fioni, H. Faust, S. Oberstedt and F.-J. Hambsch), AIP Conference Proceedings 447, Woodbury (USA) 1998, p. 330

13. M. Wöstheinrich et al. : *in Proc. of 3rd Int. Workshop of Dynamical Aspects of Nuclear Fission, Častá-Papiernička (Slovakia) 1996 (Eds. J. Kliman and B.I. Pustylnik)*, Dubna 1996, p. 231
14. M. Davi et al. : *in Proc. of the 2nd Int. Workshop on Nuclear Fission and Fission-Product Spectroscopy, Seyssins (France) 1998 (Eds. G. Fioni, H. Faust, S. Oberstedt and F.-J. Hambsch)*, to be published by AIP
15. M. Mutterer et al. : *in Proc. of 3rd Int. Workshop of Dynamical Aspects of Nuclear Fission, Častá-Papiernička (Slovakia) 1996 (Eds. J. Kliman and B.I. Pustylnik)*, Dubna 1996, p. 250

YIELDS OF CORRELATED FRAGMENT PAIRS AND NEUTRON MULTIPLICITY IN SPONTANEOUS FISSION OF ^{242}PU.

M. VESELSKÝ [1], J. KLIMAN[1], M. MORHÁČ[1], J.H. HAMILTON[2], A.V. RAMAYYA[2], J. KORMICKI[2], J.O. RASMUSSEN[3], M.A. STOYER[4], G.M. TER-AKOPIAN[5], A.V. DANIEL[2,5], G.S. POPEKO[5], YU.TS. OGANESSIAN[5], W. GREINER[6], J.D. COLE[7], R. ARYAEINEJAD[7] AND GANDS COLLABORATION

[1] *Institute of Physics of SASc, Dubravska 9, 84228 Bratislava, Slovakia*
[2] *Physics Department, Vanderbilt University, Nashville, USA*
[3] *Lawrence Berkeley National Laboratory, Berkeley, USA*
[4] *Lawrence Livermore National Laboratory, Livermore, USA*
[5] *Joint Institute for Nuclear Research, Dubna, Russia*
[6] *Institut fur Theor. Physik, J.W. Goethe Universitaet, Frankfurt a. M., Germany*
[7] *Idaho National Engineering Laboratory, Idaho Falls, USA*

Yields of correlated fragment pairs were obtained in spontaneous fission of ^{242}Pu. Charge, mass and neutron multiplicity distributions of fragment pairs were determined and compared to available data. The yield of cold fission without neutron emission was determined to about 10 % for the set of observed correlated fragment pairs. The yields of ternary fission were examined.

1 Introduction

During past decades, the characteristics of fission fragments have been studied extensively. A large number of neutron-rich nuclei, almost not accessible by other means, can be produced by nuclear fission of actinide nuclei. Systematic study of fragment yields allows one to study the dynamics of the fission process. Of special interest are data from spontaneous fission, which deliver information on the dynamics of fission in the case where no excitation energy is available. This allows one to study the influence of shell structure on the fission process.

A study of ^{242}Pu spontaneous fission characteristics, carried out at Gammasphere using the prompt gamma-gamma coincidence spectroscopy of fission fragments, is described in the present work. Mass, charge and neutron multiplicity distributions are determined from correlated fragment yields. The isotope ^{242}Pu is of particular interest for the study of cold fission without emission of neutrons. In this case the energy, released in the fission process is shared only between the total kinetic energy of fission fragments and the energy of γ-rays. In the case of deformed fragments the enhancement of

γ-emission from rotational bands could be expected.

2 Experimental setup and analysis

The coincident γ-rays have been detected with Gammasphere [1] in its early implementation stage when it consisted of 32 HPGe Compton suppressed gamma detectors with 80 % efficiency and $\approx$ 2.5 keV resolution for 1332 keV gamma rays. The plutonium source, 12.7 g of PuO_2, was doubly encapsulated in brass cylinders, 0.64 mm thick, which stopped all α and β particles. Because of impurity of ^{241}Am that gave intense low energy gamma rays, the encapsulated source was surrounded with a 0.5 mm silver foil. The source consisted of 89.56 % of ^{242}Pu, 0.74, 1.33, 6.37 and 1.62 % of lighter plutonium isotopes ^{238,239,240,241}Pu and 0.42 % of ^{241}Am, produced from the β^- decay of ^{241}Pu. The source was placed in the center of Gammasphere approximately 25 cm from the front faces of the detectors.

The spectra of triple coincidences of gamma rays have been used for the analysis. The main trick of the analysis is to reduce the background from random coincidences of fully absorbed gamma rays emitted by fission fragments with the gamma rays, emitted from background sources in the sample or detector volume, and Compton scattered gamma rays. The triple gamma coincidence data were the subject of background subtraction procedure based on the use of the background elimination algorithm [2] for multidimensional spectra. This algorithm allows to unfold true peaks from background and to obtain the values of the peak areas within acceptable precision also for relatively small peaks.

The effectivity of the algorithm is documented on Figure 1 where the one dimensional projection of the experimental triple gamma coincidence spectrum (Fig.1b) is compared to the one dimensional projection of the same spectra after application of the background elimination algorithm to the three dimensional case using 8 iteration steps (Fig.1a). The number of triple coincidences has been reduced from 5.44×10^7 to 2.17×10^7, mostly because of reduction of the smooth background, and the most significant background lines (208, 511 and 606 keV).

It is well known, that the relative intensity of the $2^+ \rightarrow 0^+$ transitions in even–even fission fragments and also the sum of all transitions feeding the ground states of the even–odd isotopes represent the relative independent yields, a fact that allows one to gate on this transition and deduce the relative yields of the partner fission fragments.

Yields of correlated fragments have been obtained using the double gating of the three dimensional spectra. For every combination of fission fragments,

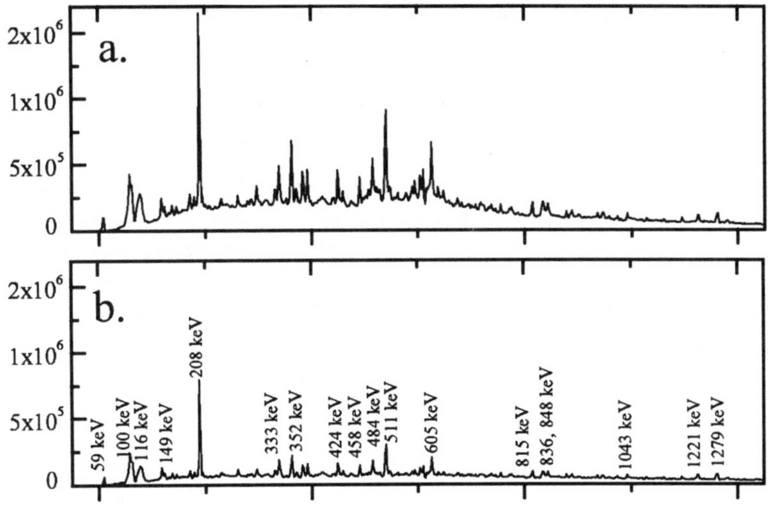

Figure 1. One dimensional projection of (a) original $\gamma - \gamma - \gamma$ spectrum of ^{242}Pu without background elimination (b) $\gamma - \gamma - \gamma$ spectrum of ^{242}Pu with eliminated background (8 iteration steps).

the most significant lines of both light and heavy fragment have been used for gating. As an additional information, also spectra gated by two gamma lines of light or heavy fragments were used . This was possible mostly for even-even nuclei, where $2^+ \rightarrow 0^+$ and $4^+ \rightarrow 2^+$ gamma rays are of moderate energy. In the case of odd nuclei many isotopes have the most significant lines with an energy below 150 keV where the background from the contaminant Pu and Am isotopes dominates.

In all cases the yields had to be corrected for the detection efficiency of both gating gamma rays and summed peak in the gated gamma spectrum. Because of the complex setup of the plutonium source, a significant suppression of low energy gamma rays occured. To take this into account, we corrected the standard efficiency calibration, obtained with the calibration sources (^{56}Co, ^{152}Eu and ^{154}Eu). The value of this correction was deduced from relative intensities of $2^+ \rightarrow 0^+$ transitions in gates of $4^+ \rightarrow 2^+$ transitions of fission fragments produced with high yields. For the correction of the spectrometer efficiency we selected the fission fragments where the energy of $2^+ \rightarrow 0^+$ transition is lower than 300 keV. Such an absolute correction includes also

Table 1. Yields of Ru - Sn fragment pairs from ^{242}Pu (S.F.) (arb. units).

	^{130}Sn	^{132}Sn
^{108,110}Ru	46±15	16±9

Table 2. Yields of Mo - Te fragment pairs from ^{242}Pu (S.F.) (arb. units).

	^{132}Te	^{133}Te	^{134}Te	^{135}Te	^{136}Te
^{102}Mo			23± 9		104± 16
^{103}Mo			28± 12		138± 21
^{104}Mo	36± 14	18± 12	84± 28		529± 91
^{105}Mo			87± 26		280± 52
^{106}Mo	22± 8	25± 11	172± 41		250± 36

Table 3. Yields of Zr - Xe fragment pairs from ^{242}Pu (S.F.) (arb. units).

	^{136}Xe	^{137}Xe	^{138}Xe	^{139}Xe	^{140}Xe	^{141}Xe	^{142}Xe
^{98}Zr				32±14	87±33	30±12	
^{99}Zr			73±31	24±14	70± 28	19±11	
^{100}Zr		19±9	84±33	113± 38	369±43	60±30	30±16
^{101}Zr		60±36	106±53	142±47	171±43	30±15	
^{102}Zr	12±7	24±10	107±24	72±24	109±22		
^{103}Zr			50±33				

the effect of internal conversion. For the energies above 300 keV the detection efficiency obtained using the calibration standards has been used. In the case of odd nuclei, where gamma transitions, used for both gating or yield determination are not of $E2$ multipolarity, an additional correction to internal conversion was also used.

The correlated yields of individual secondary fragment pairs have been determined for the charge splits Z(L)/Z(H)= 36/58, 38/56, 40/54, 42/52 and 44/50. The resulting post–neutron emission yields are given in Tables 1 - 5. The yields are given in units representing the peak area, corrected for the detection efficiency. The given errors include the statistical errors of the peak areas, errors of the subtracted three dimensional background and the uncertainty in the determination of the detection efficiency.

As may be seen from Tables 1 - 5, the number of detected fragment pairs depends on many factors. In general, the number of detected fragment pairs

Table 4. Yields of Sr - Ba fragment pairss from ^{242}Pu (S.F.) (arb. units).

	^{140}Ba	^{141}Ba	^{142}Ba	^{143}Ba	^{144}Ba	^{145}Ba	^{146}Ba	^{147}Ba
^{94}Sr			25 ± 11	22 ± 13	60 ± 24	65 ± 43	39 ± 17	12 ± 6
^{95}Sr			23 ± 12	66 ± 13	70 ± 15	97 ± 49	86 ± 31	
^{96}Sr			19 ± 11	97 ± 21	177 ± 39	90 ± 54	27 ± 10	
^{97}Sr				47 ± 24	65 ± 36			
^{98}Sr	16 ± 6		60 ± 12	45 ± 16	45 ± 15			

Table 5. Yields of Kr - Ce fragment pairs from ^{242}Pu (S.F.) (arb. units).

	^{148}Ce	^{150}Ce
^{90}Kr	22± 6	18± 6
^{92}Kr	20± 7	7± 4

Table 6. Multiplicities of neutrons emitted in spontaneous fission of ^{242}Pu.

Fragment pairs	$\bar{\nu_n}$ (this work)	$\bar{\nu_n}$ (other works)
all pairs	2.23 ± 0.20	2.10 [3]
Ru/Sn	2.16 ± 0.45	
Mo/Te	2.15 ± 0.25	
Zr/Xe	2.30 ± 0.25	
Sr/Ba	2.25 ± 0.25	
Kr/Ce	2.47 ± 0.45	

depends on the production rate of the fragment pairs with a given charge split. In the case of the Mo - Te fragment pairs the gamma transitions of heavy Te isotopes, $^{135,137,138}Te$, are unknown and cannot be identified within the available data.

3 Results and discussion

From the yields of correlated fragment pairs, several interesting quantities or distributions may be determined, which are relevant to the dynamics of spontaneous fission of ^{242}Pu. For correlated fragment pairs the mass and charge are determined unambiguously, what allows to determine the mean mass and charge of both light and heavy fragments and the mass and charge distributions of the detected fragment pairs. From the determined yields of correlated fragment pairs the values of $< A_{HF} > = 139.05 \pm 0.75$, $< A_{LF} > = 100.71 \pm 0.80$, $< Z_{HF} > = 53.78 \pm 0.45$ and $< Z_{LF} > = 40.22 \pm 0.45$ have

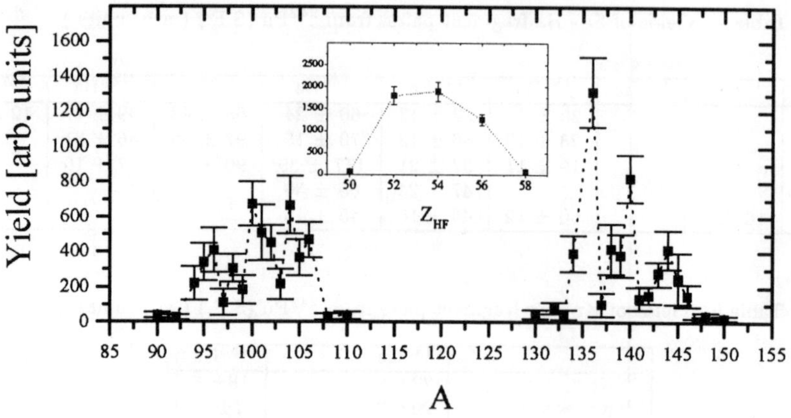

Figure 2. Mass and charge distributions of observed fission fragments in spontaneous fission of ^{242}Pu.

been determined. The mass and charge distributions of fission fragments are given in Figure 2. The obtained mass distribution of fission fragments exhibits structure which has to be attributed to the limited set of included nuclei (mostly even-even, some odd-even isotopes). The mean value of the heavy fragment mass is consistent with similar previously known more integral data [3]. Also the shape of the mass distribution of heavy fragment yields is similar to the shape given in ref. [3]. The shape of the mass distribution of the light fragments is not a mirror image to the shape of the mass distribution of the heavy fragments. Most likely, reasons for this are the limited set of observed nuclei or possible influence of shell structure.

Another type of new data extracted from the yields of fragment pairs is the multiplicity distribution of prompt neutrons emitted in the spontaneous fission of ^{242}Pu. In Figure 3 neutron multiplicity distributions are given for different charge splits of fragment pairs (open symbols) and for the whole set of fragment pairs (solid squares). In the case of ruthenium isotopes, $^{108,110}Ru$, the yields have been divided according to the fragment yields from thermal neutron induced fission of ^{241}Pu, taken from the ENDF nuclear data file [4]. The multiplicity distributions are similar for all the fragment charge splits, with a maximum in the 2n emission channel. Minor discrepancies in the yields of higher neutron multiplicities are, as in previous cases, most likely caused by a limited set of observed nuclei resp. influence of the shell structure.

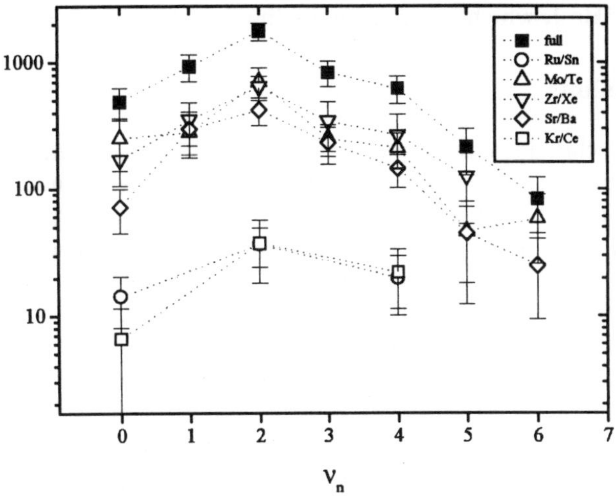

Figure 3. Neutron multiplicity distributions of ^{242}Pu(s.f.) for different charge splits of fragment pairs (open symbols) and for the whole set of fragment pairs (solid squares).

Of special interest is the yield of cold fission where no neutrons are emitted. In the present work the yield of cold fission of 9.9 ± 1.5 % was determined from the observed set of correlated fragment pairs. On the other hand, the fact that mostly the even-even nuclei were detected in the present work (more than 60 % of yield), implies that the value of cold fission yield for a complete set of correlated pairs will be slightly lower than the given value.

In the case of the most abundant cold fission fragments ^{106}Mo - ^{136}Te and ^{102}Zr - ^{140}Xe ground state bands of both fragments were examined and compared to the main channel with emission of two neutrons For both charge splits no drastic differences appeared between the 0n and 2n channel. The gamma rays of up to $10^+ \rightarrow 8^+$ (in the case of ^{136}Te up to $12^+ \rightarrow 10^+$) could be observed for both 0n and 2n channels, respectively.

As a last step of analysis, the possible yields of ternary fission were examined. The fragment pairs of ^{102}Zr - ^{134,136}Te and ^{96,98}Sr - ^{140}Xe were observed with yields of 10^{-3} of summary fission fragment yield. Such a fragment pairs may be interpreted as fission accompanied by emission of α-particle (^{102}Zr - ^{136}Te and ^{98}Sr - ^{140}Xe) or, in the latter case (^{102}Zr - ^{134}Te and ^{96}Sr - ^{140}Xe), two explantions are possible (emission of α-particle and two neutrons or emission of ^{6}He), which can not be distinguised in present analysis.

4 Conclusions

Yields of correlated fragment pairs were detected in the spontaneous fission of ^{242}Pu for five different even charge splits. Charge, mass and neutron multiplicity distributions of fragment pairs were determined and compared to available data. The determined values are in reasonable agreement with the available data on spontaneous fission of ^{242}Pu. The values of the mean number of emitted neutrons were determined for the whole set of data and for different charge splits. No significant deviations in the mean numbers of emitted neutrons or neutron multiplicity distributions were observed for different charge splits. The yield of cold fission without neutron emission was set to be about 10 % for the set of observed correlated fragment pairs. For the complete set of fragment pairs a slightly lower value might be expected. The yields of ternary fission were detected on the level of 10^{-3} of summary fission yield.

Acknowledgements

The work was partially supported by the Grant Agency of Slovak Republic through contract GAV 2/5121/98. The authors would like to thank the GAMMASPHERE collaboration, J.H. Hamilton, J.O. Rasmussen, G. M. Ter-Akopian and A. V. Daniel for the help in data preparation and valuable discussions.

References

1. GAMMASPHERE a National Gamma-Ray Facility, a proposal, March 1988, Ed.: M. A. Delaplanque and R. M. Diamond.
2. M. Morhac, J. Kliman, V. Matousek, M. Veselsky, I. Turzo, *Nuclear Instruments and Methods* **A401**, 113 (1997)
3. C. Wagemans et al., *Physics Review* **C30**, 218 (1984)
4. Evaluated Nuclear Data Files, National Nuclear Data Center, Brookhaven National Laboratory, Brookhaven.

THE MUNICH ACCELERATOR FOR FISSION FRAGMENTS (MAFF) AT THE NEW REACTOR FRM II

T. VON EGIDY, P. KIENLE, U. KÖSTER

Physik-Department, Technische Universität München, D-85748 Garching, Germany

E-mail: egidy@e18.physik.tu-muenchen.de

D. HABS, M. GROSS, O. KESTER, H.J. MAIER, P.G. THIROLF

Sektion Physik, Ludwig-Maximilians-Universität München, D-85748 Garching, Germany

AND THE MAFF COLLABORATION

In Garching near Munich the new research reactor FRM II is under construction. It is supposed to be commissioned in 2001. The reactor will have a power of 20 MW, a rather small core diameter of 25 cm and a maximum neutron flux of $8 \cdot 10^{14}$ n/cm$^2 \cdot$s. Many new powerful instruments and novel experiments are planned in material science, solid state physics, nuclear physics, fundamental physics, chemistry, biology, medicine and technology. Consequently, the applications will range from material science diffractometers and inelastic neutron scattering over neutron tomography and cancer therapy to fusion producing the heaviest elements using accelerated fission fragments. The acceleration of radioactive isotope beams (RIB) is an important new field in nuclear physics and many new facilities are planned or under construction all over the world. Nuclear fission is the most suitable method to produce neutron-rich isotopes due to the large fission cross sections of thermal neutrons and the high thermal neutron flux in modern research reactors. Consequently, the new facility MAFF (Munich Accelerator for Fission Fragments) is foreseen for the new FRM II reactor. Beams of about 10^{11} atoms/s are expected for ^{91}Kr, ^{132}Sn,^{144}Cs from a ^{235}U source in a graphite cylinder close to the reactor core placed in a through-going beam-tube. Laser ionisation and a linear accelerator with IH and RFQ structures will provide the possibility to accelerate many different isotopes up to about 6 MeV/nucleon. Neutron-rich isotope beams are most interesting, because they will allow the investigation of several exciting new areas: i)Production of heavy and superheavy elements. ii) Atomic physics and chemistry of the heaviest elements. iii) Nuclear structure of astrophysically relevant nuclides and data for the r-process nucleosynthesis. iv) Isospin dependence of nuclear reactions. v) Application of radioactive isotope beams for solid state physics, medicine and radioactive waste management.

1 The new Munich Reactor FRM II

After more than forty years of very successful operation the Munich Research Reactor FRM will be replaced by the more powerful modern reactor FRM II. Its construction started in 1996. Although the power, core diameter and maximum neutron flux of the FRM II (20 MW, 25 cm and $8 \cdot 10^{14} n/cm^2 \cdot s$), are somewhat lower than the corresponding values of the Grenoble High Flux Reactor of the ILL (57 MW, 42 cm and $1.5 \cdot 10^{15}$ n/cm $^2 \cdot$ s), it is tried to compensate this with most modern, efficient and innovative instrumentation. The FRM II is cooled with light water and moderated with heavy water. For best efficiency the reactor will be operated with highly enriched ^{235}U (93%). The reactor building is nearly finished and has a wall thickness of 1.8 m to protect the reactor against airplane crashes. All safety measures played an important role in the construction. Between 30 and 35 experiments are planned. The reactor is scheduled to be operational in 2001.

2 Instrumentation of the FRM II

The instrumentation will be built and operated by user groups from universities, research institutes and industry. Consequently, dedicated neutron beams will be used for research and education at universities, for testing new ideas and for a large spectrum of applied purposes. Neutrons are very complementary to x-rays, as they mainly interact with the nucleus, not the electron shell. They are easily scattered at light elements including hydrogen. Various precise neutron diffractometers will help to determine the crystal and molecular structure of many technical materials, new compounds and organic, biological and medical substances. Inelastic neutron scattering provides information on dynamic properties of matter. Medical application includes clinical cancer therapy with MeV neutrons, new radiopharmaca and boron capture therapy. The properties of the neutron itself will be investigated with ultracold neutron beams ($v \approx 5$ m/s). An extremely intense positron source ($10^9 e^+$/s) will be constructed using pair production with neutron-capture gamma-rays. The most important instrument for nuclear physics will be the Munich Accelerator for Fission Fragments (MAFF) which will be discussed below. Industry is very interested in Si doping with neutron transmutation, irradiation for various tests and analyses and tomography with cold and fast neutrons which is specially suited to see thin layers of organic material.

3 Radioactive isotope beams (RIB)

Research with accelerated radioactive isotopes is a new and exciting development in nuclear physics. Several new facilities are planned and one is already in operation (Table 1). Reactions with extremely neutron-deficient or neutron-rich radioactive nuclei allow the investigation of new regions of the chart of nuclides. Since hot nuclei evaporate mainly neutrons, neutron-deficient nuclei

Place	Primary beams accelerators	Target/ Ion source	Post-acceleration	Energy MeV/u	Beams	Status
ARENAS Louvain-la-Neuve	Cyclone 30 30 MeV p, 200 μA Cyclone 110, 80 MeV p, light ions, 20 μA	solid/molten ECR	Cyclone 110 New cyclotron (1997)	0.56-27 0.2-0.8	A < 40 *close to stability*	operational since 1987
SPIRAL GANIL/Caen	Cyclotrons light and heavy ions 25 - 95 MeV/u, 10^{12} pps	solid targets ECR	Cyclotron	4-40	n-rich and p-rich of many elements *gaseous elements favoured*	under construction target/ECR tests first beams 1999
REX-ISOLDE CERN/Geneva	CERN PS-Booster 1 GeV p, 2 μA	solid/molten Surface Plasma Laser	Penning trap EBIS LINAC	**0.8-2.2** upgrade possible	n-rich and p-rich *non refractory elements*	**60 keV beams exist** post accelerator under construction first beams 1999
EXCYT Catania	Super conducting cyclotron light and heavy ions 20 - 100 MeV, 10^{13} pps	molten/solid plasma/ECR surface charge exchange	Tandem 15 MeV	(Q+1) *15 MeV Q < 11	n-rich and p-rich of elements with *negative ions*	under construction 2000?
FRM-II Munich	FRM-II reactor n-induced fission 10^{14} n/cm²s	Uranium-Carbide	ECR-charge state breeder LINAC	3-6	n-rich A = 80 - 150	under construction
SPIRAL-2 GANIL/Caen	LINAC, deuterons 100 MeV/u, 10^{15} pps	liquid U-target ECR	Cyclotron	4-40	n-rich A = 80 - 150	planned
RAL Rutherford Appleton Lab.	proton synchrotron 800 MeV, 200 μA	solid	LINAC	4-15	n-rich and p-rich	target test facility

Table 1. European projects on ISOL based radioactive isotope beams.

and nuclei close to the proton drip-line have been frequently studied. It is more difficult to produce neutron-rich nuclei which deserve on the other hand special attention, because fusion of neutron-rich nuclei leads to more stable and eventually to the heaviest nuclei. Moreover, neutron-rich nuclides play an important role in the astrophysical r-process. Nuclear fission is the best reaction to obtain neutron-rich nuclei.

Figure 1 shows the cross sections for the production of rubidium isotopes using different reactions [1]. The fission cross sections with thermal neutrons are several orders of magnitude larger than fragmentation or spallation cross sections. The high in-pile neutron flux of the FRM II makes the fission reaction very attractive.

Table 2 compares the yields of some isotopes from MAFF with accelerator based RIB facilities and demonstrates that MAFF is the most powerful method. P. Armbruster and H. Nifenecker had proposed in 1990 to produce an intense beam of fission fragments from a special uranium target in a beam tube of the ILL reactor in Grenoble and guide this beam 400 m to the SARA accelerator complex of the adjacent Institut des Sciences Nucléaires. This PIAFE project found large international support but was terminated in 1998 due to insufficient funding. Consequently, the interested physicists concen-

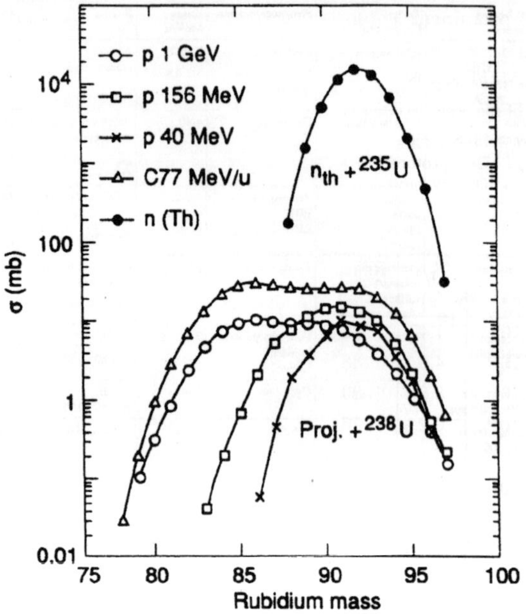

Figure 1. Cross sections for the production of rubidium isotopes by various nuclear reactions [1].

trate their effort now on MAFF. An international workshop took place in October 1996 at Benediktbeuern south of Munich on "Research with Fission Fragments" [2] which provided a comprehensive survey of the new physics to be done and the planned facilities.

4 The MAFF installation at the FRM II

The fission fragment source (Fig. 2) consists of a graphite cylinder (6 cm length, 1.5 cm diameter, with a central bore hole, loaded with $1-2\,g$ ^{235}U. It is encapsulated in a rhenium cylinder with a 2 mm diameter outlet where the fission fragments diffuse out. An electrical lens system accelerates singly charged ions to 30 keV and guides them to the mass separator and installations in the reactor hall. The source is placed in the highest neutron flux ($1.5 \cdot 10^{14}\,n/cm^2 \cdot s$) of a through-going beam tube 60 cm from the reactor centre where the flux gradient is rather small. The fragment beam is extracted on one side while the source mounting and changing and most of the conventional

Table 2. Yields of prominent isotopes at MAFF compared to accelerator based RIB facilities.

The values given for MAFF assume a total fission rate of $1.0 \cdot 10^{14}$ s^{-1}.

The primary beams of the other facilities are as follows:

SPIRAL: 10 μA of 1.15 GeV ^{12}C
ISOLDE: 2.5 μA of 1 GeV protons (600 MeV protons for the SC)
ISL: 100 μA of 100 MeV protons

Isotope	Prod. rate (s^{-1})	$T_{1/2}$ (s)	Release (%)	Ioniz. (%)	**MAFF** (s^{-1})	SPIRAL (s^{-1})	ISOLDE (s^{-1})	ISL (s^{-1})
^{78}Zn	$2.3 \cdot 10^9$	1.47	36	10	$\mathbf{4 \cdot 10^8}$	$2 \cdot 10^6$	$1 \cdot 10^6$ (SC)	$6 \cdot 10^6$
^{82}Ge	$1.2 \cdot 10^{11}$	4.6	48	10	$\mathbf{6 \cdot 10^9}$		$2 \cdot 10^5$ (SC)	$1 \cdot 10^8$
^{91}Kr	$3.3 \cdot 10^{12}$	8.6	89	15	$\mathbf{4 \cdot 10^{11}}$	$2 \cdot 10^9$	$2 \cdot 10^9$	$4 \cdot 10^{10}$
^{94}Kr	$1.1 \cdot 10^{11}$	0.2	16	15	$\mathbf{3 \cdot 10^9}$	$3 \cdot 10^7$	$4 \cdot 10^6$	$6 \cdot 10^9$
^{97}Rb	$3.2 \cdot 10^{10}$	0.17	10	80	$\mathbf{2 \cdot 10^9}$	$5 \cdot 10^8$	$3 \cdot 10^8$	$7 \cdot 10^9$
^{132}Sn	$7.0 \cdot 10^{11}$	39.7	89	10	$\mathbf{6 \cdot 10^{10}}$	$6 \cdot 10^6$	$8 \cdot 10^7$ (SC)	$1 \cdot 10^{10}$
^{142}Xe	$5.2 \cdot 10^{11}$	1.24	7	25	$\mathbf{1 \cdot 10^{10}}$	$6 \cdot 10^7$	$2 \cdot 10^8$	$4 \cdot 10^{10}$
^{144}Xe	$9.3 \cdot 10^9$	1.15	7	25	$\mathbf{2 \cdot 10^8}$	$1 \cdot 10^6$	$5 \cdot 10^6$	$2 \cdot 10^{10}$
^{144}Cs	$4.3 \cdot 10^{11}$	1.0	47	80	$\mathbf{2 \cdot 10^{11}}$	$2 \cdot 10^{10}$	$3 \cdot 10^{10}$	$6 \cdot 10^{10}$

vacuum pumping is provided from the other side. The source temperature of 2400° C will be achieved mainly by the fission process in the source and stabilized by additional electrical heating. A He cryo-pumping system in the beam tube provides the high vacuum and helps to localize gaseous fission fragments at cryo panels.

Singly ionized ions of alkali and alkaline earth elements are produced in the ion source by surface ionization. An efficient way to ionize many elements which are only poorly thermally ionized is the ionization with resonant laser light via one ore more intermediate atomic levels [3]. Two or three laser beams have to be guided into the ion source. This technique is not only element selective already within the ion source, but it might also be used to separate in special cases isotopes and even isomers. Efficiencies of 5-15% can be achieved.

The overall layout of the facility is shown in Fig. 3 with the reactor hall and the special experimental hall for MAFF. The fission fragments enter a magnetic mass separator in the reactor hall and are transferred into the experimental hall. After charge breeding to high ionization levels of $q/A \geq 0.16$ they enter a rather compact linear accelerator consisting of RFQ and IH structures. Energies up to 6 MeV/nucleon are foreseen. The beam can be switched to several experimental areas.

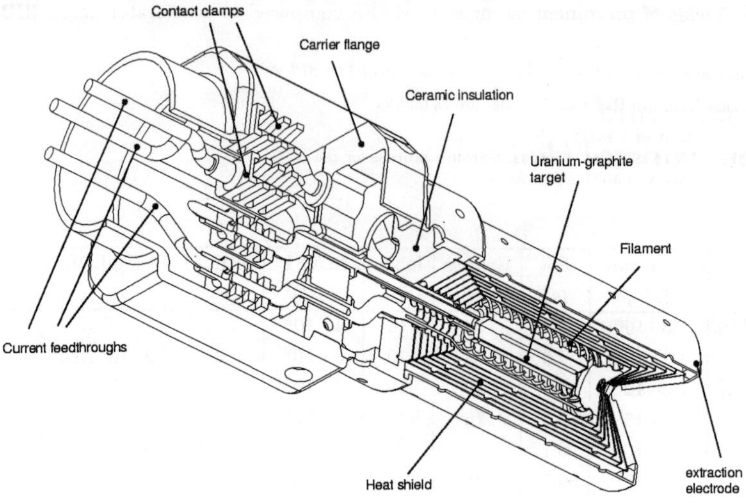

Figure 2. Fission fragment source.

5 Fusion with neutron-rich projectiles

The heaviest elements have recently been produced at GSI, Darmstadt, although the cross sections of these fusion reactions were only a few pb [4]. The systematics of the measured cross sections are shown in Fig. 4 [5]. A remarkable result is the increase of the cross section for element 110 from 3.5 pb with ^{62}Ni (^{269}110) to 15 pb with ^{64}Ni (^{271}110).

This fact indicates that the production of element 114 could be feasible with very neutron-rich projectiles [5]. Moreover, careful predictions with shell corrections [6] show that the isotope ^{294}114 is expected to be more stable (see Fig. 5) and this nuclide can only be produced with projectiles such as ^{144}Xe:

$$^{150}\mathrm{Nd} + ^{144}\mathrm{Xe} \rightarrow ^{294}114. \qquad (1)$$

Other similar reactions to superheavy nuclei are for instance [7]:

$$^{150}\mathrm{Nd} + ^{132}\mathrm{Sn} \rightarrow ^{282}110, \qquad (2)$$
$$^{208}\mathrm{Pb} + ^{82}\mathrm{Ge} \rightarrow ^{290}114, \qquad (3)$$
$$^{208}\mathrm{Pb} + ^{94}\mathrm{Kr} \rightarrow ^{302}118. \qquad (4)$$

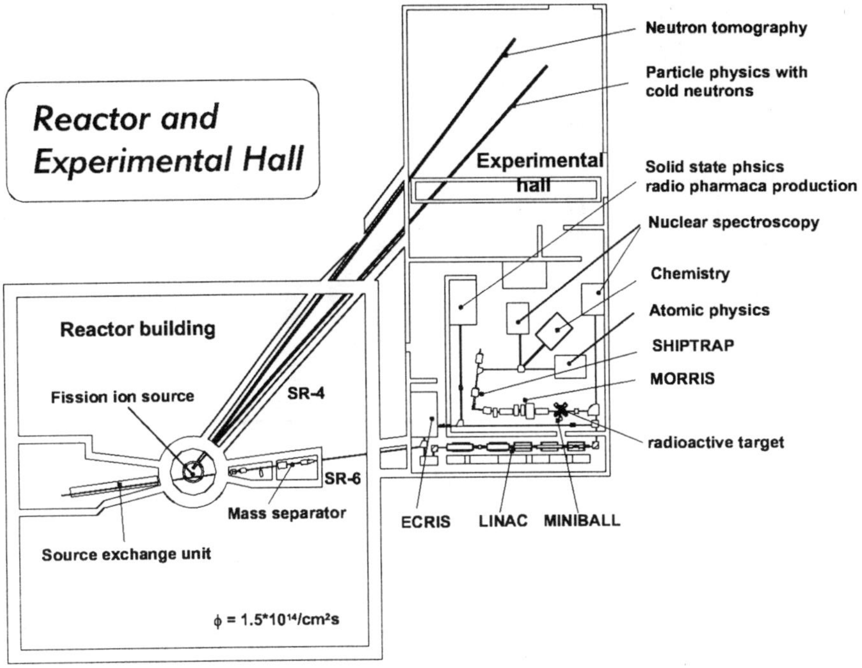

Figure 3. Overview of the reactor building and the accelerator hall.

Another challenging reaction which has not yet been performed until now is "inverse fission" of uranium or fermium. The understanding of such processes would unify fission and fusion [7], e.g.

$$^{132}\mathrm{Sn} + ^{132}\mathrm{Sn} \rightarrow ^{264}\mathrm{Fm}, \tag{5}$$

$$^{138}\mathrm{Ba} + ^{94}\mathrm{Kr} \rightarrow ^{232}\mathrm{U}, \tag{6}$$

$$^{100}\mathrm{Mo} + ^{132}\mathrm{Sn} \rightarrow ^{232}\mathrm{U}, \tag{7}$$

$$^{87}\mathrm{Rb} + ^{145}\mathrm{Cs} \rightarrow ^{232}\mathrm{U}. \tag{8}$$

We would like to note that recent investigations showed that fusion reaction with magic target or projectile nuclei are strongly enhanced.

"A compact apparatus for mass selective resonance ionization spectroscopy of transuranium elements in a buffer gas cell" has been developed in Mainz [8]. This will allow laser spectroscopy of the heaviest elements which offers the possibility for the investigation of the unknown atomic, chemical and

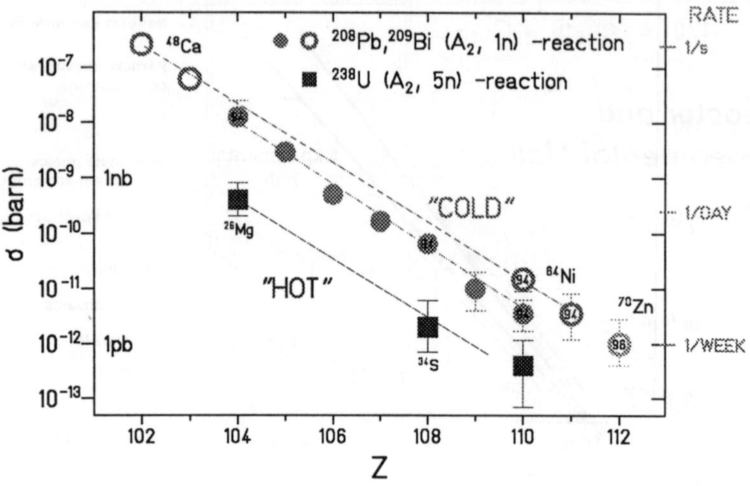

Figure 4. Systematics of measured cross sections for the production of the heaviest nuclei.

nuclear properties of these nuclides. The high sensitivity and short extraction time (1.5 ms) opens a large field of new research. For instance, the electron binding energies are known only up to Cf (Z = 98) and could be determined for superheavy elements. Single-molecule chemistry with these elements might also be possible.

6 Study of the r-process with neutron-rich nuclides

The fission process produces nuclides which are close to the neutron r-process path and approach the neutron drip-line more than those achieved by other reactions (see Fig. 6). Consequently, all informations on the structure, on masses, on cross sections, on lifetimes, on level densities etc. are very valuable for astrophysical calculations. There are indications that the shell structure known for nuclei close to the valley of stability changes considerably for very neutron-rich nuclei. This fact modifies very much the calculations of isotope abundances. The necessary information can be obtained with various methods: β- and γ-spectroscopy of fission fragments, direct reactions in inverse kinematics [9] such as d(^{132}Sn,p)^{133}Sn or p(^{144}Xe,p')^{144}Xe or Coulomb excitation [10].

Atomic masses and nuclear binding energies contain fundamental information on the nuclear structure. However, present mass predictions are still

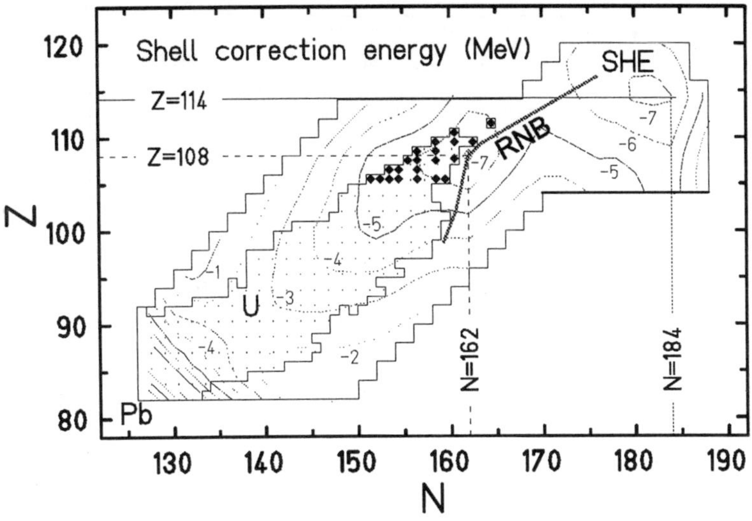

Figure 5. Contour map of shell corrections to the binding energies. Diamonds denote the heaviest nuclides synthesized up to now.

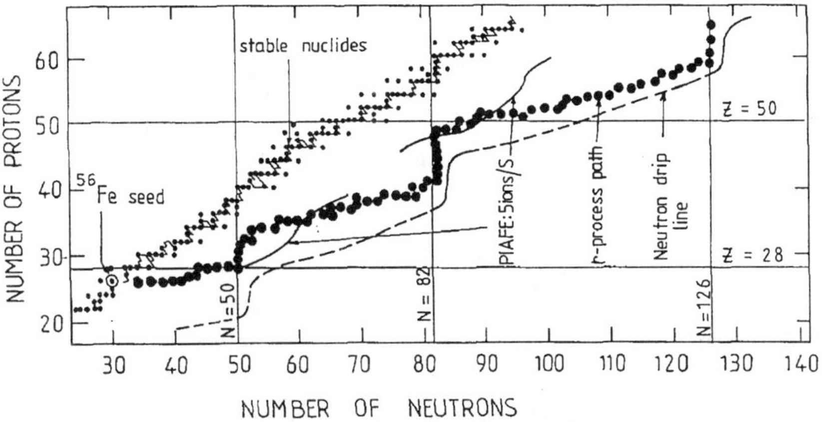

Figure 6. Comparison of the PIAFE beam (similar to MAFF) at a rate of 5 ions/s with the neutron r-process path and the neutron drip-line.

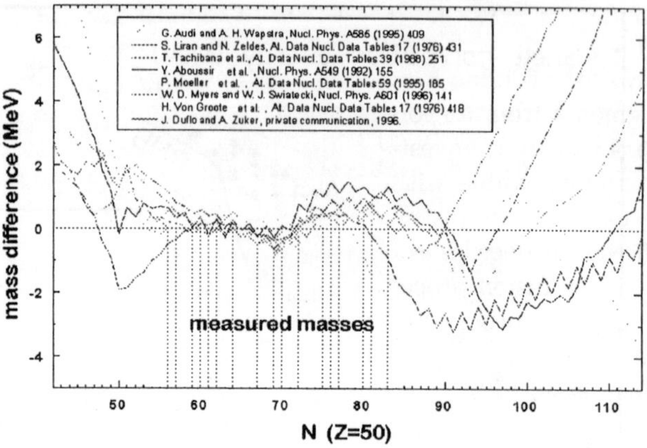

Figure 7. Differences between various mass models for the tin (Z = 50) isotopes (relative to the values of Duflo and Zuker) [11].

limited and extrapolations beyond the known region show large divergences as shown in Fig. 7 for tin nuclei [11]. Newly improved methods such as the Smith spectrometer MISTRAL [11] will allow precisions of $5 \cdot 10^{-7}$.

The predicted and observed neutron skin of neutron rich nuclei and their different isospin can change reaction cross sections and dynamics. Studies of the nucleon exchange in deep-inelastic collisions and of giant isovector dipole resonances are therefore of special interest [12].

7 Applications

There is a wealth of applications for solid state physics which can be done with radioactive fission fragments, such as Mössbauer spectroscopy, perturbed angular correlation, NMR and emission channeling. Other techniques comprise radio-tracers for semiconductor physics, migration and implantation studies. The last methods are also important for storage studies and management of nuclear waste. Also medical applications, e.g. radiopharmaca, could play a role.

8 Summary

The feasibility of the production, extraction, mass separation and acceleration of fission fragments from an ion source in in-pile position has been demonstrated by various careful investigations. Higher yields of neutron-rich nuclei are expected than possible with other methods. The large number of new and exciting experiments justifies the great effort for MAFF. The research with neutron-rich projectiles will substantially enlarge our knowledge of the heaviest nuclei, the fusion process, isospin effects in nuclear structure and reactions, astrophysical processes and even chemistry of heavy elements. Many application in solid state physics and medicine complement the attractivity of this new facility.

References

1. D. Habs et al.: *in* Research with Fission Fragments *(Eds. T. von Egidy et al.)*. World Sci., Singapore 1997, p. 18
2. Research with Fission Fragments *(Eds. T.von Egidy, D. Habs, F.J. Hartmann, K.E.G. Löbner, H. Nifenecker)*. World Sci., Singapore 1997
3. V.I. Mishin et al. *Nucl. Instr. Meth. B* **73**, 550 (1993)
4. S. Hofmann et al. *Z. Phys. A* **350**, 281 (1995); *ibid.* **354**, 229 (1996)
5. D. Habs et al. *Nucl. Phys. A* **616**, 39c (1997)
6. H.A. Ghergescu et al.: *in* [2], p. 116.
7. I. Lazarev, I. Oganesian: *in* [2], p. 314.
8. H. Backe et al.: *in* [2], p. 80.
9. P. Egelhof: *in* [2], p. 178.
10. H.J. Wollersheim: *in* [2], p. 167.
11. M.D. Lunney et al.: *in* [2], p. 225.
12. F. Hanappe et al.: *in* [2], p. 340;
    H. Nifenecker et al.: *ibid.*, p. 360.

NUCLEAR WASTE INCINERATION

T. Bolognese, A. Pluquet
CEA–Saclay
DSM/DAPNIA/SPP

In the context of one of the French programs, INCA, on nuclear waste incineration using an ADS (accelerator driven system), fundamental physics aspects are presented: accelerator studies, spallation experiments, ^{241}Am and ^{99}Tc cross section measurements, code development and hybrid systems design considerations.

1 Introduction

Radio-toxicity reduction of nuclear waste produced by civil and military nuclear reactors is one of the big goals for the coming 50 years. Solving the waste problem will contribute to make nuclear energy production socially more acceptable.

Nuclear waste policies are rather different around the world, depending mainly on the present use of spent fuel. In some countries (like USA, Sweden, Switzerland), unloaded fuel will be directly stored . This simple and maybe cheaper solution suffers from safety uncertainties of the storage over hundred of thousand years, a time scale often not understandable for most of the population. In other countries (like France, UK, Japan, Belgium) spent fuel is reprocessed to extract uranium and plutonium for further use, as MOX fuel in REP reactors.

In Table 1 [1] we compare the actinide annual production in three fuel cycles scenarios of equal power production (60 GWe, 400 TWhe per year):

- Open cycle : UOX fuel and no re-processing of the spent fuel;
- MOX1 (this is the present French situation): one step recycling: 11.7% of the UOX spent fuel is reprocessed to MOX fuel to be used in REP reactors;
- MOXE (E for equilibrium): multi-recycling in which all the UOX spent fuel is continuously recycled and, at equilibrium, all of the Pu is kept in the fuel cycle.

It has to be noted that, even if the multi-recycling allows to eliminate the plutonium and to reduce considerably the total volume of the waste to be stored, the counterpart is an increase of minor actinide production. The radio-toxicity of the waste is considerably reduced for several hundred years

Table 1: Actinide waste annual productions in three fuel cycles scenarios

Element	Open cycle tons/year	MOX1 (1 recycle) tons/year	MOXE (multi–recycling) tons/year
Pu	11.720	7.200	0.000
Am	0.559	1.146	2.900
Cm	0.116	0.256	0.500
Np	0.840	0.752	0.730

but after 400 years the Am will bring up again the radio-toxicity to the level of the open cycle. So minor actinides have to be burnt if we want a long term radio–toxicity reduction. Furthermore, more long lived fission products are also produced, in particular technetium and iodine (~1.2 tons/year for the multi-recycling).

In December 1991 the French government voted a law relative to the research to be performed to solve the nuclear waste problem. One of the articles of this law deals with the separation and transmutation of nuclear long lived elements present in nuclear waste. Because of that law and for the afore mentioned reasons, part of the French research community concentrates on minor actinides and long lived fission product burning. One of the possible solutions to burn nuclear waste is the so called ADS (accelerator driven system); a schematic view of such a system is shown in figure 1. It is based on an intense neutron source, consisting of a high intensity proton accelerator and a spallation target, surrounded by a sub-critical blanket containing actinides and selected fission products to be destroyed.

In our department we do essentially fundamental research and technology applied to fundamental research. Engineering and technological problems are treated in other departments of CEA and CNRS.

This report deals essentially with fundamental aspects of nuclear waste incineration and in the following section we present our activities:

- section 2: high intensity accelerator research;
- section 3: spallation physics experiments at Saturne;
- section 4: neutron cross sections measurements on actinides and fission products, and
- section 5: code development and ADS conceptual design.

2 IPHI: high Intensity Proton Injector project

High power proton or deuteron accelerators are more and more considered as ideal sources of neutrons. High neutron fluxes, broad energy spectra in continuous working (CW) or pulsed modes are needed for several applications. Nuclear waste incineration, tritium production are the industrial applications of this kind of accelerators but they can be used for material studies, spallation physics and possibly neutrino oscillations.

Nuclear waste incineration necessitates a continuous working proton accelerator with a beam power greater then 20 MW. Currently, cyclotrons go up to 5 MW and several years of research and development are needed in order increase this limit. Our department has an extensive experience in linear accelerators so our efforts have been concentrated on high–intensity continuous working linear accelerators. They have to be designed as a part of an industrial facility and this requirement puts more stringent constraints compared to a research facility.

In particular beam losses must be limited to an extremely low level to allow hand on maintenance. Beam losses will activate structural materials and require long times before intervention to reduce the radioactivity to an acceptable level. Furthermore limiting the beam losses will reduce the shielding needed and therefore the cost. Abrupt stops have also to be reduced to an extremely low level to reduce strength on the target window which operates at high temperature. Reactor engineers quote a maximum of 3 abrupt stops (longer than 100 ms) per year, which is very far from a 1 stop per hour that cold systems can stand nowadays.

In our department a R&D program for the injector IPHI of a high intensity linear accelerator has been approved (but not completely financed). The aim is to obtain a CW beam of 10 MeV protons of 100 mA intensity. A continuous working proton and deuteron ECR source (called SILHI) has been constructed and the results of preliminary tests are very encouraging. This is a key component because its performance determines the accelerator design. The goal is to reach a 100 mA beam current at 95 keV with a rms normalized emittance better than 0.2 π.mm.mrad and a very high reliability. The results of a 100 hours test run at 100 mA gives (excluding the first conditioning day) 98.8% availability, 5h33mn of mean time between failures, 4mn44s mean time to repair (which will be considerably improved by automatic restart) and no damage to the HV extraction electrodes after more than 300 hrs CW at 100 mA. The beams performances achieved are shown in table 2.

This work has been done in collaboration with CNRS-IN2P3 and Thomson. Further developments are foreseen for the other parts of the injector that are

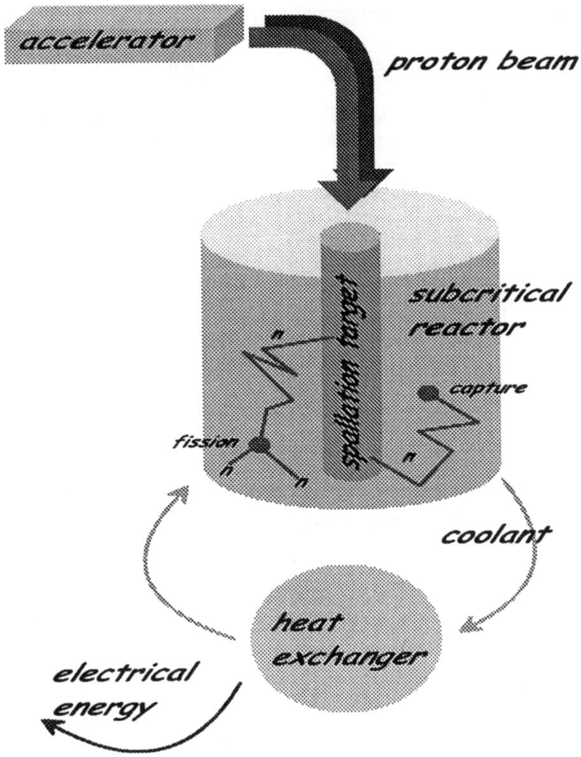

Figure 1: Schematic view of an ADS

the RFQ , DLT and the diagnostic line.

3 Spallation physics

As shown in figure 2, the proton beam hits a heavy metal target to produce a large amount of neutrons by spallation processes. Materials such as Pb or Bi are envisaged because of their high cross section on proton and their low cross section on neutron. In order to get the maximum amount of neutrons leaving the target, the geometry has to be optimized, taking into account the beam energy. It has been noticed that the number of neutrons produced will not increase at energies larger than 1.5 GeV.

Spallation reactions are modeled by a two step process. The first step, in which proton-neutron collisions are taking place, is described by intra-nuclear

Table 2: SILHI Beam characteristics achieved during a first test in 1997 at 100 mA

Parameter	Requested	Status
Energy(KeV)	95	95
Intermediate elec.(KV)	65	45
Proton current (mA)	100	98
Total current	110	122.5
Proton fraction(%)	¿90	80
Extraction diameter(cm)	1	1
Current density (mA/cm^2)	140	156
RF forward power (W)	1200	1200
Duty factor (%)	100	100
H_2 gas flow (sccm)	¡10	$\approx$3.2
Beam noise (%)	$\pm$1	N.A($\pm$2)
rms norm. emittance (π.mm.mrad)	0.20	0.21

cascade codes and leads to the production of energetic light particles and of an excited residual nucleus. In the second step, this nucleus cools down by particle evaporation. If the target is thick enough, the emitted particles will induce secondary reactions . This is the inter-nuclear cascade. These processes produce a large variety of residual nuclei. The knowledge of their distribution is necessary for different reasons. The residual nuclei yield a radio-toxicity that has to be evaluated. Some of this nuclei could modify the neutronic equilibrium in the reactor and induce corrosion and damage to structural materials. Finally experimental data on thin and thick targets measuring angular and energy cross sections distributions and residual nuclei yields will contribute to the understanding of the theoretical descriptions of these reaction mechanisms.

In order to test the nuclear physics part (intra-nuclear, fission, evaporation) of the available Monte-Carlo simulation codes, it is necessary to measure the double differential cross section for neutron production for thin targets , as a function of the incident proton energy . These measurements have been performed using the SATURNE accelerator which produces protons of energy between 0.8 and 1.6 GeV. The energy of the neutrons, produced by the interactions of protons on thin targets, is measured between 1.5 MeV and 1 GeV using two methods:

- at low energy (up to 400 MeV) a time of flight between the tagged beam particles and the neutrons detected by a thick liquid scintillator is used;
- at high energy a second method is used based on the charge exchange reaction pn$\rightarrow$pn in a secondary hydrogen target; a magnetic spectrometer with trajectory reconstruction and time of flight allows to identify and evaluate the momentum of the emitted proton.

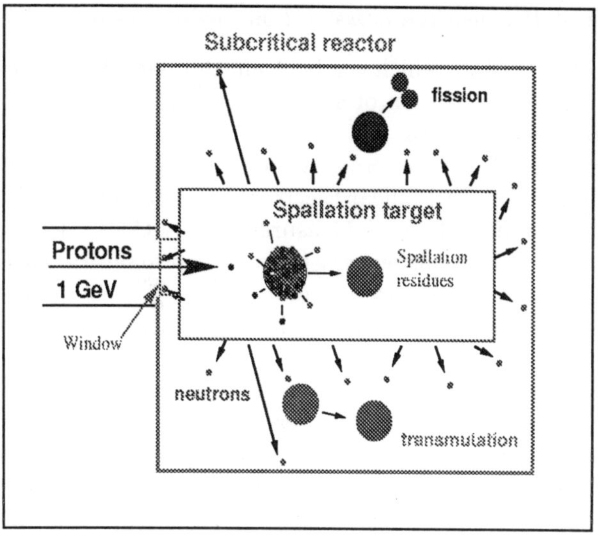

Figure 2: Neutron production by spallation reactions in an ADS system.

The two methods overlap in the range 200-400 MeV.

The neutron angular distribution is measured using six large volume liquid scintillator detectors (DEMON) and four low threshold liquid scintillator detectors; they allow the measurement of the low energy part of the spectrum; the charge exchange detection system can be rotated from 0° to 85°. Thin foils of Pb, Al, Fe, Zr, W, and Th have been analyzed. In Figure 3 experimental results obtained at SATURNE accelerator of double differential cross section (energy and angle) for neutron production on Pb at 1.2 GeV proton energy, are compared with two models predictions [4,5]. The Cugnon model [5] seems to fit better the data points in particular at high energy. See reference [2] for more experimental details.

4 Neutron cross section measurements

The lack of reliable and sufficient data on neutron cross sections of long–lived fission products and minor actinides is a major problem in the design of nuclear waste transmutation systems. For this reason several measurements of neutron cross sections are going on, in our department, for Tc, Np and Am and in the near future for other long-lived isotopes.

4.1 *High resolution neutron cross section measurement for ^{99}Tc and ^{237}Np*

The fission product ^{99}Tc, which is a beta-emitter with 200000 years of half life, goes to ^{100}Tc after absorption of a neutron. ^{100}Tc then decays in 16 seconds into the stable isotope ^{100}Ru.

Measurements of neutron cross sections (total and capture) of ^{99}Tc have been performed at the Geel Linear Electron Accelerator GELINA facility of the Institute for Reference Material Measurements (IRMM, Belgium). Neutrons with a spectrum from thermal energy (25 meV) up to several MeV are produced by an electron beam hitting an uranium target. The energy of the neutrons, collimated into beam tubes leading to the experimental area, is determined by time of flight.

In the capture experiment events are detected by means of the emitted gamma rays of the ^{99}Tc(n,γ)^{100}Tc reaction and measured with two C6D6 detectors placed at 90° on both sides of the sample, located itself 30 m from the neutron source. The spectrum of the neutron flux was measured at the same time with a boron chamber located 60 cm upstream of the sample.

In the transmission experiment one measures the neutron flux as a function of the energy with and without a sample in the neutron beam. At 25 m from the neutron source, the ^{99}Tc sample was placed in an automatic sample changer. In order to determine the background, this device also contains isotopes with large resonance filtering the neutron flux at specific energies. The neutrons passing through the samples were detected by a NE919 lithium–glass detector located 50 m from the neutron source. The measured capture and transmission yields are shown in figure 4. These two measurements allow to improve several resonance parameters. Data have been taken in the energy range from 1 to 600 eV and from 0.1 to 100 keV for ^{99}Tc and for ^{237}Np. Data analysis is almost final for ^{99}Tc [6] and under way for ^{237}Np.

4.2 *Transmutation of Am*

Among the minor actinides present in nuclear waste, Am is the most abundant (see table 2). The quality of the available nuclear data on the different Am isotopes is not always adequate to carry out calculations. For instance a sizable discrepancy of over a factor of 20 is present on the ^{242}Am^{gs} thermal neutron flux cross section data from JEF-2 (5511 barns) and ENDF-B/VI (252 barns) libraries. In order to determine the capture cross section of ^{242}Am^{gs}, a series of measurements have been carried out at the high flux reactor of the institute Laue–Langevin (ILL) in Grenoble). Experimental details and preliminary results have been presented at this conference [7]. Other experiments are planned to measure Am and Cm cross section in thermal, epithermal and

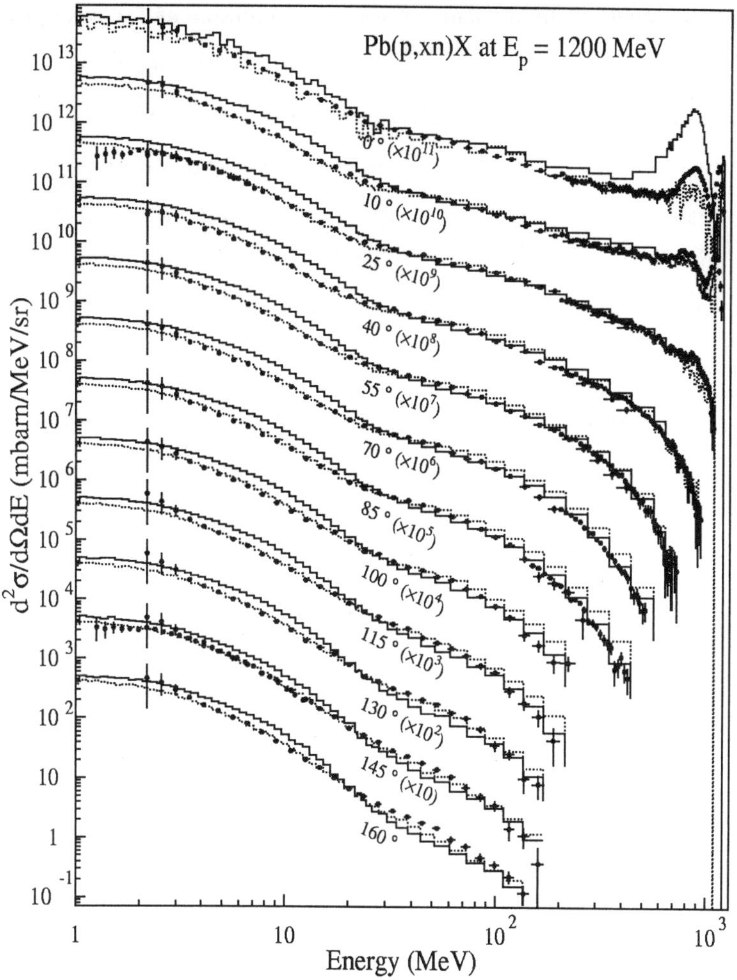

Figure 3: Neutron production cross section measured on a 2 cm thick Pb target at 1200 MeV. The histogram represent TIERCE simulation [3] using Bertini [4](full line) or Cugnon [5] (dotted line) cascade models.

possibly fast neutron fluxes using the same technique as for reference [7] with improved statistics and systematics.

5 Code development and ADS conceptual design

5.1 An Object Oriented code for simulation of hybrid systems (mcap)

The evaluation of the performance of an ADS system necessitates specialized tools. The minimal specifications include:

- the calculation of the neutronics of the sub-critical blanket (neutron flux as a function of position and energy, sub-criticity level, power),
- the prediction of the evolution of the isotopic composition of the blanket under spontaneous and induced decay for significant durations,
- the possibility of varying the many parameters that describe the system (cross section databases, geometry, materials, operating mode, etc.,).

At the time this study began and after a few attempts, it seemed to us that no software was fulfilling the above specifications. We discarded the coupling of different tools, which required the realization of tricky interfaces between programs using different conventions or physical hypotheses. We decided instead to create a new and integrated software, *mcap*, based on Monte Carlo techniques for the physical behavior. Detailed description of the physical model and of the computation schema used can be found in reference[8]. In contrast to the very poor ergonomics of previous codes and in order to safely control the calculation inputs and outputs, it was agreed that an user-friendly interface had to be developed. These concerns led to an object oriented design of the code.

mcap accepts:

- any isotope, as long the data exists in the databases (314 isotopes in JEF, 320 in ENDF, 337 in JENDL)[9], allowing comparisons or sensitivity studies.
- any material. A material is a mixture of isotopes and can be defined by many ways (weight, atomic or stoichiometric fractions with automatic conversions, mixture of known materials,...).
- any geometry. Spherical, parallelepipedic or cylindrical sub–volumes can be freely distributed in space. The inclusion tree is automatically built, and inconsistencies are reported.

- any spallation source. Simple parameterizations of the source are proposed, or, for detailed studies, external sources can be imported via a simple ASCII file.

- any operating mode. The load, unload and homogenization operations can describe many situations.

A complete graphical user interface helps the physicist to organize, visualize and control its input parameters. At the output, the program produces an easy to read document and lays emphasis on the physical behavior of the reactor.

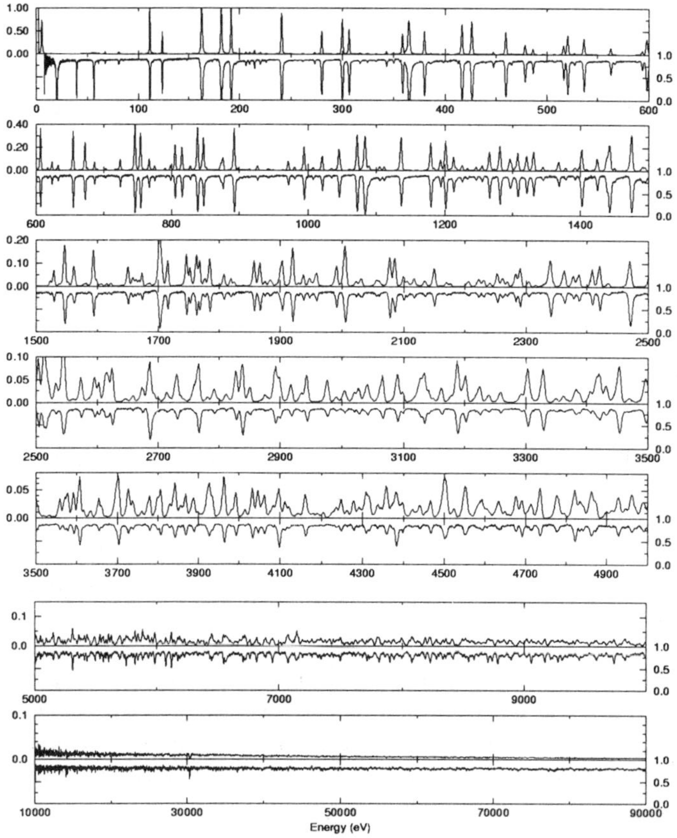

Figure 4: The capture yields and transmition factor derived from the measurements up to 90 keV on the same energy scale.

mcap is certainly not the first program dedicated to neutronics or evolution calculation. But, to our knowledge, it was the first to offer the essential functionalities in a single and coherent package, whose strong points are:

- a transparent interface to the nuclear databases. These sets of data come from the Nuclear Energy Agency of the OECD [9].They contain a huge amount of data which are evaluated by different international groups of experts. The principal drawback of these databases is an awkward format, mostly due to historical reasons. *mcap* allows simulations with any combination of isotopes available in the JEF, ENDF and JENDL databases;

- neutronics and evolution calculation are fully and automatically interconnected, sparing the user the trouble of translating data between more or less compatible programs.

Figure 5 and 6 are Examples of neutronic and evolution calculations performed by *mcap*.

5.2 Hybrid system conceptual design

Our group in mainly interested in conceptual design of hybrid systems. We have divided our study in three successive stages:

- exploratory phase. The huge number of a priori conceivable systems prevents us from simulating all of them. An analytical model [1] reduces the solution space to permitted regions (figure 7), in which the appropriateness of the selected hybrid systems is ensured,

- simulation phase of the selected hybrid systems. For this use, the level of precision reached by *mcap* and its flexibility have been found perfectly suitable [1];

- detailed study phase of the final design. Crossing the 1% precision limit will require the use of specialized softwares [10] or new developments in *mcap*, which is, by comparison, still in its early days. Preliminary studies show that improvements can be expected from a more detailed description of the neutron interactions (parameterization of the anisotropy in the high energy range, non linearities in the mean number of neutrons produced per fission, etc.)

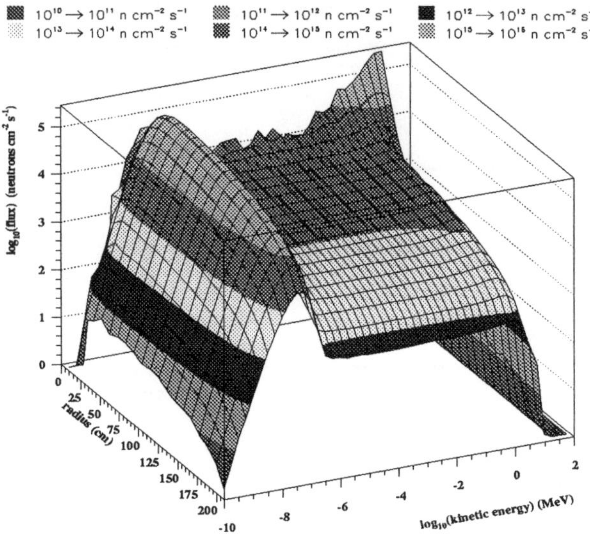

Figure 5: Example of neutronic calculations in a thermal hybrid system. The spallation target (top right) is clearly visible with a high flux of fast neutrons. On the left a thermal peak remains intense from the center to the limits of the reactor.

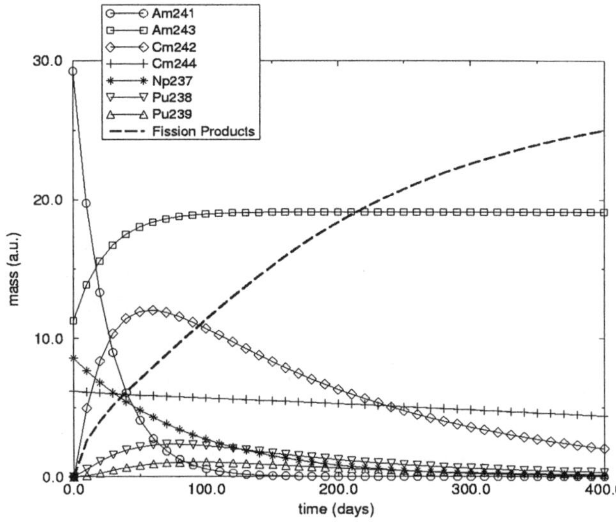

Figure 6: Example of evolution calculation: a sample composed of the mixture of Am, Np and Cm is irradiated in a high thermal flux. The ^{241}Am is efficiently destroyed

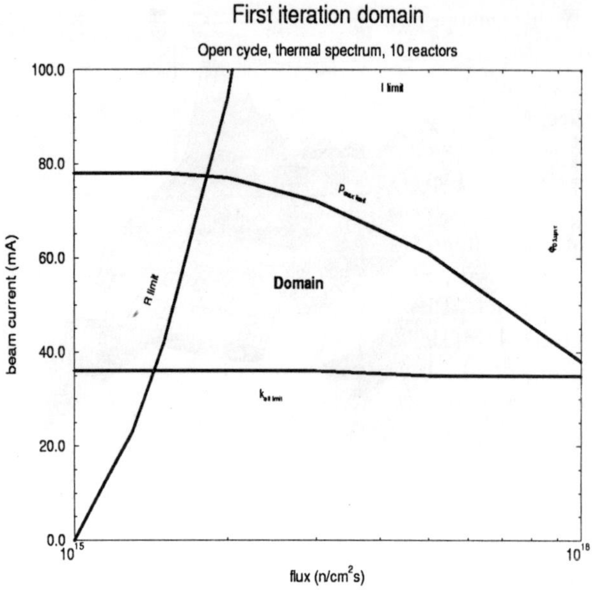

Figure 7: Allowed domains for a thermal hybrid system. The domain is limited by constraints on sub-criticality, radius and power density. Favorite solutions are located at the up left corner and correspond to low concentration systems.

6 Conclusions

Nuclear waste is among the worst inheritance we can leave to future generations. It is our duty to find solutions to burn this waste and fundamental scientists can bring an important contribution. In this paper we have shown some of the fundamental physics aspects of nuclear waste incineration. Other contributions will hopefully come if the scientific community and politicians continue to support this research.

Acknowledgments

We wish to thank all members of the INCA Saclay group.

References

1. F. Lelievre, Thesis: Stratgies pour l'incinération des déchets nucléaires dans des réacteurs hybrides (Orsay 11 décembre 1998) and M. Delphech

et al. Multirecyclage de Plutonium en REP et RNR- Impact réacteur et scénario, note technique CEA DRN/DER/SRPC/LEDC/96-461, 1996

2. X. Ledoux et al., Spallation neutron production by 0.8–1.2 GeV protons on Pb target. Submitted to Phys. Rev. Letters and International Conference on the Physics of nuclear science and technology. Long Island ; New York (USA) October 5-8, 1998
3. O. Bersillon, 2nd International Conference on Accelerator Driven Transmutation Technologies, Kalmar, Sweden, 5-7 June, 1996
4. H.W. Bertini et al., Phys. Rev. **131**, 1801 (1963)
5. J. Cugnon, Nucl. Phys. **A462**, 751 (1985) // J. Cugnon et al., Nucl. Phys. **A620**, 475 (1997)
6. F. Geunsing et al., Investigation of the neutron cross sections of ^{99}Tc in the resonance region. 6th International Seminar on Interactions of nucleon with nuclei, JINR, Dubna (Russia), 13-16 May, 1998
7. T. Bolognese, this proceedings.
8. A. Pluquet et al., An object oriented software for the simulation of sub–critical hybrid systems. Submitted to Nuc. Inst. and Meth. and Computer in Physics.
9. `http://www.nes.fr/html/dbdata/nds\_evaluated\_htm` and `http://t2.lanl.gov/data/data.html`
10. `http://www.nes.fr/html/dbprog`

PRELIMINARY RESULTS ON ^{241}Am TRANSMUTATION IN A HIGH THERMAL FLUX

T. Bolognese, G. Fioni, M. Cribier, F. Marie
DSM / DAPNIA, CEA/Saclay, 91191 Gif-sur-Yvette, France

M. Aubert
DCC / DPE, CEA/Saclay, 91191 Gif-sur-Yvette, France

S. Ayrault
DSM / DRECAM - CNRS, CEA/Saclay, 91191 Gif-sur-Yvette, France

J.-M. Cavedon, F. Chartier, F. Doneddu
DCC / DPE, CEA/Saclay, 91191 Gif-sur-Yvette, France

H. Faust
Institut Laue-Langevin, Grenoble, France

A. Gaudry
DSM / DRECAM - CNRS, CEA/Saclay, 91191 Gif-sur-Yvette, France

F. Gunsing
DSM / DAPNIA, CEA/Saclay, 91191 Gif-sur-Yvette, France

Ph. Leconte
Institut Laue-Langevin, Grenoble, France

F. Lelievre, J. Martino
DSM / DAPNIA, CEA/Saclay, 91191 Gif-sur-Yvette, France

R. Oliver
Institut Laue-Langevin, Grenoble, France

A. Pluquet, M. Spiro, C. Veyssiere

DSM / DAPNIA, CEA/Saclay, 91191 Gif-sur-Yvette, France

The possibility to transmute ^{241}Am by a high intensity thermal neutron flux depends essentially on the value of the ^{242gs}Am capture cross section. As the recommended values given in the two most widely used nuclear data libraries ENDF-B/VI and JEF-2.2 differ by more than a factor of 20, an experiment has been carried out at the Institut Laue-Langevin in Grenoble (France) to fix this discrepancy. The obtained results and their impact on the present understanding of nuclear waste incineration systems will be discussed.

1 Introduction

The transmutation of radioactive nuclear waste issued from conventional nuclear power plants, constitutes a considerable challenge for the scientific community.

The INCA (Incineration by Accelerator) project [1] of the Directorate for Science of Matter of the French Atomic Energy Authority (CEA/DSM), aims to outline the ideal physical conditions to transmute minor actinides (mainly $^{241-243}$Am, ^{237}Np and $^{244-245}$Cm) and some long living fission products (^{99}Tc, ^{129}I and ^{135}Cs) in an intense neutron flux obtained by an accelerator driven spallation facility coupled to a sub-critical system.

In this framework, a sizeable discrepancy of a factor of 20 was pointed out on the ^{242gs}Am thermal neutron capture cross section from JEF-2.2 [2] (5511 barns). and ENDF-B/VI[3] (252 barns) libraries. For the ^{242gs}Am data, JEF-2.2 is based on the JENDL-3.2 library [4] which rely on a Japanese evaluation [5]. No experimental data are actually available for this reaction and the recommended values are based on theoretical evaluations.

This great discrepancy has no major impact on the present commercial nuclear reactors, where the ^{242gs}Am is present at very low concentration. This is due to the absence of ^{241}Am in the starting fuel[a], to the low neutron flux and to the very short half-life of the ^{242gs}Am (16.02 h). On the other hand, in a minor actinide incineration system the initial fuel will contain mostly ^{241}Am and the high neutron flux will contribute to form a large amount of ^{242gs}Am. The quantitative impact of its capture cross section on the behaviour of an incineration system based on thermal neutrons, can be obtained by looking at the number of neutrons required to incinerate by nuclear fission an atom of a specific nuclear fuel. This parameter, also known as the D-factor, is plotted in Fig. 1 as a function of the thermal neutron flux [6]. A mixture of minor actinides composed by 43% ^{241}Am, 27% ^{243}Am, 17% ^{237}Np and 10% ^{244}Cm, was assumed in the calculations. This composition corresponds to the minor actinides which are supposed to be present after multi-recycling of a MOX (Mixed Oxide) fuel. The continuous and the dashed lines in Fig. 1 are obtained respectively by using the JEF-2.2 and the ENDF-B/VI libraries. A negative value of the D-factor corresponds to a gain of neutrons during the transmutation process and it should be obtained to have a transmutation system economically acceptable. In reality the "break-point", where the system becomes "critical", is situated at about -0.3 to take into account the neutron leakage from the system. From Fig.1, it is therefore evident that the transmutation of a mixture of minor actinides issued from multi-recycling. of MOX

---

[a] ^{241}Am is mainly formed by β-decay of ^{241}Pu during the fuel burn-up.

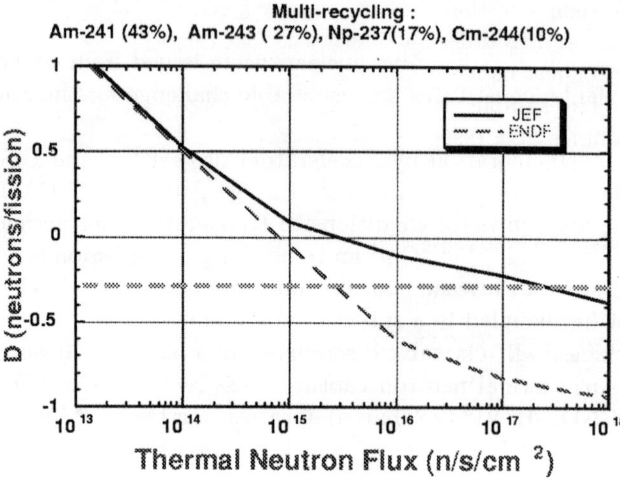

Figure 1: Number of neutrons required to incinerate by nuclear fission . one atom of the given isotope mixture as a function of the thermal neutron flux intensity. The continuous line is calculated using the JEF-2.2 library, while the dashed line is obtained with ENDF-B/VI. The starting mixture corresponds to the average composition of the minor actinides after multi-recycling of a MOX-fuel.

fuels, is possible with a thermal neutron flux of about 2×10^{15} n/s/cm^2 only if the value of the ^{242gs}Am capture cross section is lower than a few hundred barns.

A correct determination of this cross section is indeed crucial for the present understanding of minor actinides incineration systems.

2 Experimental procedure

In order to assess experimentally the transmutation efficiency of ^{241}Am in a high thermal neutron flux and to estimate the ^{242}Am capture cross sections, a total of 12 samples containing each about 30 μg of ^{241}Am oxide canned in quartz ampoules have been irradiated at the High Flux Reactor of the Institut Laue-Langevin in Grenoble (France). This single highly enriched fuel element reactor is cooled and moderated by D_2O and provides the highest purely thermal neutron flux available in the world. The combination of a very high neutron flux and the absence of epithermal neutrons, has been essential for the success of the following experiment. The experimental method consists in the measurement of the isotopic evolution of several samples irradiated for different times and in different neutron fluxes. The H9 beam position, which provides the highest neutron flux easily available at ILL, was used for 8 of the 12 samples. The sample changer of the Lohengrin spectrometer was used to bring the samples at a distance of about 60 cm from the fuel element, where

Table 1: Table I. Details of the irradiation programme. In three cases, two ^{241}Am samples were irradiated simultaneously to permit two different mass spectrometry analyses, as described in section 3.3. The neutron flux at the H9 position is about 6×10^{14} n/s/cm^2, while on T4 is about 2×10^{13} n/s/cm^2.

^{241}Am Mass [μg]	Irradiation duration [days]	Irradiation Position
29.52	30 minutes	H9
25.94	0.95	H9
30.21	2.01	H9
27.89	5.02	H9
31.52	10.98	H9
25.24	10.98	H9
25.32	19.03	H9
38.14	19.03	H9
37.92	5.01	T4
27.95	11.01	T4
29.81	24.0	T4
29.85	24.0	T4

the neutron flux approaches 6×10^{14} n/s/cm^2 with a thermalization coefficient of more than 98%. The T4 irradiation position located on the top of the ILL-reactor and giving access to a neutron flux of about 2×10^{13} n/s/cm^2 , was used for four samples. In Table 1 the experimental programme is given with the precise mass of the sample and the irradiation duration.

The main reason for using two different fluxes resides on having two independent sets of data to be used in the final fit and especially to verify that the epithermal component in H9 could be really neglected in the data analysis. In fact, the T4 position is outside the heavy water reactor vessel at about 1.3 meters from the fuel element, where the neutron flux is perfectly thermalized.

The isotopic chains formed by thermal neutron irradiation of ^{241}Am are shown in Fig. 2. Off-line gamma-ray spectroscopy and mass spectrometry were used to determine the evolution of each irradiated sample as described in section 3.3.

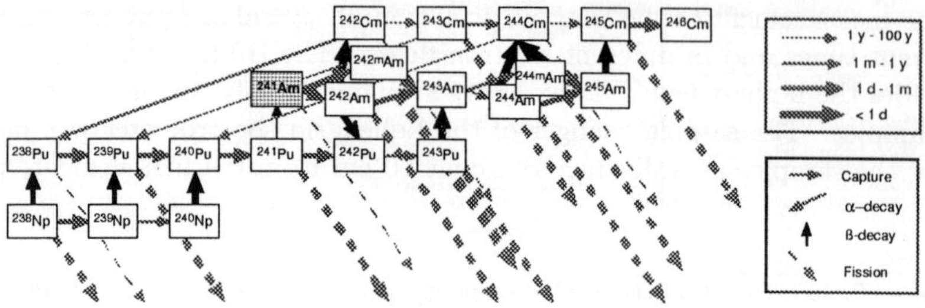

Figure 2: First steps of the isotopic chain formed during the irradiation of ^{241}Am by thermal neutrons.

3 Off-line measurements

The very short half-life of ^{242gs}Am and its very low concentration, make impossible on-line experiments. Mass spectrometry techniques must then be applied. A number of ancillary measurements are nevertheless necessary to obtain a reasonable precision on the final results.

3.1 Neutron Flux determination

A 20 mg natural iron wire, canned in a quartz ampoule, was irradiated together with each ^{241}Am sample to measure the average thermal neutron flux.

During the irradiation of the iron sample, the isotope ^{59}Fe is formed through neutron capture of ^{58}Fe. ^{59}Fe then decays with a half-life of $T_{1/2}$ = 44.5 days to the stable ^{59}Co through beta decay followed by gamma-ray emission. By measuring the absolute gamma-ray intensity of the 1291.6 keV line of the irradiated iron samples, the average neutron flux has been determined [7] with an accuracy of about 4%, which is satisfactory for the present experiment.

The choice of natural iron as a flux monitor was made in consideration of the experimental conditions. The very high temperature inside the H9 beam tube makes the use of any aluminium alloy impossible and the high neutron flux impose a very "diluted" sample to avoid to work with a too high radioactivity.

In Fig. 3 the average neutron flux is plotted as a function of the irradiation time, where the 0 value corresponds to the beginning of the irradiation campaign. We observe a steady decrease of the neutron flux as a function of time, probably due to the movement of the reactor control rod during the cycle. The reactor stability is better than 2% over the considered time interval. For the

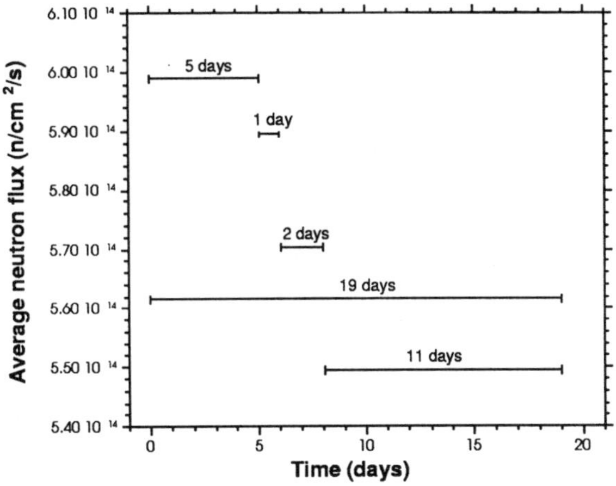

Figure 3: Average thermal neutron flux at the H9 position of the ILL reactor as a function of the time, where 0 corresponds to the beginning of the irradiation. It should be noted that after 6 days of irradiation, the reactor thermal power was decreased from 57 to 52 MW.

two initial values the reactor was operated at the nominal power, before the decrease at 52 MW from the sixth day on.

3.2 X and γ-ray spectroscopy of irradiated samples

The accuracy of the final results depends strongly on the determination of the capture cross section of ^{241}Am and the corresponding branching ratio towards the ground and the metastable states of ^{242}Am.

To determine these values, dedicated X and gamma-ray spectrometry measurements have been carried out using two samples irradiated respectively 30 minutes and 1 day on H9. The principle relies on the determination of the X and gamma activities after 1, 2 and 273 days from the end of the irradiation. In Fig. 4 a schematic view of the first steps of the ^{241}Am evolution is given. ^{242gs}Am is formed either directly by ^{241}Am(n,γ) or by decay of the ^{242m}Am by internal conversion. The ^{242gs}Am then undergoes β-decay to ^{242}Cm (83%) and electron capture to ^{242}Pu (17%).

The decay rate of ^{242gs}Am can then be determined by looking at the K_α lines of ^{242}Pu at 99.5 and 103.7 keV, and at the 42.1 keV γ-line. Complemen-

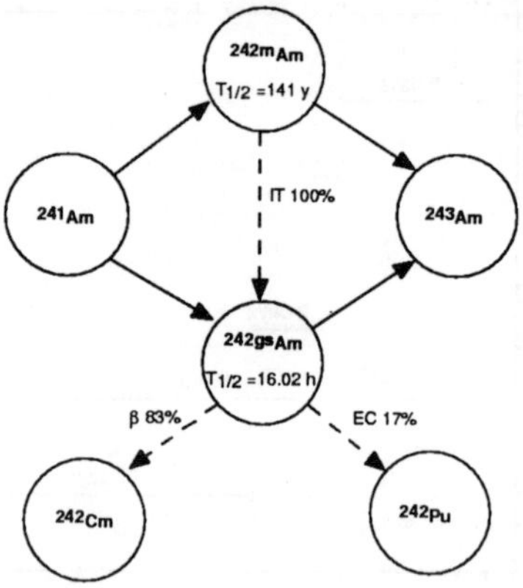

Figure 4: Schematic view of the first steps of the ^{241}Am transmutation chain.

tary information can be obtained by the α-decay of ^{242}Cm to 238Pu observed by looking at the associated 44.1 and 101.9 keV γ-lines.

At short cooling times, the decay of ^{242gs}Am formed by ^{241}Am(n,γ) dominates, with the contribution coming from ^{242m}Am decay being negligible. X and γ-spectroscopy measurements will then provide directly the capture cross section for ^{241}Am(n,γ)^{242gs}Am.

These measurements require a very short irradiation (30 minutes) in order to avoid a huge background coming from the X-rays emitted from the fission fragments.

At long cooling times, the "original" ^{242gs}Am has disappeared ($T_{1/2}$ = 16.02 h) and the ^{242gs}Am present in the sample is formed only by the Internal Conversion of ^{242m}Am ($T_{1/2}$ = 141 y). In this case, X and γ-spectroscopy measurements of a sample irradiated for 1 day on H9 will provide the capture cross section for ^{241}Am(n,γ)^{242m}Am.

The X and γ spectra after a cooling time of 1.9 and 273 days were obtained. The γ-transition at 42.1 keV is strongly suppressed after 273 days as a consequence of the disappearing of the ^{242gs}Am formed during the irradiation.

The final results are :

^{241}Am(n,γ) = (69530) barns

and

$$\frac{\sigma_c(^{241}\mathrm{Am}\rightarrow{}^{242gr}\mathrm{Am})}{\sigma_c^{tot}} = 0.91 \pm 0.01 \ .$$

While the branching ratio is in agreement with the literature value, the total capture cross section is 16% higher than the recommended JEND-3.2 valuation and 22% lower than a recent Japanese measurement [9].

3.3 Mass spectrometry analysis

All samples have been analyzed by Thermal Ionization Mass Spectrometry (TIMS). With this technique, the Am isotope ratios for all samples, as well as those of Cm and Pu for some samples, have been measured on some tens of ng of each element, with a precision ranging from 0.5 % to some percents, depending on the value of the isotopic ratios. For some samples, the isotope dilution analysis with tracers has been applied for the determination of Cm and Pu amounts relative to the initial ^{241}Am quantity.

Due to the isobars coming from the different elements, it was necessary to perform a separation between these elements before TIMS analyses. These separations have been carried out by High Performance Liquid Chromatography (HPLC), so that the composition of different Am, Cm and Pu isotopic chains has been determined with very high accuracy and precision.

In Fig. 5, the isotopic ratio ^{243}Am/^{241}Am is plotted as a function of the irradiation time. The continuous and the dotted lines are respectively calculated for a nominal neutron flux of 5×10^{14} n/s/cm^2 from JEF-2.2 and ENDF-B/VI libraries. The uncorrected experimental data are plotted with a square marker connected by a dashed line to guide the eye. It is evident from these raw data that the value of the ^{242gs}Am cross section is of the order of the ENDF-B/VI evaluation.

A qualitative and semi-quantitative analysis (15 to 20 % uncertainty) has also been carried out by Inductively Coupled Plasma Mass Spectrometry (ICP-MS) to obtain the fission fragment yields. At present the results are still under evaluation and a comprehensive article will be submitted within few months.

4 Results and conclusions

The mass spectrometry data have been analyzed to determine the capture cross sections of ^{242gs}Am. The measured isotopic ratios (^{243}Am/^{241}Am and ^{242m}Am/^{241}Am) for all samples, as well as the Cm, Pu and fission products contents for two samples, have been compared in a global fit with the prediction obtained by solving the Bateman equations describing the evolution of the samples. The values of the ^{241}Am(n,γ) and the corresponding branching ratio were taken from the γ-ray spectroscopy measurements described in the previous section. The preliminary result is:

$^{242gs}\mathrm{Am}(n, \gamma) = (210 \pm 50)$ barns.

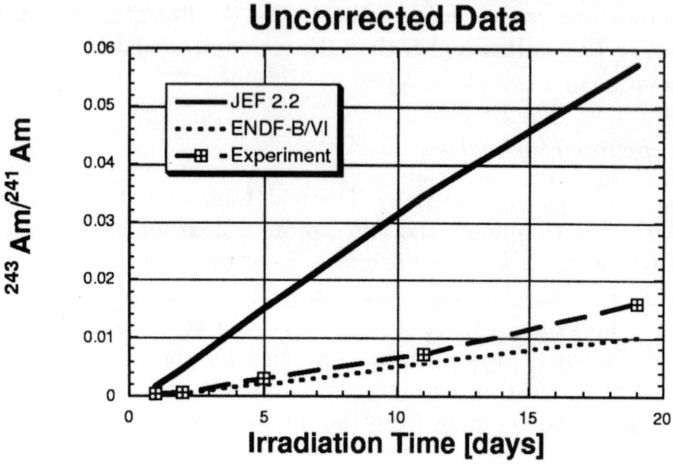

Figure 5: Isotopic ratio ^{243}Am/^{241}Am plotted as a function of the irradiation time. The continuous and the dotted lines are respectively calculated for a nominal neutron flux of 5×10^{14} n/s/cm^2 from JEF-2.2 and ENDF-B/VI libraries. The uncorrected experimental data are plotted with a square marker connected by a dashed line to guide the eye.

An accuracy of 25% is given as the result is still preliminary and further data analysis is still in progress. The preliminary result is in agreement with both the ENDF-B/VI value and with a recent Belarussian evaluation [10]. The option to transmute ^{241}Am together with the minor actinides issued from multi-recycling of a MOX fuel by an intense flux of thermal neutrons is still theoretically open.

Acknowledgments

This experiment would have not been possible without the co-operation and the essential help of the personnel of the ILL-Grenoble. We wish to acknowledge the ILL reactor department in the person of M. Bauer for the competent and efficient support.

References

1. M. Cribier, G. Fioni, R. Legrain, F. Lelievre, S. Leray, A. Pluquet, H. Safa, M. Spiro, Y. Terrien, Ch. Veyssiere: *Report CEA, DAPNIA 97-01*

(1997)

2. *JEF-2.2. The evaluated neutron nuclear data library of the NEA Data Bank. Report IAEA-NDS-129, Rev. 3.* IAEA Vienna, 1996
3. *ENDF/B-VI. The US evaluated nuclear data library for neutron reaction data. Report IAEA-NDS-100, Rev. 6.* IAEA Vienna, 1995
4. *JENDL-3.2. The Japanese Evaluated Nuclear Data Library. Report IAEA-NDS-110, Rev. 5.* IAEA Vienna, 1994
5. T. Nakagawa, S. Igarasi: *Report JAERI M8903.* (1980)
6. Y. Terrien: *private communication*
7. F. Gunsing, G. Fioni: *Report CEA DAPNIA/SPhN 98-07.* (1998)
8. F. Marie, T. Bolognese, M. Cribier, G. Fioni, F. Gunsing, J. Martino, A. Pluquet, M. Spiro: *Gamma Spectroscopy of Neutron Irradiated* ^{241}Am *Samples*, Eds. G. Fioni *et al.*, Amer. Inst. Phys. Conf. Series *(in press)*
9. N. Shinohara, Y. Hatsukawa, K. Hata, N. Kohno, J. Nucl. Sci. Technol. **34**, 613 (1997)
10. V.M. Maslov: *Report INDC(BLR)-007.* IAEA Vienna, 1997

MEASUREMENT OF ^{239}PU(N,F) FRAGMENTS' MASS AND KINETIC ENERGY IN THE RESOLVED RESONANCE REGION

L. DEMATTÈ, F.-J. HAMBSCH AND H. BAX
CEC-JRC, Institute for Reference Materials and Measurements, Retieseweg, 2440 Geel, Belgium

A new measurement on the ^{239}Pu(n,f) fragments' mass and energy characteristics in the resolved neutron resonance region has been started, in order to check the existence of fluctuations of fission fragment energy and mass characteristics, as shown by the $\overline{\nu}_p$ measurement. The experimental set-up is described and some first results are presented and discussed.

1 Introduction

^{239}Pu is belonging to the major actinides and its importance is growing due to its use in mixed oxide (MOX) fuel elements. In the past, fluctuations of the average number of prompt neutrons, $\overline{\nu}_p$, have been observed as a function of resonance neutron energies [1]. In a previous experiment [2] on ^{235}U(n,f) fluctuations in the average pre-neutron total kinetic energy ($\overline{\mathrm{TKE}}$) of the fragments, related to two-dimensional mass and TKE distribution changes, were also observed and were associated to the observed fluctuations in $\overline{\nu}_p$ of that system. A similar finding in the ^{239}Pu case would be of great interest since it would enhance the reliability of the evaluated data.

The ^{239}Pu(n,f) reaction at low neutron energy seems also to be a good candidate for the investigation of possible influences of the transition states (Bohr's channels) on the fragment characteristics. This is due to the fact that the only possible low-energy spin states ($J^\pi=0^+$, 1^+) belong to two well separated (about 1.25 MeV according to Ref. [3]) transition state bands, thus possibly having a much weaker mixing compared to other systems like, e.g. ^{235}U(n,f). However, previous experiments [3,4] being performed with surface barrier detectors with a lower intrinsic energy resolution compared to ionization chambers (IC), showed controversial results. Hence, an experiment using the IC technique seems to be more appropriate to look for the expected small differences in the fission fragment mass and kinetic energy distributions.

In addition, the ^{239}Pu(n,f) shows a relatively strong (n,γf) competition effect on the average prompt neutron ($\overline{\nu}_p$) and gamma ($\overline{\nu}_\gamma$) multiplicities, and on the average gamma energy ($\overline{E}_\gamma$) [1]. It seems very interesting to investigate whether a similar effect can be seen also on other fragment observables.

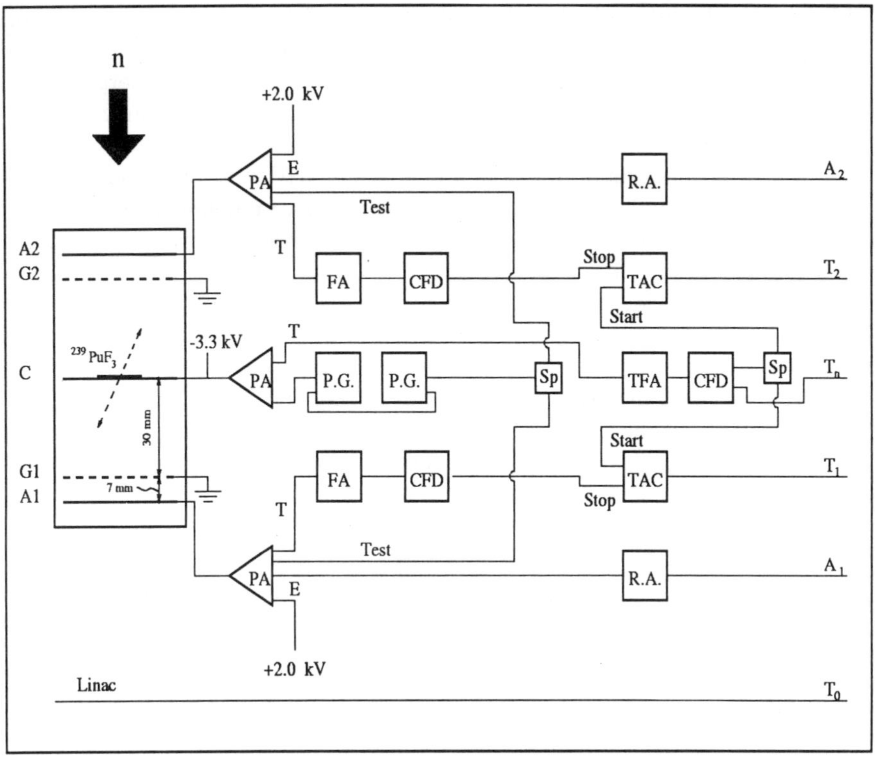

Figure 1. Experimental set-up.

In the following, the experimental set-up together with the data analysis is presented and first results are discussed.

2 Experimental set-up

A sample consisting of about 500 μg of Pu (^{239}Pu enrichment: 99.99%) evaporated on a very thin gold coated polyimide backing was prepared by the Sample Preparation Group of IRMM, and was placed on the common cathode of a twin IC. As ionizing gas, it was chosen to use pure CH_4 slightly above atmospheric pressure (1.2×10^5 Pa), constantly flowing at a rate of 0.1 l/min.

The chamber was placed at about 9.4 m from the rotating U target of the Geel Electron LINear Accelerator (GELINA) of the IRMM, which provides a pulsed neutron beam, thus allowing the determination of the neutron energy by time-of-flight (TOF) technique with a very high energy resolution. Since the neutron beam has a white spectrum, the fragments' energies of the resolved resonances are being measured simultaneously thus ensuring exactly the same experimental conditions. A schematic view of the IC together with the associated electronic devices can be seen in Fig. 1. The very high α-activity of the sample (~1MBq), did not allow us to implement in the set-up the pile-up rejection scheme described in Ref. [5]. This, however, will not introduce a systematic error in the fragment energy measurement at the various resonances since the system is calibrated by means of the ^{239}Pu(n_{th},f) reaction using the sample itself, and the fraction of fission events affected by α pile-up depends only on the activity. A paraffine block of 4 cm of thickness was used to thermalize the neutron beam during the calibration runs.

In addition to the anode signals, carrying the information on the fragment energy, also the time difference between cathode and anode signals is recorded, since it allows the determination of the fragments' emission angles [9].

The slow signals A_1, A_2, T_1 and T_2 are digitally converted by four SILENA 7423 UHS ADC's, while the two fast signals needed for the TOF measurement (T_n,T_0) are processed by a time digitizer especially developed at IRMM. Data are acquired in list-mode by means of a MPI 8000 data acquisition system of SEND GmbH. The on-line monitoring and transfer of data to a SUN workstation is accomplished by the PV-WAVE based LISA package [6].

3 Data analysis

As usual in a double energy measurement, the pre-neutron energies of the fission fragments can be deduced from the momentum- and mass-conservation relations. To this goal, firstly one has to introduce some corrections for grid inefficiency, energy loss in sample and backing and neutron evaporation (for a detailed discussion see refs. [2,7]; neutron emission data are taken from ref. [8]). The question concerning the existence of an intrinsic mass-dependent pulse-height-defect (PHD) in methane seems still controversial [9,10] since no measurement, like e.g. Ref. [11] in the case of 90%Ar+10%CH_4, have been made up to now. For this reason we decided to analyse the data, for the time being, without introducing a PHD correction. As it will be said in the next section, such correction, however, seems to be necessary. On the other hand, since the resonance data are renormalized to the thermal case, the influence of the PHD on the global fragment's characteristics is expected to be negligible.

As said in section 2, the calibration of the IC was obtained using the ^{239}Pu(n_{th},f) and transforming the centroids of the light and heavy fragment

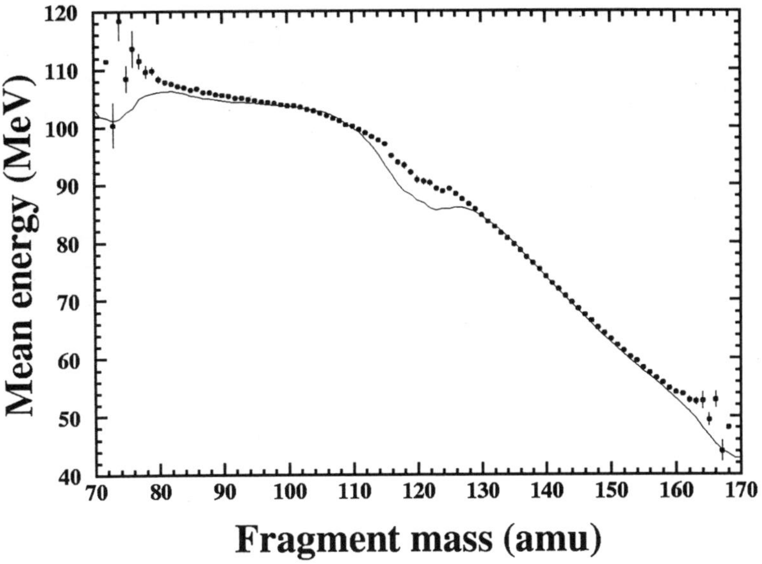

Figure 2. Mean fragment energy as a function of fragment mass compared with the results of ref. [13] (solid line).

channel spectra, into the corresponding mean light and heavy post-neutron energies reported in literature [9]. As a final remark it must be noted that this calibration must be repeated regularly in order to ensure the long-term stability of the measurement.

4 Results and discussion

The average properties of the thermal fission of ^{239}Pu are reported in Table. 1.

It can be noted that our results are very close to literature values[a]: only the mass width seems to be somewhat larger than expected. This discrepancy arises to a higher degree, also in the mass-energy correlations, as it can be

[a]The values of the pre-neutron energies do not correspond *exactly* to the literature ones used as a reference (see Table 1) because the calibration was performed on the post-neutron energies, i.e. before to perform the correction for neutron emission.

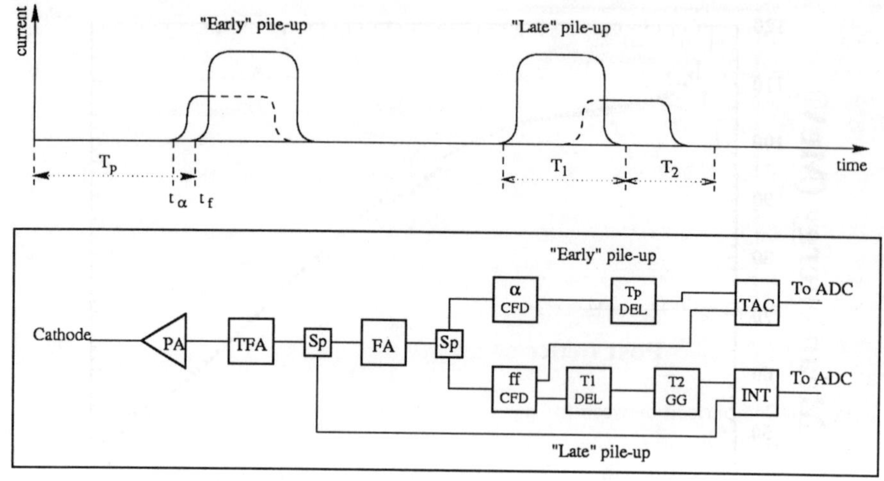

Figure 3. Schematic view of the α pile-up rejection scheme.

Table 1. Global characteristics of ^{239}Pu(n_{th},f).

	This work	Ref. [9]	Ref. [13]	Ref. [14]
$\langle TKE^* \rangle$	177.94±0.03	177.97	177.35	177.7
σ_{TKE^*}	10.91	11.00	11.79	11.09
$\langle M_H^* \rangle$	139.39±0.02	139.26	139.48	139.66
σ_{M^*}	6.63	5.88	6.50	6.01
$\langle E_L^* \rangle$	103.20±0.02	103.13	102.92	103.2
$\sigma_{E_L^*}$	5.81	5.42		
$\langle E_H^* \rangle$	74.75±0.02	74.84	74.43	74.5
$\sigma_{E_H^*}$	8.08	7.76		

seen from Fig. 2, and is most probably due to the combined effect of the PHD neglection, that causes the change in slope clearly visible in the light fragment region, and the strong α-background. While the first discrepancy is expected to disappear when introducing a PHD correction, the latter problem can be solved only applying an efficient pile-up rejection system. Such a system is currently being under development and it is sketched in Fig. 3. The cathode signal is split and analysed separately looking for "early" pile-up and "late" pile-up with respect to a fission fragment event. The first kind of pile-up can

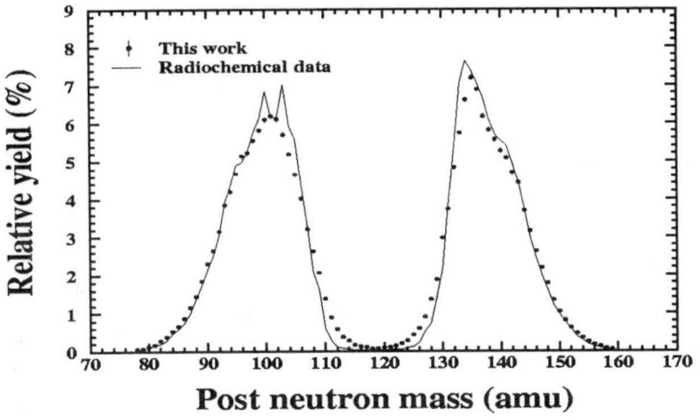

Figure 4. Comparison between the deduced post neutron mass distribution and radiochemical data (from [8]).

be discriminated recording the time difference between a start signal triggered by a fission event and a (delayed) stop signal triggered by an α particle. The second kind of pile-up can be discriminated by integrating the cathode current signal for a time T_2, starting from a time T_1 after the beginning of a fission fragment event. The amount of delay to be introduced in the stop signal and the sum $T_1 + T_2$ should be of the order of the pulse pair resolution time of the system. The actual discrimination is performed off-line. Preliminary tests seem to indicate that the amount of pile-up can be decreased by a factor of five or more. The efficiency, however, decreases to about 0.35. These new data will soon be analyzed in order to check whether this pile-up suppression system can improve the resolution of our system.

In Fig. 4 the deduced post-neutron mass distribution is compared with radiochemical data [8] showing, despite of the discrepancies outlined above, a nice agreement.

The mean mass and energy characteristics were obtained for twenty low energy resonances ranging from 7 to 60 eV (see Fig. 5). It has to be noted that the resonances at 41.46 and 41.74 eV, and also at 49.58 and 50.14 eV cannot be individually resolved. Numerical results and associated statistical uncertainties are reported in Table 2 (resonance data are taken from JEF-2.2) and are plotted in Fig. 6 and 7. Despite of the still large statistical errors one can observe the presence of fluctuations of the order of 1 MeV for $\overline{\mathrm{TKE}}$ and 0.5 amu for $\overline{\mathrm{M_H}}$. Looking at the first 9 resonances (i.e. those with presently better

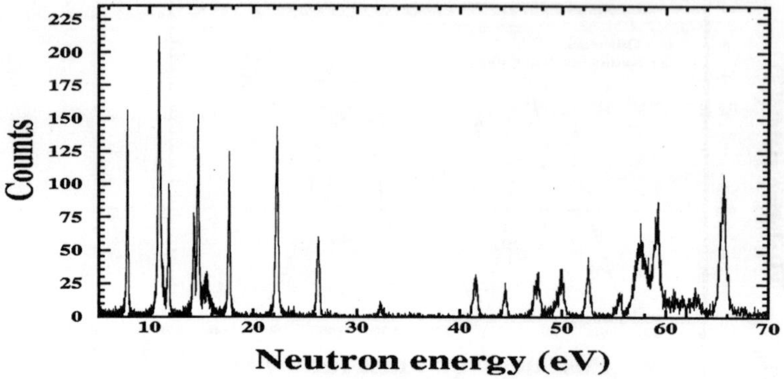

Figure 5. Neutron energy spectrum of ^{239}Pu(n,f) between 5 and 70 eV.

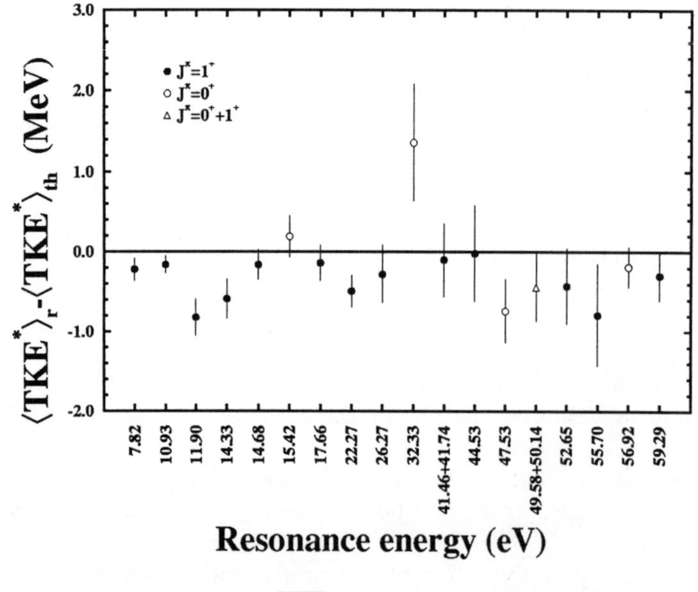

Figure 6. Fluctuations of $\overline{\mathrm{TKE}}$ plotted as the difference with the thermal value.

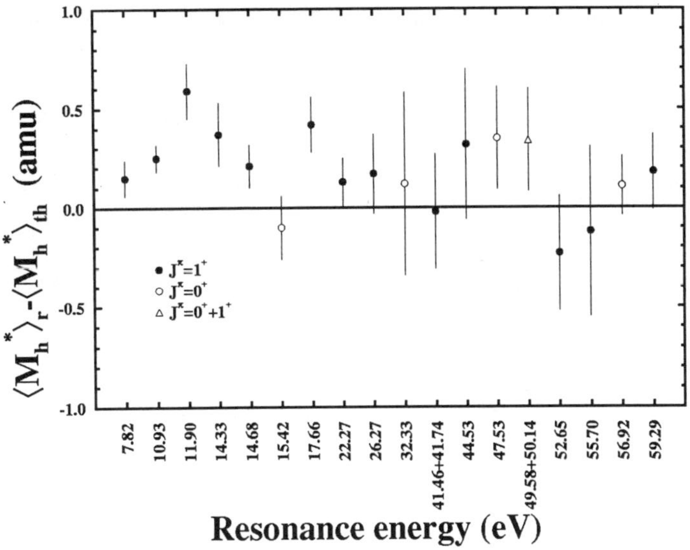

Figure 7. Same as Fig. 6 but for $\overline{M_H}$.

statistics), it appears that fluctuations in total kinetic energy are associated with fluctuations in mass asymmetry. This could be a first indication of the existence of a fission mode competition, namely Standard I (less asymmetric and with a higher TKE) and Standard II (more asymmetric and with a lower TKE). A similar interpretation was already given for $^{235}U(n,f)$ [2] and for the series of spontaneously fissioning Pu isotopes [13]. A more quantitative analysis however, needs a much better statistics, and hence the data taking is going on.

5 Summary

A new measurement of the fragment mass and energy characteristics of the $^{239}Pu(n,f)$ in the resolved resonance region has been started at the IRMM of Geel. The adopted experimental technique is the double energy measurement with a double Frisch-gridded ionization chamber. A high resolution neutron beam is provided by the GELINA accelerator. First results on twenty low-energy resonances, seems to indicate the existence of fluctuations both in $\overline{TKE}$

Table 2. Mean pre-neutron total kinetic energy and heavy fragment mass of 20 resolved resonances.

Res. energy (eV)	J^π	$\langle$TKE$\rangle$	$\langle M_H \rangle$	Res. energy (eV)	J^π	$\langle$TKE$\rangle$	$\langle M_H \rangle$
7.82	1^+	177.72(14)	139.54(09)	41.46	1^+	177.84(46)	139.37(29)
10.93	1^+	177.78(11)	139.64(07)	41.74	1^+		
11.90	1^+	177.12(23)	139.98(14)	44.53	1^+	177.92(60)	139.71(38)
14.33	1^+	177.35(25)	139.76(16)	47.53	0^+	177.20(40)	139.74(26)
14.68	1^+	177.78(19)	139.60(11)	49.58	0^+	177.50(43)	139.73(26)
15.42	0^+	178.13(26)	139.29(16)	50.14	1^+		
17.66	1^+	177.80(22)	139.81(14)	52.65	1^+	177.51(47)	139.16(29)
22.27	1^+	177.45(20)	139.52(12)	55.70	1^+	177.15(64)	139.27(43)
26.27	1^+	177.66(36)	139.56(20)	56.92	0^+	177.75(25)	139.50(15)
32.33	0^+	179.30(72)	139.51(46)	59.29	1^+	177.64(31)	139.57(19)

and $\overline{M}_H$, in analogy with the experimental findings on ^{235}U(n,f) [2]. Due to the limited statistics an estimation of the influence of the transition states is up to now impracticable.

References

1. J. Fréhaut and D. Shackleton in *Proc. III IAEA Symp. on Physics and Chemistry of Fission*, vol. II (1974) p. 201.
2. F.-J. Hambsch *et al.*, Nucl. Phys. **A491** (1989) 56.
3. R.L. Walsh *et al.* in *Proc. IV IAEA Symp. on Physics and Chemistry of Fission*, vol. II (1980) p. 129.
4. C. Wagemans *et al.*, *ibid.*, p. 143.
5. C. Budtz-Jørgensen *et al.*, Nucl. Instr. & Meth. **A258** (1987) 209.
6. A. Oberstedt and F.-J. Hambsch, Nucl. Instr. & Meth. **A340** (1994) 379.
7. G. Barreau *et al.*, Nucl Phys. **A432** (1985) 411.
8. A. Wahl, At. Data & Nucl. Data Tables **39** (1988) 1.
9. G. Simon *et al.*, Nucl. Instr. & Meth. **A286** (1990) 220, G. Simon, Thèse de doctorat, Université de Paris Sud (1990).
10. F.-J. Hambsch *et al.*, Nucl. Instr. & Meth. **A351** (1995) 257.

11. P.T. Nielsen and N.Rud, University of Aarhus, private communication (1984).
12. P. Geltenbort, Ph.D. thesis, University of Tübingen (1985).
13. L. Demattè *et al.*, Nucl. Phys. **A617** (1997) 331, L. Demattè, Ph.D. thesis, University of Gent (1996).
14. J. Neiler *et al.*, Phys. Rev. **149** (1966) 894.

SIGNATURES OF FISSION DYNAMICS DEDUCED FROM HIGHLY EXCITED FRAGMENTS IN PROTON-INDUCED REACTIONS

J. Benlliure [a]
University of Santiago de Compostela, E-15706 Santiago de Compostela, Spain

P. Armbruster, T. Enqvist, F. Farget, K.-H. Schmidt
Gesellschaft für Schwerionenforschung, Planckstrasse 1, 64291 Darmstadt, Germany

M. Bernas, L. Tassan-Got, C. Stéphan
Institut de Physique Nucléaire, F-91406 Orsay Cedex, France

A. Boudard, R. Legrain, S. Leray, C. Volant
DAPNIA/SPhN, CEA/Saclay, F-91191 Gif sur Yvette Cedex, France

J.P. Dufour
CENBG, F-33175 Gradignan Cedex, France

Interactions of ^{197}Au projectiles of 800 A.MeV with protons which lead to fission are investigated in order to study nuclear viscosity of highly excited nuclear matter. The formation cross sections of all fission residues which are fully identified in nuclear charge and mass are measured. From the width of the fission-fragment charge distribution, the mean excitation energy at saddle is deduced. The influence of nuclear viscosity on the fission dynamics is clearly observed

1 Introduction

Nuclear fission constitutes the most clear example of a large-scale collective motion of nuclear matter. The study of the dynamics of this process at high excitation energy is expected to provide information about the nature and magnitude of nuclear dissipation. This problematic was first pointed out by Kramers [1] when he proposed that the fission probabilities could be affected by the nuclear viscosity, in contrast to the approach of Bohr and Wheeler [2].

From an experimental point of view, different techniques have been used to evidence the influence of the nuclear viscosity on the fission process. γ emission of the giant dipole resonance, pre-scission neutron, or light-particle evaporation, or Coulomb proximity effects indicate that fission is a slow process [3].

[a] E-mail: benlliur@fpddux.usc.es

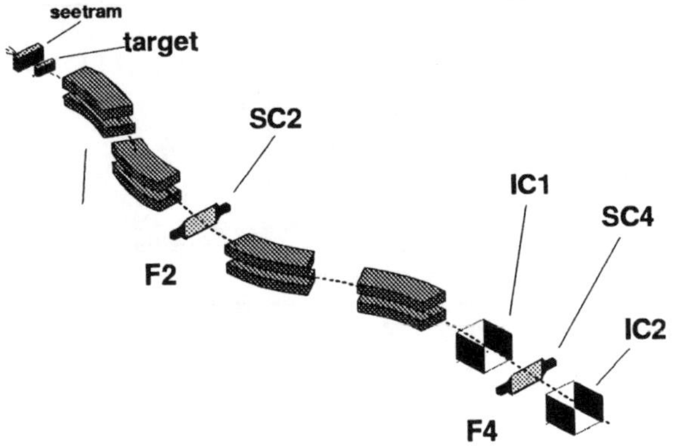

Figure 1: Schematic plot of the projectile-fragment separator (FRS) at GSI with its standard detection set up.

The experimental technique used in this work allows to identify all the reaction products in mass number A and nuclear charge Z by using inverse kinematics and a magnetic in-flight separation. This very specific information is used to characterize the fissioning nuclei at saddle and to deduce signatures of the nuclear dissipation. The selected reaction ^{197}Au (800 A.MeV)+p allows to study the nuclear fission process at high excitation energy and low angular momentum. Furthermore, the fact that the configurations at saddle and at scission are rather similar for nuclei in the region of ^{197}Au simplifies the description of the fission dynamics.

2 Experiment

The experiment was performed at the SIS synchrotron accelerator of GSI Darmstadt. A ^{197}Au beam was accelerated up to 800 A MeV with a typical intensity of $1 \cdot 10^7$ ions/s. The beam impinged on a liquid H_2 target of 85 mg/cm^2 thickness located at the entrance of the projectile-fragment separator (FRS). A second dummy target was used to subtract the contributions from the H_2 target windows. A secondary-electron current monitor allowed to determine the intensity of incoming projectiles during the whole experiment.

The projectile-fragment separator was tuned in its achromatic mode [4] in order to determine the magnetic rigidity of the particles passing through from their positions at the two image planes (F2 and F4 in fig.1). These positions were recorded by means of two position-sensitive plastic scintillators (SC2 and SC4). The nuclear charge and the mass number of any particle passing through

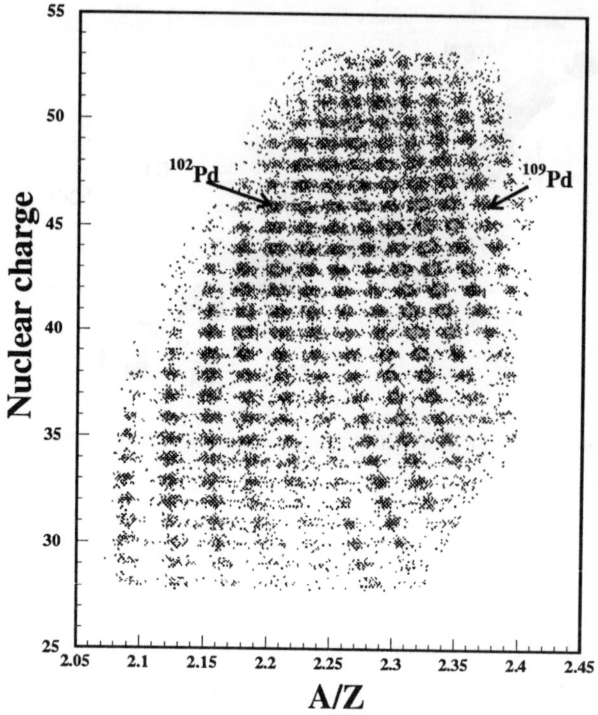

Figure 2: Z-A/Z scatter plot of fission residues observed for a typical tuning of the fragment separator for the reaction ^{197}Au (800 A MeV) + p.

the FRS could be determined from its magnetic rigidity, the time-of-flight (ToF), provided by the scintillators, and the energy loss, measured with a multi-sample ionization chamber (IC1) located at the exit of the FRS.

In figure 2 we illustrate the high Z and A resolution obtained for a typical tuning of the FRS used to identify fission residues.

The measurement of the magnetic rigidity provides also the velocity of the transmitted particles. In the case of the fission residues, and due to the angular acceptance of the spectrometer ($\theta < 15$ mrad), only the forward and backward emitted fragments can be observed. In figure 3 we represent the velocity distribution of Zr isotopes in the center-of-mass frame as a function of

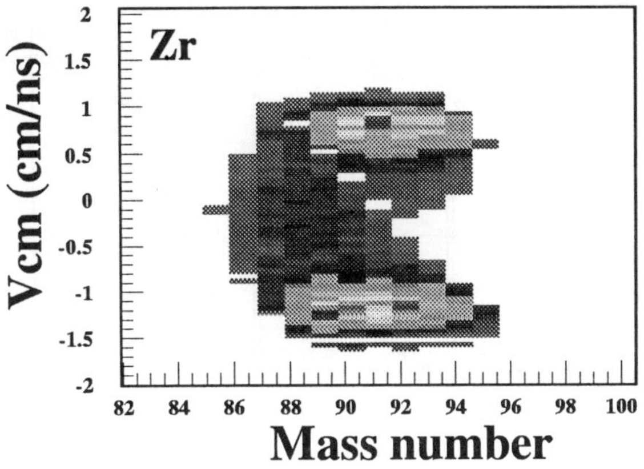

Figure 3: Velocity distribution in the frame of the projectile as a function of the neutron number for all the Zr isotopes produced in the reaction ^{197}Au(800 A MeV) + p.

their neutron number. In this picture we can disentangle clearly forward fission at positives velocities, backward fission at negatives ones, and fragmentation residues around the center-of-mass velocity.

3 Results

In order to determine the absolute production cross sections of the reaction residues, the measured yields were corrected for dead time, secondary reactions in the scintillator at the intermediate image plane and transmission of the particles through the FRS.

In figure 4 we have report on a chart of the nuclides all the isotopes measured as resulting from the interaction between the ^{197}Au at 800 A MeV and the proton. In order to understand how the residual fragments are distributed on the chart of the nuclides first we should consider these proton-induced reactions can be described as a two-step process; first, a fast interaction between the proton and the nucleus leading to the formation of a highly excited prefragment and, secondly, the slow deexcitation of this prefragment by particle evaporation or fission. In the upper part of the chart of the nuclides we can identify the evaporation residues which survive fission while in the lower part

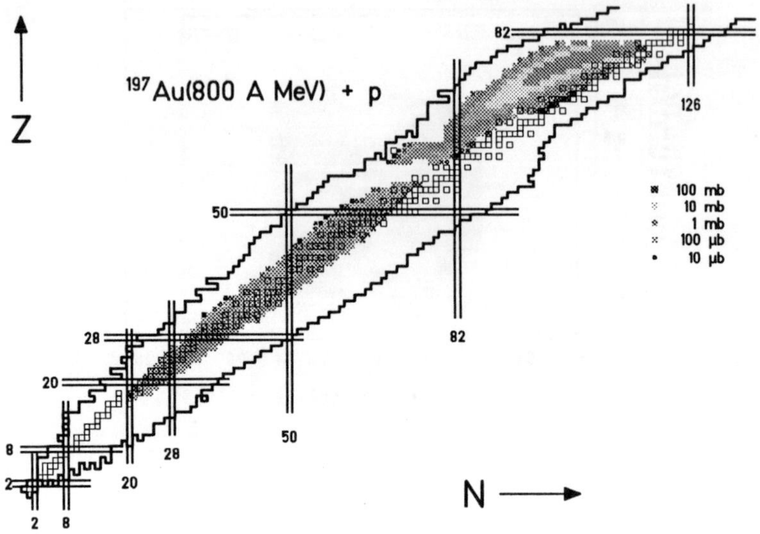

Figure 4: Measured reaction residues from ^{197}Au(800 A MeV) + p collisions on a chart of the nuclides.

we observe the fission residues. In the following we will concentrate on the fission residues. A more detailed description of the evaporation residues can be found in reference [5].

In contrast to the fission of actinides, in this case the fission residues are characterized by a symmetric mass and charge distribution. Another important feature is that the mean A/Z ratio of the fission residues is considerably lower than the one of the projectile. Taking into account that the prefragments produced in the first interaction between the proton and the ^{197}Au nucleus preserve the same A/Z ratio of the projectile, we can conclude that the fission process for these collisions is associated with the formation of highly excited prefragments since many neutrons are emitted. Now we have two possible scenarios. Maybe these highly excited prefragments fission and then the fission residues deexcit by evaporating neutrons, or the fission is inhibited at high excitation energy, then the prefragments deexcite by evaporating neutrons and finally they fission at lower energy. In the following, we will try to answer this question from the detailed analysis of our data.

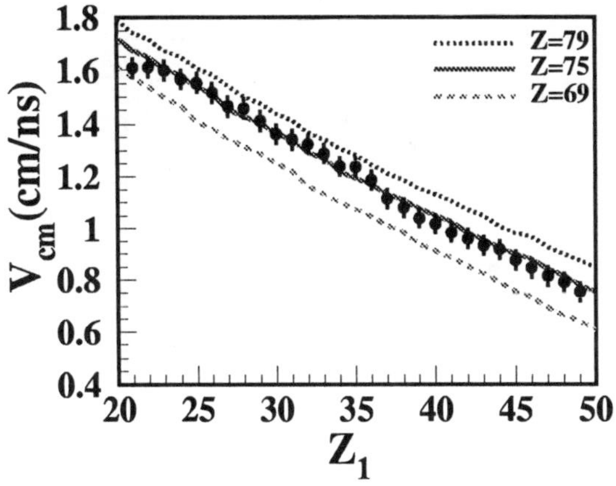

Figure 5: Mean velocity in the center-of-mass as a function of the nuclear charge for the fission residues produced in the reaction ^{197}Au(800 A MeV) + p.

4 Characterization of the fissioning nuclei at saddle

The fission process can be characterized by the properties of the fissioning system at saddle and its dynamical evolution down to scission. In our case, saddle and scission point appear at rather similar deformations, and thus we can neglect the effects of this evolution. From the analysis of our experimental observables we can deduce the following properties of the fissioning system.

The final velocity of the fission residues is directly related to the charge of the fissioning nucleus at scission. In figure 5 we report the mean measured velocity of the fission residues as a function of their nuclear charge. These data are compared with different Coulomb repulsion calculations according to the expression:

$$E_c = \frac{e^2 Z_1 Z_2}{R_1 + R_2 + 2fm} \tag{1}$$

where $Z_2 = Z - Z_1$, Z_1 is the charge of the observed fragment and Z is the charge of the fissioning system at scission. The radii of the nuclei are defined as:

$$R_i = r_o A_1^{1/3} \left(1 + \frac{2\beta}{3}\right) \tag{2}$$

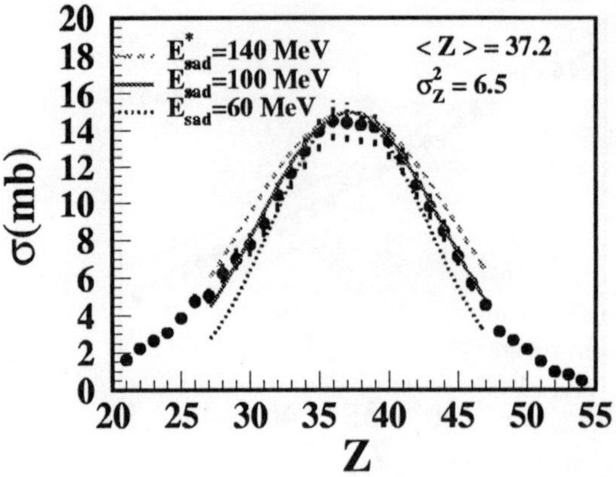

Figure 6: Charge distribution of fission residues from the reaction ^{197}Au(800 A MeV) + p. The experimental data are compared with calculated charge distributions assuming different excitation energies above the barrier (see text).

where r_o= 1.16 fm and β = 0.625. The parameters have been adjusted to measured total kinetic energy values of a variety of fissioning nuclei [6].

The comparison between the measured and the calculated velocities for different initial charges shows the sensitivite of the velocities to the charge of the system at scission. From this comparison we can deduce that the charge of the fissioning system at scission, and from the previous considerations, at saddle is around Z=75 with a precision of about 1 unit. Another important conclusion that we obtain from this picture is that the experimental velocities are well described in the full range of charge considered assuming the same charge at scission. This observation indicates that the charge distribution of the fissioning nuclei is quite narrow, in other words, only few different nuclei seem to fission. Otherwise one would expect heavier fission fragments to be produced by heavier fissioning nuclei and lighter fission fragments by lighter fissioning nuclei.

Because of the previous results indicating that only few different nuclei are fissioning we can relate the width of the charge distribution of fission residues to the excitation energy of the fissioning nuclei at saddle according to the expression:

$$\sigma_Z^2 = \frac{1}{2}\frac{\sqrt{E^*_{sad}}}{\sqrt{a}C_{mac}} \qquad (3)$$

where E^*_{sad} is the excitation energy above the barrier, a is the level-density parameter and C_{mac} is the curvature of the potential energy at the saddle point as a function of the charge asymmetry as defined in reference [6]. From figure 6, we obtain that the mean excitation energy above the barrier (E^*_{sad}) is around 100±20 MeV. The total excitation energy of the fissioning nucleus will be obtained by taking into account the fission barrier (E^*_o = 120±20). This information is crucial in order to conclude about the role of nuclear dissipation in fission. The sensitivity of the method is shown in figure 6 by comparing the data with the expected width of the charge distributions at different excitation energies. Since the sum of the mean charges of the final fission fragments (Z =74.4) is very close to the calculated charge at saddle (Z_{sad}=75), the influence of the post-scission proton evaporation process on the width of the charge distribution can be disregarded.

Finally, we can deduce the mass of the fissioning system at saddle from the measured mean value of the mass of the final fission residues by taking into account the post-scission particle evaporation due to the excitation energy at saddle. From these considerations we obtaine a value for the mass of the fissioning system at saddle around A_{sad}=181±2.

Because of the correlation between the excitation energy and the mass of the fissioning system at saddle, both parameters are determined by means of an iterative procedure.

5 Dissipation in high-energy fission

In order to conclude about the fission dynamics and the role of nuclear dissipation we can compare our experimental information with model calculations.

Proton-induced reactions with heavy nuclei can be described by means of an intranuclear-cascade model which describes the fast interaction between the proton and the heavy nuclei, and the later deexcitation step including the description of particle evaporation and fission.

In our case we use the Cugnon cascade [7] while the statistical evaporation is described by the Weisskopf formalism according to references[8,9]. The fission width is described by the Bohr-Wheeler model[2] but we introduce a dynamical approach following the ideas of Kramers and Grangé[1,10]. This model describes fission as a dissipative process where the introduction of the nuclear viscosity

Table 1: Comparison between the experimental information which characterize the fission process: initial excitation energy (E_i^*), mean excitation energy, Z and A of the fissioning nuclei (E_o^*(MeV), Z_{sad}, A_{sad}), mean total numbers of protons and nucleons in both final fission fragments (Z_f, A_f) and total fission cross section (σ_f), with three model calculations based on the Bohr-Wheeler model (Γ_{BW}) and including the fission hindrance due to the nuclear viscosity with two different time dependences (Γ_{step}, Γ_{exp}). β=0 indicates that the Bohr-Wheeler approach was used.

	E_i^*(MeV)	E_o^*(MeV)	Z_{sad}	A_{sad}	Z_f	A_f	σ_f(mb)
data		120±20	75±1	181±2	74.4±1	170±2	105±22
Γ_{BW} β=0	298	247	78	189	75	170	530
Γ_{step} β=1	260	126	76	181	76	171	29
Γ_{exp} β=2	300	200	77	185	76	172	92

leads to a delay of the fission channel and then a suppression of fission at high excitation energy.

Two different forms of the time dependence of the fission decay width have been used. One is a simple step function as used in reference [12], which opens the fission decay channel only at a time τ after the original formation of the system. The second is an exponential ingrowth function of the form $\lambda(t) = \lambda[1 - exp(-t/\tau)]$ as in reference [13].

Finally, and in order to compare with our data, we introduce also a description of the fission-fragment properties and in particular its mass distribution and A/Z ratio following reference [14].

The results from the calculations including the different descriptions of the fission decay width: Bohr-Wheeler model (Γ_{BW}), dissipative step function time dependence (Γ_{step}) and dissipative exponential function time dependence (Γ_{exp}), are shwon in table 1.

The comparison with the experimental values clearly indicates that the most sensitive parameters to the different model calculations are the excitation energy of the fissionning system (E_o^*) and the total fission cross section (σ_f). From this comparison we can also conclude that the calculation based in the pure statistical Bohr-Wheeler model is far to describe the data while the inclusion of the fission hindrance by means of the nuclear dissipation gives a better description. The same conclusion can be drawn from figure 7 where we compare some measured isotopic distributions of fission residues with two model calculations where the fission hindrance is included (Γ_{exp}) or not (Γ_{BW}). The overestimation of the fission channel when the nuclear viscosity is not in-

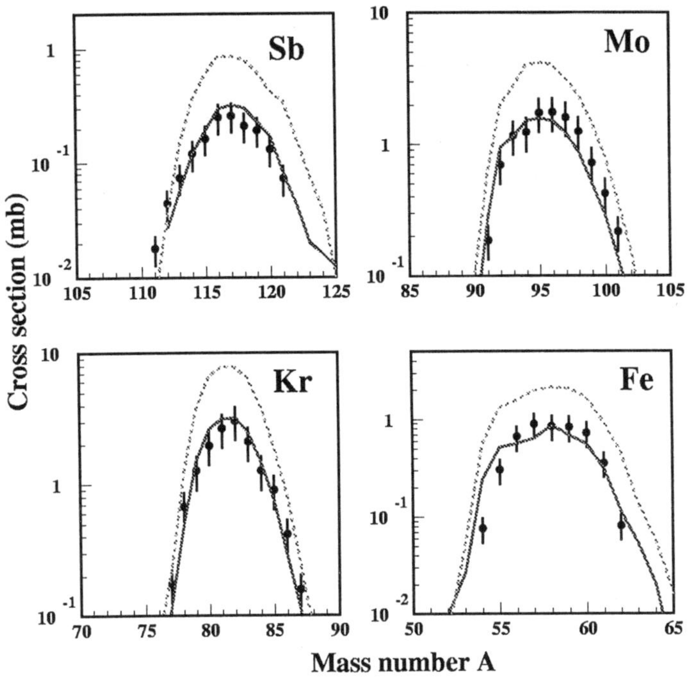

Figure 7: Measured isotopic distribution of different fission fragments (black points) compared with two fission model calculations using the Bohr-Wheeler model (dashed line) and including the nuclear dissipation (Γ_{exp})in the description of the fission dynamics (solid line)

cluded in the description of the fission channel is evident.

Another important feature that we obtain from the comparison of the experimental data with our model calculations is that the fission channel appears in a relatively small range in excitation energy. This effect can easily be understood in terms of the nuclear fission barrier and the role of the nuclear viscosity. Fission at low excitation energy is not observed because the fissioning nuclei have rather large fission barriers ($B_f > 20$ MeV). On another hand, the nuclear viscosity causes a suppression of the fission channel at high excitation energy.

Although we have a clear evidence for the influence of the nuclear viscosity in high-energy fission, it is a more difficult task to characterize the time dependence of the fission decay width ,which will allow to obtain quantitative information about the nuclear viscosity. The dependence of the absolute value of the nuclear viscosity (β) on the time evolution of the fission decay width is clear from the calculations perfomed with the two different time approaches (Γ_{step}, Γ_{exp}). From this comparison in table 1 we can deduce that the exponential dependence for the time evolution of the fission decay width (Γ_{exp}), does not seem to be realistic because the fission hindrance at high-excitation energy is not sufficient. The step-function approach seems to be more realistic although the final fission cross sections are underestimated. However, this discrepancy may be traced back to the influence of angular momentum which was neglected in our calculation.

6 Conclusion

Fission dynamics at high excitation energy and low angular momentum was investigated by means of ^{197}Au on proton induced reactions at 1 A.GeV. More than 370 fission fragments were identified in atomic number and nuclear charge.

For the first time, the width of the charge distribution together with the kinematical properties of the fission residues were used to characterize the fissioning system at saddle. Comparison with model calculations indicates a clear evidence for the supression of the fission decay channel at high excitation energy. From this comparison we can conclude that 60% of the excitation energy of the fissionning nuclei is evacuated by pre-scission particle evaporation indicating that fission occurs in a small range of excitation energy as a consequence of the large fission barriers and the role of the nuclear viscosity.

The deduced excitation energy of the fissioning system at saddle seems to be an important constraint in order to conclude about the time dependence of the fission decay width and then about the absolute value of the nuclear viscosity.

References

[1] H.A. Kramers Physica VII 4 (1940) 284
[2] N. Bohr, J.A. Wheeler Phys. Rev. 56 (1939) 426
[3] D. Hilscher, H. Rossner Ann. Phys. Fr. 17 (1992) 471
[4] H. Geissel et al. Nucl. Instrum. Methods B70 (1992) 286
[5] L. Tassan-Got et al.Proc. of the International Conference on the Physics of the Nuclear Science and Technology, October 1998, Island Mariott Long Island, USA
[6] C. Böckstiegel et al. Phys. Lett. B398 (1997) 259
[7] S. Mulgin et al. Nucl. Phys. A640 (1998) 375
[8] J. Cugnon et al. Nucl. Phys. A620 (1997) 475

[9] J.J. Gaimard, K.-H. Schmidt Nucl. Phys. A531 (1991) 709
[10] A.R. Junghans et al. Nucl. Phys. A629 (1998) 635
[11] P. Grangé et al. Phys. Rev C27 (1983) 2063
[12] A.V. Ignatyuk et al. Nucl. Phys. A593 (1995) 519
[13] R. Butsch et al. Phys. Rev C44 (1991) 1515
[14] J. Benlliure et al. Nucl. Phys. A628 (1998) 458

INVESTIGATION OF FISSION TIME DISTRIBUTIONS BY THE BLOCKING TECHNIQUE IN SINGLE CRYSTALS

M. MORJEAN*, M. CHEVALLIER†, C. COHEN‡, D. DAUVERGNE†, J. DURAL§, J. GALIN*, F. GOLDENBAUM*, D. JACQUET¶, R. KIRSCH†, A. L'HOIR‡, E. LIÉNARD*, B. LOTT*, Y. PÉRIER*, J.C. POIZAT†, G. PRÉVOT‡, J. REMILLIEUX†, D. SCHMAUS‡, M. TOULEMONDE§

*GANIL DSM/CEA, IN2P3/CNRS ,BP 5027, 14076 Caen Cedex 5, France

†Institut de Physique Nucléaire de Lyon, IN2P3/CNRS, Univ. Cl. Bernard, 43 Bd. 11 Novembre 1918, F-69622 Villeurbanne Cedex, France

‡GPS,2 place Jussieu, 75251 Paris Cedex 05, France

¶Institut de Physique Nucléaire d'Orsay,BP 1, F-91406 Orsay Cedex, France

§CIRIL, BP 5133, 14040 Caen Cedex, France

The blocking dips associated with fission fragments of uranium-like nuclei with initial excitation energies smaller than 250 MeV point to fissions occurring with sizeable cross-sections at very low residual excitation energies.

1 Introduction

Immediately after the discovery of blocking phenomena in single crystals [1], their application to lifetime measurements was suggested [2]. This technique has been applied to fission lifetime measurements [3,4,5,6], leading at low excitation energies ($E^* < 10$ MeV) [3] to results in very satisfactory agreement with the Bohr and Wheeler's statistical model predictions. However, at higher excitation energies, analyses of the blocking patterns associated with fission fragments concluded [4,5] to the existence of long lifetime components in the fission time distributions. The experimental evidence for long lifetime components arose from the observed discrepancy between the measured blocking patterns for fission fragments and the one measured for elastically scattered nuclei, after a scaling has been performed to take into account the respective energies and charges. In the experiment reported here, the evolution with the initial excitation energy of the blocking pattern associated with fission fragments has been measured during the very same experiment, thus permitting a reliable determination of the actual blocking technique limit for short fission times. Blocking pattern simulations have permitted an investigation of the fission time distributions for excitation energies smaller than 250 MeV.

2 Experimental techniques

2.1 Experimental set-up

A detailed description of the experimental set-up can be found in references [7,8]. A silicon single crystal has been bombarded by a ^{238}U beam accelerated at 24 MeV/nucleon. The energies of the coincident fission fragments have been measured and their atomic numbers identified. The blocking patterns have been measured at 7° and at 1° with respect to the beam axis for fission fragments and elastically scattered nuclei, respectively. The excitation energy has been inferred event-by-event from the neutron multiplicity measured by a 4π detector.

2.2 The blocking technique

The principle of time measurement by the blocking technique [9] can be briefly summarized as follows. When a single crystal is used as a target, blocking effects give rise to dips in the measured angular distributions around the directions of the crystal axes and planes. The dips arise from the repulsive potentials due to the atomic rows or planes that modify the trajectories within the crystal.

In the case of fission processes, the distributions of the scission distances from the axis or plane of the crystal determine the blocking patterns. In the present experiment, for fission fragments detected at 7° around the $< 110 >$ axis direction, the range of sensitivity of the blocking technique is between 3×10^{-19}s and 10^{-16}s assuming exponential fission time distributions. The lower and upper time limits have two different origins. On the one hand, if the scissions occur on the average within the thermal vibration domain of the atoms of the crystal, the potential experienced by the fission fragments is smeared and the dips in the angular distributions do not depend anymore on the fission times, thus leading to the lower time limit. On the other hand, if the fissioning nuclei has paths through the crystal that are, on the average, much larger than the spacing between two adjacent rows or planes, no dip is observed in the angular distribution, thus leading to the upper time limit.

3 Results

The shape of the blocking patterns associated with fission fragments is essentially determined by the distribution of fission times. These distributions are highly dependent on the initial friction and on its evolution with deformation or temperature. Therefore, simulations of the blocking patterns should

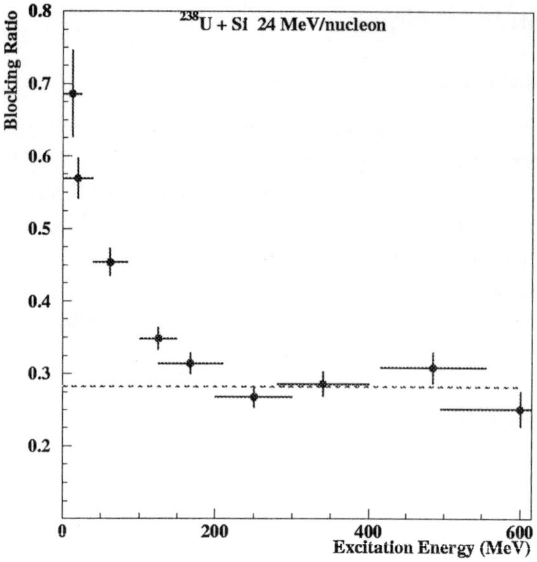

Figure 1. Blocking ratio as a function of the initial excitation energy of the fissioning nucleus.

provide some important pieces of information on the actual shapes of these distributions and on the friction and dissipation mechanisms involved during the fission process.

3.1 Evolution with excitation energy

A first order analysis can be performed by defining the blocking ratio B, a typical quantity associated with the blocking dips. The blocking ratio represents the filling of the angular distribution in the angular range where blocking dips are observed [8]: for $B = 1$ no dip is observed in the angular distribution whereas, for $B = 0$, there is no count in the dip. B is presented in Fig. 1 as a function of the initial excitation energy E^* of the fissioning nucleus. A selection on fissioning nuclei with atomic numbers $Z = 92 \pm 5$ has been applied. The error bars correspond to statistical uncertainties on the B values and to uncertainties arising from the E^* determination from the measured neutron multiplicity. As already stressed in references [7,8], the strong evolution of B

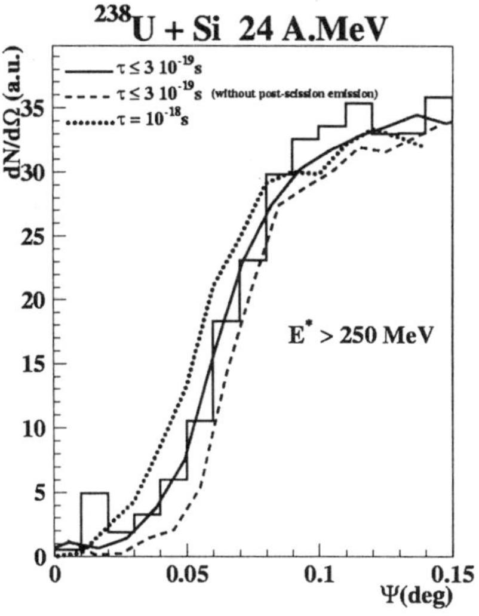

Figure 2. Measured blocking dip for initial excitation energies $E^* > 250$ MeV (histogram). The lines correspond to results of simulations with various hypotheses.

observed for $E^* < 250$ MeV indicates sizeable components in the fission time distributions corresponding to lifetimes of the fissioning nuclei longer than 3×10^{-19}s. For $E^* \geq 250$ MeV, no further evolution is observed, indicating either that the weight of the components with lifetimes longer than 3×10^{-19}s becomes negligible or that it becomes constant.

3.2 Effect of post-scission neutrons

The blocking dip associated with the < 110 > axis of the silicon crystal is shown for $E^* > 250$ MeV as a histogram in Fig. 2. The curves correspond to simulations in which the atomic collisions between fission fragments and atoms of the lattice have been followed in a Monte Carlo code. This code reproduces in a very satisfactory way the measured dip associated with elastically scattered nuclei at 1°. Nevertheless, in the case of the fission process, the effects of neutron emission by the recoiling nucleus before its scission and by the detected fission fragment have to be considered. Pre-scission emission has only a small effect on the recoil velocity of the fissioning nucleus, thus leading to a quite negligible error on the distance actually covered by the projectile-like nucleus before scission. Therefore, the effect of pre-scission emission will be neglected in the following. On the opposite, post-scission neutrons, despite their small momentum, can modify in a quite sensitive way (considering the very small angles associated with the axial blocking dips in the present experiment) the fission fragment trajectories. From unfolding procedures of the neutron energy and angular distributions [10], about 4 neutrons are known to be emitted, on the average, by the fission fragments of uranium-like nuclei, with a very weak dependence (if any) on the initial excitation energy. Therefore, simulations of the blocking patterns have been performed assuming that two neutrons are emitted by each fission fragment with kinetic energies and emission times estimated from a statistical code applied to typical fission fragments. With this correction, a very satisfactory agreement between the simulation and the data is achieved whatever the assumed shape of the fission time distribution, provided the latter has no component corresponding to lifetimes longer than 3×10^{-19}s. The full line in Fig. 2 corresponds to a simulation assuming an exponential fission time distribution with any average value $\tau \leq 3 \times 10^{-19}$s. For the sake of comparison, the result of the same simulation is shown as a dotted line for $\tau = 10^{-18}$s.

The dashed line in Fig. 2 corresponds, for an exponential fission time distribution with $\tau \leq 3 \times 10^{-19}$s, to the result of a simulation in which post-scission emission has not been taken into account. The importance of the correction for post-scission emission can be estimated by comparing in Fig. 2 the full curve to the dashed one and to the dotted one, respectively. This rather strong effect has not been considered in most of the previous blocking experiments, probably leading to overestimates of the components at very long times in the fission time distributions [4,5]. The effect is maximum for scissions occurring within the thermal vibration domain of the crystal atoms. The magnitude of this effect decreases progressively for longer fission times and it becomes quite negligible, in the present experiment, for components in the fission time distribution corresponding to lifetimes longer than 10^{-18}s.

3.3 Investigation of the fission time distributions

As already stressed by Fig. 1, for $E^* < 250$ MeV the relative weight of the components of the fission time distribution corresponding to lifetimes larger than 3×10^{-19}s increases when E^* decreases. The measured dip for $E^* = 120$ MeV is presented as a histogram in Fig. 3. The dashed curve corresponds to the best fit achieved with the simulation code assuming an exponential shape for the fission time distribution and considering the average time τ as an ad-

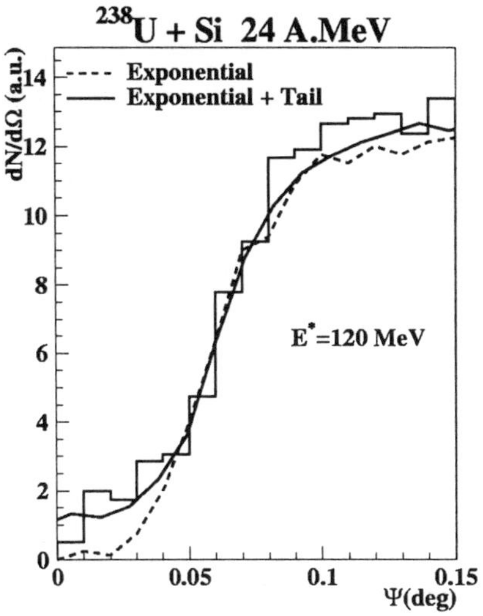

Figure 3. Measured blocking dip for initial excitation $E^* = 120MeV$ (histogram). The lines correspond to results of simulations with various hypotheses.

justable parameter. The best fit is achieved for $\tau = 10^{-18}$s. A satisfactory agreement is observed for angles larger than about 0.05° but this simulation fails in reproducing the dip for smaller angles. The only way to fill in the dip at small angles is to consider components in the fission time distribution corresponding to lifetimes much larger than $\tau = 10^{-18}$s. Such long lifetime components in the fission time distribution, essential in order to reproduce the blocking pattern at the smallest angles, implies components corresponding to lifetimes shorter than $\tau = 10^{-18}$s in order to keep a satisfactory agreement for larger angles. Therefore, the actual fission time distribution has to be much broader than an exponential distribution. This behaviour is observed for any excitation energy between a few tens of MeV and 250 MeV. Unfortunately, taking into account the statistical uncertainties, very satisfactory fits can be achieved for any excitation energy $E^* < 250$ MeV assuming very different shapes. As an example, the full curve in Fig. 3 corresponds to the best fit achieved assuming a fission time distribution with two components (a component with an exponential shape, with $\tau \leq 3 \times 10^{-19}$s, and a component with a constant probability density between 3×10^{-19}s and 6×10^{-17}s) and adjusting the relative weights of these components. However, whatever the assumed shape of the fission time distribution, the inferred average values τ are found much longer than the ones inferred from pre-scission multiplicities [10,11] due to the long lifetime components that can be probed only by the blocking technique. Such components seem to arise from fissions occurring at very low residual excitation energies. A large reduction of the fission widths as compared to the predictions of Bohr and Wheeler's statistical model seems essential to permit the excited nucleus to reach such low excitation energies before undergoing fission.

4 Conclusions

Blocking dips associated with fission fragments have been measured as a function of the initial excitation energy of fissioning uranium-like nuclei. Simulations of the blocking dips have shown the important effect of post-scission emission on the shape of the blocking dips when approaching the lower time limit of the blocking technique. For initial excitation energies lower than 250 MeV, the shapes of the blocking dips point to broad fission time distributions that can only be probed by the blocking technique in single crystals. For these initial excitation energies, a sizeable part of the fission processes seem to occur at very low residual energies, thus implying large reductions of fission widths due to high friction coefficients.

References

1. B. Domeij and K. Bjorkqvist, *Phys. Lett.* **14**, 129 (1965)
2. D.S. Gemmell and R.E. Holland, *Phys. Rev. Lett.* **14**, 945 (1965)
3. J.U. Andersen et al., *Nucl. Phys.* **A241**, (1975) 317
4. J. U. Andersen et al., *Phys. Rev. Lett.* **36**, 1539 (1976)
5. J. S. Forster et al., *Nucl. Phys.* **A464**, 497 (1987)
6. S.A. Karamyan et al., *Sov. J. Part. Nucl.* **4**, 196 (1973)
7. M. Morjean et al., *Nucl. Phys.* **A630**, 200c (1998)
8. F. Goldenbaum et al., *Proc. XXXVI Int. Winter Meeting on Nucl. Phys., Bormio*, ed I. Iori (Univ. Milano, 1998)
9. W. M. Gibson, *Ann. Rev. Nucl. Sci.* **25**, 465 (1975)
10. D. Hilscher and H. Rossner, *Ann. Phys. Fr.* **17**, 471 (1992)
11. P. Paul and M. Thoennessen, *Ann. Rev. Part. Nuc. Sci.* **44**, 65 (1994)

PROTON AND ALPHA PARTICLE EMISSION IN THE FISSION AND EVAPORATION RESIDUE CHANNELS FROM NUCLEI WITH A≃140

E. VARDACI, G. LA RANA, A. BRONDI, M. A. DI MEO,
R. MORO, A. PRINCIPE

INFN and Dipartimento di Scienze Fisiche dell'Universitá di Napoli, I-80125 Napoli, Italy
E-mail: vardaci@na.infn.it

D. FABRIS, G. NEBBIA, G. VIESTI

INFN and Dipartimento di Fisica dell'Universitá di Padova, I-35131 Padova, Italy

G. PRETE

INFN, Laboratori Nazionali di Legnaro, I-36020 Legnaro (Padova), Italy

Charged-particles in coincidence with fission fragments (FF) and evaporation residues (ER) have been measured for the system 180-MeV ^{32}S + ^{109}Ag leading to the composite system ^{141}Eu at the excitation energy $E_x \simeq$90 MeV. Energy spectra and multiplicities in both channels have been compared to the predictions of the statistical model by using the code PACE2. The same multiplicity analysis has been employed in the system 905-MeV ^{121}Sb + ^{27}Al. For both reactions, it was not possible to find a unique set of input parameters which could result in a simultaneous fit of the multiplicities in both channels. A time delay was still necessary in both systems to reproduce the respective pre-scission multiplicities and fission cross-section alone. Charged particle multiplicities in the ER channel are largely overestimated by the model in both reactions and implications of this finding on the extraction of the fission timescale are briefly discussed.

1 Introduction

A large variety of studies[1-7] on the light particles pre-scission emission from composite nuclei with A≃200-250 have shown a rapid increase of the neutron, proton and alpha pre-scission multiplicities with increasing excitation energy not predicted by the standard statistical model (SM). The overall picture suggested by these studies is that the excess of the n, p and α pre-scission multiplicities observed is due to a slowed-down dynamical evolution of the composite system from the equilibrium to the scission point because of the nuclear viscosity. Moreover, light charged particles seem to be emitted predominantly in the equilibrium-to-saddle path, whereas neutrons are favoured in the saddle-to scission one[1,3].

A dynamical fission time scale τ_f of $\simeq$ (35 $\pm$ 15) x 10^{-21} has been estimated [1] by using the pre-scission neutrons as a "clock" and the SM which includes a pre-saddle delay τ_d.

An estimate of the equilibrium-to-saddle and saddle-to-scission time scales, τ_d and τ_{ssc}, respectively, has been achieved in studies[3-6] involving light charged particles pre-scission multiplicities combined to a modified version of the SM which assumes the evaporation of particles to occurr at two fixed points in the deformation space, chosen in the pre-saddle region and the post-saddle one. The total fission time scale is defined as $\tau_f = \tau_d + \tau_{ssc}$. It turns out[3-6] that $\tau_d \simeq 10 \times 10^{-21}$ and $\tau_{ssc} \simeq (20-50) \times 10^{-21}$. Compared to neutrons, pre-scission charged particles, and especially alpha particles, offer the advantage of being sensitive to the emission barrier as well as to the yrast line at high angular momentum. Therefore, besides the fission time scale, charged particles can provide information on the average deformation of the nucleus in route toward the fission. The trend generally observed[3-6,8] is that the alpha average energies are overestimated by ~2 MeV in the SM calculation, whereas the proton average energies are quite well reproduced.

The estimates of the time scales are, however, quite controversial, and blurred by the fact that different sets of input parameters can result in equally good fits[2-6]. Since the extraction of dynamical effects in the fission process relies on SM calculations, it seems mandatory the use of as many observables as possible to constraint the relevant parameters of the model, and, in particular, to reproduce the variety of observables with a unique set of input parameters.

Composite systems of intermediate fissility can help to persue this goal. These systems are characterized by an evaporation residue cross-section σ_{ER} comparable or larger than the fission cross section $\sigma_{fission}$, and by a shorter path in the deformation space from the saddle-to-scission point [9]. Consequently, the role of the pre-saddle dynamic relative to the saddle-to-scission one is enhanced [9], and SM input parameters can be further constrained by the multiplicities of the light particles in the evaporation residues channel.

In this work we propose the measurement of protons and alpha in coincidence with the evaporation residues (ER), in conjunction with particles in the fusion-fission (FF) channel, as a mean to impose further constraints on the SM calculations.

We have measured energy spectra and multiplicities of the charged particles in the fusion-fission and fusion-evaporation channels for the system 180 MeV $^{32}S + ^{109}Ag$, leading to the composite system ^{141}Eu (fissility of 0.56) at the excitation energy $E_x \simeq$90 MeV. Furthermore, the same SM analysis of the multiplicities has been carried out for the same measured observables in the reaction $^{121}Sb + ^{27}Al$ at 905 MeV of bombarding energy [11].

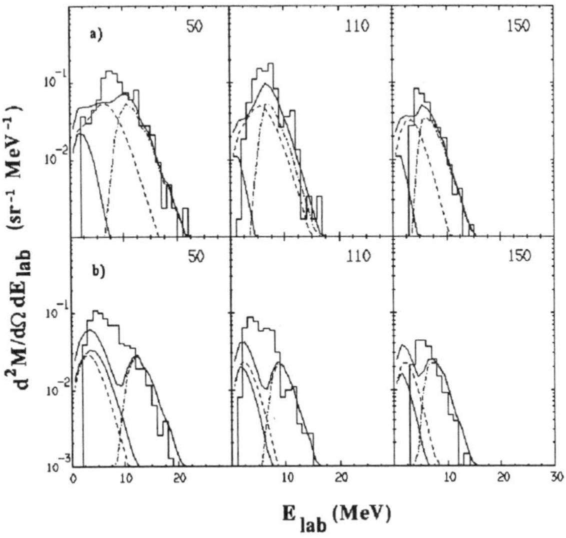

Figure 1. Fragment-proton coincidence spectra (histograms) at three laboratory angles. (a) In-plane FF trigger. (b) Out-of-plane FF trigger. The curves are from reaction simulation of evaporation from the trigger fragment (solid line), the undetected partner fragment (dashed line), and the composite system prior to fragment separation (dot-dashed line). The dark solid line is the sum of the components.

2 Experimental Procedures and Results

The experiment was performed at the XTU Tandem of the Laboratori Nazionali di Legnaro using a 180 MeV beam of ^{32}S on ^{109}Ag target $\approx$900 $\mu g/cm^2$ thick.

Light charged particles were detected in coincidence with fission fragments (FF) and evaporation residues (ER) by using three solid state telescopes (SST) at laboratory angles $\theta_{lab} = 50°$, 110° and 150°. Fission fragments were detected in two gas telescopes both positioned at 50° with respect to the beam, one in the same plane as the SSTs, and the other in the perpendicular plane to detect in-plane and out-of-plane particles. A value of the fission cross section $\sigma_{fission} = 113\pm22$ mb was measured. Evaporation residues were detected by three 7-strip silicon detectors after being separated from the beam by an electrostatic deflector.

In the sharp cut-off approximation it is possible to associate the ER cross section σ_{ER}, and $\sigma_{fission}$ to a corresponding window of entrance channel spin.

This information is necessary as input in the procedure used to extract the multiplicities, and for the SM analysis. From the measured $\sigma_{fission}$ and from the value of the fusion cross section $\sigma_{FUSION} = \sigma_{ER} + \sigma_{fission} = 800$ mb, taken from literature [10], we obtained $0 \div 59\hbar$ for the ER spin window and $59 \div 64\hbar$ for the FF spin window.

2.1 Fragment-Particle Coincidence

The laboratory multiplicity spectra for the protons and alpha particles in coincidence with the FF are shown as histograms in Figure 1 and Figure 2. In each figure, the particles detected in the same plane of the fragment trigger (in-plane configuration) are in part (a) of the figure, and the particles detected in the plane orthogonal to the one of the fragment trigger (out-of-plane configuration) are in part (b) of the figure. The lines are the results of the calculations performed with the code GANES [12]. It is possible to observe for both particles a strong dependence of the shapes upon the plane of observation. This is mainly due to the kinematic shifts of the particles emitted from the rapidly moving fission fragments.

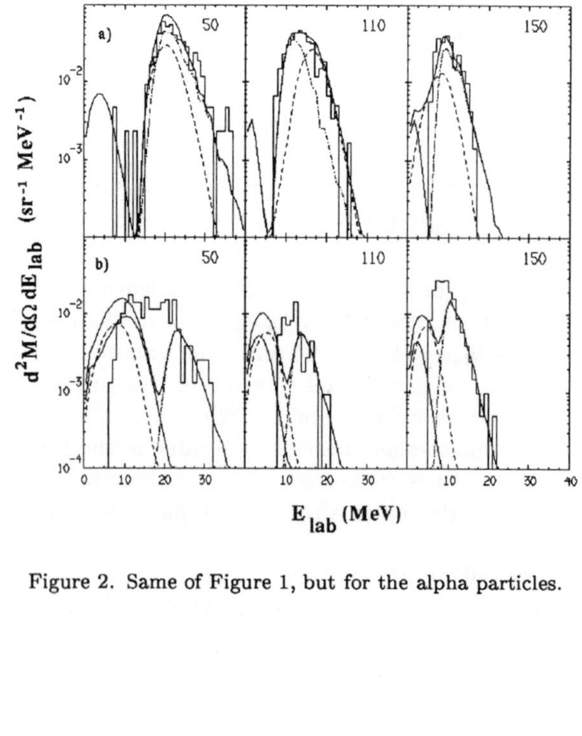

Figure 2. Same of Figure 1, but for the alpha particles.

For a specific fragment trigger, the particle spectra can arise from several sources. In order to extract the pre- and post-scission integrated multiplicities it is necessary to unfold the contributions in the spectra due to the three possible evaporative sources: the composite nucleus prior to scission, the trigger fragment, and the undetected partner fragment. The first will be also referred to as Composite Emission (CE), and the second and third to as Fragments Emission (FE).

We have used a well established procedure which employes the Monte Carlo statistical code GANES [12, 13]. Proton and alpha particle spectra were calculated, for each source of emission, for the trigger configurations defined in the experiment. The energy and charge distributions of the fission fragments obtained from GANES simulations have been matched with the experimental ones. Furthermore, emission from deformed composite systems and from spherical fragments was assumed. The normalization of the simulated spectra to those measured provides the determination of the integrated multiplicities for each simulated emission source. A large deformation of the composite system prior to scission was necessary to fit simultaneously the energy spectra and the angular distributions.

Although the fits are in general good, there are some important deviations which indicate contributions from processes not accounted for by the simulations. In the out-of-plane spectra it is quite evident a surplus of particles with energies intermediate between those for FE and CE which is also slightly observed in the in-plane data for the protons. This same pattern was already found for a similar system [13] and has been attributed to the near scission emission [14].

¿From the fitting of the experimental spectra by the calculated CE and FE contributions, we have derived the proton and alpha particle integrated multiplicities. These are presented in Table 1.

2.2 Evaporation Residue-Particle Coincidence

The extraction of the integrated proton and alpha multiplicities in the ER channel, M_p^{ER} and M_α^{ER}, respectively, was accomplished by means of the SM code Lilita_N95 [15, 16]. The constraints of the experimental setup (SSTs and deflector system) were included in the code. The integration of the multiplicities distributions of the particles in coincidence with the ER obtained from Lilita_N95, normalized to the experimental ones, has provided M_p^{ER} and M_α^{ER} which are shown in Table 1.

Table 1. Measured multiplicities from FF-particle and ER-particle coincidences[a].

	M^{PRE}	M^{POST}	M^{ER}
FF-p	0.104±0.023	0.182±0.041	
FF-α	0.068±0.015	0.080±0.018	
ER-p			1.3±0.3
ER-α			0.6±0.1

[a] M^{PRE}, M^{POST} and M^{ER} denote the pre-scission multiplicity, the post-scission multiplicity and the multiplicity in the ER channel, respectively.

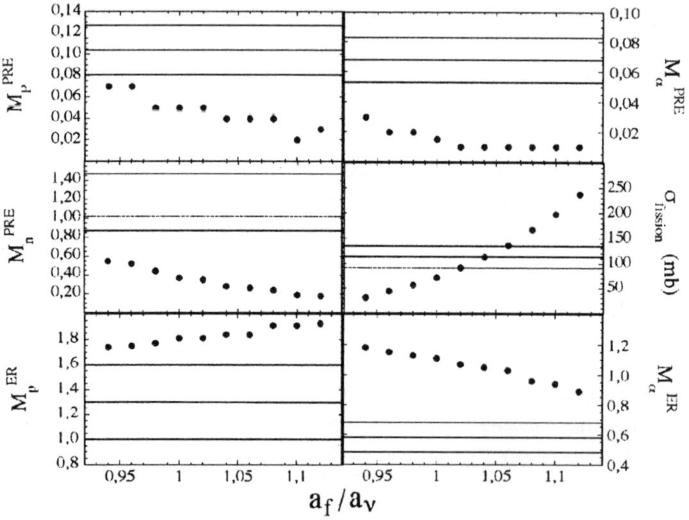

Figure 3. Measured particle multiplicities in the ER and FF channels, and fission cross section (horizontal lines) for the reaction ^{32}S + ^{109}Ag compared to the predictions of the model (dots) for different values of a_f/a_ν and τ_d=0.

3 Statistical Model Analysis

We have analysed the measured multiplicities and the spectral shapes of the particles in the FF and ER channels on the basis of the SM as implemented

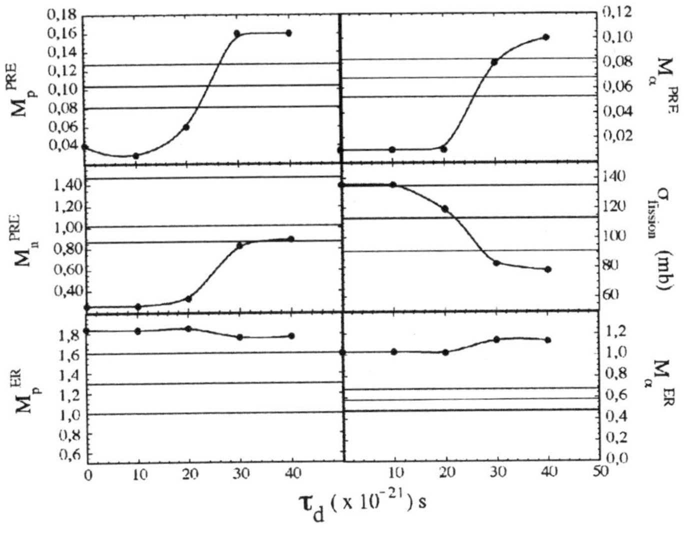

Figure 4. Measured particle multiplicities in the ER and FF channels, and fission cross section (horizontal lines) for the reaction $^{32}S + {}^{109}Ag$ compared to the predictions of the model for different values of τ_d and a_f/a_ν=1.06 (dots); solid lines are to guide the eye.

in the code PACE2 [17]. We have used the default transmission coefficients [17], fission barriers from Sierk model [18] and the value of the level density parameter for particle emission a_ν= A/8.09 [19]. With this value of a_ν the code PACE2 reproduces the high energy part of the energy spectra of the protons and alphas in the ER channel and constitutes a constraint for the SM calculations. The code PACE2 was also modified to include a fission delay τ_d in such a way that the fission probability is zero up to the time τ_d and has the full statistical value subsequently. Therefore, this time delay should be considered only as a pre-saddle delay.

Since we expect that the multiplicities are very sensitive to the value of the ratio a_f/a_ν, we first performed PACE2 calculations for a range of values of a_f/a_ν between 0.94 and 1.12 and without fission delay. The results are shown in Figure 3. The measured values with errors are indicated by full lines. The experimental value of the neutron pre-scission multiplicity was taken for a similar system at nearly the same excitation energy [7] and is shown just for reference. The trend of the calculations is the one expected: the increase of the ratio a_f/a_ν results in the raise of $\sigma_{fission}$ and in the fall of the pre-

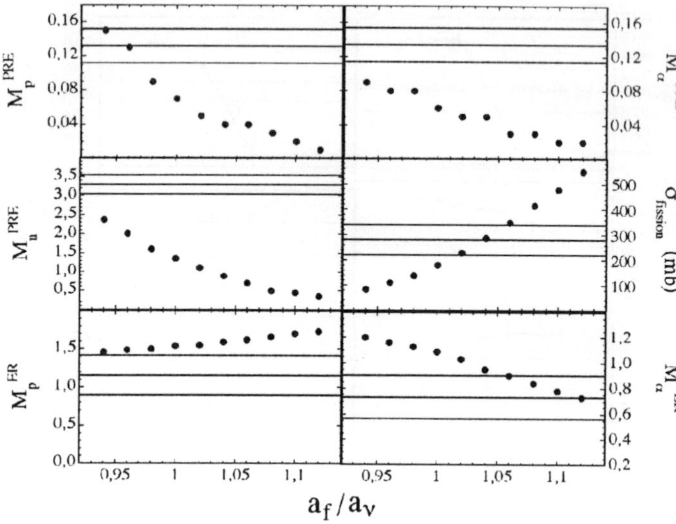

Figure 5. Same as Figure 3, but for the reaction $^{121}Sb + ^{27}Al$

scission multiplicities. In the ER channel, the decrease of the M_α^{ER}, as an angular momentum effect, results in a slight increase of M_p^{ER}. However, there is no single value of a_f/a_ν that allows the reproduction of the whole set of observables in the FF channel, even if the multiplicities in the ER channel are not considered. On the other side, the computed average energies are not much sensitive upon a_f/a_ν. The code is able to reproduce all the average energies except for the α particle in the pre-scission channel where they are overestimated by about 2 MeV.

The use of a fission delay is expected to enhance the particle pre-scission multiplicities and to reduce $\sigma_{fission}$. Therefore, one can suppose that there might be a combination of both a_f/a_ν and τ_d for which the pre-scission multiplicities and $\sigma_{fission}$ will be reproduced simultaneously.

We performed the whole grid of calculations for $0.94 \leq a_f/a_\nu \leq 1.12$ and $10 \leq \tau_d \leq 40$ zs. The multiplicities in the ER channel turned out to be always overestimated, and the best agreement, at least for M_p^{PRE}, M_α^{PRE} and $\sigma_{fission}$, was found for $a_f/a_\nu = 1.06$ and $21 < \tau_d < 23$ zs. This is shown in Figure 4. The neutron multiplicity is not expected to be reproduced because the saddle-to-scission path [3] is not accounted for by the model, but it is certainly less important for charged particles and intermediate fissility

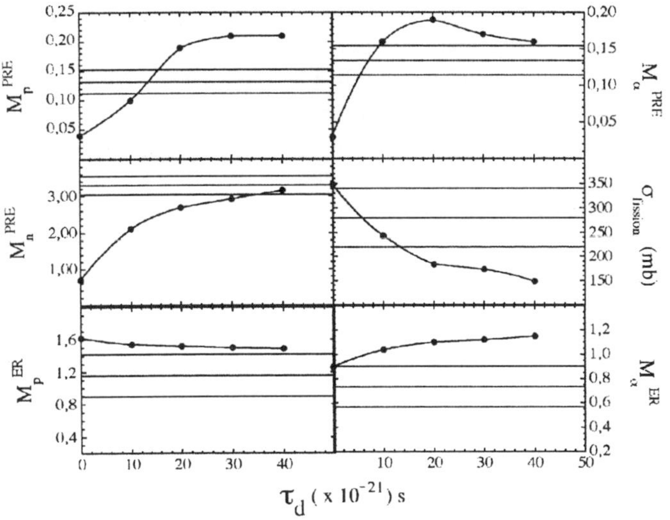

Figure 6. Same as Figure 4, but for the reaction $^{121}Sb + ^{27}Al$

systems [9].

In order to test the same analysis procedure on a common ground, the reaction $^{121}Sb + ^{27}Al$ at 905 MeV of bombarding energy [11] was considered. This reaction produces the nucleus ^{148}Gd (fissility of 0.56) at the excitation energy $E_x \simeq$135 MeV. The fusion-fission spin zone is 64-75 $\hbar$. Figure 5 and Figure 6 summarize the PACE2 calculations. A delay of 10-14 zs was necessary to reproduce simultaneously M_p^{PRE}, M_α^{PRE} and $\sigma_{fission}$. Once again M_p^{ER} is not reproduced, and M_α^{ER}, which is reproduced without the fission delay and a large value of a_f/a_ν (Figure 5), turns out to be overestimated when the fission delay is introduced.

4 Summary and Conclusions

We have measured the energy spectra and the multiplicities of protons and alpha particles in coincidence with FF and ER for the system 180 MeV ^{32}S + ^{109}Ag leading to the composite system ^{141}Eu at the excitation energy $E_x \simeq$90 MeV. Pre-scission particle energy spectra and multiplicities have been obtained by means of the code GANES. The data in the ER and FF channels have been analysed in the framework of the SM using a modified version of the code PACE2 which includes the fission delay.

A set of calculations has been performed for different values of a_f/a_ν and of the fission delay τ_d using a_ν=8.09. The best overall agreement for M_p^{PRE}, M_α^{PRE} and $\sigma_{fission}$ was obtained for a_f/a_ν=1.06 and $21 < \tau_d < 23$ zs. The same analysis has been applied to the system $^{121}Sb + ^{27}Al$. A fission delay of 10-14 zs is necessary also in this case.

These results are in substantial agreement with the findings of Lestone et al. [3] and Saxena et al. [5], even though the choice of some of the relevant parameters of the model, like a_ν, a_f/a_ν and the fission barriers, follows different criteria.

The overprediction of the protons and alpha multiplicities in the ER channel, for both reactions, raises a question on the reliability of the method of extracting quantitative estimates of the fission timescale from the pre-scission multiplicities predicted by the SM. By reviewing recent literature, we have found several cases in which the charged particles multiplicities in the ER channel were overpredicted [21]. We suspect that these discrepancies derive from the failure of the SM to account properly for the competition between particle emissions: protons and alpha are enhanced whereas neutron are suppressed, contrarily to what is measured [21]. Thus, it might be reasonable to conclude that the overprediction of M_p^{ER} and M_α^{ER} indeed supports the inclusion of a fission delay: if the same overprediction would occur also in the pre-scission channel, the amount of fission delay one would need to reproduce M_p^{PRE}, M_α^{PRE} and $\sigma_{fission}$ toghether, should be even larger than what is usually found [5], and this might also bring the larger fission delays obtained with GDR studies [22] to agree with the ones obtainable from pre-scission particle multiplicities alone.

This study supports the importance of measuring energy spectra and multiplicities of the charged particles in the ER channel, in conjunction with the same observables in the pre-scission channel, in order to impose more constraints on the relevant parameters of the model, and therefore, to extract a more significative picture on the presence of dynamical effects.

References

1. D.J. Hinde, D. Hilsher, H. Rossner, B. Gebauer, M.Lehmann and M. Wilper, *Phys. Rev.* C **45**, 1229 (1992);
2. H. Ikezoe, N. Shikazono, Y. Nagame, Y. Sugiyama, Y. Tomita, K. Ideno, I. Nishinaka, B.J. Qi, H.J. Kim and A. Iwamoto, T. Ohtsuki,*Phys. Rev.* C **46**, 1922 (1992)
3. J.P. Lestone, J.R. Leigh, J.O. Newton, D.J. Hinde, J.X. Wei, J.X. Chen,

S. Elfström and M. Zielinska-Pfabé, *Nucl. Phys.* A **559**, 277 (1993)
4. J.P. Lestone, *Phys. Rev. Lett.* **70**, 2245 (1993)
5. A. Saxena, A. Chatterjee, R. Choudhury, S. S. Kapoor, D. M. Nadkarni,*Phys. Rev.* C **49**, 932 (1994)
6. A. Chatterjee, A. Navin, S. Kailas, P. Singh, D. C. Biswas, A. Karnik, S. S. Kapoor, *Phys. Rev.* C **52**, 3167 (1995)
7. L. Fiore, G. D'Erasmo, D. Di Santo, L. Celano, N. Colonna, E. M. Fiore, A. Pantaleo, V. Paticchio, G. Tagliente, G. Viesti, D. Fabris, G. Nebbia, M. Cinausero, E. Fioretto, G. Prete, A. Brondi, R. Moro, E. Vardaci, F. Lucarelli, P.F. Bortignon, *Nucl. Phys.* A **620**, 71 (1997)
8. E. Vardaci, G. La Rana, A. Brondi, R. Moro, A. Principe, D. Fabris, G. Nebbia, G. Viesti, M. Cinausero, E. Fioretto, G. Prete, F. Lucarelli, E. M. Fiore, L. Fiore, V. Paticchio, P.F. Bortignon, *Eur. Phys. J.* A **2**, 55 (1998)
9. M. G. Itkis, A. Ya. Rusanov, *Phys. Part. Nucl.* **29**, 160 (1998)
10. J. R. Birkelund, L.E. Tubbs, J.R. Huizenga, J.N. DE, D. Sperber, *Phys. Rep.* **56**, 107-166 (1979)
11. W. E. Parker, M. Kaplan, D. J. Moses, J.M. Alexander, J. T. Boger, R. A. Lacey, D. M. Castro Rizzo, *Nucl. Phys.* A **568**, 633 (1994)
12. N.N. Ajitanand, G. La Rana, R. Lacey, D. J. Moses, L. C. Vaz, G. F. Peaslee, D. M. Castro Rizzo, M. Kaplan, and J.M. Alexander, *Phys. Rev.* C **34**, 877 (1986)
13. R. Lacey, N.N. Ajitanand, J.M. Alexander, D. M. Castro Rizzo, G.F. Peaslee, L. C. Vaz, M. Kaplan, M. Kildir, G. La Rana, D. J. Moses, W. E. Parker, D. Logan, M.S. Zisman, P. DeYoung, L. Kowalski, *Phys. Rev.* C **37**, 2540 (1988)
14. E. Duek, N.N. Ajitanand, J.M. Alexander, D. Logan, M. Kildir, L. Kowalski, L.C. Vaz, D. Garreau, M.S. Zisman, M. Kaplan, and D. J. Moses, *Phys. Lett.* B **131**, 297 (1983);
15. J. Gomez del Campo and R.G. Stokstad, Report ORNL TM-7295,1981;
16. G. La Rana and M. Kaplan, private communication. The code has been renamed to distinguish it from several other versions of LILITA.
17. A. Gavron, *Phys. Rev.* C **21**, 230 (1980)
18. A.J. Sierk, *Phys. Rev.* C **33**, 2039 (1986)
19. J. Töke and W.J. Swiatecki, *Nucl. Phys.* A **372**, 141 (1981)
20. F. Pühlhofer, *Nucl. Phys.* A **280**, 267 (1977)
21. see for example: J. M. Alexander, G. Auger, M. Kaplan, L. Kowalski, R. Lacey, G. La Rana, M. T. Magda, G. F. Peaslee, in Nuclear Dynamics and Nuclear Disassembly, World Scientific, 1990, p.211, and references therein; B. John and S. K. Kataria, B. S. Tomar, A. Goswami, G. K.

Gubbi, and S. B. Manohar *Phys. Rev.* C **56**, 2582 (1997)
22. P. Paul and M. Thoennesen, *Annu. Rev. Nucl. Part. Sci.* **44**, 65 (1994)

DYNAMIC EVOLUTION IN THE DISASSEMBLY OF HOT NUCLEI

J. CIBOR, K. HAGEL, R. WADA, M. LUNARDON[1], N. MARIE[2], R. ALFARO[3], B. XIAO, W. SHEN[4], Y. ZHAO, J. LI, B. A. LI[5], M. MURRAY AND J. B. NATOWITZ

Cyclotron Institute, Texas A&M University, College Station, Texas 77843-3366

Z. MAJKA[6] AND P. STASZEL[6]

Institute of Physics, Jagellonian University ul. Reymonta 4, 30-059 Krakow, Poland

Detailed studies of light charged particle emission have been carried out for intermediate energy heavy ion reactions. For reactions at 47A MeV, application of coalescence model analyses to the observed particle yields provides evidence for increasing expansion of the hot nuclei produced as the projectile mass increases. Temperatures are derived for these expanded nuclei and the role of dynamics in determinations of caloric curves is discussed.

To employ intermediate energy heavy ion collisions as a tool to probe the nuclear equation of state, a detailed understanding of the reaction dynamics in well- characterized collisions is essential. When a central collision of two heavy nuclei leads to multifragmentation [1,2,3,4,5] of an expanded and equilibrated nucleus the thermal shock and compression which results can cause the nucleus to expand to low density, cluster and disassemble.

Figure 1 presents a schematic picture in which central collision of two heavy nuclei leads to multifragmentation, the process which we are attempting to isolate and study. The schematic similarity of this evolution to that observed in lower energy binary fission processes such as those extensively discussed in this meeting is readily apparent.

The collision produces a thermal shock and some compression. Reacting to this, the nucleus expands to low density, clusters and disassembles. This stage is labeled "freezeous" in the figure. In an ideal situation this disassembly would be that of a thermally and chemically equilibrated nucleus. In practice this ideal state may not be reached and the final product distribution may include fragments and particles originating from non-equilibrium processes and reflecting correlations already present in the separated projectile and target nuclei [6,7]. Distinguishing between these different production mechanisms of fragment and light particle production is difficult but essential to our understanding of the multifragmentation process.

The light particle emission, which occurs throughout the evolution of the

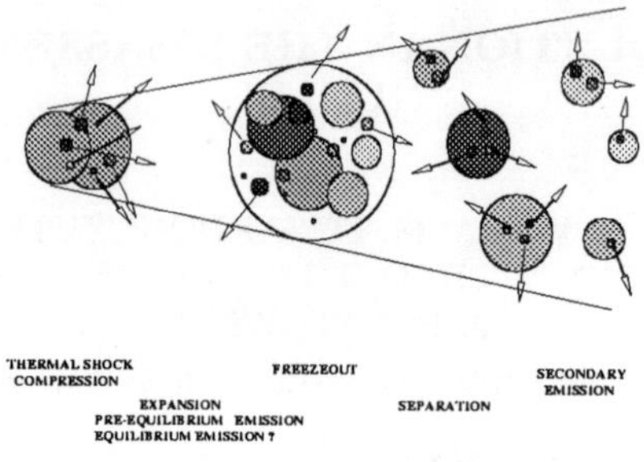

Figure 1. Schematic View of Multifragmentation

system, carries essential information on the early dynamics and on the degree of equilibration at each stage of the reaction. As a first step toward exploiting the light particle information to follow the dynamical and thermodynamical evolution of multifragmentation reactions of medium to heavy mass nuclei we recently applied particle-fragment correlation techniques to determine the last stage yield of secondary decay particles in central collisions of ^{129}Xe with ^{124}Sn at 50A MeV [8]. We found that such emission accounted for 28% of the total light charged particle multiplicity. The data allowed reconstruction of the atomic numbers and average excitation energies of the primary fragments as shown in Figure 2. The results indicate an excitation energy of 3 MeV/u for all fragments and provides evidence for the equilibrated nature of the disassembling system.

Our next goal has been to develop techniques which will allow us to obtain more specific information on the very early dynamic and thermal evolution of multifragmenting system. As a first step in pursuing such studies we recently carried out a series of measurements of reactions induced by 47A MeV projectiles, ^{4}He, ^{12}C, ^{22}Ne, ^{40}Ar and ^{64}Zn, on medium and heavy targets using a combined CsI ball-Neutron Ball detection system. Figures 3 and 4 show

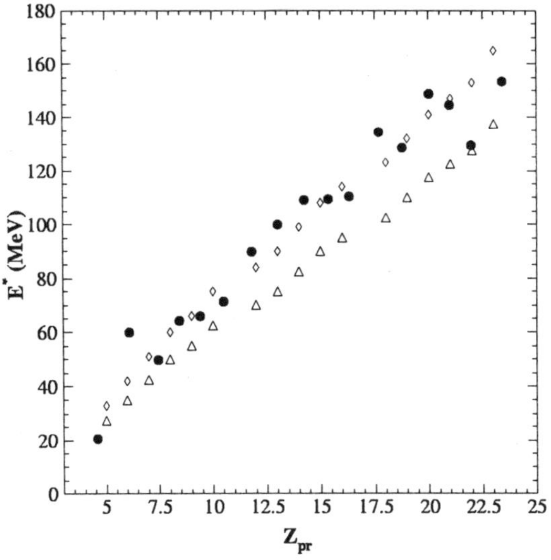

Figure 2. Average total excitation energies for primary fragments as a function of Z_{pr}. Black dots represent experimental data. Open symbols represent reconstructed excitation energies using a statistical decay model and two different assumptions for excitation energies. Triangles - 2.5 Mev/u, Diamonds - 3 MeV/u.

results of calculations for the medium mass system carried out with the QMD code, CHIMERA [9].

As indicated in the figures, the particular set of target and projectile combinations used in our experiment are predicted to lead, after pre-equilibrium emission of particles, to excited composite nuclei of very similar mass, A 120. Compression and excitation energy are predicted to increase monotonically as the projectile mass is increased. The combined thermal and compressional energies which are deposited in the collisions lead to increasing expansions of the composite nuclei. At the maximum predicted expansions, excitation energies range from 1 to 7 MeV/u and densities range from normal density, ρ_0 to $\approx 0.4\rho_0$. In the case of both Ar and Zn projectiles, the minimum densities are such that the expanded nuclei are expected to enter the region of mechanical instability known as the spinodal region.

In Figure 5 invariant velocity plots of the intensity of light charged par-

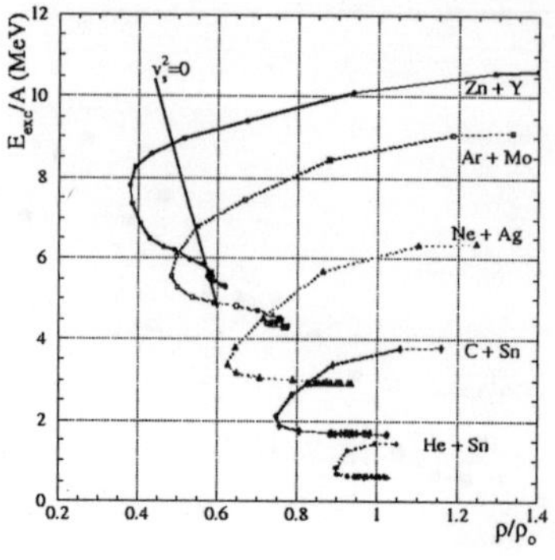

Figure 3. Excitation energy per nuclion as a function of normalized density of the composite system for five different reactions at 47 A MeV. Results are for central collisions.

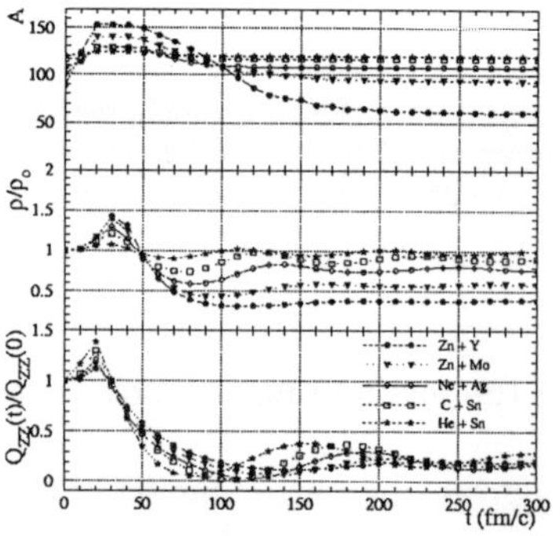

Figure 4. Average mass, normalized density and Qzz of the compound system as a function of time for five reactions.

ticles emitted in the reactions of 47A MeV ^{12}C with Ag and 47A MeV ^{64}Zn with ^{89}Y are presented as a function of their parallel and transverse velocities in the laboratory. An immediately striking feature of the plots for a given particle species is the strong similarity for the two different systems, reflecting a fairly intense emission of particles from a "source" which has a velocity close to one half of the projectile velocity. For ^{64}Zn + ^{89}Y this source is so dominant that emission from the target like source is only weakly evident.

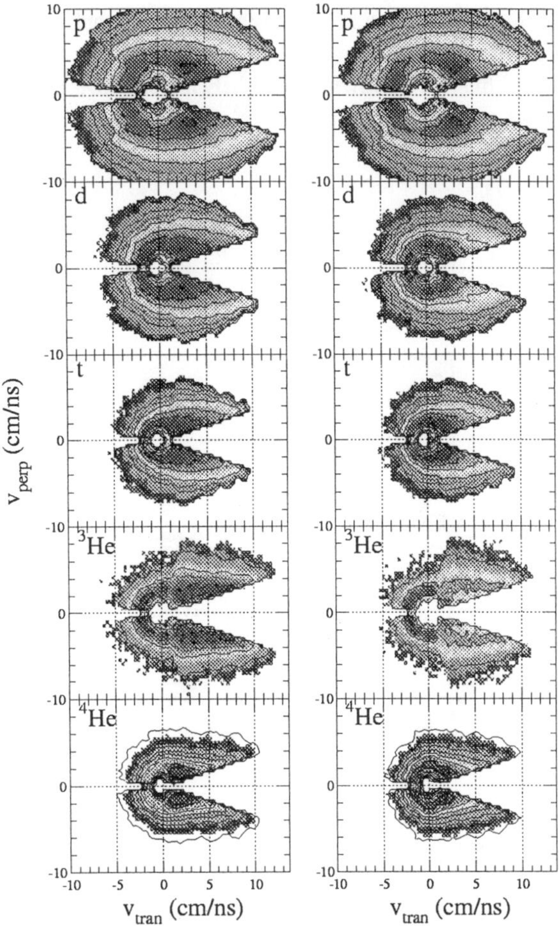

Figure 5. Invariant velocity plots for light particles emitted in reactions with ^{64}Zn %(left) and ^{12}C (right) projectiles.labelfig:radish

For ^{12}C + ^{116}Sn the emission from the target like source is easier to see as the relative intensity of emission from the intermediate source is less than in the Zn + Y reaction.

To emphasize the similarities of the higher energy portions of the energy spectra in the two different reactions we present comparisons of the particle spectra in Figure 6. These are shown as dM/dE for the10% highest multiplicity events as selected above.

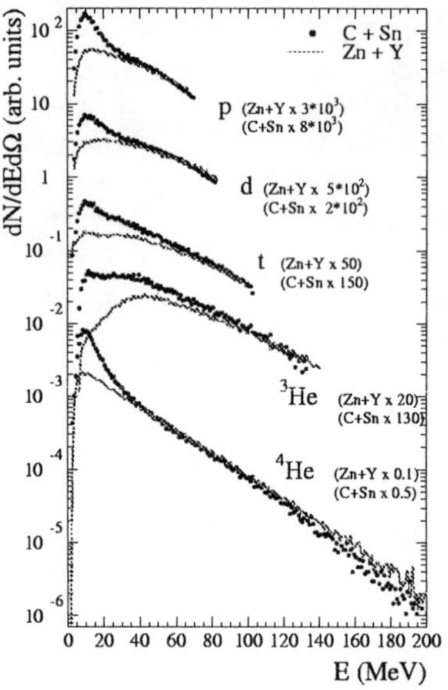

Figure 6. Energy spectra of light particles emitted in the ^{12}C and ^{64}Zn induced reactions at 45 degrees in the laboratory.fig4:radish

Recently it has been suggested, based on an expanding emitting source model, that the observed difference of the spectra from those of other particles reflects the strong weighting of the probability of thermal ^3He evaporation towards early times in the emission cascade and the very low emission probability at later stages [25]. Our results support the arguments that ^3He is predominately emitted in the early stages of the reaction, however, the strong similarities of the ^3He spectra for all systems studied and the fact that for all systems a large fraction of the ^3He appears to, be emitted from a source having a velocity close to that of the NN frame indicates that thermal emission accounts for only a small fraction of the ^3He observed in this work.

Since coalescence models have been applied to learn about source sizes over a wide range of projectile energies [10,11,12,13,14] we are currently attempting to employ coalescence model analyses of light particle emission to probe the size of the emitting system in these collisions. The very similar formal structures of such models applied in both non-equilibrium and equilibrium conditions suggest that they can provide a natural framework for this purpose. Under suitable conditions extraction of coalescence radii provides size information the analogous to that obtained in particle-particle correlation measurements. There are, however, several important questions which must be addressed to make this possible. Even though the coalescence models have enjoyed considerable success, the non- equilibrium models are based upon a coalescence relationship between the observed light clusters and *primary* neutron and proton yields while in the thermal coalescence models the coalescence relationship is between observed light clusters and the *observed* neutron and proton yields [10]. If, as the system evolves, the coalescence mechanism shifts from non-equilibrium to thermal, this must be recognized and taken into account in determination of the system size.

The CHIMERA QMD calculations indicate that the first light particles are emitted at $\approx$50 fm/c after contact and that the system reaches minimum density by $\approx$ 100 fm/c where thermal equilibrium appears to be established. During this time period a significant mass loss can occur. An inspection of the calculated energy spectra of the emitted particles over this time range reveals that there is a monotonic decrease of the average energy with increasing time and suggests that this relationship might be exploited to follow the time evolution of the system in more detail than has been attempted in previous works, i.e., if the particle energies are well correlated with emission time, and secondary emission contributions contribute primarily at the lower energies, yields of higher energy particles may be relatively uncontaminated by secondary emission processes and can be used to probe the evolution of the system. A related suggestion has recently been made by Friedman who has noted that, in thermal models, the cooling of the system leads to progressively softer particle spectra and therefore the extraction of double isotope yield ratio temperatures as a function of particle energy might be used to follow the temperature of the system as it cools [15]. In the transport calculations no a-priori assumption of thermal equilibrium is made.

To determine the coalescence parameter, P_0, we have adopted the coulomb corrected coalescence model formalism of Awes *et al.* [13]. In principle coalescence models require knowledge of both neutron and proton yields. Because measurements which include neutron information are relatively rare, it is assumed in most coalescence model analyses that the neutron yields are simply N/Z times the proton yields and the neutron energy spectra are identical to the (coulomb corrected) proton spectra. These assumptions can introduce

proj	V_{surf} cm/ns	N/Z	P_0 MeV/c	T MeV	R fm
C	7	3.34	122.30	7.00	7.37
	4	3.58	126.61	5.80	7.45
Ne	7	3.44	108.56	8.53	7.25
	4	3.52	102.18	6.82	8.07
Ar	7	3.29	115.94	10.41	7.28
	4	3.20	106.26	7.62	8.35
Zn	7	2.32	121.20	9.78	7.02
	4	2.11	102.58	6.67	8.88

Table 1.

some error into the determination of the coalescence radius. In this work we have used our data for ^{3}H and ^{3}He emission to determine the effective N/Z ratio. It is significantly different from that of the combined system as seen in Table I where ratios ^{3}H to ^{3}He yields observed at 45 degrees in the laboratory are presented for several different coulomb corrected surface velocities [13].

Also shown in the table are values of the P_0 parameters derived for tritons at those same velocities. We emphasize that the data at 45 degrees have been selected to minimize contributions to the particle emission from projectile like sources and further, that contributions from target like sources (determined from source fits) have been subtracted from the observed yields. Our results indicate that this technique is suitable for the velocity range represented in the table.

For a low density nuclear system at chemical and thermal equilibrium the temperature of the emitting system can be derived directly from the *first chance emission* double isotope yield ratios of two adjacent isotopes of two different elements [16,17]. Since the dynamic transport calculations indicate that thermal equilibrium is established only after some particle emission occurs care must be taken in the interpretation of isotope ratio temperatures but, suitably applied, this technique is quite promising [15,18,19]. Here we have extracted such temperatures as a function of velocity from yields of d,t, ^{3}He and ^{4}He particles, corrected for contributions from late stage emission. The results are presented in Table I. The average emission times derived from the relationship between emission energy and time ≈85 fm/1c at 7cm/m and 105 fm/c at 4 cm/m.

Understanding the system evolution in detail allows us, in principle, to derive the nuclear caloric curve for a well-defined low-density nuclear source with A   120. In our earlier work on heat capacities and the nuclear caloric curves we found evidence for a limiting temperature near 6.8 MeV for medium mass nuclei with A   125 excited to energies near 5 MeV/u [20]. We suggested that this limit might reflect an emission dynamics equivalent to a gentle boiling of the nuclear system in which nucleons with kinetic energies above the binding energy per nucleon are emitted from the system [21]. The temperatures derived here are in good accord with these earlier results. They are significantly higher than derived from time and energy integrated particle yields [18,19,22,23].

Our results are close to those obtained in temperature dependent Hartree-Fock calculations [24] or temperature dependent Thomas-Fermi calculations [25]. Results of such calculations are sensitive to the assumed nucleon-nucleon interactions and surface energies [26]. In addition, the recent Thomas-Fermi results indicate that the caloric curve is sensitive to flow dynamics [25]. Thus observed differences in caloric curves extracted from different reaction systems may reflect the particular dynamic evolution of the system being studied and great care must be taken to understand this dynamics.

Acknowledgments

The authors appreciate many useful conversations with Dr. B. A. Li. This work was supported by The Robert A. Welch Foundation and the United States Department of Energy (Grant No. DE-FE05-86ER40256) and the Polish Scientific Research Committee (Grant No. 2 2392 91 02).

[1]Present address INFN, Via Romea 3, I-35020 Legnaro, Italy

[2]Present address GANIL, BP 5027, F-14021, Caen Cedex, France

[3]Present address UNAM, A.P. 20-364, 01000 Mexico, D.F.

[4]Present address Shanghai Institute of Nuclear Research, Chinese Academy of Science, P.O. Box 800204, 201800 Shanghai, China

[5]Present address Arkansas State University, Dept. of Chemistry and Physics, P.O. Box 419, State University, Arkansas 72467-0419

[6]Present address Institute of Nuclear Physics, U.L. Radzikouskiego 152, Krakow, Poland

References

1. L. G. Moretto and G. Wozniak, Ann, Rev. Nucl. Sci **43**, 379 (1993).
2. J.A. Hauger *et al.*, Phys. Rev **57**, 764 (1998).
3. J. Hubele *et al.*, Z. Phys. A **340**, 263 (1991); Phys. Rev. C **46**, R1577 (1992).
4. N. Marie *et al.*, Phys. Lett. B **391**, 15 (1997).
5. R. Bougault *et al.*, Proceedings of the XXXV International Winter Meeting on Nuclear Physics, Bormio , 251 (1997)
6. P.-B. Gossiaux and J. Aichelin, Subatech Preprint , (1998).
7. Ph. Eudes *et al.*, Subatech Preprint , 98-05 (1998).
8. N. Marie *et al.*, Phys. Rev. C **58**, 256 (1998).
9. J. Lukasik and Z. Majka, Acta. Phys. Pol. **24B**, (1993 1959).
10. L. P. Csernai and J. I. Kapusta, Phys. Rep. **131**, 223 (1986).
11. A. Z. Mekjian, Phys. Rev. C **17**, 1051 (1978).
12. H. Sato and K. Yazaki, Phys. Lett. **98B**, 153 (1981).
13. T. C. Awes *et al.*, Phys. Rev. C **24**, 89 (1981).
14. W. J. Llope *et al.*, Phys. Rev. C **52**, 2004 (1995).
15. H. Xi *et al.*, Michigan State University Preprint MSUCL-1073 , (1997).
16. S. Albergo *et al.*, Nuovo Cimento **89**, 1 (1985).
17. A. Kolomiets *et al.*, Phys. Rev. C **55**, 1376 (1997).
18. F. Gulminelli and D. Durand, Nucl. Phys. A **615**, 117 (1997).
19. Z. Majka *et al.*, Phys. Rev. C **55**, 2991 (1997).
20. R. Wada *et al.*, Phys. Rev. C **39, 497**, (1988).
21. A. Bonasera, Contribution to the Les Houches Conference(1998).
22. J. Pochodzalla *et al.*, Phys. Rev. Lett. **75**, 1040 (1995).
23. M. B. Tsang *et al.*, Phys. Rev. C **53**, R1057 (1996).
24. P. Bonche *et al.*, Nucl. Phys. **A427**, 278 (1984).
25. S. K. Samaddar *et al.*, Phys. Rev. Lett. **79**, 4962 (1997).
26. H. Q. Song and R. K. Su, Phys. Rev. C **44**, 2505 (1991).

PARTICLE EMISSION AS A PROBE FOR DYNAMICS OF FISSION OF HEATED NUCLEI

W.H. TRZASKA, J. ÄYSTÖ, Z. RADIVOJEVIC, V.A. RUBCHENYA
Department of Physics, University of Jyväskylä, P.O.Box 35, FIN-40351, Jyväskylä, Finland
E-mail: trzaska@phys.jyu.fi

I.D. ALKHAZOV, S.V. KHLEBNIKOV, A.V. KUZNETSOV, A. EVSENIN, V.G. LYAPIN, O.I. OSETROV, G.P. TIOURIN, D.N. VAKHTIN
V.G. Khloplin Radium Institute, 194021, St.Petersburg, Russia

A.A. ALEXANDROV, YU.E. PENIONZHKEVICH, YU.V. PYATKOV, YU.G. SOBOLEV
Joint Institute for Nuclear Research, 141980, Dubna, Russia

M. MUTTERER, J. VON KALBEN
Institut für Kernphysik, Technische Hochschule, D-64289, Darmstadt, Germany

K. BRINKMANN
Institut für Kern- und Teilchenphysik, Technische Universität, D-01069, Dresden, Germany

New data on neutron, proton and alpha particle multiplicity measured for the reactions ^{58}Ni (340 MeV) + ^{112,118,124}Sn, ^{20}Ne (160 MeV) + ^{162}Er, ^{40}Ar (243 MeV) + ^{142}Nd, and ^{58}Ni (320 MeV) + ^{124}Sn are presented. The results are compared with new model calculations. New results on proton induced very asymmetric division of ^{238}U at intermediate energies are discussed. A short description of experimental set-up HENDES for investigation of fusion-fission processes is given.

1 Introduction

Understanding of dynamical processes governing nuclear matter is still far from complete. There is no coherent model to explain the mechanism of nuclear reactions much beyond what was already known, for example, from optical model or from statistical model predictions. Perhaps it was this lack of a major break through that, in the past, has diminished the interest in nuclear reaction research. However, as the quest for evermore-exotic nuclei reaches the fringes of the chart of nuclei, the lack of knowledge in the field of nuclear reactions becomes a serious obstacle. Production of radioactive ion beams (RIB) of high intensity, synthesis of super heavy elements, or population of exotic isotopes far from the stability line are just some of the topics that urgently need solutions. Having similar problems in mind, we have initiated nuclear reaction program at the cyclotron facility at the Department of Physics of the University of Jyväskylä. The main emphasis of our program is on the study of fusion- fission reactions using correlated measurements of light particles with fission fragments. We believe that particle emission is the most direct way to extract the information on the dynamical processes taking place following collision of heavy nuclei during their de-excitation and decay. For instance, particles emitted at different stages of fission process carry information on the temperature and on the time scale of nuclear processes. One can say that they work as a thermometer and a clock. By measuring several kinds of particles simultaneously one has several independent clocks and thermometers gauging the same process. This substantially increases the reliability of the measurement and of the subsequent interpretation of the data. Naturally, in addition to the measurements, we conduct thorough model analysis of the results. For that purpose, we have developed a modified version of the statistical model that includes dynamical hindrance of fission that depends on nuclear dissipation. We hope that, by applying modern equipment and improved model calculations, we can make a meaningful contribution towards the solving of at least some of the important problems of fission research.

2 Description of HENDES

HENDES stands for High Efficiency Neutron Detection System [1,2]. It consists of a set of fission fragment detectors, neutron detectors and light charged particle (LCP) detectors assembled in and around reaction chamber. As this set-up is constantly evolving, we would like to describe here shortly its main features. The device has been built and is operated by a joint Finnish-Russian-German collaboration since 1994 and is located at the K=130 cyclotron in Jyväskylä, Finland. Inside the thin stainless steel spherical reaction chamber of 0.8m diameter, there are two large stop detectors - position sensitive avalanche counters (PSAC) - and one or two start detectors. When two start detectors are used they are placed on the opposite sides of the target, each in front of the corresponding PSAC. Presence of a fast coincidence between the two PSACs and at least one of the start detectors is the required trigger condition. Our start detectors are mostly parallel plate avalanche counters but we have also used, with good success, micro channel plate detectors. They provide reference for all time-of-flight measurements. The remaining space inside the chamber is taken by various LCP detectors; mostly PIN diodes but also, depending on the application, by Si(Li) detectors, CsI(Tl) scintillators, proportional counters and so on.

Position sensitive neutron detectors (PSND), filled with NE213 liquid scintillator, are located outside the chamber at about 60cm from the target. Each PSND has 1m long active length and delivers 10 to 20cm position resolution; depending on neutron energy. Currently, two detector types are in use: cylindrical, with 6 cm diameter and 2.6 liter volume [3] and rectangular with 10 x 10 cm base and 10 liter volume [4]. There is a big flexibility as to the number and placement of PSNDs. However, there is seldom need for more than 8 detectors. To accommodate more and to prevent increased cross-talk one would have to increase the distance from the target thus ending up with only a small net gain in efficiency but with considerably more complex set-up. Similar arguments apply to the number of PSAC detectors. Even if there is a provision for at least 5, we seldom use more than 2 PSACs to leave space for LCP detectors and to minimize neutron scattering and interaction by reducing the amount of material inside the chamber.

3 HI induced fusion-fission reactions

After successful investigation of the reaction ^{180}Hf(^{40}Ar, fission) at beam energies E_{lab} = 180 - 249 MeV [5,6], we have started a comprehensive study of Ni induced fusion- fission reaction on Sn isotopes. The reasons for this choice are quite obvious. The abundance of stable Sn isotopes allows us to reach not only very neutron deficient compound systems but also those close to the stability line. At the same time, while Coulomb forces in the entrance channel remain the same, fission barrier height and light charged particle binding energy change significantly. Also, our calculations predict that, in addition to neutron emission, we should observe high multiplicity of protons and alpha particles. This is the necessary condition to make our approach of multiple clocks and thermometers possible.

While being more interesting, Ni + Sn reactions pose a more serious experimental challenge. The major difficulty results from the relative symmetry between beam and target nuclei. Naturally, fusion-fission is not the only reaction channel that is open; and certainly it is not the dominant one. At the bombarding energy of around 6 MeV per nucleon, in addition to elastic and quasi-elastic processes, we also have a whole range of deep inelastic reactions, including fast fission. Unfortunately, for this beam- target combination, these competing processes can not be easily distinguished from the proper (slow) fission of the compound nucleus. The only way to sort out the influence of the other reactions is by changing parameters of entrance channel leading to the same compound nucleus and the same excitation energy. For instance, to reach ^{182}Pt at about 105 MeV excitation energy, one can use the following reactions:

- ^{58}Ni + ^{124}Sn, $E_{lab} = 320$ MeV
- ^{40}Ar + ^{142}Nd, $E_{lab} = 243$ MeV
- ^{20}Ne + ^{162}Er, $E_{lab} = 160$ MeV

For each of them the contribution of deep inelastic and fast fission will be different. Therefore, it should be possible to extract time scale for the compound nucleus fission and for fast fission. As the excitation energy will remain constant, the data should also yield the dependence of particle multiplicity on the mean introduced angular momentum.

Recently, we have measured all these reactions as well as ^{58}Ni + ^{112,118,124}Sn at $E_{Ni} = 340$ MeV and ^{58}Ni + ^{112}Sn at $E_{Ni} = 320$ MeV. Although the data is not yet fully analyzed and some measurements need to be repeated, we would like to present here our preliminary results.

To get an indication of the emission source, during the analysis all particle spectra are kinematically separated into the ones emitted from the source moving with the velocity of the compound nucleus and those emitted from each of the fully accelerated fission fragment. There are later called pre- and post-scission particles respectively. Since the separation is strictly based on kinematics, we have no way to determine at which point of fission process particle emission took place. This can only be answered by calculations. As the multiplicity of pre-scission particles depends on the excitation energy and lifetime of the compound system, it is a good indicator of the time scale of the fission process (the clock) while the slope of the energy spectra provides information of nuclear temperatures (the thermometer). Table I summarizes our preliminary experimental results.

Table 1. Preliminary values of the observed neutron, proton, and α- particle multiplicity. The minus sign indicates that the collected statistics was not enough for a meaningful analysis.

Projectile	^{58}Ni	^{58}Ni	^{58}Ni	^{58}Ni	^{58}Ni	^{40}Ar	^{20}Ne
Energy	340	340	340	320	320	243	160
Target	^{112}Sn	^{118}Sn	^{124}Sn	^{112}Sn	^{124}Sn	^{142}Nd	^{162}Er
Compound	^{170}Pt	^{176}Pt	^{182}Pt	^{170}Pt	^{182}Pt	^{182}Pt	^{182}Pt
M_n^{pre}	0.42(10)	1.59(10)	2.33(30)	0.41(20)	1.30(20)	2.45(20)	3.85(30)
M_n^{post}	2.65(20)	2.97(20)	2.86(20)	2.01(25)	2.80(30)	3.37(30)	2.28(20)
M_n^{tot}	3.07(22)	4.56(22)	5.19(36)	2.42(32)	4.10(36)	5.82(36)	6.13(36)
M_p^{pre}	-	-	-	2.1	1.9	0.22	0.19
M_p^{post}	-	-	-	<0.1	<0.1	<0.1	<0.1
M_α^{pre}	0.87(20)	0.82(21)	0.74(19)	0.52	0.44	0.12	0.14
M_α^{post}	0.06(5)	0.06(5)	0.07(4)	<0.1	<0.1	<0.1	<0.1

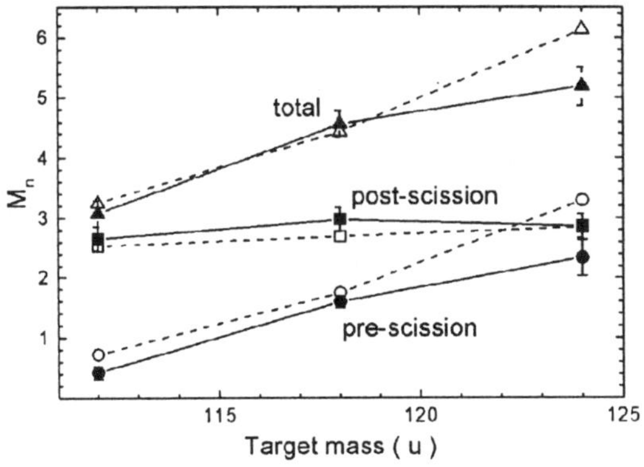

Figure 1. Pre-, post-scission, and total neutron multiplicities from 340 MeV ^{58}Ni induced fusion-fission reaction on ^{112,118,124}Sn. Solid symbols with error bars indicate measured data. The calculated values are shown with open symbols. The connecting lines are to guide the eye only.

There is a reasonably good agreement of the measured values with our model calculations [6]. The calculations were made using the new version of time-dependent statistical model which includes the main dynamical effects in fission: retardation of the fission degree of freedom; emission of light particles during the saddle-to-scission transition; de-excitation process of fission fragments. The reduced friction coefficient is parameterized as a function of the excitation energy to describe consistently the pre-scission characteristics and evaporation residue cross-sections.

Figure 1 shows the comparison of experimental (solid symbols) and calculated (open symbols) neutron multiplicities measured in the reaction 340 MeV ^{58}Ni on ^{112,118,124}Sn.

The measured post-scission values overlap perfectly with the predicted values. The pre-scission multiplicity for ^{112}Sn and for ^{118}Sn is also in agreement with the calculated value, however, the measured pre- scission multiplicity for ^{124}Sn is smaller (by about one unit) than the calculated value. If it is a real effect, such a drop of neutron multiplicity should be balanced by increased LCP multiplicity to fulfill energy conservation. As it can be seen from Figure 2, it is not the case.

Alpha particle multiplicity does not show any abnormality for ^{124}Sn target. It agrees well with our model predictions. The only discrepancy is for

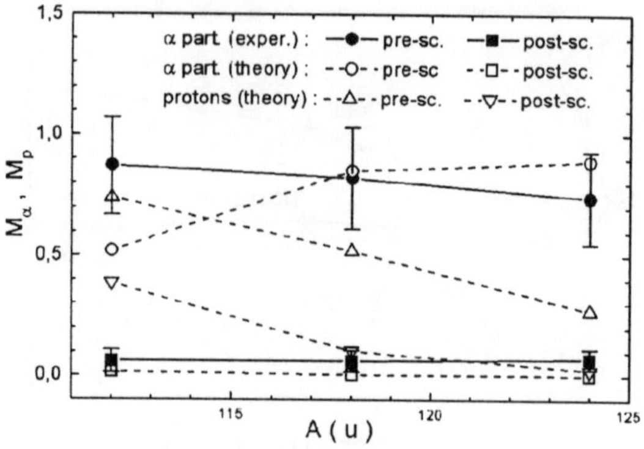

Figure 2. Pre-, and post-scission multiplicities of protons and alpha particles from 340 MeV ^{58}Ni induced fusion-fission reaction on ^{112,118,124}Sn. Solid symbols with error bars indicate measured data. The calculated values are shown with open symbols. The connecting lines are to guide the eye only.

^{112}Sn and it is reasonably within the accuracy of the experiment. Proton data was not taken at that time due to temporary lack of suitable detectors.

Figures 3 and 4 show, respectively, neutron and LCP multiplicities compared to the calculated values for the ^{182}Pt compound nucleus populated with the same excitation energy in three different reactions. The multiplicities were plotted as a function of effective fissility parameter x_{eff}:

$$x_{eff} = (4Z_P Z_T/(A_P^{1/3} A_T^{1/3}(A_P^{1/3} + A_T^{1/3}))/(Z^2/A)_{crit}$$

The agreement of measured and calculated neutron multiplicities for ^{20}Ne and ^{40}Ar induced reactions is excellent but there is a big discrepancy for the ^{58}Ni values. Similar situation is for LCP data (see Figure 4). Even if the experimental points have substantial uncertainties, there is a huge and unexpected jump in proton pre-scission multiplicity observed with ^{58}Ni beam. All the other values are reproduced reasonably well by the calculations.

There are several possible explanations of this discrepancy. First of all one can question the validity of experimental results. Indeed, we would like to repeat Ni irradiation to verify our measurements. However, the present results seem sufficiently self-consistent to be taken seriously. The increased multiplicity of charged particles is well compensated by the drop in neutron multiplicity. This balance is not only qualitative but also quantitative: the drop in total kinetic energy taken by neutrons is compensated by higher than expected share of kinetic energy carried out by LCP. It is therefore quite likely that with 320 MeV ^{58}Ni beam we do not yet obtain equilibrium of

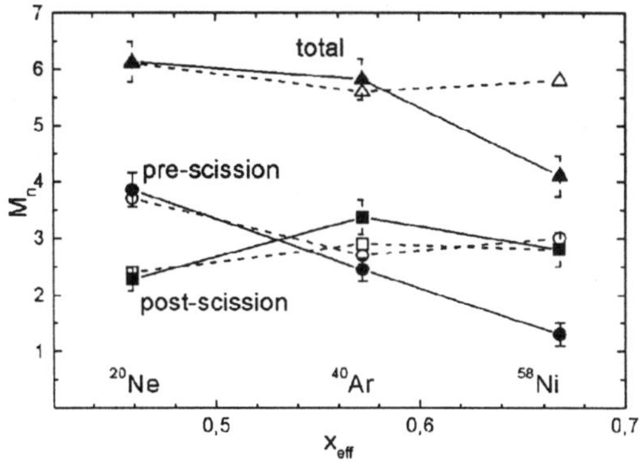

Figure 3. Pre-, post-scission, and total neutron multiplicities from 3 different reactions leading to the same compound nucleus ^{182}Pt at the same excitation energy of 105 MeV: ^{20}Ne (160 MeV) + ^{162}Er, ^{40}Ar (243 MeV) + ^{142}Nd, and ^{58}Ni (320 MeV) + ^{124}Sn. Solid symbols with error bars indicate measured data. The calculated values are shown with open symbols. The connecting lines are to guide the eye only. The values are plotted as a function of the effective fissility parameter.

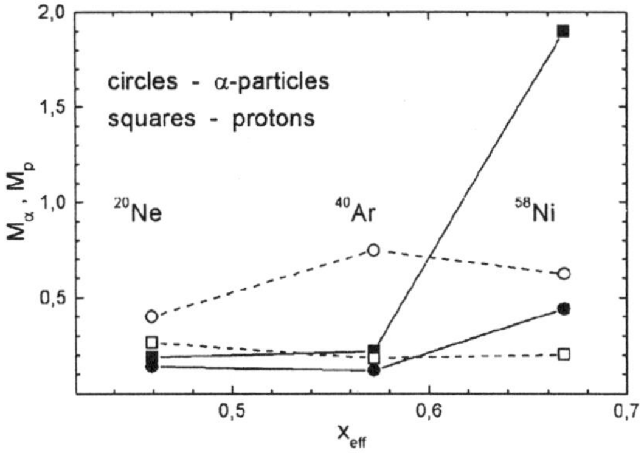

Figure 4. Pre-scission multiplicity of protons and alpha particles from 3 different reactions leading to the same compound nucleus ^{182}Pt at the same excitation energy of 105 MeV: ^{20}Ne (160 MeV) + ^{162}Er, ^{40}Ar (243 MeV) + ^{142}Nd, and ^{58}Ni (320 MeV) + ^{124}Sn. Solid symbols indicate measured data. The calculated values are shown with open symbols. The connecting lines are to guide the eye only. The values are plotted as a function of the effective fissility parameter.

the compound nucleus. Instead, particle emission takes place from the highly deformed, rotating system. Such a system would have increased LCP emission resulting from the lowering of the Coulomb barrier at the extremities. We may get a better agreement by modifying or updating the data on ^{58}Ni + ^{124}Sn reaction. Perhaps it would also improve our description of pre-scission neutron multiplicity on Figure 1 and change the value of the average angular momentum brought to the compound system as plotted in Figures 3 and 4.

At the higher bombarding energy of 340 MeV, particle emission conforms well to our calculations. It will be worthwhile to investigate closer the energy dependence of the fusion-fission process around this energy.

4 Search for super-asymmetric fission mode

The main limitation in the study of exotic, neutron-reach nuclei produced through fission, lies in very small yields. For instance, measured fragment mass distribution of fission products from ^{235}U(n_{th},f) [7] shows 4 orders of magnitude drop in the yield between mass 80 and mass 70. To search, in that mass range, for new isotopes and to investigate their properties one needs much higher yields. Hence, the renewed interest in fission mass distributions, especially for the extreme mass division. If a major super-asymmetric fission mode would ever be found for some beam/target combination, it could give an enormous boost to the way we produce exotic, neutron- reach nuclei. Unfortunately, by so far, all the earlier claims have been found to result from experimental artifacts that disappear under stringent analysis and/or sufficient statistics [8].

Available experimental data seem to indicate that the very asymmetric fission modes of heavy nuclei are prohibited in low energy fission. Better yields are achieved at moderate excitation energies. Indeed, an experiment at IGISOL facility in Jyväskylä has shown [9] that for proton induced fission on ^{238}U the drop in yield between mass 80 and 70 is only two orders of magnitude as opposed to 4 orders of magnitude that was observed in thermal neutron induced fission [7].

As HENDES is very well suited for efficient measurements of fission fragment mass distributions with mass resolution of up to 3 a.m.u. (FWHM), we have decided to verify the results obtained by our colleagues at IGISOL and extend the search for super-asymmetric fission modes to proton and deuteron induced fission on uranium and thorium targets at beam energies available at K=130 cyclotron in Jyväskylä. Preliminary analysis of the proton experiment confirms the IGISOL results. In addition, no appreciable change in the yield with the increase of proton energy from 20 to 50 MeV has been noticed. At mass 70 the yield at 50 MeV is not more than factor of 2 higher from the 20 MeV value (see Figure 5).

Thanks to the relatively high efficiency of HENDES we were able to accu-

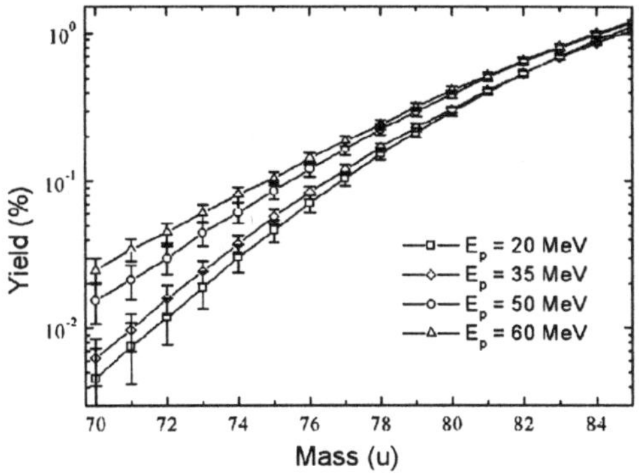

Figure 5. Fission fragments mass yields in the super asymmetric mass region for the reaction $^{238}U(p,f)$ measured at 4 proton energies: 20, 35, 50, and 60 MeV

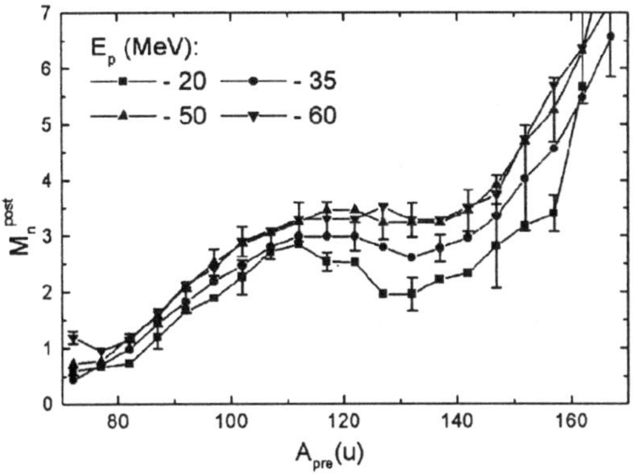

Figure 6. Post-scission neutron multiplicity as a function of fragment mass prior to the particle emission. Measurements were done for the reaction $^{238}U(p,f)$ at 4 proton energies: 20, 35, 50, and 60 MeV.

mulate sufficient data to investigate dependence of neutron multiplicity emitted from fission fragments (post-scission multiplicity) as a function of the mass number of the fragment for the $^{238}U(p,f)$ reaction. Figure 6 shows this dependence measured for 4 proton energies: 20, 35, 50 and 60 MeV. It is interesting to notice that the 50 MeV and the 60 MeV curves lie practically on top of each other. It would indicate that with 50 MeV protons we have already reached the saturation. It would be very interesting to extend this curve further into the asymmetric mass region but it would require substantially longer measurements.

In the latest experiments we have measured fission fragment mass distribution and neutron multiplicities from deuteron induced fission of ^{238}U. With the aid of LCP telescopes we have also looked for the ^{238}U(d,pf) mode to get some indication on mass distribution from fast neutron induced fission. Further experiments are planned at IGISOL facility in Jyväskylä and at other European laboratories.

Acknowledgments

Financial support of Academy of Finland, Center for International Mobility, and Nordic Council of Ministers for Research in Finland is gratefully acknowledged. This work has been supported by the Access to Large Scale Facility program under the Training and Mobility of Researchers program of the European Union.

References

1. W.H.Trzaska et al. *Proc. XV Nuclear Physics EPS Div. Conf., St.Petersburg, April 1995* , 246 (1995)
2. W.H.Trzaska et al. *Proc. XIV Int. Conf. Appl. Accel. Res. and Industry, Denton, Texas, November 6–9, 1996, AIP Conference Proc. 392, edited by J.L.Duggan and I.L.Morgan, AIP Press, New York 1997* , 1059 (1997)
3. A.V.Kuznetsov et al. *Nucl. Instr. Meth.* **A346**, 259 (1994)
4. K.Brinkmann: *private communication*
5. A.V.Kuznetsov et al. *Z. fur Phys.* **A354**, 287 (1996)
6. V.A.Rubchenya et al. *Phys. Rev.* **C58**, 1587 (1998)
7. J.L.Sida et al. *Nucl. Phys.* **A502**, 233c (1989)
8. F.-J.Hambsch, S. Oberstedt *Nucl. Phys.* **A617**, 347 (1997)
9. M.Huhta et al. *Phys. Lett.* **B405**, 230 (1997)

FISSION FRAGMENT MASS-ENERGY DISTRIBUTIONS AND GAMMA-RAY MULTIPLICITIES IN MULTIMODAL FISSION OF LIGHT THORIUM ISOTOPES

G. G. Chubarian[+,†], M.G. Itkis[†], N. A. Kondratiev[†], E. M. Kozulin[†], Yu. Ts. Oganessian[†], I.V. Pokrovsky[†], V. S. Salamatin[†], V. V. Pashkevich[‡], A.Ya. Rusanov[§], B. J. Hurst[+], D. O'Kelly[+], R. P. Schmitt[+], F. Hanappe[++], E. Liatard[ff], A. Huck[f], L. Stuttgé[f], L. Calabretta[&], C. Maiolino[&], K. Lukashin[&], C. Agodi[&], G. Bellia[&]

[+]Cyclotron Institute, Texas A&M University, TX 77843, USA;
[†]FLNR, JINR, Dubna, Russia 141980;
[‡]LTF, JINR, Dubna, Russia 141980;
[§]INP, Alma-Ata, Kazakhstan;
[++]ULB, PNTPM CP229, av. F.D. Roosewelt, B1050, Brussels, Belgium;
[ff]ISN, 53 avenue des Martyrs, 38026, Grenoble, Cedex, France;
[f]CRN, 23, rue du Loess, 67037 Strasbourg, Cedex 2, France
[&]LNS-INFN, 57 Corso Italia, I-95100, Catania, Italy;

The fission fragment mass-energy distributions and the γ-rays from the multimodal fission of the $^{220,224,226}Th$ was investigated. The corresponding excitation energies at the saddle point, E^*_{sp}, ranged from 16 to 40 MeV. As E_{sp}^* decreases, an asymmetric mass component becomes visible on the predominately symmetric mass distribution. The contribution of the asymmetric mode is characterized by the total yield ratio Y_s/Y_a, which decreases rapidly for the heavier isotopes of thorium. For all isotopes studied, the subtracted asymmetric fission component, $Y_a = Y_t - Y_s$, exhibits a complex structure. The average γ-ray multiplicities and relative γ-ray energies as a function of the mass of the fission fragments also exhibit a complex structure and strong variations. Such strong variations have never been previously observed in heavy ion-induced fusion-fission reactions. Obtained results may be explained with the influence of shell effects on the properties of the fission fragments.

1. Introduction

It is well known that dramatic changes occur in the mass and energy distributions of certain spontaneously fissioning Fermium isotopes and some other transuranic elements [1,2]. For the isotopes $^{258,259}Fm$ and ^{260}Md, fission produces narrow symmetric mass distributions and fragments with relatively high kinetic energies. Only small changes in the nucleonic composition radically alter this picture.

Dramatic changes in the symmetric/asymmetric character of the fission probabilities can be accounted for by theories, which incorporate a micro-macroscopic approach including the effects of nuclear shell structure on the formation of the fragments. Detailed calculations of the nuclear potential energy surface indicate the possibility of a co-existence of various (at least two) distinct fission modes [3]. Multimodal fission phenomena have also been observed in the Pb-At region in proton and alpha induced reactions [4], as earlier predicted [5].

Unfortunately, with light charged particles it is not possible to extend this work into the largely unstudied At-Th region due to the absence of suitable targets. In this region dramatic changes are expected to occur in the fission properties as a function of the nucleonic composition and the excitation energy. The transition form symmetric to asymmetric fission along with the accompanying changes in the barrier height and the saddle to scission times can be expected to strongly influence the fission mass and energy distributions, the fission cross sections, the fragment γ-ray multiplicities as well as the pre- and post-fission neutron multiplicities.

A new strategy for probing the At-Th region was developed several years ago utilizing heavy ion beams in near- and sub-barrier fusion-fission reactions [6]. Similar experiments have been performed at GSI to investigate Coulomb fission reactions of secondary radioactive beams [7].

The current paper focuses on the results of studies of the multimodal structure of the fission fragment mass and energy distributions of the neutron deficient nuclei ^{220}Th, ^{224}Th and ^{226}Th formed via ^{16}O and ^{18}O reactions at near and sub-barrier energies, and on the results of studies of the γ-ray multiplicities from the fission fragments of ^{226}Th isotope formed in reactions 78 and 75 MeV ^{18}O+^{208}Pb. Additional aspects of the work including pre- and post-fission neutron multiplicities will be presented in a future when the analysis has been completed.

2. Experimental procedures

The main problem in studying low energy fission ($20 < E^* < 30$ MeV) in the above mentioned transition region using heavy ion beams is the very small fusion-fission cross section near the Coulomb barrier. By choosing the appropriate target-projectile combination, one can achieve the "lowest possible excitation energy," $E^* = E_{cm} + Q$, in the compound system when $E_{cm} \approx B_C$ where Q is the fusion Q-value. This is, of course, the strategy used in cold fusion reactions for producing heavy nuclei [8]. As an illustration, Fig. 1 shows the fission excitation function for the reaction ^{208}Pb(^{16}O,f)^{224}Th. The range of bombarding energies which produce compound systems with $E^* = 20 - 30$ MeV corresponds to $72 < E_p(^{16}O) < 83$ MeV. Over this energy interval, the fission cross section varies from 10^{-3} to 10^{2} mb, placing severe constraints on the types of measurements which can be successfully conducted.

The experiments were carried out using time-of-flight spectrometer DEMAS and his following modification [12,]. This spectrometer was based on large area parallel-grid avalanche counters (PGAC) that provided a reliable and precise system for measuring mass and energy distributions of correlated fragments. The time-of-flight spectrometer consists of four PGAC's ("stop" detectors) with 200x300 mm^2 active area each and two small parallel plate avalanche counters (PPAC's) with areas of 30x40 mm^2 ("start" detectors). Each arm had a flight path about 40 cm. The spectrometer was calibrated with fission fragments from a thin ^{252}Cf source. The measured peak-to-valley ratio for the reconstructed masses is greater than 20, indicating a mass resolution $\Delta M/M \approx 3\%$ (results were compared with [13]).

The angular distributions, time-of-flight and energy spectra of neutrons were measured using eight elements of the DEMON array [14]. The γ-ray multiplicities were measured with an array of six 63x63 mm NaI detectors arranged symmetrically around the target.

3. Results and analysis

The characteristics of the reactions studied are summarized in Tab. 1. This table lists the entrance channels, the bombarding energies, E_i, the compound nucleus excitation energies, E^*, the fission barriers, E_f, the saddle point excitation energies, E^*_{sp}, the average total kinetic energies, <TKE>, the TKE dispersions, σ^2_{TKE}, and the mass dispersions, σ^2_M. The last column is related to the relative ratios of different fission modes (see below).

Figure 2 shows contour plots of fission events in the TKE versus mass plane for the systems ^{220}Th, ^{224}Th and ^{226}Th along with the projected total mass distributions (normalized to 200%). The average kinetic energies as a function of fragment mass are indicated by the circles superimposed on the contour diagrams. These plots clearly show that at similar excitation energies there is an increasing tendency to fission asymmetrically as the N of the compound system increases.

Mass and kinetic energy distributions for $^{224,226}Th$ at different bombarding energies are shown on Fig. 3. Evidence for the presence of different fission modes is also readily apparent in the mean kinetic energies as a function of fragment mass.

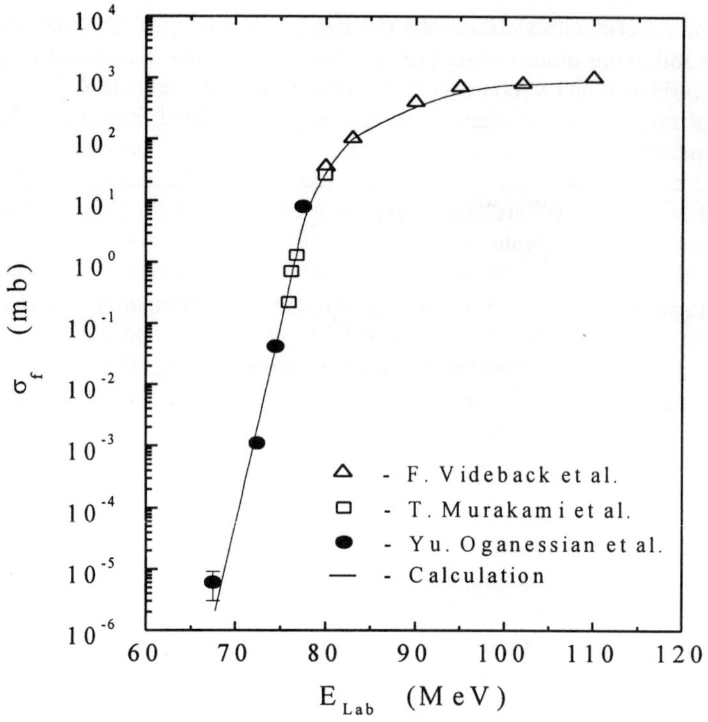

Figure 1. Fission excitation function for the system $^{208}Pb + ^{16}O$. Calculations and • - [9], □ - [10], Δ - [11].

Table 1. Parameters For the Various Fission Reaction.

Reaction	E_i, MeV	E^* MeV	E_f MeV	E^*_{sp} MeV	TKE MeV	σ^2_{TKE} MeV^2	σ^2_M amu^2	$\Sigma Y_s/\Sigma Y_a$
$^{208}Pb+^{18}O$	78	26.1	6.9	19.2	165.4±1.1	137±6	280±15	2.8±0.5
	75	23.3		16.4	165.2±2.0	150±11	291±22	2.4±0.8
$^{208}Pb+^{16}O$	108	53.8	7.2	46.6	162.4±1.0	137±9	224±15	-
	85	32.4		25.2	164.0±1.1	113±7	193±13	8.0±1.0
	77	25.0		17.8	163.6±1.2	98±6	187±12	4.8±0.5
	75	23.2		15.9	164.7±1.4	94±6	202±13	3.3±0.4
$^{204}Pb+^{16}O$	108	53.6	8.8	46.8	162.1±1.0	151±10	236±15	-
	85	34.3		25.5	163.1±1.2	118±7	188±12	7.6±0.8
	77	26.9		18.1	163.6±1.2	102±6	166±11	6.0±0.7
	75	25.0		16.2	163.7±1.4	103±6	162±11	6.3±1.0

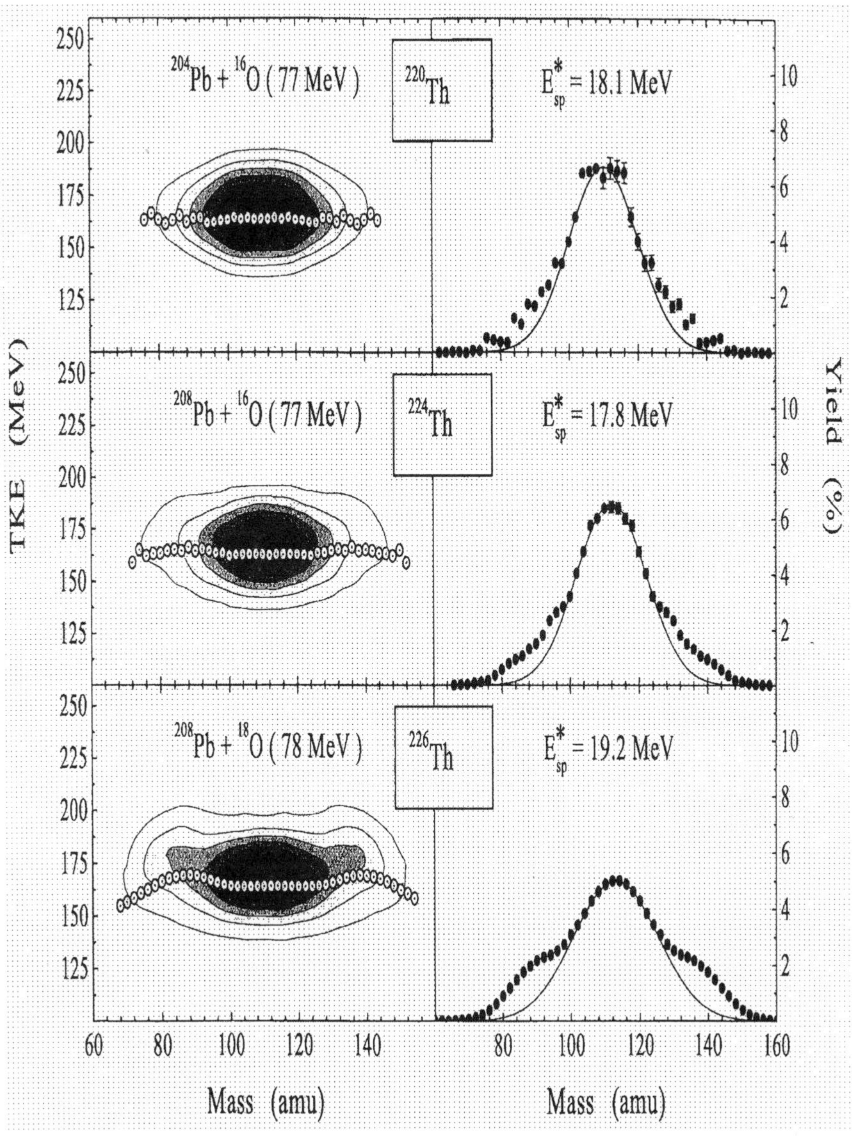

Figure 2. (Left) Contour plots showing the distribution of fission events in the TKE-mass plane for the reactions $^{204,\,208}Pb$ + ^{16}O (77 MeV), ^{208}Pb + ^{18}O (78 MeV). The open circles indicate the average TKE-values. (Right) Corresponding mass yield distributions for the reactions above. The curves reflect fits to the symmetric fission component.

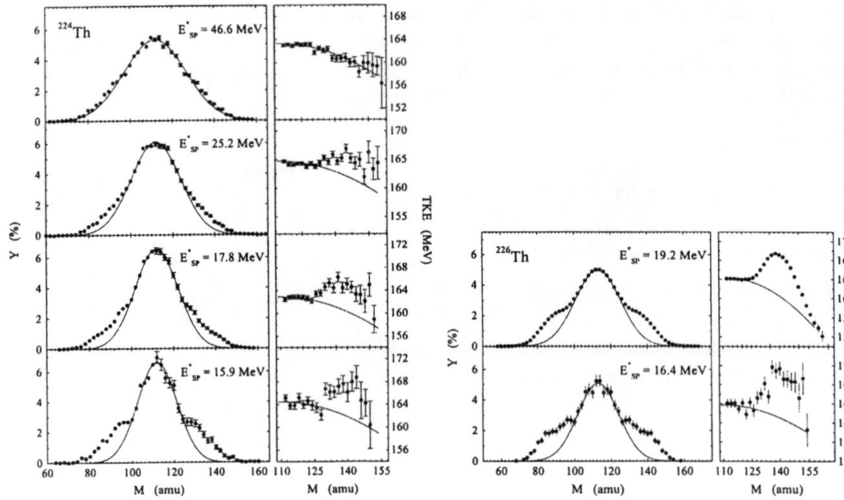

Figure 3. Mass yield distributions for fission following the reaction ^{208}Pb + $^{16,18}O$ at different bombarding energies and corresponding average TKE-values as a function of fission fragment mass.

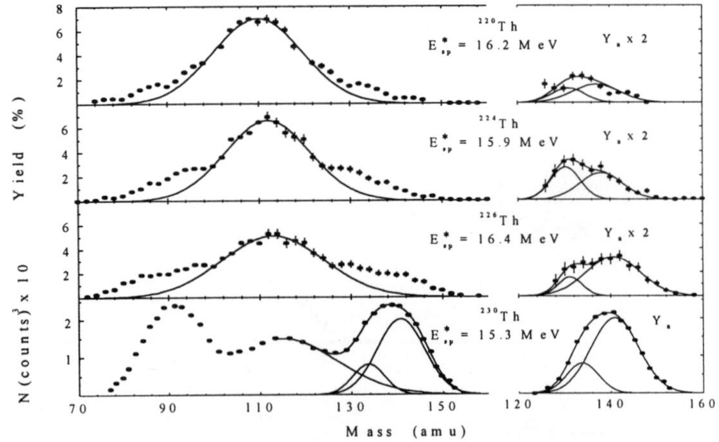

Figure 4. Mass distributions for fission of $^{220,224,226}Th$ and ^{230}Th [16] at excitation energies $E^*_{sp} \approx 16$ MeV. The curves depict fits used to decompose the asymmetric component into the contributions from the fission modes a_o and a_1 (see text).

According to the macroscopic model [15], the mean TKE is expected to exhibit a parabolic mass dependence. This behavior is clearly indicated at the highest excitation energies, ≈ 46 MeV. However, at the lowest bombarding energies, deviations from this smooth parabolic dependence as large as 15 MeV are visible in the mass range 132 - 140.

The data for Y(M), TKE(M), and $\sigma^2_{TKE}(M)$ suggest the presence of two asymmetric modes for M > 130 designated by a_o and a_1. These appear to peak around the mean masses M_{a1} = 132 and M_{a0} = 140. To estimate the contribution from each of the three modes (the symmetric mode and two asymmetric modes), the total mass distributions were analyzed assuming that each component is described by a Gaussian.

Thus, the total mass distributions, Y(M), normalized on 200% were subjected to least square fits with a total of five Gaussians.

The fits to the total and asymmetric portions of the mass distribution for ^{220}Th, ^{224}Th, ^{226}Th and ^{230}Th [16] for E^*_{sp} ≈ 16MeV are shown in Fig. 4. The left part of the figure presents the total mass distributions. The fits to the asymmetric portions of the mass distributions are shown on the right. The curves show the yield distributions of $Y_{a1}(M)$ and $Y_{a0}(M)$, which are centered about M = 132 and 140, respectively, and their sum. It is interesting to note that the Y_{a1} component decreases in yield as the compound system becomes more neutron deficient.

The last column of Tab. 1 shows the Y_s^t/Y_a^t ratio from the five-component fits

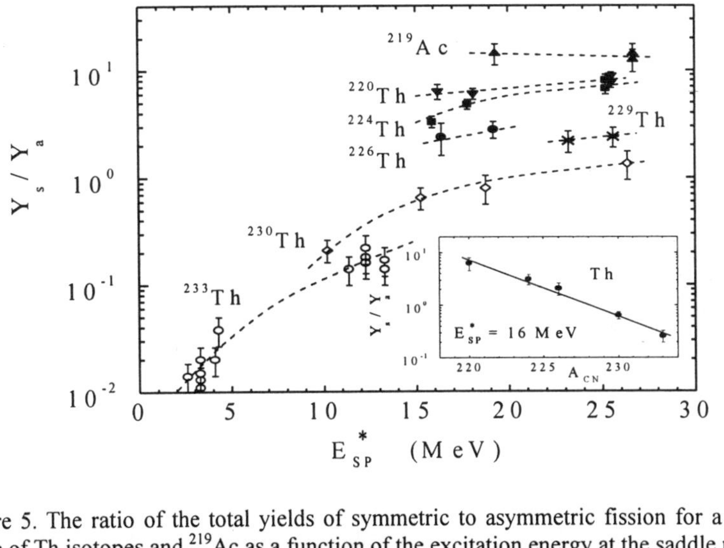

Figure 5. The ratio of the total yields of symmetric to asymmetric fission for a wide range of Th isotopes and ^{219}Ac as a function of the excitation energy at the saddle point. ✳ - [16], ✧ - [16, 17], ○ - [18-23], ▲, ▼, ■, ● - present work. (Insert) The same ratio as a function of compound nucleus mass for fixed excitation energy.

to the ^{220,224,226}Th mass distributions. The dependence of ratio Y_s^t/Y_a^t on the excitation energy of the fissioning nuclei at the saddle point is shown in Fig. 6. Additional results are also shown from analyses of the mass distributions for other Th isotopes studied in reactions induced with neutrons, γ-quanta, and ^{4}He ions [17-23].

The values of Y_s^t/Y_a^t for all nuclei studied at the same excitation energy, $E^*_{sp} \approx 16$ MeV, are shown in insert in Fig. 5. The main point of this figure is to demonstrate the strong dependence of Y_s^t/Y_a^t on the neutron number of the fissioning nucleus.

Dramatic effects are observed in the mass distributions when they are gated on different TKE values. Figure 6 shows the energy gated differential mass distributions for ^{224}Th and ^{226}Th at compound nucleus excitation energies of E^*=26 MeV. It is readily apparent in the case of ^{226}Th that changing the total kinetic energy TKE from 146 MeV to 194 MeV produces strong transformations in the mass distributions.

Thus, if the N of the fissioning nucleus is changed by just two neutrons, one observes a transition from symmetric to asymmetric fission when the fission excitation energy, Q-TKE, approaches 0.

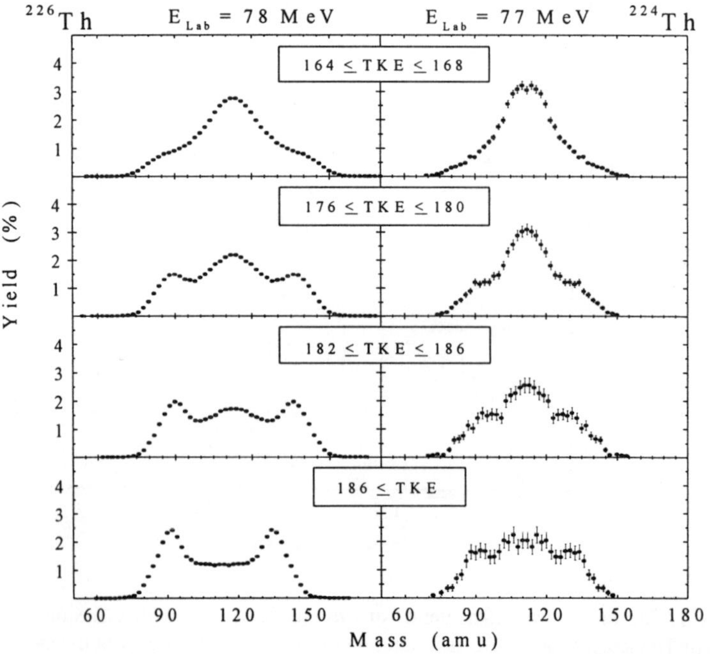

Figure 6. Mass yield distributions for ^{226}Th (Left) and ^{224}Th (Right) for four TKE windows.

Further understanding of the characteristics of the fission modes can be achieved by analyzing the differential total kinetic energy spectra at different mass asymmetries. Figure 8 shows the experimental spectra, Y(TKE), for ^{226}Th (E_i=78MeV) for three different mass regions: M=A/2=112-114, M=134, and M=142. Both the average total kinetic energy, and the corresponding dispersion, σ_{TKE}, vary with mass cut. When one of the fission modes is dominant (i.e., symmetric fission for M=112-114, or asymmetric fission for a_0 at M=142), the spectra are essentially Gaussians. The energy dispersions σ_{TKE} are also smaller than those seen in the spectra for M=134 where there are comparable contributions from all the fission modes (s, a_0 and a_1).

The TKE spectra for M=132 and 134 have been fitted with three Gaussians. Both the dispersion and the area for one of the Gaussians had a fixed value. Since the $\sigma^2_{TKE}(M)$ for symmetric fission does not depend on fragment mass, it was assumed that σ^2_{TKE} (M=A/2), which has been obtained from the fit to the TKE spectrum with M=112-114. The relative contribution of Y_s^t to Y^t was determined from the decomposed mass yield. The results of the analysis are illustrated in the middle panel of Fig. 7. This shows the experimental energy spectra, the simulated distributions for

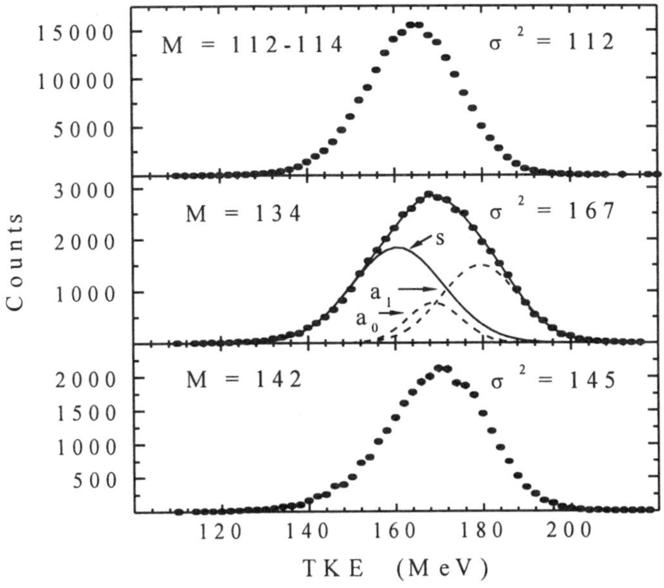

Figure 7. Energy spectra of fragments from the fission of ^{226}Th for three mass cuts. (Center) The curves show the results of fits to the three components. Energy spectra for all symmetric fission s (solid curve), and for the two asymmetric fission modes a_0 and a_1 (dashed curves).

the symmetric fission mode (solid curve), and difference spectra (dashed curves), which should correspond to both the asymmetric modes a_0 and a_1. A similar analysis was carried out for M=132. These observations again provide strong evidence for independent fission modes.

4. Gamma–ray multiplicities

In the current section we will discuss the results of studies of the γ-ray multiplicities from the fission fragments of the isotope ^{226}Th formed in reactions 78 and 75 MeV $^{18}O+^{208}Pb$.

Figure 8(a) shows relative average γ-ray energies, $\langle E_\gamma \rangle$ (top, solid circles), $\langle M_\gamma \rangle$, (middle, open circles) as a function of the mass of fragments for the reaction ^{208}Pb(^{18}O,f) at E_i=78 MeV. The mass distribution of the fission fragments is presented in the bottom of the figure. The variation of the $\langle M_\gamma \rangle$ with fragment mass is dramatic. Near symmetry $\langle M_\gamma \rangle$ is about 12. As one deviates from symmetry, $\langle M_\gamma \rangle$ stays approximately constant over a small mass interval and then plunges almost precipitously. Around A ≈ 94 and 132, small shoulders are observed and then $\langle M_\gamma \rangle$ falls to a minimum value of about 8 around A ≈ 86 and 142. At larger mass asymmetries, the multiplicity rises and then appears to fall.

Such a structure and strong variation in $\langle M_\gamma \rangle$ have never been previously observed in heavy-ion induced fission. The dependence of $\langle E_\gamma \rangle$ on mass also exhibits a strong variation. Near symmetry it has wide minimum and increases sharply while approaching the range of asymmetry in mass distribution. Around masses mentioned above, $\langle E_\gamma \rangle$ has well pronounced local maximums. For the largest masses, approximately in the range where $\langle M_\gamma \rangle$ rises again, $\langle E_\gamma \rangle$ appears to fall. The total variation of $\langle E_\gamma \rangle$ is about 30%. It seems likely that the marked structure in $\langle M_\gamma \rangle$ and $\langle E_\gamma \rangle$ is due to strong shell effects in the cold fission fragments, where predominantly rare, but high-energy γ-ray transitions are possible.

It is interesting to compare the present results with liquid drop calculations [24]. The dashed curve in fig.8(a) shows the results of such calculations assuming total fragment spin $\ell=15\hbar$ and $S_T = 2(\langle M_\gamma \rangle-6)$ [25]. The calculations predict a maximum at symmetry, however, the calculated values are larger than the experimental values. This is not particularly surprising since the assumed orbital angular momentum is a rough estimate. Furthermore, no corrections have been made for the spin removed by evaporated neutrons. The calculations do not include the influence of shell effects, which should have a significant impact on the fragments shape and the relevant temperatures.

As it was shown in higher, TKE may be a convenient tool for separation of distinct fission modes. Figure 8(b-d) shows $\langle M_\gamma \rangle$ as a function of the fragment mass for three selected ranges of TKE. Mass distributions for the same TKEs' are also presented. For low TKEs', where the input of the asymmetric masses is small, $\langle M_\gamma \rangle$ reaches its highest values, in the mean time the variation of $\langle M_\gamma \rangle$ is the smallest. On the contrary, at the highest TKE where the yield of asymmetric fission fragments is dominating, variation of the $<M_\gamma>$ is the largest, but has the lowest absolute value.

Gamma-ray multiplicities as a function of the TKE for different ranges of mass are shown on the figure 9(a-c). For all selected ranges (total, symmetric and asymmetric) $<M_\gamma>$ exhibits same behavior, but absolute values are different. $<M_\gamma>$ is

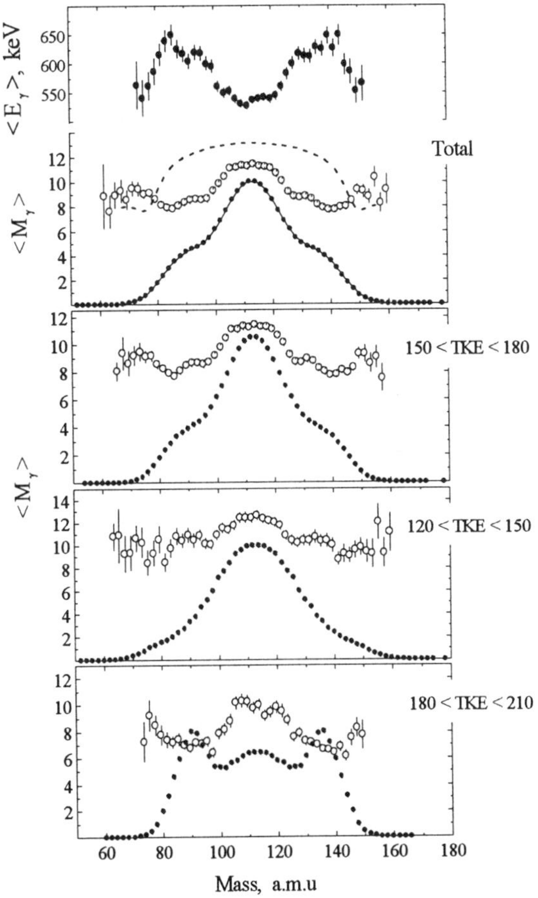

Figure 8. $<M_\gamma>$ and $<E_\gamma>$ as a function of the fission fragment mass. See details in text.

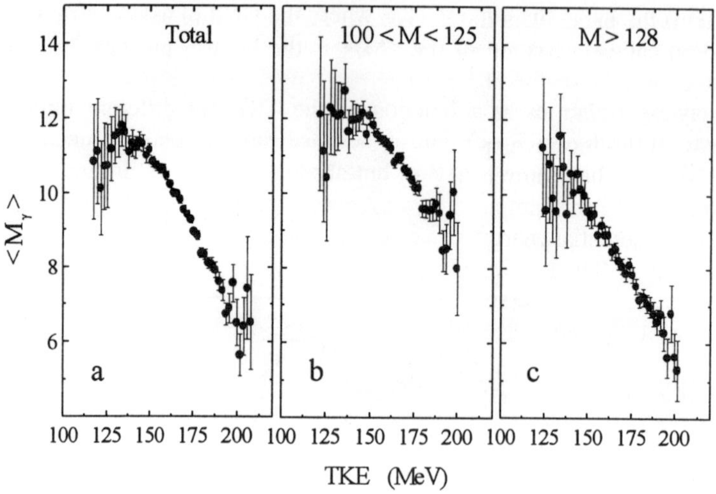

Figure 9. <M_γ> as a function of fission fragment total kinetic energy. See details in text.

approximately 30% higher for symmetric masses than for asymmetric part of the mass distribution, where the influence of shell structure on the fission fragments is the strongest.

5. Discussion and results

The potential energy surfaces for the various Th isotopes have been calculated as a function of the mass asymmetry. Cassinian ovals [5] were used to describe the nuclear shapes. In the calculations, shell corrections were taken into account using spherical harmonics up to 7th order. For a given shape, a Woods-Saxon single-particle potential was generated as a function of the distance $l(\vec{r})$ from the origin to the surface [26].

The methods used to calculate the single-particle spectrum [26, 27], the shell corrections [28], the liquid drop parameters [29], and the potential [30] are the same as those described in ref. 3. In some cases, the surface diffuseness was taken into account using the "Yukawa plus exponential" model [31] with the parameters given in ref. 36. The results of calculations for ^{220,226}Th are shown in Fig. 10. Here the potential energy is given as a function of the elongation parameter α and the mass asymmetry parameter α_3. At each point on the surfaces, a minimization was carried out with respect to the parameters α_4-α_7. The surfaces are completely symmetrical with respect to α_3, so only one of the two asymmetric valleys is presented here. Because of the choice of scales, the ground state, the second minimum, and the intervening barrier (usually called A) are not visible in Fig. 10. For all three systems, a symmetric valley develops at $\alpha \approx 0.7$. This valley runs almost parallel to the asymmetric valley. The occurrence of these two valleys obviously influences the final mass asymmetry. For example, in ^{220}Th one sees that as the system evolves towards scission the potential energy surface favors a shift to the valley with smaller mass asymmetry. In ^{226}Th the two valleys are separated by populated from the primary asymmetric valley by passing

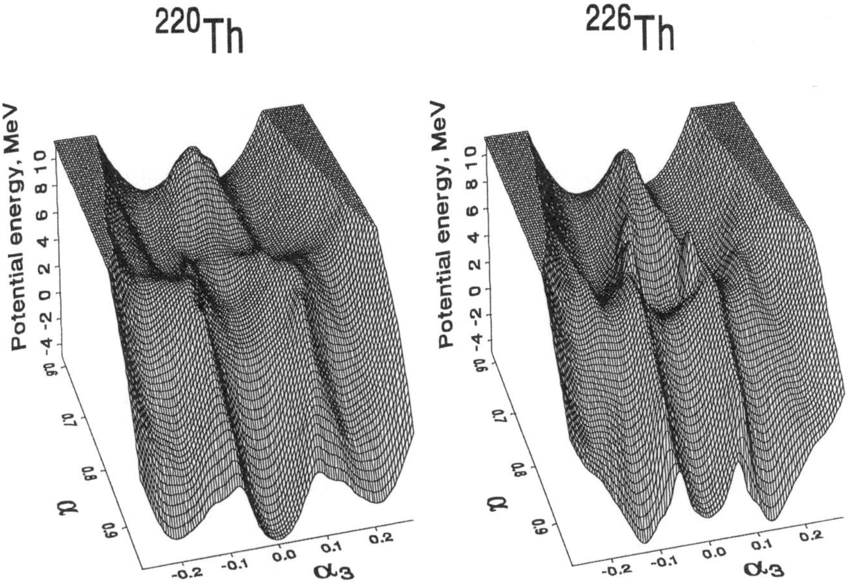

Figure 10. Calculated potential energy as a function of the elongation parameter, α, and the asymmetry variable, α_3, for the fission of ^{220}Th, and ^{226}Th.

over the very thin ridge that separates them. This ridge, apparently due to the shell effects, should quickly disappear as the excitation energy is increased.

The same type of potential energy surface calculations was performed for all the thorium isotopes from ^{220}Th and ^{232}Th. All of the surfaces show the same basic valley structure. The differences lie in the quantitative details of the surfaces, i.e., the depths of the asymmetric and symmetric valleys, the height of the third fission barrier (E_C), the energy at the exit of the asymmetric valley, and the energy of the pass (E_S) between the asymmetric and the symmetric valleys. The latter quantity is expected to play a crucial role in determining the ratio of asymmetric to symmetric fission.

6. Summary

The mass and the energy spectra of the fission fragments of the neutron deficient isotopes ^{220,224,226}Th have been investigated using near- and sub-barrier fusion reactions. The data gathered in this work provide a detailed view of the evolution of the mass and the energy distributions with the excitation energy and the nucleonic composition of the fissioning system.

The measured mass and energy distributions of the fission fragments show that the ratio of symmetric to asymmetric fission depends very strongly on the neutron number of the Th nucleus. These distributions also exhibit features associated with multimodal fission phenomena.

Calculations of the fission potential energy surfaces have been presented. The predicted features of these surfaces are in good agreement with the observed experimental features.

The properties of γ-rays from the fission fragments of the neutron deficient isotope of ^{226}Th have been investigated using near- and sub-barrier fusion-fission reactions. The data gathered in this work provide a detailed view of the evolution of the $\langle M_{\gamma} \rangle$ and $\langle E_{\gamma} \rangle$ with the mass of the fission fragment.

Complex structure and strong variation of the $\langle M_{\gamma} \rangle$ and $\langle E_{\gamma} \rangle$ as a function of mass and TKE may be explained with the influence of shell structure of the fission fragments on the yield and energy of the γ-rays.

For accurate fit of $\langle M_{\gamma} \rangle$ dependencies one needs shell corrected theoretical calculations of the shapes and temperatures of the fission fragments at the scission point.

This was supported by the Russian Fundamental Research Foundation, under grant 96-02-17743A, and the US Department of Energy, under grant DE-FG03-93ER40773, and the US National Science Foundation, under grant PHY96-03143.

References

1. Hullet E.K et al., *Phys. Rev. Let.* 56, 313 (1986).
   Hullet E.K. et al., *Phys. Rev.* C 40, 779 (1989).
2. Hoffman D., Proc. of the 4th Int. Symposium on the Phys. and Chem. Of Fission, Julich, Germany,
   1979, 1, 275 (IAEA, Vienna, 1980).
3. Möller P., Nix R.J. and Swiatecki W.J, *Nucl. Phys.* A 492, 349 (1989);
   Pashkevich V.V., *Nucl. Phys.* A 477, 1 (1988);
4. Itkis M.G., Okolovich V.N.,Rusanov A.Ya. , Smirenkin G.N., *Z. Phys.* A 320,

433 (1985);
Itkis M.G., Okolovich V.N., Smirenkin G.N., *Nucl. Phys.* A 502, 243 (1989) 243;
5. Pshkevich V.V., *Nucl. Phys.* A 169, 275 (1971).
6. Itkis M.G. et al., JINR Rapid Commun. 3 (1994) 19;
7. Shmidt K.-H. et al., *Phys. Lett.* B325, 313 (1994).
8. Oganessian Yu.Ts. et al, In Proc. of the Robert A. Welch Conf. On Chemical Research, Houston,
159, Texas, 1990.
9. Oganessian Yu.Ts. et al., *JINR Rapid Commun.*, 1(75), 123 (1996).
10. Murakami T. et al.,, *Phys. Rev.* C 34, 1353 (1986).
11. Videback F., Goldstein R.B., Grodzins L., and Steadman S.G., *Phys. Rev.* C 15, 954 (1977).
12. Chubarian G.G. et. al., *Sov. Journal of Nuclear Phys.* 56(3), 3 (1993).
13. Whetstone S.L., *Phys. Rev.* 131, 1232 (1963).
14. Tilquin I. et al., *Nucl. Instr. And Meth. In Phys. Res.* A 365, 446 (1995).
15. Cohen S., Swiatecki W.J., *Ann. Phys.* 22, 406 (1963).
16. Unik J.P., Huizenga J.R., *Phys. Rev.* 134, B390 (1964).
17. Britt H.C., Wegner H.E., Gursky J.C., *Phys. Rev.* 129, 2239 (1963).
18. Pfeiffer E., *Z. Phys.* 240, 403 (1970).
19. Dubrovina S. M. et al., *Sov. Journal of Nuclear Phys.* 17, 470 (1973).
20. Gonnenwein F., Pfeiffer E., *Z. Phys.* 207, 209 (1967).
21. Glendenin L.E., *Phys. Rev.* C 22, 152 (1980).
22. Vorobjeva V.G., Djachenko N.P., Kuzminov B.D., In Proc. of the All Union Conf. On Neutron
Physics, Kiev, USSR, April 1977, Part 3, 183, Atomenergoizdat, Moscow, 1997.
23. Holubarsch W, Pfiffer E., Gonnenwein F., *Nucl. Phys.* A 171, 631 (1971).
24. Schmitt R.P. et.al., *Z. Phys.* A 321, 411 (1985).
25. Schmitt R.P. et.al., *Nucl. Phys.* A 487 370 (1988).
26. Damgaard J., Pauli H.C., Pashkevich V.V., Strutinsky V.M., *Nucl. Phys.* A 135, 432 (1969).
27. Pashkevich V.V. and Strutinsky V.M., *Sov. Journal of Nuclear Phys.* 9, 56 (1969).
28. Strutinsky V.M., *Nucl. Phys.* A 95, 420 (1967).
29. Myers W.D. and Swiatecki W.J., *Avk. Fys.* 36, 343 (1967).
30. Ivanova S.P., Komov A.L., Malov L.A., Soloviev V.G., *Particles and Nuclei* 7(2), 450 (1976).
31. Krappe H.J., Nix J.R. and Sierk A.J., *Phys. Rev.* C 20, 992 (1979).

STRUCTURE OF SUPERHEAVY NUCLEI

A. SOBICZEWSKI, I. MUNTIAN, Z. PATYK
Soltan Institute for Nuclear Studies, Hoża 69, PL–00–681 Warsaw, Poland and GSI, D-64291 Darmstadt, Germany

Recent theoretical studies of structure and properties of superheavy nuclei are shortly reviewed. Properties such as deformations, single-particle spectra, shell correction to mass, rotational energies and half-lives with respect to main decay modes are discussed. Even-even nuclei with proton number $Z = 82 - 120$ and neutron number $N = 126 - 190$ are considered.

1 Introduction

The objective of this paper is to give a short review of recent theoretical studies on the structure and properties of superheavy nuclei. Properties such as deformation, single-particle energies, shell correction to energy (mass), rotational spectra, main decay modes (α decay and spontaneous fission) and respective half-lives are discussed. Even-even nuclei with proton number $Z = 82 - 120$ and neutron number $N = 126 - 190$ are considered. The paper is an extension of the work [1], to include more recent results.

We concentrate on the studies performed in a macroscopic-microscopic approach (e.g. [2,3,4,5,6,7], cf. also the reviews [8,9,10]), which seem to be, at the moment, most complete. Some properties of heaviest nuclei have been, however, also studied by fully microscopic methods, like Hartree–Fock–Bogolubov and Relativistic Mean Field approaches (e.g. [11,12,13,14]).

The theoretical studies are closely connected with, and motivated by, an intensive experimental activity in this field (e.g. [15,16,17,18,19]).

2 Essential role of shell effects

Shell effects are important for all nuclei. Their role for the heaviest nuclei is, however, essential, as many of them would not exist at all without these effects [20].

The analysis of shell effects, performed in [20], has shown that these effects elongate the α-decay half-lives T_α of known even-even heavy nuclei by up to about 5 orders of magnitude, and the spontaneous-fission half-lives $T_{\rm sf}$ by up to about 15 orders of magnitude.

The mechanism by which practically whole half-life of a very heavy nucleus is made up by shell effects is illustrated in Fig. 1. The figure shows the spontaneous-fission barrier of the nucleus $^{264}108$ (^{264}Hs), i.e. the dependence of the ground-state energy of this nucleus on its quadrupole-deformation parameter β_2. For each β_2, the energy is minimized with respect to the hexadecapole-deformation parameter β_4. The total fission barrier (Y+SHELL), including shell effects, is shown by solid line and its smooth part (obtained by the Yukawa-plus-exponential model (Y) [21]), by dashed line. The smooth barrier obtained by another macroscopic model (liquid drop, LD [22]) is also shown (dotted line), for comparison.

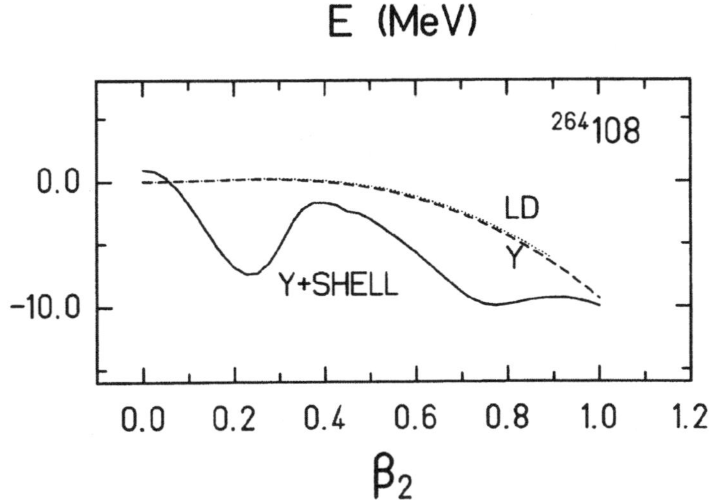

Figure 1. Total fission barrier (Y+SHELL) and its macroscopic (smooth) part obtained by the Yukawa-plus-exponential (Y) and liquid-drop (LD) models, for the nucleus [20] $^{264}108$.

One can see that a significant height (about 6 MeV) of the fission barrier is obtained only after the inclusion of shell effects. Without them, no fission barrier (Y and LD) appears.

A particular feature of the considered region of nuclei is that some deformed nuclei show shell effects which are similarly strong as the effects observed in spherical magic nuclei, i.e. that we observe deformed shells in these nuclei. Specifically, effects of the deformed neutron shell at the neutron number $N = 152$ are experimentally observed for a long time. There is also, however, an increasing experimental evidence for the existence of even stronger deformed shells at $N = 162$ and $Z = 108$, predicted theoretically. The nucleus

$^{270}108$ (^{270}Hs) is expected theoretically [23,2] to be a doubly magic deformed nucleus.

3 Theoretical model

As already stated in the Introduction, we concentrate on the macroscopic-microscopic approach. In this approach, the energy (mass) of a nucleus is composed of two parts: macroscopic and microscopic. The macroscopic part is usually described by the Yukawa-plus-exponential model [21]. The microscopic part is the Strutinski shell correction, based on a model for the internal (microscopic) structure of a nucleus. As this model, we take the Woods-Saxon single-particle potential [24].

Equilibrium deformation of a nucleus is calculated by minimization of its energy in a multidimensional deformation space [25]. A study performed in [2,3] has shown that the 4-dimensional deformation space $\{\beta_\lambda\}$, $\lambda = 2, 4, 6, 8$, is sufficient for the heaviest nuclei considered in this paper. Here, β_λ are the usual deformation parameters, appearing in the expression for nuclear radius (in the intrinsic frame of reference) in terms of spherical harmonics.

To obtain the energies of the ground-state rotational band, the moment of inertia of a nucleus is calculated. This is treated in the cranking approximation (cf. e.g. [26,27,28]).

The $\alpha-$decay half-lives are described by the phenomenological formula of Viola and Seaborg, but with its free parameters readjusted to account for recent data. Details of the calculations are given in [2,3].

The spontaneous-fission half-lives are analyzed in a dynamical way, with the mass tensor (describing the inertia of a nucleus with respect to its deformation) taken into account. Details of the calculations are given in [4].

In the study of the effect of the nuclear surface diffuseness on nuclear properties, the macroscopic part of energy of a nucleus is treated in the Thomas-Fermi approach [7].

4 Basic results

4.1 Equilibrium deformations

The equilibrium deformations of heaviest nuclei have been studied in [3], as has been already mentioned in sect. 3. The analysis has been performed in the 8-dimensional deformation space $\{\beta_\lambda\}$, $\lambda = 2, 3, ..., 9$. In accordance with [25], it has been found, however, that the odd-multipolarity components β_λ^0, $\lambda = 3, 5, 7$, of the deformation are different from zero for only a very few of the nuclei (which are situated around radium). Thus, almost all nuclei considered here are reflection symmetric. Due to this, we only show the results for even-multipolarity deformations β^0_λ , $\lambda = 2, 4, 6, 8$. Contour maps of these deformations are given in Fig. 2, taken from [3]. One can see that most of the investigated nuclei are deformed. Only two rather small regions of spherical nuclei, situated around the doubly magic spherical nuclei ^{208}Pb and ^{298}114, appear: a smaller one around ^{208}Pb and a slightly larger region around ^{298}114. The main, quadrupole, component of the deformation, β^0_2, is largest and is positive in the whole region of deformed nuclei. Its value ($\beta^0_2 \approx 0.24$) is about constant in a large part of the region (around its center) and it rapidly decreases as one moves towards the boundaries of this region. The higher-multipolarity components are smaller and they change sign as one moves through the region. The largest β^0_6 (in their absolute values) are obtained for nuclei around ^{254}102, they are negative and equal to about -0.05. The values of β^0_8 are even smaller in their absolute magnitude. The largest β^0_8 are equal to about 0.03 and are obtained for nuclei which are close to the nucleus ^{270}108 (^{270}Hs).

4.2 Shell correction to energy (mass) of a nucleus

Shell correction to the ground-state mass of a heavy nucleus gives us a first orientation in the stability of this nucleus. Figure 3, taken from [3], shows the shell correction, $E_{\rm sh}$, calculated for the large region of nuclei under consideration. One can see that $E_{\rm sh}$ has three minima in this region. The first one, which is the deepest ($E_{\rm sh}$=–14.3 MeV), is obtained for the doubly magic spherical nucleus ^{208}Pb. The second one ($E_{\rm sh}$=–7.2 MeV) appears at the nucleus $^{270}108_{162}$, which is predicted [2] to be a doubly magic deformed nucleus. The third minimum, with the same depth ($E_{\rm sh}$=–7.2 MeV) as that of the second minimum, is obtained for the nucleus $^{296}114_{182}$, which is close to the nucleus $^{298}114_{184}$ predicted [29,30] to be a doubly magic spherical nucleus, the next one to the last experimentally known ^{208}Pb.

One can see in Fig. 3 that some of the already synthesized nuclei profit by 6-7 MeV in their binding energy from the shell correction. Without this profit they could not exist, as discussed in [20].

The appearance of the region of nuclei around the second minimum (deformed superheavy nuclei) constitutes the main change in our view of stability of heaviest nuclei in recent years. Before, it was believed for a long time that spherical superheavy nuclei, predicted to be situated around the third minimum, would constitute an island, separated from the usual peninsula of relatively long-lived nuclei by an "ocean" of full instability. After the appear-

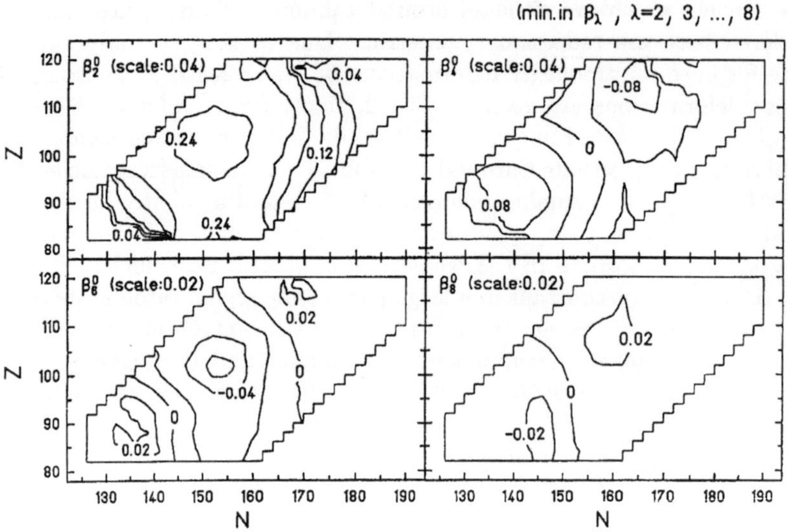

Figure 2. Contour maps of the equilibrium deformations β^0_λ, $\lambda = 2, 4, 6, 8$, plotted as functions of the proton Z and the neutron N numbers. Numbers at the contour lines give the values of the deformations. The difference in the values between neighbouring contour lines is specified by the scale [3].

ance of deformed superheavy nuclei, however, the peninsula is expected to be extended, to include also the spherical superheavy nuclei.

4.3 Single-particle structure

It is interesting to see the single-particle spectra of the doubly magic nuclei: ^{208}Pb, ^{270}Hs and ^{298}114, for which the three minima of the shell correction $E_{\rm sh}$ have been obtained in Fig. 3.

The spectra are shown in Figs. 4 and 5, for protons and neutrons, respectively. One can see in Fig. 4 large energy gaps at $Z = 82$ (in the spectrum of ^{208}Pb), at $Z = 108$ (in ^{270}Hs) and at $Z = 114$ (in ^{298}114). In the spectrum of ^{270}Hs, a low density of levels extends from $Z = 106$ to $Z = 110$. For neutrons (Fig. 5), besides a large energy gap obtained at $N = 162$ in the spectrum of ^{270}Hs, a smaller gap at $N = 152$ is also seen. All this finds its reflection in the shell correction to energy, $E_{\rm sh}$, shown in Fig. 3.

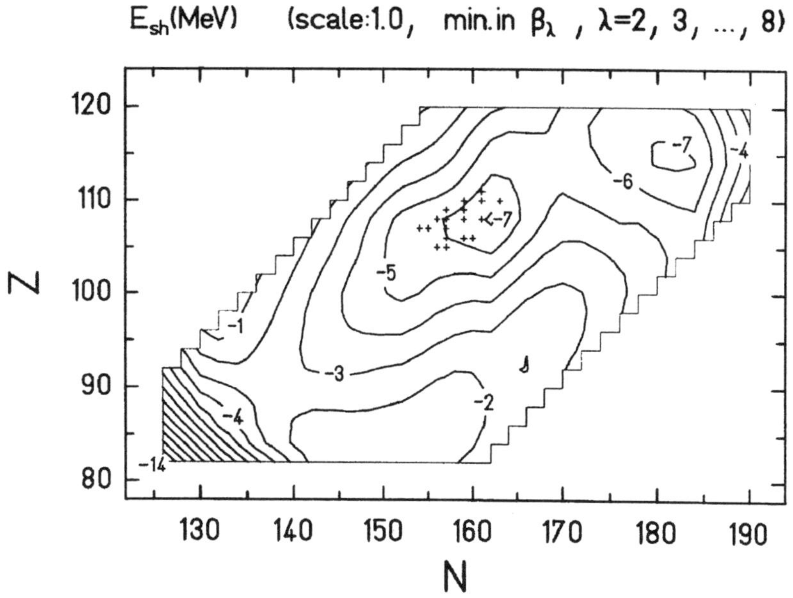

Figure 3. Contour map of the shell correction to energy, E_{sh}. Crosses denote the heaviest nuclides synthesized up to now [3].

5 Properties of deformed superheavy nuclei

5.1 Half-lives

Theoretical half-lives for nuclei situated around the doubly magic deformed nucleus ^{270}Hs (deformed superheavy nuclei), i.e. around the second minimum in Fig. 3, have been given and extensively discussed in [2,3,4]. Here, we only illustrate a comparison of these results with the experimental values obtained for the element 110. This is shown in Fig. 6. The experimental values are taken from [31] for ^{257}110, from [32] for ^{259,261}110 and from [33] for ^{263}110. One can see that the measured values are rather close to the predicted ones. In particular, they seem to confirm the existence of the predicted neutron deformed shell at $N = 162$. They also confirm the prediction that T_{sf} is larger than T_α for respective nuclei, as only α-decay has been observed for them.

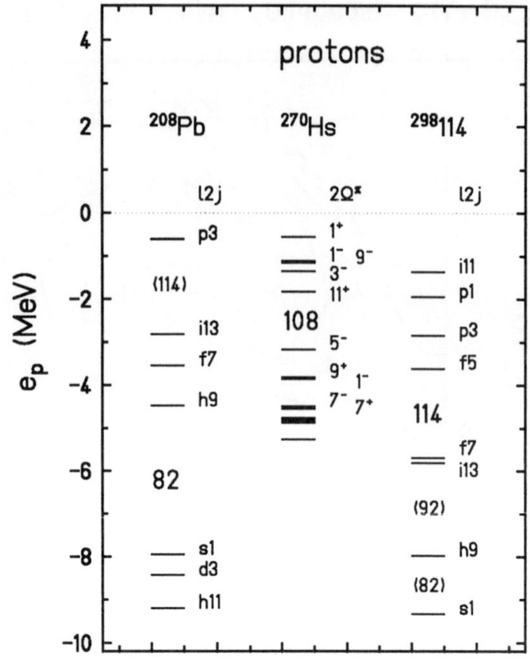

Figure 4. Proton single-particle levels calculated for the doubly magic nuclei: ^{208}Pb, ^{270}Hs and ^{298}114. Spectroscopic symbol for the orbital angular momentum l and total spin (multiplied by two) $2j$ are given at each level of the spherical nuclei ^{208}Pb and ^{298}114. Projection of spin (multiplied by two) 2Ω and parity π are shown at each level of the deformed nucleus ^{270}Hs [10].

5.2 Rotational energies

Nuclei situated around ^{270}Hs, in nuclear chart, and discussed in this section are called "deformed". But the only indication for this are theoretical results. There is no direct experimental evidence for their deformed shapes. Thus, it would be important to obtain such an evidence.

A direct evidence for the stable deformation of a nucleus is the observation of a rotational band in its spectra. The heaviest nuclei, for which such bands have been observed up to now, are the isotopes of fermium: ^{254,256}Fm [34]. Very recently, a rotational band has been also seen for ^{254}No [35]. Here, we would like to show what rotational energies may be expected for superheavy nuclei, up to those of the element 114.

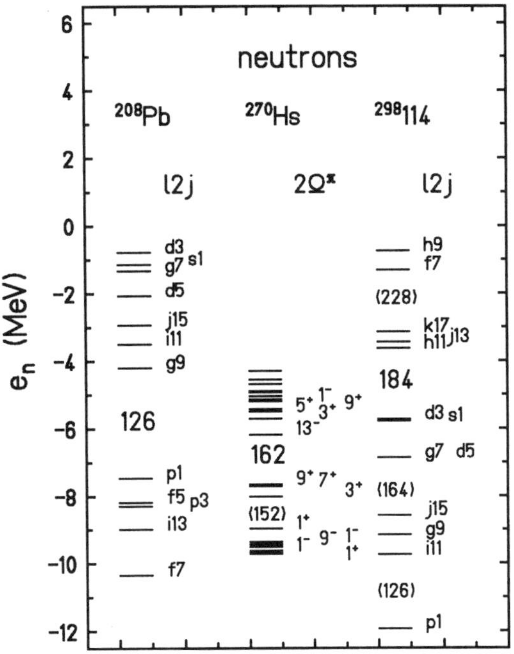

Figure 5. Same as in Fig. 4, but for neutrons.

To specify the energies of the ideal rotational band, it is enough to give the energy of any state of the band. For even-even nucleus, this may be e.g. the first state: 2+. To obtain this energy, we calculate the moment of inertia of the nuclei. As we are interested in the ground-state rotational bands, we analyze the ground-state moments of inertia. This quantity is calculated in almost the same way as used and described e.g. in [26,27,28], the differences concerning only some details of the calculations. For example, we use the Woods-Saxon single-particle potential, instead of the Nilsson potential taken in those papers. The results obtained for the energy of the first 2+ state, E_{2+}, are shown in Fig. 7.

One can see in Fig. 7 that two minima of E_{2+} are obtained for the considered nuclei. One of them (41.7 keV) is obtained for the nucleus ^{254}No, and the other (40.0 keV) for ^{270}Hs. These two minima should be connected with the two local maxima of the shell correction to energy (in their abso-

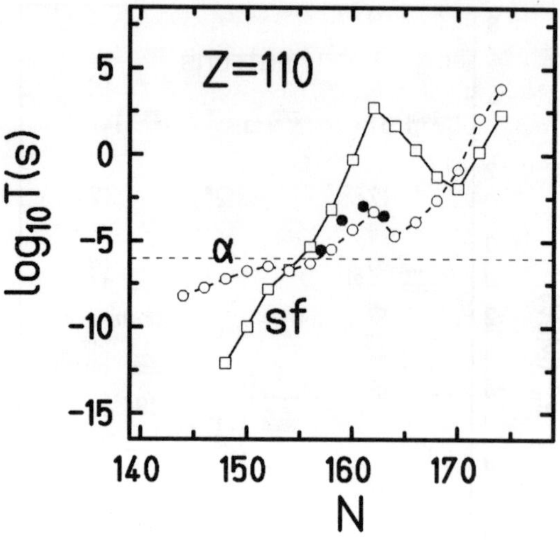

Figure 6. Comparison between predicted theoretically (open cirles) and measured (full circles) α-decay half-lives, for isotopes of the element 110.

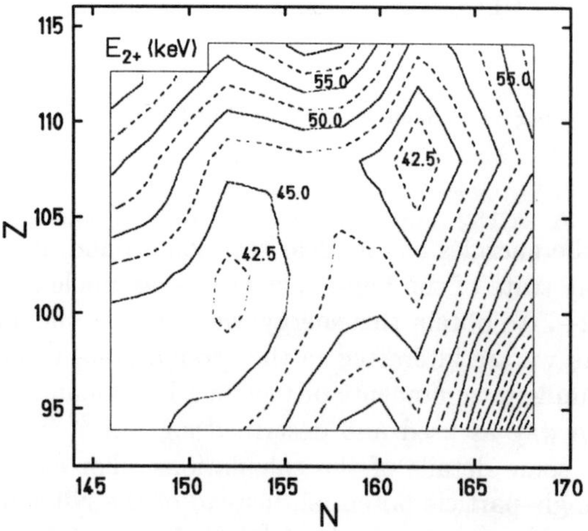

Figure 7. Contour map of calulated energy E_{2+} of the first rotational state 2+ [6].

lute values) obtained for nuclei around ^{254}No and ^{270}Hs, as discussed e.g. in [2]. A maximum of the shell correction corresponds to a large energy gap in the single-particle spectrum, which results in a weak pairing correlation. The latter is directly illustrated in Fig. 8 for neutrons, where a map of the pairing-energy gap Δ_n is shown.

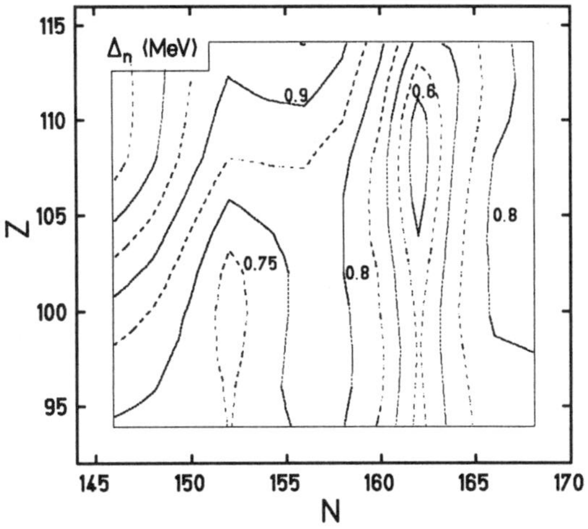

Figure 8. Contour map of the neutron pairing-energy gap Δ_n [6].

One can see in Fig. 8 that clear minima of Δ_n are obtained for the neutron shell closures at $N = 152$ and 162. As small pairing correlations correspond to large values of the moment of inertia of a nucleus (cf. e.g. [27]), these two minima of Δ_n result in the two minima of the rotational energy E_{2+}. Thus, we interpret the two minima of E_{2+}, shown in Fig. 7, as due to the largest shell corrections to energy, which, in considered region of nuclei, are obtained for ^{254}No and ^{270}Hs.

6 Properties of "spherical" superheavy nuclei

In this section, we discuss properties of nuclei situated around the third minimum in Fig. 3, i.e. around the hypothetical doubly magic spherical nucleus ^{298}114. Certainly, only nuclei close to this nucleus are expected to be spherical; this is the reason for which we put the word "spherical" in the title of the section into quotation marks.

6.1 *Half-lives*

Half-lives of "spherical" superheavy nuclei have been recently calculated in [36,37] in a dynamical way, using 3-dimensional deformation space $\{\beta_\lambda\}$, $\lambda = 2, 4, 6$. The results are shown in Fig. 9, taken from [37]. The isotopes with the neutron number N=176–184 of the elements 110–118 are considered in the figure. Only for the element 114, the isotope with N=186 is also shown, to see the behaviour of the half-lives above the shell closure at N=184. The role of the dimension of the deformation space admitted in the analysis of spontaneous fission has been discussed in [38].

One can see that the fission half-lives are longer than those of α-decay for all considered isotopes of the elements 112–118. The opposite relation is only obtained for the three lightest isotopes of the element 110. As illustrated for the element 114, the half-lives decrease above the shell closure at N=184, with the fission half-lives decreasing especially fast.

One can also see in Fig. 9 that the longest half-lives, with respect to both decay modes, are expected to appear for nuclei around the nucleus $^{292}110$ and to be of the order of one hundred years. As the nucleus $^{292}110$ is expected to be β-stable (e.g. [39]), these longest half-lives are expected to be the total half-lives. Nuclei with so long half-lives could be accumulated (in distinction to those situated around the doubly magic deformed nucleus ^{270}Hs, which have short half-lives). This would give a chance for extensive studies of physical and chemical properties of these exotic nuclei and elements. Certainly, on the condition that cross sections for their synthesis appear sufficiently large.

Fission half-lives of "spherical" superheavy nuclei have been also considered recently in [40] applying the approach used in [4], i.e. the dynamics in two-dimensional deformation space. They have been also studied in [41].

6.2 *Effect of nuclear surface diffuseness on the properties of spherical superheavy nuclei*

Let us first illustrate the dependence of the shell structure of a superheavy nucleus on the surface diffuseness.

Figure 10 shows the dependence of the proton single-particle energies e_p on the surface diffuseness z of the Woods-Saxon potential for protons, calculated for the nucleus $^{310}126$. For each value of z, radius of the proton potential is chosen so that the number of protons ($Z = 126$), contained in the potential up to the Fermi surface, is kept constant. Density of protons is obtained in the Thomas-Fermi approximation.

One can see in Fig. 10 that shell structure of a nucleus strongly depends

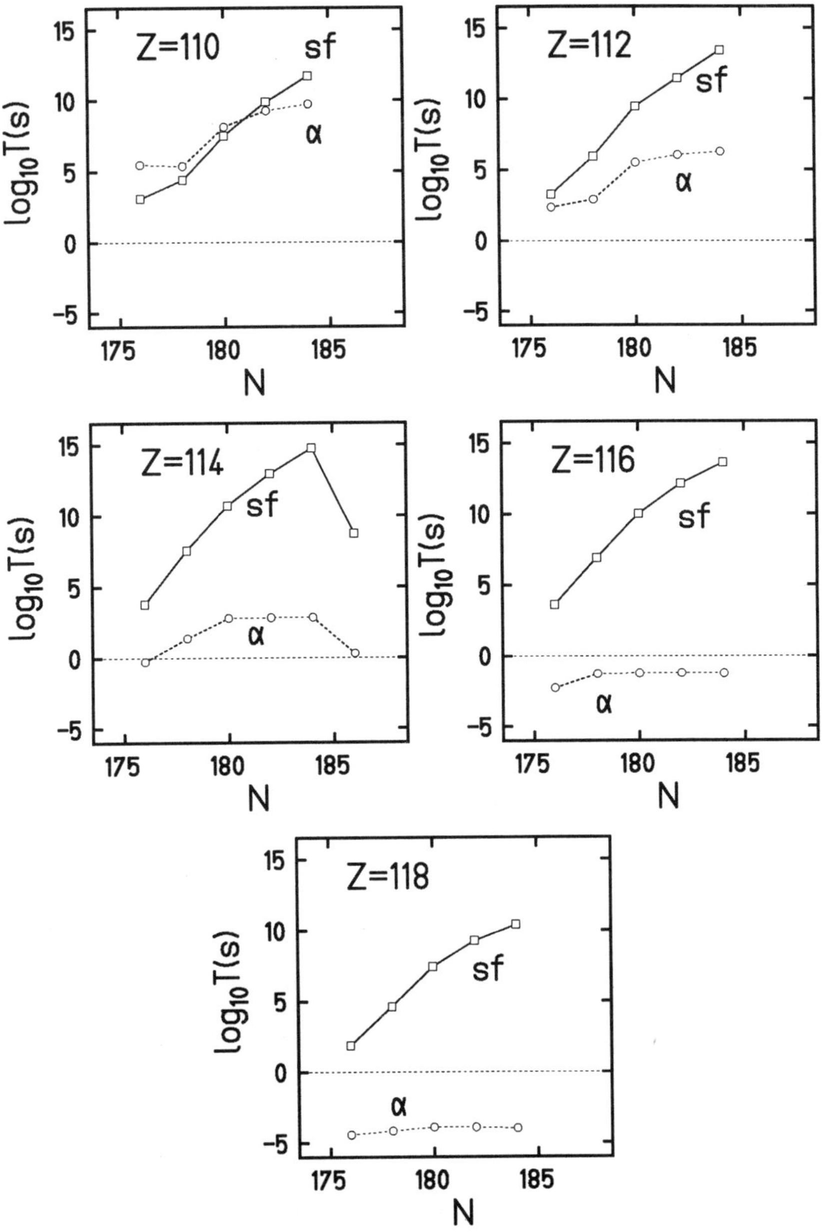

Figure 9. Logarithm of the calculated spontaneous-fission (sf) and α-decay (α) half-lives (given in seconds), as function of the neutron number N, for the elements: 110–118 [37].

on z. In particular, the energy gap at $Z = 114$ decreases , while the gap at $Z = 126$ increases with increasing z. For an illustration, the Figure is plotted for a large range of z. In reality, the equilibrium value of z, obtained for protons in the nucleus $^{310}126$ is: $z_{eq} = 0.78$ fm, i.e. it is only about 12% larger than the value: $z_0 = 0.70$ fm, usually used by us in earlier calculations (e.g. [2,3]). This increase of z does not change the gap at $Z = 114$, while it increases about three times the gap at $Z = 126$.

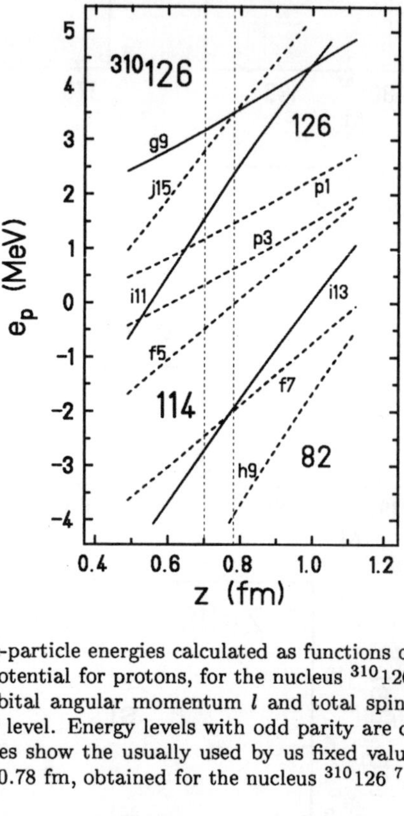

Figure 10. Proton single-particle energies calculated as functions of the surface diffuseness z of the Woods-Saxon potential for protons, for the nucleus $^{310}126$. Spectroscopic symbol corresponding to the orbital angular momentum l and total spin (multiplied by two) $2j$ are given at each energy level. Energy levels with odd parity are denoted by dashed lines. Thin dashed vertical lines show the usually used by us fixed value $z_0 = 0.70$ fm and the equilibrium value $z_{eq} = 0.78$ fm, obtained for the nucleus $^{310}126$ [7].

Figure 11 shows shell correction, E_{sh}, to the ground-state energy, calculated for 9 even-even nuclei with the magic neutron number $N = 184$. According to the calculations, these nuclei are spherical. The calculations are done in two cases: one, when the surface diffusenesses are treated as free parameters (TF), and the other, when they are fixed (Y).

One can see in Fig. 11 that the free surface diffusenesses increase (in

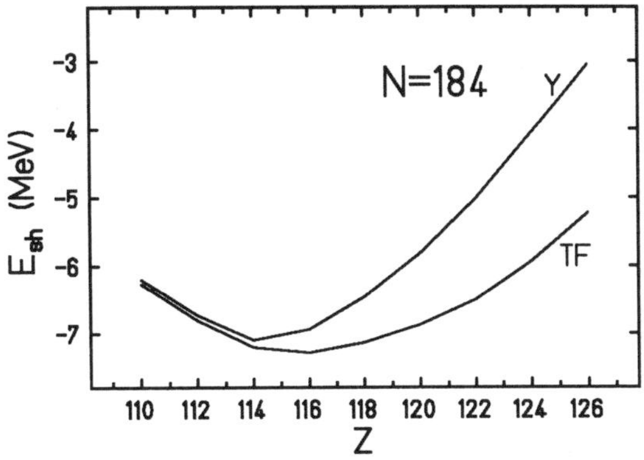

Figure 11. Shell correction energy, E_{sh}, calculated when the surface diffusenesses are treated as free parameters (TF) and when they are fixed (Y) [7].

absolute value) the shell correction energy (and, thus, also the stability) of the nuclei, especially those of heavier elements. The shell correction is only a little larger for the nucleus $^{298}114$, while it is by more than 2 MeV larger for the nucleus $^{310}126$, in comparison to the case of fixed diffusenesses. The complete result for the nucleus $^{310}126$ is: z_{eq} is increased by 12% for protons and by 4% for neutrons, shell correction is increased (in absolute value) by 2.18 MeV and the macroscopic energy (also in absolute value) by 2.34 MeV. Thus, the total binding energy of this nucleus is increased by 4.52 MeV, with respect to the case of the diffusenesses z fixed at the value $z_0 = 0.70$ fm.

The authors would like to thank P. Armbruster, S. Hofmann, T.L. Khoo, M. Leino, G. Münzenberg and Yu. Ts. Oganessian for helpful discussions. Support by the Polish Committee for Scientific Research (KBN), grant no. 2 P03B 117 15, is gratefully acknowledged.

References

1. A. Sobiczewski, I. Muntian and Z. Patyk, *Acta Phys. Slov.* **49**, 83 (1999).
2. Z. Patyk and A. Sobiczewski, *Nucl. Phys.* **A533**, 132 (1991).
3. R. Smolańczuk and A. Sobiczewski, Proc. XV Nucl. Phys. Conf.:*Low Energy Nuclear Dynamics*, St. Petersburg (Russia) 1995, eds. Yu.Ts. Oganessian, W. von Oertzen, R. Kalpakchieva (World Scientific, Singapore 1995) p. 313.

4. R. Smolańczuk, J. Skalski and A. Sobiczewski, *Phys. Rev.* **C52**, 1871 (1995).
5. Z. Patyk, J. Skalski, R.A. Gherghescu and A. Sobiczewski, *APH N.S., Heavy Ion Physics* **7**, 13 (1998).
6. I. Muntian, Z. Patyk and A. Sobiczewski, *Acta Phys. Pol.* **B30**, 000 (1999).
7. Z. Patyk *et al.*, *Acta Phys. Pol.* **B30**, 000 (1999).
8. P. Möller and J.R. Nix, *J. Phys.* **G20**, 1681 (1994).
9. A. Sobiczewski, *Usp. Fiz. Nauk* **166**, 943 (1996); *Physics–Uspekhi* **39**, 885 (1996).
10. A. Sobiczewski, Proc. of the R.A. Welch Foundation 41st Conf. on Chemical Research: "The Transactinide Elements", Houston (USA) 1997 (The R.A. Welch Foundation, Houston, 1997) chap. 13.
11. J.-F. Berger *et al.*, Proc. 24th Intern. Workshop: *Extremes of Nuclear Stucture*, Hirschegg (Austria) 1996, eds. H. Feldmeier, J. Knoll and W. Nörenberg (GSI, Darmstadt 1996) p. 43.
12. G.A. Lalazissis, M.M. Sharma, P. Ring and Y.K. Gambhir, *Nucl. Phys.* **A608**, 202 (1996).
13. S. Ćwiok *et al.*, *Nucl. Phys.* **A611**, 211 (1996).
14. K. Rutz *et al.*, *Phys. Rev.* **C56**, 238 (1997).
15. S. Hofmann, *Rep. Prog. Phys.* **61**, 639 (1998).
16. A. Türler *et al.*, *Phys. Rev.* **C57**, 1648 (1998).
17. S. Hofmann, *Acta Phys. Pol.* **B30**, 000 (1999).
18. Yu.Ts. Oganessian, Proc. 4th Intern. Conf. on Dynamical Aspects of Nuclear Fission, Casta-Papiernicka (Slovakia) 1998 (World Scientific, Singapore, 1999)
19. F.P. Hessberger, *Acta Phys. Slov.* **48**, 000 (1999).
20. Z. Patyk, A. Sobiczewski, P. Armbruster and K.-H. Schmidt, *Nucl. Phys.* **A491**, 267 (1989).
21. H.J. Krappe, J.R. Nix and A.J. Sierk, *Phys. Rev.* **C20**, 992 (1979).
22. W.D. Myers, W.J. Świątecki *Ark. Fys.* **36**, 343 (1967).
23. Z. Patyk and A. Sobiczewski, *Phys. Lett.* **B256**, 307 (1991).
24. S. Ćwiok *et al.*, *Comput. Phys. Commun.* **46**, 379 (1987).
25. A. Sobiczewski, Z. Patyk, S. Ćwiok and P. Rozmej, *Nucl. Phys.* **A485**, 16 (1988).
26. A. Sobiczewski, S. Bjørnholm and K. Pomorski, *Nucl. Phys.* **A202**, 274 (1973).
27. K. Pomorski *et al.*, *Nucl. Phys.* **A205**, 433 (1973).
28. K. Pomorski and A. Sobiczewski, *Acta Phys. Pol.* **B9**, 61 (1978).
29. A. Sobiczewski, F.A. Gareev and B.N. Kalinkin, *Phys. Lett.* **22**, 500

(1966).
30. H. Meldner, *Ark. Fys.* **36**, 593 (1967).
31. A. Ghiorso *et al.*, *Phys. Rev.* **C51**, R2293 (1995).
32. S. Hofmann *et al.*, *Z. Phys.* **A350**, 277 (1995); **A350**, 281 (1995).
33. Yu.A. Lazarev *et al.*, *Phys. Rev.* **C54**, 620 (1996).
34. Table of isotopes, 8th edition, eds. R.B. Firestone and V.S. Shirley, vol. 2 (J. Wiley, New York, 1996).
35. T.L. Khoo and M. Leino, private communication (1998).
36. R.A. Gherghescu, Z. Patyk and A. Sobiczewski, *Acta Phys. Pol.* **B28**, 31 (1997).
37. R.A. Gherghescu, Z. Patyk, J. Skalski and A. Sobiczewski, Proc. Intern. Workshop: *Research with Fission Fragments*, Benediktbeuern (Germany) 1996, eds. T. von Egidy, D. Habs, F.J. Hartmann, K.E.G. Löbner and H. Nifenecker (World Scientific, Singapore 1997) p. 116.
38. R. Smolańczuk, H.V. Klapdor-Kleingrothaus and A. Sobiczewski, *Acta Phys. Pol.* **B24**, 685 (1993).
39. W.D. Myers and W.J. Świątecki, *Nucl. Phys.* **A601**, 141 (1996).
40. R. Smolańczuk, *Phys. Rev.* **C56**, 812 (1997).
41. A. Staszczak *et al.*, Proc 3rd Int. Conf.:*Dynamical Aspects of Nuclear Fission*, Casta-Papiernicka (Slovakia) 1996, eds. J. Kliman and B.I. Pustylnik (JINR, Dubna, 1996) p. 22.

THE ROLE OF QUASI-FISSION IN THE SUPERHEAVY ELEMENT SYNTHESIS

V.V. VOLKOV

Joint Institute for Nuclear Research, 141980 Dubna, Russia

The role of quasi-fission in the synthesis of transfermium and superheavy elements within the framework of the dinuclear system concept is discussed. The models of competition between complete fusion and quasi-fission elaborated on the basis of this concept are used to calculate the probability of the compound nucleus formation in the cold and warm fusion reactions. It is shown that in cold fusion reactions quasi-fission is the main factor of decreasing the superheavy element production cross section with increasing the element's atomic number. Quasi-fission plays an important role in warm fusion reactions where a ^{48}Ca beam is used. The very intensive quasi-fission channel in symmetric nuclear reactions between massive nuclei prevents using these reactions for the synthesis of superheavy elements.

1 Introduction

In my talk I would like to discuss the role of quasi-fission in the reactions of synthesis of transfermium and superheavy elements (SHE). This problem is analyzed within the framework of the dinuclear system concept (DNS-concept) of the complete fusion process. This concept was proposed and elaborated at Dubna [1,2]. It is based on the information on the interaction of two nuclei in close contact which has been obtained in the study of deep inelastic transfer reactions [3]. The main result of this analysisis: quasi-fission is a very important factor in the process of complete fusion of massive nuclei and it is a serious danger for the synthesis of SHE.

2 The difficulties of traditional models of complete fusion in the description of fusion of massive nuclei

It is well known that the production cross section of transuranium elements drops with increasing the atomic number of an element. The cause of such a drop is fission of a compound nucleus during its deexcitation. In the region of SHE the fission barriers are increased, and this inspires the hope that the production cross section will be not too small. Figure 1 shows calculation results of the production cross section of elements 104, 108 and 110 in the cold fusion reactions. The calculations were made by Boris Pustylnik at Dubna [4]. He used the optical model for the calculation of the capture cross section - σ_c and a statistical model for the calculation of the survival probability of the compound nucleus – W_{sur}. The calculated cross section grows with increasing the atomic number of an element. There is agreement between the calculated and experimental data for element 104. However, in the case of elements 108 and 110 one can see dramatic disagreement. The experimental data [5] lie three-five orders of magnitude lower than the calculated ones. It is not possible to explain this discrepancy within the framework of existing models of complete fusion. One can think that this discrepancy may be the result of quasi-fission of the initial nuclear system. The system can decay into two fragments without the compound nucleus formation. However the existing theoretical models provide no possibility to calculate the competition between complete fusion and quasi-fission. Indeed, the optical model [6] and the surface friction model of Fröbrich [7]

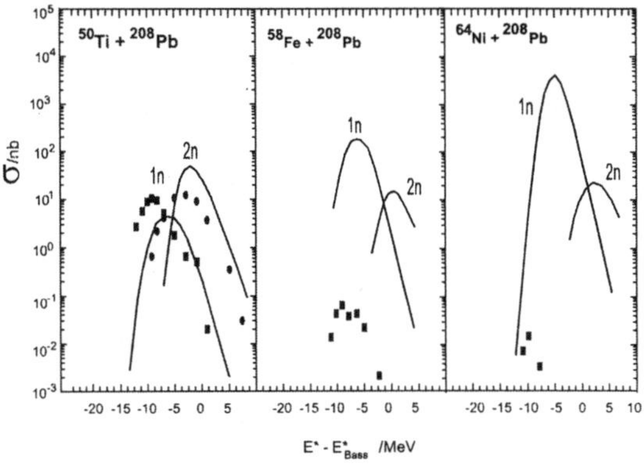

Figure 1. The production cross section of elements 104, 108, 110 in cold fusion reactions (HI,1-2n), the points are experimental data [5], the lines are results of calculations [4]. E^* – the excitation energy of the compound nucleus, E^*_{Bass} – the excitation energy for the collision on the Bass barrier.

permit calculation of only the capture cross section in reactions between massive nuclei. In the macroscopic dynamical model by Swiatecki (MDM) [8], complete fusion or quasi-fission are determined by the bombarding energy E_i. If $E_i > B_c + E_{xx}$, nuclei fuse, if $E_i < B_c + E_{xx}$, the quasi-fission is realized. B_c is the Coulomb barrier, E_{xx} is the extra-extra push. The competition between these two channels is not possible. It is the consequence of the determinism of the classical dynamical model.

The DNS-concept is now only an approach to the complete fusion process which reveals the competition between complete fusion and quasi-fission and provides a basis for creating models of this competition.

3 The competition between complete fusion and quasi-fission within the framework of the DNS-concept

According to the DNS-concept the scenario of the complete fusion process is as follows. On the capture stage, after full dissipation of the collision kinetic energy the dinuclear system (DNS) is formed. The complete fusion is an evolution process in which the nucleons of one nucleus gradually, shell by shell are transferred to another nucleus. The nuclei of the DNS conserve their individuality until the end of the fusion process. This important peculiarity of the DNS evolution is the consequence of their nucleon and shell structure.

Figure 2 illustrates the compound nucleus formation process within the framework of the DNS-concept and within the framework of a popular macroscopic dynamical model by Swiatecki. In the MDM the fused nuclear drops lose their individuality very quickly as the result of the neck formation. In the DNS-concept the fused nuclei retain their individuality until the end of complete fusion process.

As is known from the deep inelastic transfer reactions [3], the evolution of the DNS is determined by the potential energy of the system as a function of charge (mass) asymmetry and the angular momentum of collision. Figure 3 shows the potential energy of the DNS which is formed in the reaction $^{110}Pd + ^{110}Pd$ [2]. Z is the atomic number of one of the DNS nuclei. L is the angular momentum of collision which transforms into the

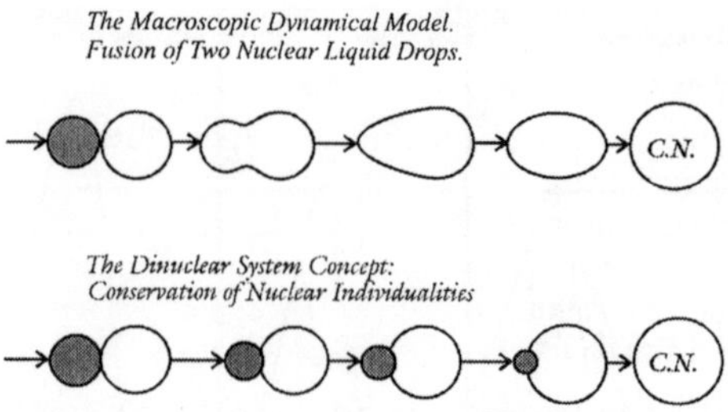

Figure 2. Nuclear fusion from the viewpoint of the macroscopic dynamical model and the dinuclear system concept.

DNS spin. The potential energy of the compound nucleus including its rotation is taken as zero. The initial DNS is situated in the minimum of the potential energy. It is similar to the gigantic nuclear molecule. For the realization of complete fusion and formation of the compound nucleus the initial DNS has to overcome the inner fusion barrier – B^*_{fus}. The asterisk indicates that the energy for the barrier overcoming is taken from the DNS excitation energy - E^*. The initial DNS is also situated in the local minimum of the nucleus-nucleus potential V(R), at the bottom of the potential pocket (Fig. 4). The quasi-fission barrier B_{qf} preserves the DNS from the decay. The excited DNS has two ways of changing. It may evolve in the direction of increasing its charge asymmetry and after overcoming the barrier B^*_{fus} it may transform into a compound nucleus. It may alternatively overcome the quasi-fission barrier B_{qf} and decay into two nearly equal fragments. It is the quasi-fission. The competition between complete fusion and quasi-fission in the DNS-concept arises naturally as a consequence of the statistical nature of the complete fusion process.

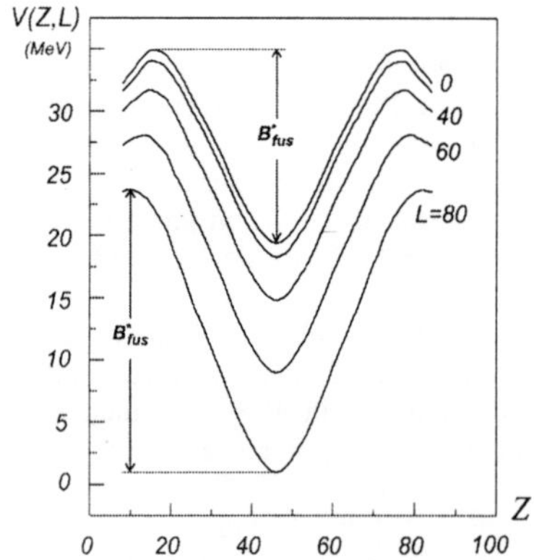

Figure 3. Potential energy of DNS formed in the reaction ^{110}Pd + ^{110}Pd. Z is the atomic number of one of DNS's nuclei, L is the collision angular momentum. The inner fusion barriers B^*_{fus} are indicated. The potential energy of the compound nucleus is taken as 0.

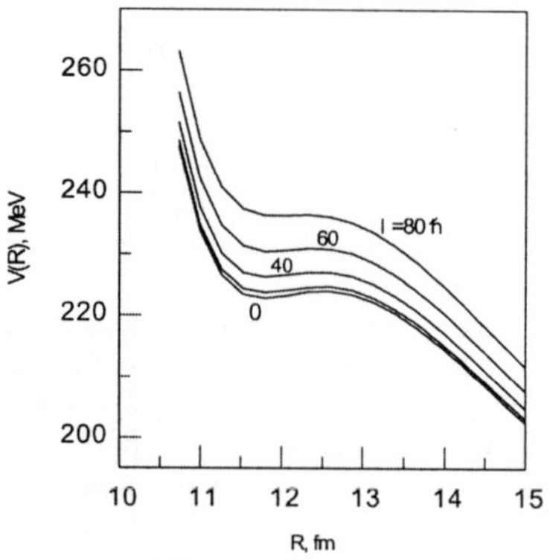

Figure 4. Nucleus-nucleus potential in the reaction ^{110}Pd + ^{110}Pd, R is the distance between centres of the nuclei, L is the angular momentum of the collision.

4 The model of the competition between complete fusion and quasi-fission in symmetric nuclear reactions

Our first model of the competition between complete fusion and quasi-fission was elaborated for symmetric nuclear reactions between massive nuclei [2]. In these reactions the initial DNS is situated in the minimum of the potential energy V(Z, L) and in the minimum of the nucleus-nucleus potential V(R). It means that the DNS is in a quasi-equilibrium state. One can suppose that the probability of going to the complete fusion channel or to the quasi-fission channel is proportional to the level density of the DNS on top of the fusion barrier $B^*_{fus} - \rho_{B^*fus}$ and on top of the quasi-fission barrier $B_{qf} - \rho_{qf}$. The probability of the compound nucleus formation after the realization of capture is defined by the ratio:

$$P_{cn} = \frac{\rho_{B^*_{fus}}}{\rho_{B^*_{fus}} + \rho_{qf}} \tag{1}$$

The density of the DNS states was taken from Nërenberg's proposal [9].

In traditional models the evaporation residue production cross section – σ_{ER} is determined by the capture cross section - σ_c and by the survival probability of the compound nucleus during its deexcitation - W_{sur}.

$$\sigma_{ER} = \sigma_c \cdot W_{sur} \tag{2}$$

The DNS-concept adds to these two factors the third one - P_{cn} - the probability of compound nucleus formation in the competition with quasi-fission:

$$\sigma_{ER} = \sigma_c \cdot P_{cn} \cdot W_{sur} \tag{3}$$

Using this model we have calculated the σ_{ER} value in the reaction $^{110}Pd + {}^{110}Pd$. The results are presented in figure 5 by the solid line [2]. The squares represent experimental data by Moravek *et al.* [10] obtained at Darmstadt. The dashed line shows the results of the σ_{ER} calculation using the MDM [2]. The MDM does not take into account the competition between complete fusion and quasi-fission. But in this reaction P_{cn} is equal to ~10^{-3}. It means that the quasi-fission channel very strongly dominates over the complete fusion channel.

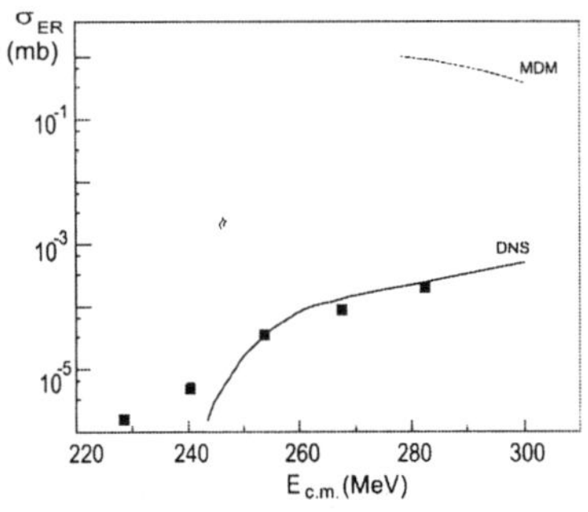

Figure 5. Cross sections for the evaporation residues σ_{ER} in the reaction $^{110}Pd + {}^{110}Pd$ as a function of bombarding energy. The filled squares represent the experimental data from [10], the solid curve – the σ_{ER} calculations using DNSC, the dashed curve – the σ_{ER} calculations using MDM.

5 The models of the competition between complete fusion and quasi-fission in asymmetric nuclear reactions

In the asymmetric reactions the DNS has two ways of evolution (see Fig. 6) [11]. One way leads to the compound nucleus formation. On this way the DNS has to overcome the inner fusion barrier B^{*}_{fus}. The second way leads to the symmetrisation of the DNS form. On this way the DNS evolution does not meet any obstacles. The symmetric DNS is in the minimum of the potential energy, i.e. in the quasi-equilibrium state. However in the symmetric form the Coulomb repulsion between DNS nuclei has the maximum value, and the DNS decays into two nuclear fragments. The maximum value of the DNS excitation energy in the symmetric form favours the decay.

In the asymmetric nuclear reactions the initial DNS is in the unequilibrium state and for the calculations of P_{cn} we cannot use the model discussed in 3. We have elaborated two new models for asymmetric reactions.

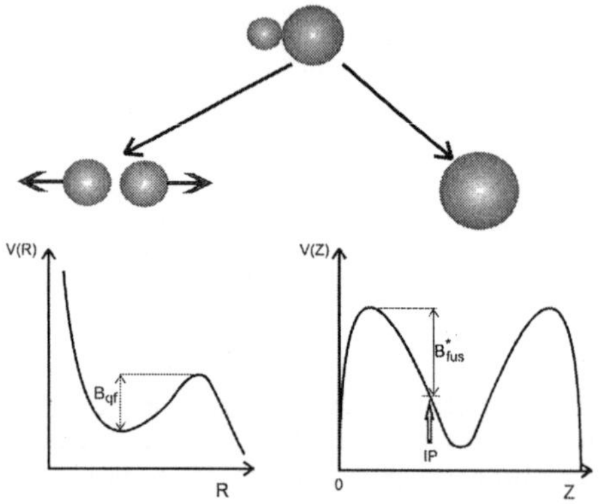

Figure 6. Two ways of evolution of a massive asymmetric DNS, V(R) is the nucleus-nucleus potential, V(Z) – the potential energy of DNS, Z – the atomic number of one of DNS nuclei, B_{qf} – the quasi-fission barrier, B^{*}_{fus} – the inner fusion barrier, I.P. – the reaction injection point.

5.1 The model in which the Monte-Carlo method is used

In the first model the Monte-Carlo method is used for the calculation of the DNS evolution. Some simplifications of the DNS evolution have been introduced [12]. From any configuration the DNS can only go into a configuration neighboring in Z and A. It means that one proton and one or two neutrons are transferred from one nucleus to another. The cluster transfer is excluded. The probability of the nucleon transfer is proportional to the DNS level densities in the neighboring configuration. The level density is determined by the DNS excitation energy. It is calculated according to the ratio proposed by Nörenberg [9].

The DNS evolution proceeds along a large number of trajectories in the Z and A space of the DNS nuclei. In the model [12] all these trajectories are substituted by one trajectory which goes along the potential energy valley. The calculation of the DNS evolution is carried out for different angular momenta L. It is assumed that the DNS which crossed the top of B^{*}_{fus} goes irreversibly into the complete fusion channel. The DNS which reached the symmetric form goes irreversibly into the quasi-fission channel.

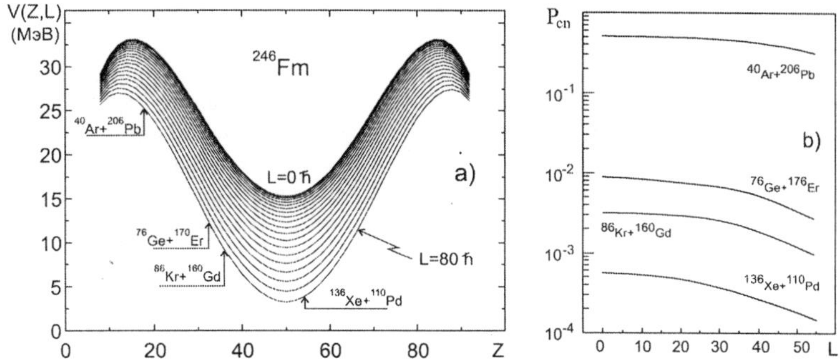

Figure 7. a) Potential energy of DNS formed in four reactions, the same compound nucleus ^{246}Fm being formed. b) Production probability for the compound nucleus ^{246}Fm as a function of the collision angular momentum.

The model was used in the analysis of four different reactions producing the same compound nucleus ^{246}Fm [13]. Figure 7 shows potential energy of the DNS's which are formed in these reactions. The arrows indicate the injection points of the reactions. One can see that in the reaction ^{40}Ar + ^{206}Pb the fusion barrier B^{*}_{fus} is equal to a few MeV only, but in the reaction ^{136}Xe + ^{110}Pd the B^{*}_{fus} value increases up to 20 MeV.

In the right part of figure 7 the calculated values of P_{cn} are shown as a function of the angular momentum of collision. In the reaction with ^{40}Ar ions, P_{cn} is equal to $\sim 5 \cdot 10^{-1}$ and it drops to $\sim 5 \cdot 10^{-4}$ for the reactions with ^{136}Xe ions. In the experiment performed by Gäggeler *et al.* [13] at Darmstadt the fermium isotope ^{244}Fm was registered. That isotope is a product of the two-neutron evaporation from the compound nucleus ^{246}Fm. It decays by spontaneous fission. Using our data for P_{cn} it is possible to reproduce experimental data for the production cross section of ^{244}Fm in the reactions with ^{40}Ar and ^{76}Ge ions [12]. The low values of P_{cn} in the reactions with ^{86}Kr and ^{136}Xe ions explain why the experimenters did not observe ^{244}Fm in the reactions with these ions.

5.2 The model in which the Fokker-Planck equation is used

In the second model the evolution of the DNS is considered as a diffusion process along the two collective coordinates [14]. The diffusion along the mass asymmetry coordinate $\eta=(A_1-A_2)/(A_1+A_2)$ leads to the complete fusion. The diffusion along the R-coordinate (R is the distance between the two centres of DNS nuclei) leads to quasi-fission. A quasi-stationary solution to the two-dimensional Fokker-Planck equation is used for the description of the DNS evolution. The parameter of the model is the DNS viscosity. The stationary flow of the probability through the inner fusion barrier B^{*}_{fus} and the quasi-fission barrier B_{qf} defines the value of P_{cn}:

$$P_{cn} = \frac{\lambda_\eta}{\lambda_\eta + \lambda_R} \qquad (4)$$

The model was tasted in nuclear reactions between massive nuclei where P_{cn} could be calculated using experimental data for σ_{ER} and the corresponding data for σ_c and W_{sur}.

Then the model was applied to the calculations of P_{cn} in the reactions of synthesis of transfermium and superheavy elements.

The dependence of P_{cn} on the atomic number of the compound nucleus for cold fusion reactions is presented in figure 8. For element 104 P_{cn} is equal to $\sim 5\cdot 10^{-2}$. However for elements 112 and 114 P_{cn} drops to $\sim 10^{-6}$ and 10^{-7}. The drop of the P_{cn} value for the most heavy elements is in agreement with the drop of the production cross sections. In going from element 110 to element 112, P_{cn} decreases by ten times. The experimental production cross sections for these elements are 12 and 1 picobarn, respectively [15].

The calculations of the P_{cn} values indicate that quasi-fission is the main factor in decreasing the production cross section with increasing Z of the element in the cold fusion reactions.

Quasi-fission plays an important role in the warm synthesis of SHE also when using ^{48}Ca ions. Figure 9 shows the calculated data for P_{cn} in the reactions ^{232}Th, ^{238}U and ^{244}Pu + ^{48}Ca [16]. The value of P_{cn} is given in dependence on the excitation energy of the compound nucleus. In these reactions the reaction channel with the two or three-neutron evaporation produces the largest yield. The excitation energy of the compound nucleus is nearly 25–35 MeV for these channels. One can see that P_{cn} drops considerably in going from element 110 to element 114.

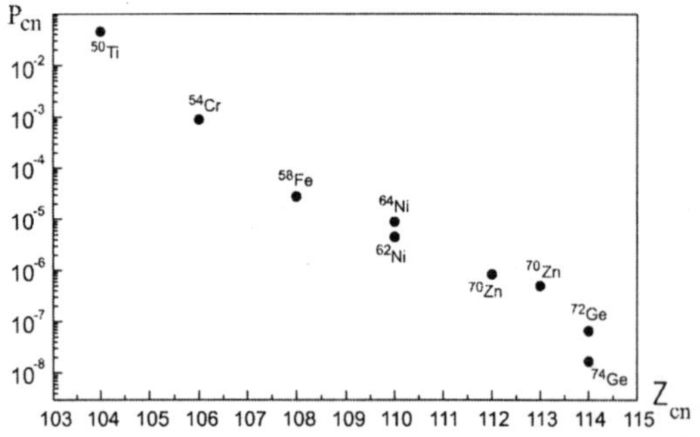

Figure 8. Production probability for the compound nucleus in cold fusion reactions (H.I.,1n) used to synthesise TUE and SHE. Z_{cn} is the atomic number of the compound nucleus, the bombarding ions are indicated.

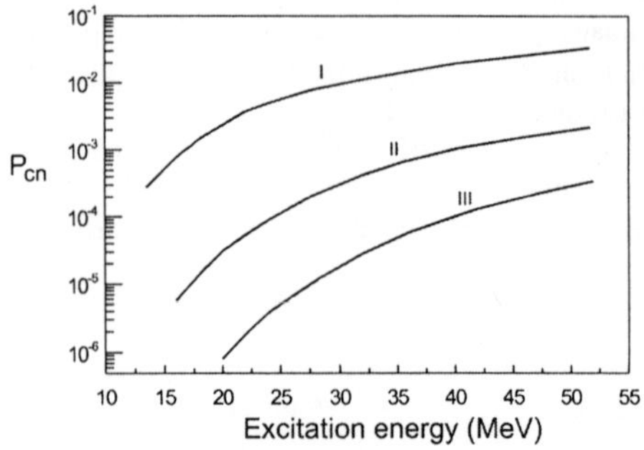

Figure 9. Production probability for the compound nuclei of elements 110, 112 and 114 as a function of the excitation energy of the compound nucleus for the ^{232}Th (I), ^{238}U (II), ^{244}Pu (III) targets and ^{48}Ca ions.

6 The prospects for using symmetric nuclear reactions for the SHE synthesis

Symmetric nuclear reactions between two massive nuclei are considered to be one of possible ways for the synthesis of SHE. Due to the large negative values of Q the compound nucleus in these reactions should have a very low excitation energy at a level of several MeV.

We considered the fusion of two ^{136}Xe nuclei into the nucleus ^{272}108 within the framework of the DNS-concept [17]. Figure 10 shows the potential energy of the DNS which is formed in this reaction in a head-on collision. One can see that the inner fusion barrier B^*_{fus} has the value of about 30 MeV. To overcome the fusion barrier the DNS must have an excitation energy of no less than 30 MeV. The compound nucleus will have nearly the same excitation energy. With increasing Z of the compound nucleus, B^*_{fus} increases also. This means that it is not possible to produce a cold nucleus of SHE in symmetric nuclear reactions.

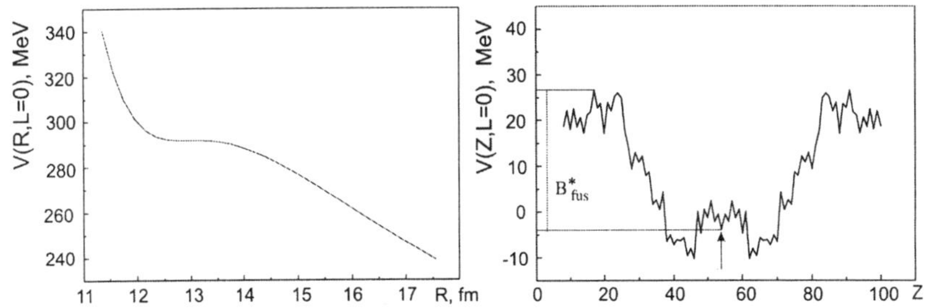

Figure 10. The DNS potential energy and nucleus-nucleus potential in the reaction $^{136}Xe + ^{136}Xe \rightarrow ^{272}108$.

The symmetric massive DNS has a vanishingly low quasi-fission barrier. The high fusion barrier and the low quasi-fission barrier make the quasi-fission channel to dominate strongly over the complete fusion channel. The P_{cn} value in the reaction ^{136}Xe + ^{136}Xe is equal to a few units of 10^{-9} [17]. Thus the DNS-concept indicates that symmetric nuclear reactions are not promising for the synthesis of SHE.

During many years the fission of an excited compound nucleus was considered as the main danger for the successful synthesis of transfermium and SHE. The DNS-concept reveals a new serious danger on the way to SHE. This new danger is the quasi-fission of the DNS, which is formed on the capture stage. Figure 11 illustrates the two dangers, encounted by experimenters on the way to the SHE.

7 Conclusion

1. The DNS-concept reveals an important role of quasi-fission in the complete fusion of massive nuclei.
2. The DNS-concept becomes the basis for the models of the competition between complete fusion and quasi-fission. These models permit calculation of the probability of the compound nucleus formation in reactions between massive nuclei.
3. The DNS-concept reveals a new serious danger – quasi-fission which the experimenters encounter in the synthesis of SHE.

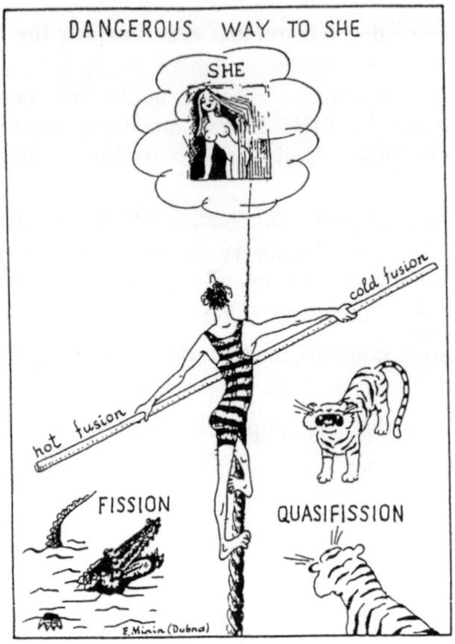

Figure 11. The two dangers, encounted by experimenters on the way to the superheavy elements.

References

1. V.V.Volkov, *Izv. AN SSSR ser. fiz.* **50** (1986) p.1879; in *Proc. Int. School-Seminar on Heavy Ion Physics (Dubna, 1986)* D7-87-68, 1987, p.528; in *Proc. 6th Int. Conf. on Nuclear Reactions Mechanism* (Varenna, Italy 1991), ed. E.Gadioli (Ricerca Scientifica) p.39.
2. N.V.Antonenko, E.A.Cherepanov, A.K.Nasirov, V.B.Permjakov and V.V.Volkov, *Phys. Lett* **B319** (1993) p.425; *Phys. Rev.* **C51** (1995) p.2635.
3. V.V.Volkov, *Phys. Reports* **44** (1978) p.94.
4. B.I.Pustylnik, in *Dynamical aspects of nuclear fission*, ed. J.Kliman, B.Pustylnik, (JINR E6,7-97-49, Dubna, 1996) p.121.
5. S.Hofmann, in *Proc. on XV Nucl. Phys. Conf. LEND-95,* St.Petersburg, April 18-22, 1995, ed. Yu.Oganessian, R.Kalpakchieva and W.von Oertzen (World Scientific, Singapore) p.305.
6. A.S.Iljinov, Yu.Ts.Oganessian and E.A.Cherepanov, *Sov. J. Nucl. Phys.* **36** (1982) p.118.
7. P.Fröbrich, *Phys. Report* **116** (1984) p.337.
8. W.J.Swiatecki, *Phys. Scripta* **24** (1981) p.113; S.Bjornholm, W.J.Swiatecki, *Nucl. Phys.* **A391** (1982) p.471; J.P.Blocki, H.Feldmeier, W.J.Swiatecki, *Nucl. Phys.* **A459** (1986) p.145.
9. S.Ayik, B.Schurman and W.Nörenberg, *Z. Phys.* **A277** (1978) p.299.
10. W.Morawek *et al., Z. Phys.* **A341** (1991) p.75.
11. V.V.Volkov, E.A.Cherepanov, N.V.Antonenko, A.K.Nasirov, in *Proc. on XV Nucl. Phys. Conf. LEND-95,* St.Petersburg, April 18-22, 1995, ed. Yu.Oganessian, R.Kalpakchieva and W.von Oertzen (World Scientific, Singapore) p.336.

12. E.A.Cherepanov, V.V.Volkov, N.V.Antonenko, A.K.Nasirov, in *Heavy Ion Physics and Its Applications,* ed. Y.X.Luo, G.M.Jin, J.Y.Liu (World Scientific, Singapore 1996) p.272.
13. H.Gäggeler *et al., Z. Phys.* **A316** (1984) p.291.
14. G.G.Adamian, N.V.Antonenko, W.Scheid, V.V.Volkov, *Nucl. Phys.* **A627** (1997) p.361.
15. S.Hofmann *et al., Z. Phys.* **A350** (1995) p.277; **A350** (1995) p.281; **A354** (1996) p.229.
16. G.G.Adamian, N.V.Antonenko, V.V.Volkov, W.Scheid, E.A.Cherepanov, *RAN, Izv. AN ser. fiz* (in press).
17. E.A.Cherepanov, G.G.Adamian, N.V.Antonenko, V.V.Volkov, A.K.Nasirov, *RAN, Izv. AN ser. fiz.* **61** (1997) p.2213.

UPGRADE OF THE DUBNA GAS-FILLED RECOIL SEPARATOR FOR THE SYNTHESIS OF SUPERHEAVY NUCLEI IN ^{244}PU+^{48}CA REACTION

YU.TS. OGANESSIAN, V.K. UTYONKOV, YU.V. LOBANOV, F.SH. ABDULLIN, A.N. POLYAKOV, I.V. SHIROKOVSKY, YU.S. TSYGANOV, S. ILIEV, V.G. SUBBOTIN, A.M. SUKHOV, G.V. BUKLANOV, A.N. MEZENTSEV, K. SUBOTIC AND O.V. IVANOV

Joint Institute for Nuclear Research, 141980 Dubna, Russian Federation
E-mail: utyonkov@sungns.jinr.dubna.su

K.J. MOODY, J.F. WILD, N.J. STOYER, M.A. STOYER AND R.W. LOUGHEED

University of California, Lawrence Livermore National Laboratory, Livermore, California 94551, USA
E-mail: moody@isd.llnl.gov

Preparing experiments on the synthesis of superheavy nuclei with Z=114 in the reaction ^{244}Pu+^{48}Ca, we modified the Dubna gas-filled recoil separator and its detection system and carried out bombardments of the lead targets with ^{48}Ca projectiles. We studied excitation functions for the reactions ^{206}Pb(^{48}Ca,1-4n) and ^{204,207,208}Pb(^{48}Ca,2n). Maximum cross sections for evaporation of 1-3 neutrons in the reaction ^{206}Pb+^{48}Ca were measured to be σ_{1n}=60 nb, σ_{2n}=500 nb, and σ_{3n}=30 nb; an upper limit of 0.2 nb was obtained for the 4n channel. In the bombardment of ^{204}Pb target by 217-MeV ^{48}Ca production cross sections for evaporation of 2 neutrons from compound nuclei ^{256}No, ^{255}No, and ^{252}No were measured to be 2.1 μb, 1.3 μb, and 10 nb, respectively. Collection efficiency of the separator was estimated to be ~45%. Spontaneously fissioning even-even isotope ^{250}No with a half-life $T_{1/2}$=46 μs was identified in this experiment. Test irradiation of ^{244}Pu target with ^{48}Ca beam was performed, aimed at estimating of all kinds of background counting rates and optimizing detection conditions for future experiments.

1 Introduction

A fundamental outcome of macro-microscopic nuclear theory is the prediction of an "island of stability" of superheavy elements. Calculations performed over more than 35 years with different versions of the nuclear shell model predict substantial enhancement of the stability of heavy nuclei when approaching the closed spherical shells Z=114 and N=184. In the recent years considerable progress was achieved in the research of the heaviest elements. Experiments extended the Chart of Nuclides up to Z=112 (see, e.g., Refs. [1-3] and references therein). These results provided test of the nuclear theory and increased confidence and accuracy of predictions for the spherical superheavy nuclei.

The heaviest isotopes of element 114 with neutron numbers from 174 to 176, close to the predicted spherical nuclei and, consequently, being relatively stable, can be produced in the fusion reaction of ^{244}Pu with ^{48}Ca ions [4]. With the doubly-magic ^{48}Ca projectile, the resulting compound nucleus ^{292}114 should have excitation energy at the Coulomb barrier of about 30 MeV. Correspondingly, nuclear shell effects are still expected to persist in the excited nucleus, increasing the survival probability of the evaporation residues (EVRs), as compared to “hot fusion” reactions ($E^* \approx 45$ MeV), which were used for the synthesis of heavy isotopes of elements with atomic numbers Z=106, 108 and 110 [3]. Additionally, the high mass asymmetry in the entrance channel should decrease the dynamical limitations on nuclear fusion arising in more symmetrical "cold fusion" reactions.

Previous attempts to synthesize new elements in ^{48}Ca-induced reactions with actinide targets gave only upper cross-section limits for their production (see, e.g., [5] and references therein) except for the result obtained in 1998 when two spontaneous fission (SF) events were assigned to the decay of a new isotope of element 112 produced in the reaction ^{238}U(^{48}Ca,3n)^{283}112 with a cross section of 5 pb [6]. Cross sections could be expected to be at a level of 1 pb for the ^{244}Pu+^{48}Ca reaction. To increase the sensitivity of future experiments we modified the Dubna gas-filled recoil separator and its detection system and carried out bombardments of the lead targets with ^{48}Ca projectiles. Test irradiation of ^{244}Pu with ^{48}Ca ions was also carried out.

2 Experimental Technique

Beam of ^{48}Ca projectile was delivered by the Dubna U400 cyclotron. The beam energy was verified with a precision of ~1 MeV, by measuring the energies of scattered ions, and by a time-of-flight technique. Targets of isotopically enriched material were arranged on a rotating wheel to lower possible heat and radiation damage by the beam. Experiments were performed with either pulsed or continuous cyclotron beams. For the pulsed-beam experiments, the beam was modulated with 150 Hz frequency so that targets were exposed to the ~2.2 ms beam pulse during each ~6.7 ms beam cycle. Targets of ^{206}Pb (95.9% ^{206}Pb, 1.34% ^{208}Pb, 2.76% ^{207}Pb), and ^{204}Pb (53.5% ^{204}Pb, 15.9% ^{208}Pb, 11.0% ^{207}Pb, 19.6% ^{206}Pb) with thickness of 0.38, and 0.20 mg cm^{-2}, respectively, were deposited by evaporation in vacuum on ~0.55 mg cm^{-2} Cu substrates.

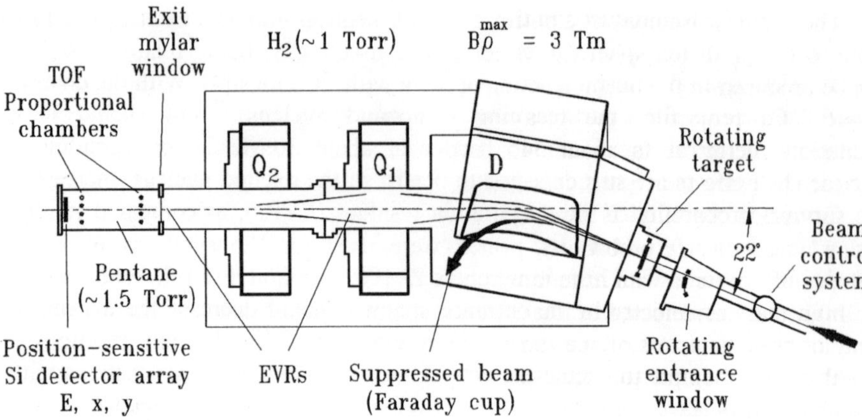

Figure 1. Lay-out of the Dubna gas-filled recoil separator (the dipole magnet D followed by the quadrupole doublet Q_1 Q_2).

The Dubna gas-filled recoil separator [7] (Figure 1) was employed to separate in flight evaporation residues (EVRs) recoiling out of the targets from beam particles and transfer-reaction products. To set the magnetic field of the separator's dipole magnet for collecting specific EVRs on the focal-plane detectors, we used our measurements of the average charge states of slow atoms moving through and exchanging charge with ~1 Torr of hydrogen [8]. Detection module at the focal plane of the separator included a time-of-flight (TOF) measuring system and a semiconductor detector array (Figure 2). A 0.5-μm Mylar window separated the hydrogen media of the separator from detector module filled with pentane at ~1.5 Torr. The TOF detector consisted of two muliwire proportional chambers operated in pentane atmosphere. The high efficiency of the TOF proportional chambers (~99.7%) allowed both distinguishing recoil-like events and efficient discrimination of background counts from various ionizing particles, entering the detector set-up from the separator. The recoil nuclei were finally implanted into a 40×120 mm^2 position-sensitive detector (PSD) array. The PSD array consisted of three 40×40-mm^2 passivated boron-implanted silicon detectors manufactured by *Canberra Semiconductor NV*, each having four 40-mm high and 10-mm wide strips. Each of the detector strips had vertical position sensitivity. We obtained horizontal positions for the reaction products from the 12 strips and vertical positions from the 40-mm high resistive layer of the detectors. To detect escaping α's eight detectors of similar type, without position sensitivity, were mounted in a box-like array around the perimeter of the focal-plane detectors.

The ^{252}No EVRs produced in the $^{206}Pb+^{48}Ca$ reaction at the excitation function maximum had an initial energy of ~41 MeV, which was reduced to ~24 MeV at implantation due to energy losses in the target, hydrogen and pentane. The detection efficiency for α particles emitted in the decays of implanted nuclei by the focal-plane detectors is about 54% of 4π, with an overall efficiency of 88% provided by using the eight side detectors.

Certain background is produced by low-ionizing particles, e.g., protons and α's, which, with low probability, pass through the separator and reach the focal-plane. These particles can have ranges exceeding the focal-plane detector thickness of 300 μ and are detected by the TOF system with less than 10% efficiency. An array of three detectors of the same dimensions, without position sensitivity, was added 15 mm behind the focal-plane detectors. With these back detectors, we detected particles that passed through the focal-plane detectors. This allowed identification of the p, ^{2}H, ^{3}H and α particles and measurement of the corresponding energy spectra. Employing the extra back detectors in "veto" mode suppressed the background in the focal-plane detectors when measuring in-beam α-particle spectra.

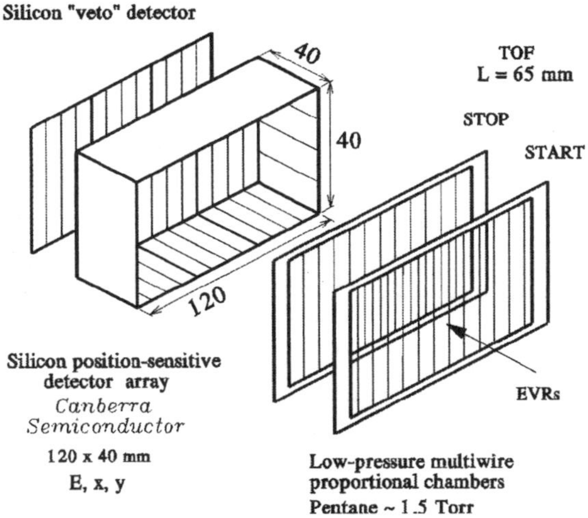

Figure 2. Detection system of the Dubna gas-filled recoil separator.

For total suppression of the background caused by various particles in the course of the heavy-ion bombardment, an electronic system for fast (~100 μs) switch-off of the cyclotron radio-frequency system was developed and tested in these experiments. This system operates in real-time mode and allows blocking the ion beam for a fixed time interval after detecting of time- and position-correlated signals, which correspond to expected EVR-α or α-α sequences. The system permits to chose corresponding energy and position characteristics of sequential signals, as well as TOF of EVRs, and time distances between them [9].

Alpha-energy calibrations were performed using α emitting Th isotopes with their daughters produced in the $^{nat}Yb+^{48}Ca$ reaction. Most of the strips of the focal-plane detectors had α-energy peak resolution of about 30 to 38 keV FWHM. When α particles escape the focal-plane detectors at varying angles they lose correspondingly a variable amount of energy traversing the detector's dead layer, the entrance window and pentane, that fills the detection module. Thus, the strips of the side detectors generally showed α-energy peak FWHM of 90 to 160 keV. By using known sequences of successive alpha decays from the products of the $^{nat}Yb+^{48}Ca$ and $^{206}Pb+^{48}Ca$ reactions, we measured the FWHM position deviation, Δpos, to be 1.0 mm for α-α sequences, 1.3 mm for No EVR-α and 1.0 mm for No EVR-SF. Thus, a chain of consecutive signals from EVR, α particles, and spontaneous fission should be registered within position window of ~2 mm with 95% probability. In this respect, the whole detector array is equivalent to more than 200 separate cells. This essentially reduces probability to include random particles in the chain and allows expanding time intervals in searching for correlated events.

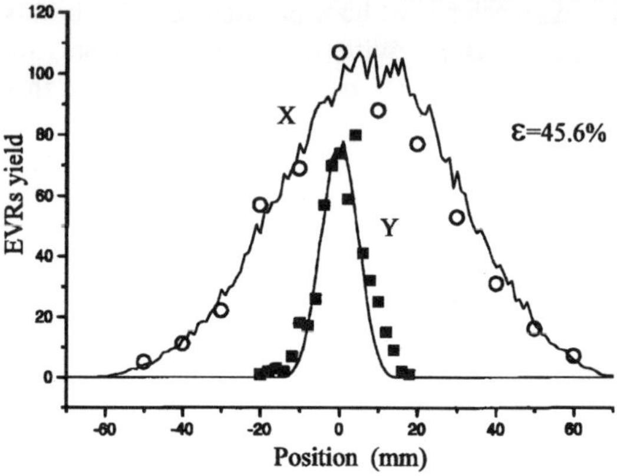

Figure 3. Horizontal (X) and vertical (Y) position spectra for ^{252}No EVRs produced in the $^{206}Pb+^{48}Ca$ reaction. Distributions calculated using ANAMARI code are shown by solid lines.

The collection efficiency of the separator was estimated by using the code ANAMARI [10]. This computer code is based on the Monte-Carlo technique. It is used to generate EVR-events and calculate their trajectories in the separator and corresponding spatial distributions with account for reaction kinematics, energy losses and multiple scattering in target and separator's media, equilibrium charge states of EVRs and magnetic-optical system of the separator. The calculations successfully reproduce the distributions of EVRs on the focal-plane detectors and their collection efficiencies for numerous reactions studied. Calculated horizontal and vertical distributions of ^{252}No EVRs produced in $^{206}Pb+^{48}Ca$ reaction in comparison with experimental data are shown in Figure 3; the collection efficiency of the separator was calculated to be 45.6%. We used collection efficiency of 45% when calculating cross sections of the reactions $Pb(^{48}Ca,xn)No$.

3 Experimental results and discussion

Cross sections for producing $^{250\text{-}254}No$ on ^{204}Pb and ^{206}Pb targets were measured for ^{48}Ca beams with energies of 213 to 219 MeV and 213 to 239 MeV, respectively. Production cross sections for $^{251\text{-}254}No$ were calculated using the nuclear decay data of [11,12] and data from the present work for ^{250}No (see below). The experimental fusion-evaporation cross sections for the reactions ^{206}Pb(^{48}Ca,1-4n) and ^{204,206,207,208}Pb(^{48}Ca,2n) are presented in Figure 4 and Table 1.

An enriched ^{206}Pb target was used to study evaporation of 1-4 neutrons from the ^{254}No compound nucleus. The 2n-evaporation product, ^{252}No, was identified by detecting α-particle energies, correlations with the α decays of the daughter ^{248}Fm,

Table 1. Summary of measured fusion-evaporation cross sections.

Reaction	Laboratory energy (MeV)	Excitation energy (MeV)	Cross section (nb) 1n	2n	3n	4n
^{48}Ca+^{208}Pb	213.5	19.6		1870^{+430}_{-330}		
	216.7	22.3		2050^{+460}_{-340}		
	219.4	24.4		1190^{+290}_{-230}		
^{48}Ca+^{207}Pb	213.5	19.8		670^{+460}_{-450}		
	216.7	22.4		1310^{+430}_{-410}		
	219.4	24.6		710±370		
^{48}Ca+^{206}Pb	212.7	19.8	29^{+25}_{-29}	100^{+20}_{-15}		
	213.5	20.4		308^{+88}_{-73}		
	216.6	23.0		500^{+110}_{-100}		
	217.0	23.2		488^{+76}_{-45}		
	217.1	23.3		515^{+80}_{-47}		
	217.4	23.6	58^{+16}_{-17}	489^{+31}_{-21}	≤0.9	
	218.1	24.1		475^{+74}_{-44}		
	219.4	25.2		327^{+91}_{-75}		
	219.9	25.7	54^{+38}_{-37}	325^{+61}_{-43}		
	223.9	28.9	≤15	166^{+37}_{-28}	$7.9^{+7.9}_{-4.3}$	
	226.2	30.7		56^{+14}_{-11}	30^{+9}_{-7}	
	233.5	36.6		$3.8^{+1.9}_{-0.9}$	18^{+9}_{-8}	
	239.1	41.2		$1.7^{+1.0}_{-0.7}$	$3.7^{+1.7}_{-1.2}$	≤0.2
^{48}Ca+^{204}Pb	213.5	20.6		$2.4^{+6.5}_{-2.2}$		
	216.7	23.2		$9.5^{+7.6}_{-5.0}$		
	219.4	25.4		$6.9^{+8.4}_{-4.6}$		

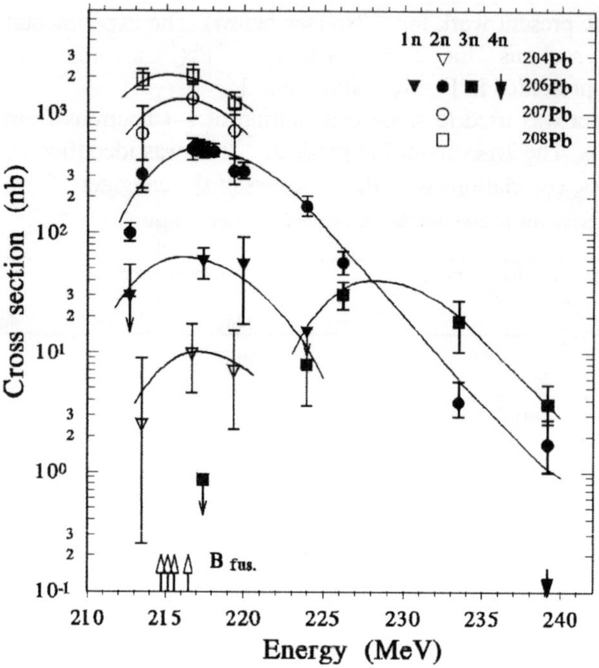

Figure 4. The experimental fusion-evaporation cross sections for the reactions ^{206}Pb(^{48}Ca,1-4n) (solid symbols) and ^{204,207,208}Pb(^{48}Ca,2n) (open symbols). Bass barriers for ^{204,206,207,208}Pb+^{48}Ca reactions are shown by open arrows. The lines are drawn to guide the eye.

and correlation of alpha-particle and spontaneous fission events with the known half-life relative to the EVR implantation time. The estimated contributions due to ^{207}Pb(^{48}Ca,3n) and ^{208}Pb(^{48}Ca,4n) reactions on target impurities did not exceed 10% at the corresponding cross section maxima (see, e.g., [13,14]) and were ignored in calculating the 2n cross section. The maximum cross section of 515^{+80}_{-47} nb was measured at 217 MeV for the reaction ^{206}Pb(^{48}Ca,2n)^{252}No. The sum in-beam energy spectrum of α particles registered by focal-plane detector in the ^{206}Pb+^{48}Ca reaction is shown in Figure 5a. Note that α particles with E_α=7.45 MeV and below 7 MeV originate from the decay of long-lived activities (^{210}Rn, ^{211}Rn, ^{211}At, and ^{211}Po) produced in the ^{nat}Yb+^{48}Ca calibration reaction.

We identified ^{251}No by the known energies of α particles correlated with α decays of the daughter ^{247}Fm and half-life. At E(^{48}Ca)=226 MeV the cross section of the ^{206}Pb(^{48}Ca,3n)^{251}No reaction reaches its maximum of 30^{+9}_{-7} nb.

In determining cross sections for the reaction ^{206}Pb(^{48}Ca,1n)^{253}No, contributions from the 2n products of the reactions with admixtures of ^{207}Pb and ^{208}Pb appeared to be considerable. Contribution from the reaction ^{207}Pb(^{48}Ca,2n)^{253}No was determined from our experimental measurement of the excitation function and the certified isotopic composition of the target material. We also considered the fact that α-particle spectrum of ^{253}No overlaps the 8.1-MeV line of ^{254}No [12], while the corresponding half-lives differ by a factor of less than two. To account for the contribution from ^{254}No, we used our measured excitation function $^{208}Pb(^{48}Ca,2n)^{254}No$ together with that from [14]. Impurities contributed up to 50% of the α-particle yield of ^{253}No at the 1n reaction maximum. Finally, we obtained maximum cross section of 58^{+16}_{-17} nb for the reaction $^{206}Pb(^{48}Ca,1n)^{253}No$ at 217 MeV.

No short-time EVR-SF correlations were detected at the ^{48}Ca energy of 239 MeV, close to the expected cross section maximum for the reaction $^{206}Pb(^{48}Ca,4n)$. With the 84-μs dead time of the electronics system, uncertainty in the radioactive half life of ^{250}No could significantly influence estimates of the $^{206}Pb(^{48}Ca,4n)$ reaction cross section. Spontaneous fission as a predominant decay mode and $T_{1/2}=250\pm50$ μs are reported in a single paper [15]. However, the data concerning ^{254}Rf [15] was not confirmed in a recent experiment [12]. With the aim of determining properties of ^{250}No we irradiated an enriched ^{204}Pb target with ^{48}Ca ions at energies covering the range expected for the peak of the 2n reaction. We assigned the observed new SF activity with $T_{1/2}=46^{+19}_{-11}$ μs to the even-even nuclide ^{250}No (see below). With the new half-life value of ^{250}No, we set a 0.2-nb cross section limit for the reaction $^{206}Pb(^{48}Ca,4n)$ at 239 MeV.

In experiments with the ^{204}Pb target, we simultaneously obtained excitation functions of $^{206,207,208}Pb(^{48}Ca,2n)$ reactions induced the substantial isotopic contaminants in the target material (see above). The maximum production cross section of ^{252}No was measured to be 500^{+110}_{-100} nb, which agrees both with the data of previous experiments and our present measurements with ^{206}Pb. Overlapping of the ^{253}No and ^{254}No α-particle spectra were taken into account when determining yields of these isotopes. The independent yield of ^{254}No was calculated from the number of detected α particles of its daughter, ^{250}Fm, taking α-branching values of 95±5% [11] and 90% [16,17] for ^{250}Fm and ^{254}No, respectively. The corresponding contribution was subtracted from the sum α spectrum of ^{253}No and ^{254}No to obtain the ^{253}No fraction. We obtained a maximum cross section of 2050^{+460}_{-340} nb for the $^{208}Pb(^{48}Ca,2n)^{254}No$ reaction, in good agreement with data of [18,14], and 1310^{+430}_{-410} nb for the reaction $^{207}Pb(^{48}Ca,2n)^{253}No$.

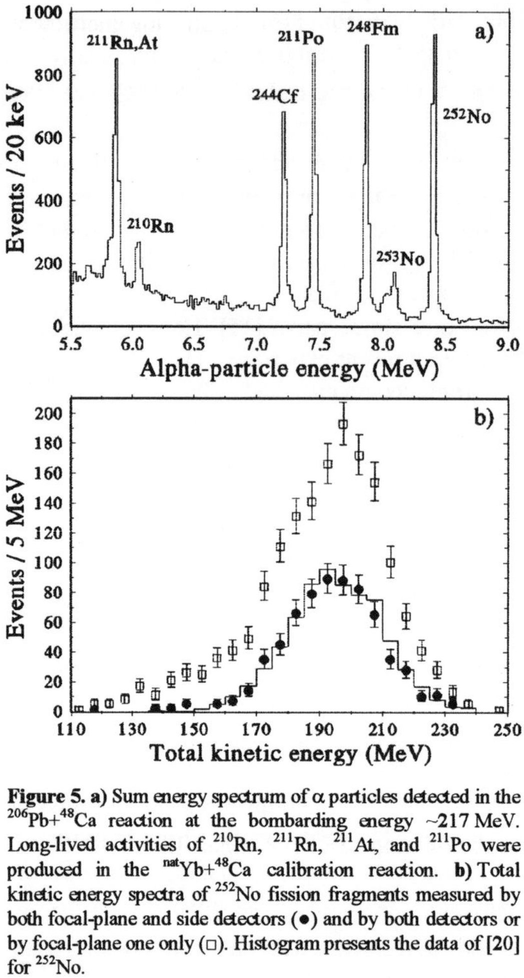

Figure 5. a) Sum energy spectrum of α particles detected in the $^{206}Pb+^{48}Ca$ reaction at the bombarding energy ~217 MeV. Long-lived activities of ^{210}Rn, ^{211}Rn, ^{211}At, and ^{211}Po were produced in the $^{nat}Yb+^{48}Ca$ calibration reaction. **b)** Total kinetic energy spectra of ^{252}No fission fragments measured by both focal-plane and side detectors (●) and by both detectors or by focal-plane one only (□). Histogram presents the data of [20] for ^{252}No.

In the present series of experiments, we detected a total of more than 3500 decays of ^{252}No. From the sum distribution of the EVR-α and EVR-SF correlations, we obtained a half-life of $2.45^{+0.05}_{-0.07}$ s for the decay of ^{252}No. This agrees with the reported values: $T_{1/2}$=2.30±0.22 s [19], $2.25^{+0.18}_{-0.16}$ s [18] and 2.44±0.12 s [20]. A spontaneous-fission branch b_{SF}=32.0±1.1% which follows from the present data is somewhat higher than b_{SF}=26.9±1.9% or 21.6±4.2% reported in [19] and [21], respectively. The corresponding partial SF half-life $T_{SF}=7.67^{+0.30}_{-0.35}$ s is included in the systematics presented in Figure 6. For ^{248}Fm, we obtained $T_{1/2}$=32.9±1.8 s from α-α correlations with ^{252}No, in agreement with [11].

For fission-energy calibration we measured an energy spectrum of fission fragments from the SF of ^{252}No implanted in the focal-plane detector. For about 40% of fission events, coincident fragments were detected by both focal-plane and side detectors. Focal-plane detectors absorbed both fragments with a probability of for about 25%, since the average fragment range exceeded the implantation depth of ^{252}No EVRs by a factor of only 3 to 4. The initial measured total kinetic energies (TKE) should be corrected for the pulse-height defect of detectors and energy losses of escaping fragments in the entrance windows, detector's dead layers and pentane. With this aim in view, the average measured kinetic energy was compared to the total kinetic energy (TKE) value of 194.3 MeV measured for SF of ^{252}No in [20]. The corrected TKE distributions for SF of ^{252}No implants are shown in

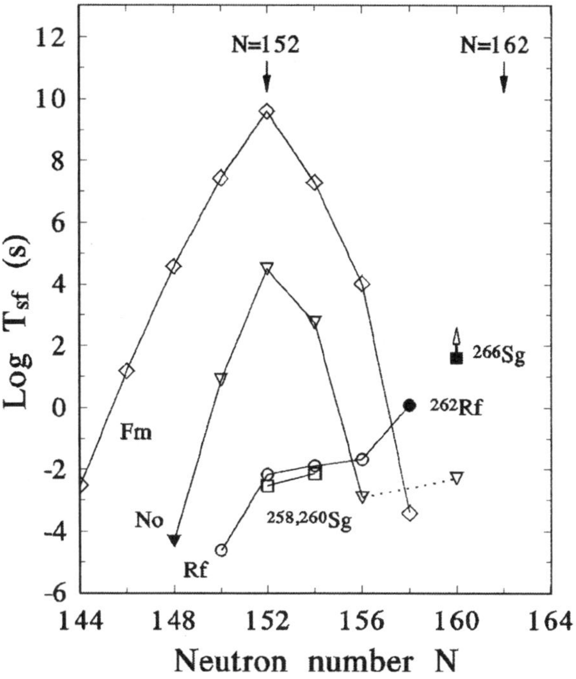

Figure 6. Partial SF half-lives of even-even nuclei with Z=100 to 106. Solid symbols show the data for ^{266}Sg, ^{262}Rf and ^{250}No from Ref.[22] and present work.

Figure 5b together with the results obtained with an external source [20]. Due to large yield of spontaneously fissioning ^{252}No in the ^{206}Pb+^{48}Ca reaction, this nuclide can be used for fission-energy calibration in the forthcoming experiment on the synthesis of superheavy nuclei in the ^{244}Pu+^{48}Ca reaction.

Analyzing the EVR-SF correlations detected in the bombardment of the enriched ^{204}Pb target with ^{48}Ca ions, we isolated a new short-lived SF activity. One sequence with a correlation time of 143 μs was detected at E(^{48}Ca)=213 MeV, seven at 217 MeV, with $T_{1/2}=41^{+24}_{-11}$ μs and three at 219 MeV ($T_{1/2}=39^{+47}_{-14}$ μs). The sum distribution results in a half-life of 46^{+19}_{-11} μs. The corresponding excitation energies of the ^{252}No compound nucleus were 21, 23 and 25 MeV, respectively, calculated with mass values of [23]. These cover the energy range expected for the ^{204}Pb(^{48}Ca,2n) reaction. The observed energy dependence of the yield of 46-μs SF activity follows excitation functions of $^{206-208}$Pb(^{48}Ca,2n) reactions measured in the same irradiation. The latter were obtained by detecting isotopes $^{252-254}$No produced on isotopic admixtures in target material and agree well with the available experimental data (see above). The corresponding EVRs are characterized by implantation energy, time of flight and ionic charge, which are in accord with those measured simultaneously for the known EVRs with Z=102. Finally, the lower yield of this activity, as compared to other No isotopes produced via 2n evaporation, its decay mode and half-life are consistent with those expected for ^{250}No. On this basis, we assign the observed SF activity to the decay of the neutron-deficient even-even nucleus ^{250}No produced in the reaction ^{204}Pb(^{48}Ca,2n) with a maximum cross section of $9.5^{+7.6}_{-4.9}$ nb at 217-

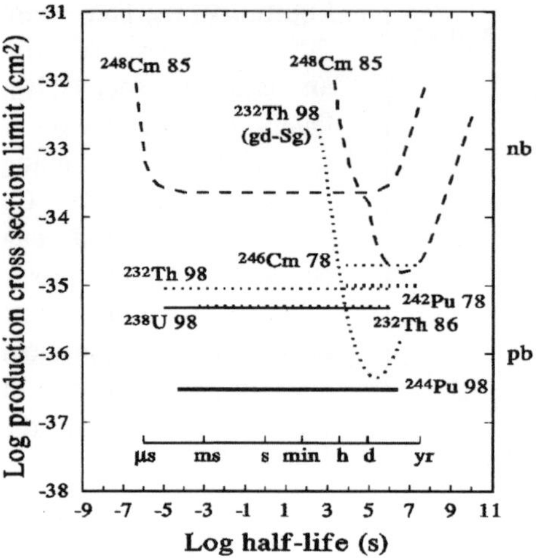

Figure 7. Sensitivity levels for the production of superheavy elements in ^{48}Ca-induced reactions with actinide targets as a function of the half-lives of the synthesized nuclei. Date of experiment and target used, are given near the corresponding curves. Expected cross section limit of 0.3 pb for the forthcoming $^{244}Pu+^{48}Ca$ experiment is also shown.

MeV (see Figures 4 and 6). The partial α-decay half-life of ^{250}No is calculated from systematics to be about 0.2 s [24], which corresponds to $b_\alpha \sim 2*10^{-4}$.

The half-life of the even-even spontaneously fissioning nucleus ^{250}No synthesized in our experiments is about 5×10^8 times shorter than that of ^{254}No, which has just four extra neutrons and is the most stable against SF No isotope. Such a significant decrease in SF half-lives, which is observed as neutron number in No isotopes is reduced from N=152, is the largest of all known even-Z nuclei. This fact can apparently be explained by steep disappearance of the macroscopic part of the fission barriers and, accordingly, by increased influence of the shell effects on stability of No nuclei in comparison with lower-Z elements [25]. Considerable decrease of the SF half-lives with moving off the N=152 shell for Fm and No isotopes was explained in [26] by the lowering of the outer fission barrier below the ground state. On the other hand, this change of the barrier shape for the nuclei with Z≥104 [26] results in weakening of the N=152 shell stabilization for these nuclei, however this influence is still active for Z=104 [12].

Together with the presented results, these experiments allowed to specify conditions for the future experiments on the synthesis of superheavy nuclei in the $^{244}Pu+^{48}Ca$ reaction. We collected new data for the average equilibrium charge states of heavy atoms moving through dilute hydrogen, in a wide range of average velocity including that expected for atoms of 114 element. Systematics of measured recoil energy vs. energy at the implantation for different heavy EVRs was extended. Based on the measured collection efficiency of the separator for the studied reactions with ^{48}Ca, we deduced that 40% of the recoiling Z=114 nuclei formed in the $^{244}Pu+^{48}Ca$ reaction would be implanted in the focal-plane detector. With the continuous projectile beam in the $^{244}Pu+^{48}Ca$ experiment, one could expect the counting rate of α-particle-like background with $E_\alpha \geq 8$ MeV to be about 1 per hour in a position window of 2 mm. The signals similar to Z=114 EVRs would be detected in the same position with a frequency of about 2 events per hour.

Such counting rates of background particles allow identification of the nuclei with lifetimes up to tens of minutes. Thus, in one month of irradiation of the ^{244}Pu target with a thickness of 0.4 mg cm^{-2} by ^{48}Ca projectiles with an intensity of 4×10^{12} pps, the total beam dose of about 10^{19} can be collected. With the separator collection efficiency of 40% and α-particle detection efficiency of 80%, the sensitivity of such experiment reaches 0.3 pb for the detection of one event. Figure 7 illustrates the reachable cross-section level of the planned experiment, compared to the previous attempts of synthesizing the new elements in the ^{48}Ca-induced reactions with actinide targets (see, e.g., [5,27,28] and Refs. therein). The sensitivity of the forthcoming experiment exceeds by 2 to 3 orders of magnitude that of the last complex experiment with the reaction ^{248}Cm+^{48}Ca [5].

4 Addendum

During November and December 1998, we carried out an experiment to synthesize nuclei with Z=114 in the vicinity of predicted spherical nuclear shells in the complete fusion reaction ^{48}Ca+^{244}Pu. We present here initial results of this experiment [28].

The average intensity of the ^{48}Ca ion beam delivered by the Dubna U400 cyclotron to the target was 4×10^{12} pps. A total beam dose of ^{48}Ca ions was 5.2×10^{18}. The targets consisted of the enriched isotope ^{244}Pu (98.6 %) in the form of PuO_2, deposited onto 1.5-μm Ti foils to a thickness of ~0.37 mg cm^{-2}. We used a ^{48}Ca bombarding energy of 236 MeV, corresponding to the calculated maximum of the 3n-evaporation channel to form the isotope ^{289}114.

One position-correlated decay chain observed in this experiment is shown in Figure 8. An α-particle was detected in the focal-plane detector 30.4 s after the implantation in the middle of the 8th strip of a recoil nucleus. The energy of the first α-particle was E_α=9.71 MeV. A second α-particle, having an energy E_α=8.67 MeV, was observed at the same location after 15.4 min. A third α-particle, escaping the front detector leaving an energy $E_{\alpha 1}$=4.04 MeV and absorbed in the side detector with $E_{\alpha 2}$=4.79 MeV (E_{tot}=8.83 MeV), was measured 1.6 min later. Finally, 16.5 min later, the SF event was observed as two coincident signals (two fission fragments) with energy deposited in the focal-plane detector E_{F1}=120 MeV and in the side detector E_{F2}=52 MeV; E_{tot}=172 MeV.

All 5 signals (EVR, α_1, α_2, α_3, SF) appeared within a position interval of 1.6 mm (Figure 8), which strongly indicates that there is a correlation among the observed decays.

To estimate the probability of the candidate event being due to random correlations 10^5 artificial fissions were inserted into the data distributed at random positions and times over the entire detector array and experiment duration by Monte-Carlo technique. The probability per fission of finding correlations of an EVR-like event followed by three α-particle signals within the energy interval of 8.5-10 MeV in the 34 minutes preceding each artificial fission was determined to be N_b=0.006. Even without further restricting the energies of sequential events or event-to-event times within the given interval, this value is quite low. Another N_b calculation was performed for strip 8 at the position in which the candidate event occurred. Thus, calculated from event rates alone, with event-to-event time distances restricted to actual measured values, the probability of this decay sequence being caused by the chance correlation of unrelated events in strip 8 is 1×10^{-4}.

All events of the decay chain are correlated in time and position. These events arise as a result of the α-decay of a parent nucleus (E_α=9.71 MeV) and continue until SF takes place. This matches the decay of a superheavy nucleus that is predicted by theory. For the whole decay chain the basic rule for α-decay, defining the relation between Q_α and T_α, is fulfilled. The detected sequential decays have larger $T_{1/2}$ vs. $E_\alpha$values than the known radioactive nuclides. The best candidate for the parent nucleus is the even-odd isotope $^{289}114$, produced in the 3n-evaporation channel. This one event corresponds to a cross section of about 1 pb. The decay properties of the observed nuclei are also in agreement with calculations [24,29], assuming reasonable hindrances for the decay of nuclei with odd neutron numbers.

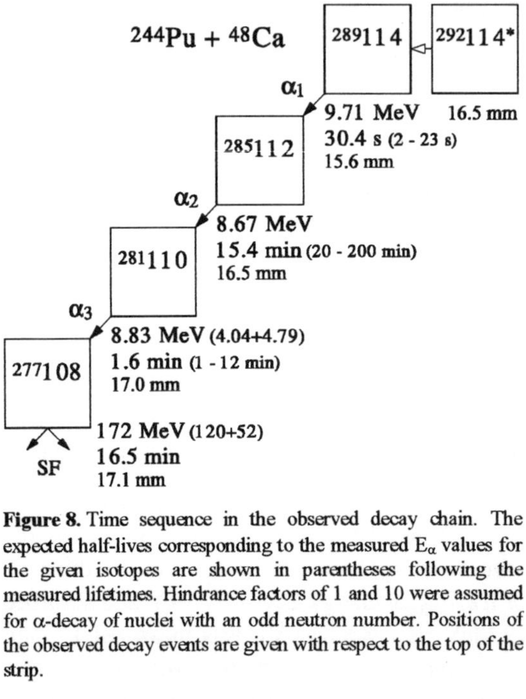

Figure 8. Time sequence in the observed decay chain. The expected half-lives corresponding to the measured E_α values for the given isotopes are shown in parentheses following the measured lifetimes. Hindrance factors of 1 and 10 were assumed for α-decay of nuclei with an odd neutron number. Positions of the observed decay events are given with respect to the top of the strip.

The lifetimes of the new isotopes, in particular $^{285}112$ and $^{281}110$, appear to be approximately 10^6 times longer than those of the known nuclei $^{277}112$ and $^{273}110$ [30,31], which have eight fewer neutrons. Therefore, the observed decay properties of the synthesized nuclides, together with the data obtained earlier for $^{283}112$ [6], can be considered the first experimental proof of the existence of enhanced stability in the region of superheavy elements.

5 Acknowledgements

We are indebted to our late colleague Dr. Yu.A. Lazarev who greatly contributed to this research at the initial stages. We express our thanks to V.I. Krashonkin, V.I. Tomin, A.M. Zubareva, and A.N. Shamanin, for their help in preparing and carrying out the experiment. We would like to express our gratitude to the personnel of the U400 cyclotron and the associates of the ion-source group for obtaining an intense ^{48}Ca beam.

The help of Dr. E.K. Hulet in the early phases of our collaboration is gratefully acknowledged.

This work has been performed with the support of the Russian Foundation for Basic Research under grant No. 96-02-17377 and of INTAS under grant No. 96-662. Much of the support for the LLNL authors was provided through the U.S. DOE under Contract No. W-7405-Eng-48. These studies were performed in the framework of the Russian Federation/U.S. Joint Coordinating Committee for Research on Fundamental Properties of Matter.

References

1. Hofmann S, In Proceedings of the VI International School-Seminar "Heavy Ion Physics", Dubna, Russia, September 22-27, 1997, (World Scientific, Singapore, 1998) p.385.
2. Lazarev Yu.A., In Proceedings of the International Workshop XXIV on Gross Properties of Nuclei and Nuclear Excitations "Extremes of Nuclear Structure", Hirschegg, Austria, January 15-20, 1996, (GSI, Darmstadt, 1996) p.11.
3. Ghiorso A. et al., Phys. Rev. C 51, (1995), p.R2293.
4. Oganessian Yu.Ts., In Proceedings of the International Conference on Nuclear Physics at the Turn of the Millennium "Structure of Vacuum & Elementary Matter", Wilderness, South Africa, March 10-16, 1996 (World Scientific, Singapore, 1997), p.11.
5. Armbruster P. et al., Phys. Rev. Lett. 54, (1985), p.406.
6. Oganessian Yu.Ts. et al., JINR Report No. E7-98-212, Dubna,1998; Eur. Phys. J. A, to be published.
7. Lazarev Yu.A. et al., In Proceedings of the International School-Seminar on Heavy Ion Physics, Dubna, Russia, May 10-15, 1993, (JINR Report E7-93-274, Dubna, 1993) Vol.II, p.497.

8. Lazarev Yu.A. et al., In Heavy Ion Physics, Scientific Report 1995-1996 (JINR Report No. E7-97-206, Dubna, 1997), p.30.
9. Tsyganov Yu.S., In Proceedings of the XVII International Symposium on Nuclear Electronics, Varna, Bulgaria, September 15-21, 1997, (JINR Report D13-98-66, Dubna, 1998) p.206.
10. Subotic K. et al., to be published.
11. Firestone R.B., Shirley V.S. (editor): Table of Isotopes eighth edition, John Wiley & sons, inc. New York, Chichester, Brisbane, Toronto, Singapore, 1996.
12. Heßberger F.P. et al., Z. Phys. A 359, (1997), p.415.
13. Flerov G.N. et al., Nucl. Phys. A 267, (1976), p.359.
14. Gäggeler H.W. et al., Nucl. Phys. A 502, (1989), p.561c.
15. Ter-Akopyan G.M. et al., Nucl. Phys. A 255, (1975), p.509.
16. Orlova O.A. et al., Yad. Fiz. 30, (1979), p.618 [in Russian] [Sov. J. Nucl. Phys. 30, (1979) p.317].
17. Türler A. et al., Z. Phys. A 331, (1988), p.363.
18. Lazarev Yu.A. et al., Physica Scripta 39, (1989), p.422.
19. Bemis C.E., Jr. et al., Phys. Rev. C 15, (1977), p.705.
20. Wild J.F. et al., J. Alloys Compounds 213/214, (1994), p.86.
21. Andreyev A.N. et al., Z. Phys. A 345, (1993), p.389.
22. Lazarev Yu.A. et al., Phys. Rev. Lett. 73, (1994), p.624.
23. Audi G. and Wapstra A.H., Nucl. Phys. A 595, (1995), p.409.
24. Smolańczuk R., Skalski J., and Sobiczewski A., In Proceedings of the International Workshop XXIV on Gross Properties of Nuclei and Nuclear Excitations “Extremes of Nuclear Structure”, Hirschegg, Austria, January 15-20, 1996, (GSI, Darmstadt, 1996) p.35.
25. Patyk Z. et al., Nucl. Phys. A 491, (1989), p.267.
26. Oganessian Yu.Ts. et al., Nucl. Phys. A 239, (1975), p.157.
27. Oganessian Yu.Ts et. al., this volume.
28. Oganessian Yu.Ts. et al., JINR Report No. E7-99-53, Dubna,1999; Phys. Rev. Lett. (submitted).
29. Smolańczuk R., Phys. Rev. C 56, (1997), p.812.
30. Lazarev Yu.A. et al., Phys. Rev. C 54, (1996), p.620.
31. Hofmann S. et al., Z. Phys. A 354, (1996), p.229.

ESTIMATION OF PRODUCTION CROSS SECTIONS OF SUPERHEAVY ELEMENTS

M. VESELSKÝ

Inst. of Physics, Slovak Acad. Sci., Dúbravská cesta 9, 842 28 Bratislava, Slovakia
E-mail: veselsky@savba.sk

The available production cross sections of the $Z \geq 102$ nuclei are compared to the results of a statistical model using the parameters taken from the $N \approx 126$ region. The hot fusion data are reproduced using a summary fission barrier reduced by 15 - 20 %. For the cold fusion reactions the parameters of the fusion channel are found, nevertheless the cross sections are not reproduced for the nuclei above $Z = 104$ where the fusion probabilities are determined. A simple statistical approach, based on the neutron multiplicity data from the spontaneous fission as a unique information on the scission configuration, is proposed and the fusion probabilities of a cold fusion reactions are reproduced using one set of parameters. With these parameters, the estimate of the production cross sections in reactions for the synthesis of superheavy nuclei is calculated. The steep decrease of estimated cross sections is predicted for the cold fusion reactions. In the case of the ^{48}Ca-induced reactions the overall decrease of production cross sections is less significant.

1 Introduction

In the recent years, the heavy elements up to $Z = 112$ have been synthesized in the cold fusion reactions using the ^{208}Pb, ^{209}Bi - targets in the channel with the emission of one neutron. The experimentalists have faced a steep decrease of the cross sections up to the picobarn level. The same level of the cross sections has been reached also in the hot fusion reactions with the emission of 5 neutrons. Such a trend is in contradiction with the expected stabilization of heavy nuclei because of the effect of shell structure. A further progress towards the doubly magic spherical shell closure $Z = 114$, $N = 184$ is complicated by the expected further decrease of the production cross sections.

2 Hot fusion

The available data on the hot fusion transfermium cross sections are rather scarce. Considerable effort to this topic has been paid in the Flerov Laboratory of Nuclear Reactions at Dubna in recent years. Using the kinematic separator, the production cross sections of evaporation residue have been measured for the transfermium nuclei in the 4-5n channels [1]. It seems to be reasonable to expect that the measured cross sections could be reproduced using the statistical model of nuclear de-excitation.

The statistical code HIVAP [2] was used for calculations. The level densities have been calculated using the well known Fermi gas model. The same level density parameters have been used for the fission and particle evaporation channels ($\tilde{a}_f/\tilde{a}_\nu = 1$). A phenomenological way of the introduction of shell effects into the level density calculation according to Ignatyuk [3], have been used in the evaporation channels. The full (summary) fission barrier has been calculated as the sum of the liquid drop and shell components, scaled by a coefficient C.

$$B_f(l) = C(B_f^{LD}(l) + \Delta B_f^{Shell}). \tag{1}$$

This formula for the fission barrier has been proved to be useful in the case of the nuclei around closed neutron shell $N = 126$, where typical values of the coefficient C were 0.65 – 0.75 independently of the ratio of macroscopic to shell components of the fission barrier. The liquid drop component of the fission barrier (B_f^{LD}) has been calculated according to the rotating charged liquid drop model of Cohen-Plasil-Swiatecki [4]. The shell component of the fission barrier (ΔB_f^{Shell}) has been taken from the calculation [5]. The principal difficulty of transfermium region is that the liquid drop barriers are small comparing to the shell corrections. This brings a new question of the dependence of the shell component of fission barrier on the angular momentum. In the region around $N = 126$ there was no essential difference when the constant shell component was taken for all values of the angular momentum or when it was scaled according to the rotating liquid drop barrier angular dependence. In the transfermium region, however, the use of the constant shell barrier component leads to unrealistic shapes of the excitation functions, while the scaled shell barrier component is more preferable and has been used in calculations. The masses of nuclei have been taken from [5] (the measured masses have been preferred wherever available). As a default, the Bass potential [6] was used for the calculation of the fusion cross sections. As an alternative, the modified WKB formula with the the quasi-gaussian barrier distribution [7] was used. Both calculations gave very similar results. The coefficient C was the only free parameter within the calculations. The optimum values of the parameter C, necessary to reproduce the experimental cross sections, are given in Fig. 1 as a function of neutron number. One can see that the optimum values of C remain stable within $Z = 102 - 110$. The value of $C = 0.8 - 0.9$ is higher when compared to $N = 126$ region and this difference could be most probably attributed to the differences of the saddle point configurations in both regions.

3 Cold fusion

For the case of the Pb-target, similar calculations as in the case of hot reactions have been carried out. In this case, the two different fusion cross section calculations give strongly different results. In the case of the Bass potential, the 1n channel appears to be deeply subbarrier. In the case of barrier distribution, no suppression by interaction barrier could be observed. Some indices on the barrier position could be obtained from the ratio of maximum cross sections of 1n and 2n channels. The available data [8] on cold fusion σ_{1n}/σ_{2n} ratio indicate that the 1n channel is subbarrier in the reaction ^{48}Ca+^{208}Pb, in the reaction ^{50}Ti+^{208}Pb both channels are of the same magnitude and for the heavier projectiles ^{54}Cr and ^{58}Fe the 1n channel is dominating. Such a tendency indicates that for the heavier projectiles the 1n channel is not suppressed and a large set of angular momenta contributes. For the further calculations the barrier distribution concept has been used with the parameter set (V_0 = 40 MeV, r_0 = 1.11 fm, d = 0.75 fm, σ_{r_0} = 3 %) which allows to reproduce the maximum cross sections of channels near the fusion barrier for the reactions from ^{16}O+^{208}Pb, ^{22}Ne+^{208}Pb up to ^{48}Ca+^{208}Pb, ^{50}Ti+^{208}Pb and to set the right positions of the maximum cross sections for heavier projectiles.

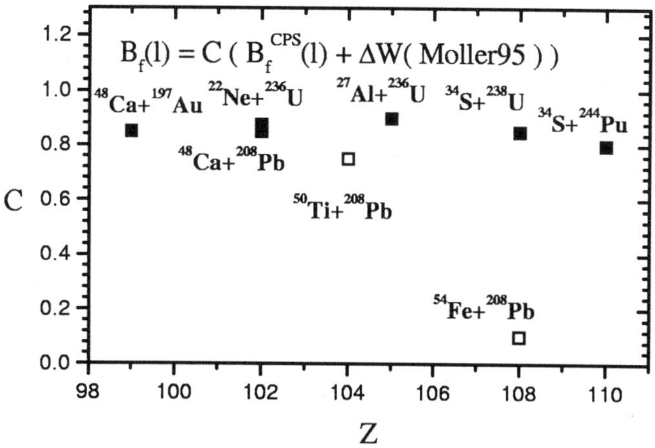

Figure 1. Optimum values of the parameter C, necessary to reproduce the experimental cross sections of hot fusion reactions (solid symbols), as a function of atomic number of residual nuclei. Open symbols - the cold fusion reactions falling out of systematics at $Z > 104$.

Reaction	σ_{1n}(exp)	σ_{1n}(calc)	P_{fus}
^{50}Ti+^{208}Pb	10 nb	40 nb	0.25
^{54}Cr+^{208}Pb	0.5 nb	45 nb	8.8×10^{-3}
^{58}Fe+^{208}Pb	60 pb	94 nb	6.3×10^{-4}
^{62}Ni+^{208}Pb	3 pb	0.28 μb	1.1×10^{-5}
^{64}Ni+^{208}Pb	15 pb	1.38 μb	1.1×10^{-5}
^{70}Zn+^{208}Pb	1 pb	4.3 μb	2.3×10^{-7}

Table 1. The fusion probability in the cold fusion reactions with ^{208}Pb target.

Using the same parameters of statistical model as in the case of hot reactions ($C = 0.85$), one can reproduce the 1-2n cross sections in the case of ^{48}Ca+^{208}Pb and only slightly overestimate the cross sections of 1-2n channels in the case of the ^{50}Ti+^{208}Pb. In the latter case the use of C=0.75 gives acceptable agreement of the experimental and calculated cross section. For the heavier projectiles a still stronger overestimation of the experimental cross sections could be observed. Such an overestimation could be corrected only by a strong decrease of C by one or more orders of magnitude what is out of the physically acceptable limits (see Fig. 1, open symbols). In the Table I are given the experimental and calculated ($C = 0.85$) values of 1n cross sections and the estimated fusion probability factors. As one may see the fusion probabilities steeply decrease and reach the value of about 10^{-7} for the heaviest projectile ^{70}Zn. This means that, unlike to the hot fusion, in the cold fusion reactions, leading to the evaporation residues with $Z > 104$, the fusion-fission competition to evaporation residue channel is not strong enough and another mechanism of fission with shorter time scale could contribute.

4 Fusion probability - Model

One can in principle assume that during the fusion process the projectile-target system can pass nearby the traditional saddle point configuration. In that case, an additional competition of the immediate fission with the complete fusion may take place and the statistical weights of the compound nucleus and scission configuration, as a final energetically allowed states, may determine the probabilities of the fusion or immediate fission. The open question is the statistical weight of scission point. Also the reaction timescale is relevant for the reaction scenario determination.

For the quantitative calculations the following formula for the fusion probability was used

$$P_{fus}^{stat} = \frac{\rho(E_{CN}^*)}{\rho(E_{CN}^*) + \rho(E_{sc,eff}^*)}, \tag{2}$$

where the level density has been calculated according to the Fermi-Gas formula. The use of such a simple level density approach without shell effects may be under discussion but in the region of $N = 126$ it gives a quite good results for the calculation of evaporation residue cross sections [9] and, in the present case, it allows to avoid the nontrivial calculation of shell correction to scission point. The effective excitation energy in the scission point could be expressed as

$$E_{sc,eff}^* = (\nu_n^{s.f.}(A_{CN}) + \Delta\nu_n(E_{CN}^*))E_n, \tag{3}$$

where $\nu_n^{s.f.}(A_{CN})$ is the linear extrapolation of the available spontaneous fission neutron multiplicity data to given A_{CN} ($\nu_n^{s.f.}(A_{CN}) = 3.316 + 0.0969(A_{CN} - 250)$), $\Delta\nu_n(E_{CN}^*)$ is the additional increase of the post-scission neutron multiplicity at given E_{CN}^* which can be expressed approximately as $\Delta\nu_n(E_{CN}^*) = 0.035E_{CN}^*$, as follows from the available post-scission neutron multiplicity data [10], and E_n is the amount of excitation energy per emitted neutron. The values of E_n have been determined for the cold fusion reactions and exhibit the linear dependence on A_{CN},

$$E_n = 3.795 + 0.04(A_{CN} - 260). \tag{4}$$

Using obtained set of parameters, the estimations of the cross sections of heavier nuclei are possible for both the cold and hot fusion.

5 Cross section estimations

For the production of the heavier nuclei approaching the spherical doubly magic shell closure at $Z = 114$, $N = 184$ several possible concepts exist. One of them considers the use of still heavier beams with the improved Pb-targets allowing the increase of beam current. For the synthesis of the elements 114 resp. 116 the beams of ^{74,76}Ge and ^{80,82}Se are considered and the 1n channel is expected to be dominant. The second considered possibility is the ^{48}Ca beam with the still heavier targets with large enough lifetime (^{232}Th, ^{238}U, ^{244}Pu and ^{248,250}Cm). In this case the dominant channel depends on the position of the fusion barrier but it is expected to sink to still lower excitation energies.

Table 2. The estimated cross sections and parameters of fusion probability model for the production of superheavy nuclei.

Reaction	E^*_{max} MeV	$E^*_{sc,eff}$ MeV	P^{stat}_{fus}	σ_{ER} systematics	σ_{ER} estimation
$^{208}Pb(^{74}Ge,1n)^{281}114$	15	32.5	10^{-8}	2.9 μb	0.03 pb
$^{208}Pb(^{76}Ge,1n)^{283}114$	15	33.9	2.5×10^{-9}	9.5 μb	0.025 pb
$^{208}Pb(^{80}Se,1n)^{287}116$	15	37.0	10^{-10}	9.8 μb	0.001 pb
$^{208}Pb(^{82}Se,1n)^{289}116$	15	38.5	4×10^{-11}	19.3 μb	0.0008 pb
$^{232}Th(^{48}Ca,3n)^{277}110$	27.5	33	6×10^{-3}	0.3 nb	1.8 pb
$^{232}Th(^{48}Ca,4n)^{276}110$	40	35	0.9	11 pb	10 pb
$^{238}U(^{48}Ca,3n)^{283}112$	33	38	9×10^{-3}	0.16 nb	1.5 pb
$^{238}U(^{48}Ca,4n)^{282}112$	38	39.3	0.22	7.3 pb	1.6 pb
$^{244}Pu(^{48}Ca,2n)^{290}114$	22	41.5	10^{-9}	18 nb	$2.\times10^{-5}$ pb
$^{244}Pu(^{48}Ca,3n)^{289}114$	24	41.6	10^{-8}	7.5 nb	$7.\times10^{-5}$ pb
$^{244}Pu(^{48}Ca,4n)^{288}114$	32	43	10^{-4}	0.15 nb	0.015 pb
$^{244}Pu(^{48}Ca,5n)^{287}114$	44	45	0.5	0.2 pb	0.1 pb
$^{240}Pu(^{48}Ca,4n)^{284}114$	36	40.5	0.023	30 nb	0.7 pb
$^{242}Pu(^{48}Ca,4n)^{286}114$	35	42	5×10^{-3}	50 nb	0.25 pb
$^{248}Cm(^{48}Ca,5n)^{291}116$	44	49	0.04	0.2 pb	0.008 pb
$^{250}Cm(^{48}Ca,5n)^{293}116$	43	50	0.01	1 pb	0.01 pb

For the reactions with the ^{208}Pb target, it is reasonable to expect that one can simply calculate the fusion probabilities using extrapolated parameters from the cold fusion data. The situation with the ^{48}Ca-induced reactions is more complicated and one has to take into account what set of angular momenta contributes to the production cross sections. This is to a large extent a question of used angular momentum dependence of the shell component of fission barrier. The scaling with the liquid drop component leads to a fact that for both types of the reactions roughly the same set of angular momenta contributes ($l \leq 20$) at the maximum of excitation function. As a consequence, the same parameters have been used for the calculation of fusion probability for both types of reactions.

The estimated cross sections of selected channels of the considered reactions are given in Table II. For all cases, the production cross sections were calculated using the hot fusion parameters taken from systematics of Fig. 1 ($C = 0.80$) and for the maxima of dominant channels the fusion probabilities were calculated. By inspection of the fusion probability formula (2) one could expect that the fusion probability would increase with increasing excitation that can lead to the shift of maxima of the excitation functions to higher excitations. The available cold fusion excitation functions do not show such an effect. It can be understood under assumption that the increase of cross section at higher excitations is compensated by the rapid decrease of fusion probability at higher angular momenta in the tails of excitation functions.

For the cold fusion reactions the increase of projectile mass brings a still lower cross sections. For the reactions of ^{48}Ca-beam there are more possibilities for the choice of the channel. For the reactions ^{48}Ca+^{232}Th and ^{48}Ca+^{238}U two channels are examined, one of them near the fusion barrier and second one which corresponds to the value of P_{fus}^{stat} near to unity. In the reaction ^{48}Ca+^{244}Pu four channels ranging from the 2n to 5n are examined. In this case the 5n channel seems to be most promising and the estimated cross sections for the channels near to the barrier decrease rapidly with decreasing excitation energy. The cross sections of the most promising 4n channels are also shown for the reactions ^{48}Ca+^{240,242}Pu. For the reactions ^{48}Ca+^{248,250}Cm, the estimated cross section are the highest for the 5n channels. Generally, the estimated production cross section decrease with the increasing mass of compound nucleus. The decrease is less strong for the hot fusion channels and the estimated values are in principle within the reach of experiments in the near future.

6 Conclusions

The proposed method of the estimation of production cross sections of superheavy nuclei indicates that the effect of shell stability will, probably, not result in the increase of production cross sections. The method further indicates that with the increasing mass of compound system one may expect a fusion hindrance at still higher excitation energies. As the unhindered the 5n channel may be considered and the lower number of emitted neutrons will result in the increasing fusion hindrance.

Acknowledgments

The work was supported by the Grant Agency of Slovak Republic through contract GAV 2/5121/98. The author would like to thank to P. Armbruster, D.D. Bogdanov, F.P. Hessberger, S. Hofmann, B. Pustylnik, R. Smolanczuk and A.V. Yeremin for valuable discussions.

References

1. Andreyev A.N. et al. *Z. Phys. A* **344**, 1992 (225); *ibid.* **345** (1993) 389; Lazarev Yu.A. et al. *Phys. Rev. Lett.* **75**, 1995 (1903); *Phys. Rev.* **C54** (1996) 620
2. W. Reisdorf *Z. Phys. A* **300**, 1981 (227)
3. A.V. Ignatyuk, G.N. Smirenkin, A.S. Tishin *Yad. Fiz.* **21**, 1975 (485)
4. S. Cohen, F. Plasil, W.J. Swiatecki *Ann. Phys. (N.Y.)* **82**, 1974 (557)
5. P. Moller et al. *Atom. Data Nucl. Data Tbl.* **59**, 1995 (185)
6. R. Bass: *Nuclear Reactions with Heavy Ions (Lect. Notes on Phys. 117)*. Springer-Verlag, Heidelberg 1980, p.281
7. W. Reisdorf et al. *Nucl. Phys. A* **444**, 1985 (154)
8. H.-W. Gaggeler et al. *Nucl. Phys. A* **502**, 1989 (561c); S. Hofmann: *GSI-Preprint-98-16*. GSI Darmstadt, 1998
9. D.Vermeulen et al. *Z. Phys. A* **318**, 1984 (157)
10. S.M. Lukyanov et al.: *in Proc. Int. School of Heavy Ion Phys., Dubna, 3-12 Oct 1989*. JINR Dubna 1990, p.225; D. Hilscher et al. *Phys. At. Nucl.* **57**, 1994 (1187)

INVESTIGATION OF THE FUSION-FISSION CROSS-SECTION AND THE MULTIMODAL NATURE OF THE FISSION PROCESS IN THE *At – Th* REGION USING THE ^{220}Ra COMPOUND NUCLEUS AS AN EXAMPLE.

I.V.Pokrovski[1], L.Calabretta[2], M.G.Itkis[1], E.M.Kozulin[1], N.A.Kondratiev[1], C.Maiolino[2], E.V.Prokhorova[1], A.Ya.Rusanov[3], S.P.Tretiakova[1]

1 Flerov Laboratory of Nuclear Reaction, Joint Institute for Nuclear Research, RUSSIA

2 Laboratori Nazionali del Sud, Istituto Nazionale di Fisica Nucleare, ITALY

3 Institute of Nuclear Physics, KAZAKHSTAN

Using a double-arm time-of-flight spectrometer CORSET, the mass and energy distributions of fission fragments from the ^{220}Ra compound-nucleus produced in the ^{12}C + ^{208}Pb reaction were measured in a wide energy range of bombarding ^{12}C ions. It was shown that with a cooling of the compound-nucleus the asymmetric component consisting of two standard fission modes with masses of ~132 and ~139 appeared on slopes of the predominant symmetric fission. In total kinetic energy distributions the three-component structure was also observed in the asymmetric mass region.

1. Introduction

It is well known from numerous experimental data that actinide nuclei with masses of up to ~256 predominantly disintegrate into two fragments asymmetrically, that is mass and energy distributions (MED) of the fragments have a two-humped shape with a minimal yield for disintegrating into two equal parts [1,2]. On the contrary, nuclei in the lead region predominantly disintegrate symmetrically but have a small contribution of the asymmetric component (less than 0.5%) which decreases with decreasing A in the disintegrating nucleus [3]. The fission simultaneously through both these ways is common to the ^{225}Ra-^{228}Ac region in the case of intermediate excitation energies. But in the case of energies near the Coulomb barrier the symmetric way is a predominant one [4]. The region A=214-224 that can be named the intermediate region has not been so far studied from the point of view of the fission modes. There were attempts to study MED of fission fragments in the reactions ^{7}Li + ^{209}Bi and ^{12}C + ^{208}Pb at intermediate excitation energies in [5]. But in [5] no any information was presented on the asymmetric fission of these nuclei. Nonetheless, the irregularities in energy characteristics (total kinetic energy distributions and its dispersions) were observed in the fragment mass region which is common for the asymmetric mode.

Some years ago experimental studies of different characteristics of fission in the intermediate region were started by our group at Dubna and continued in Catania, Grenoble and Strasbourg. MED of the ^{219}Ac fission fragments and $^{220,224,226}Th$ compound-nuclei produced in reactions with $^{16,18}O$ ions at energies near and below the Coulomb barrier [6-9] were measured. The multiplicities of pre- and post-neutrons and gamma-quanta emitted from the fission fragments were measured for the ^{226}Th compound-nucleus [10,11].

At the same time the experimental studies of the charge distributions of the fission fragments from the secondary beams of nuclei from ^{214}Ra to ^{234}U [12-14] were started at GSI (Darmstadt).

The experiments [6-14] showed that in this region of nuclei with low excitation energies the transition between the predominantly symmetric to predominantly asymmetric fission (with increasing A) took place.

The present work continues a series of investigations devoted to the fission process characteristics in the intermediate region of nuclei, and we report here some experimental results concerning the MED of fission fragments and fission cross-section for the ^{220}Ra compound-nucleus produced in the reaction ^{12}C + ^{208}Pb at bombarding energies of ^{12}C ions of higher and deep below the Coulomb barrier.

2. The experiment

The measurements were performed on an external beam of the TANDEM accelerator (LNS, INFN, Catania, Italy) at bombarding energies of ^{12}C ions E_{lab}=54, 56, 57, 59, 62, 65, 73 and 90 MeV. The velocities and coordinates of the binary fission fragments were measured by the kinematics coincidence method with a double-arm time-of-flight position-sensitive spectrometer CORSET [15]. This spectrometer consists of four position-sensitive microchannel plate detectors providing registration of the reaction products within the solid angle of 360msrad and two microchannel plate detectors with electrostatic mirrors providing a start signal for the time-of-flight measurements.

Two start detectors allow us to measure two different times of flight for each particle registered by stop detectors. In other words, each timing parameter has a controlling parameter. To avoid saturation in the electronic channel we measured the T_{zero} parameter, which is the difference between the timing signal from the stop detector and the same signal transmitted through a coincidence unit. Taking into account this parameter allows us to improve the time resolution of the spectrometer.

The mass resolution of the spectrometer was 2-3amu and position resolution was 0.1°. The target consisted of a ^{208}Pb layer 220 μg/cm^2 in thickness delivered onto a carbon foil 50 μg/cm^2 in thickness.

The fission cross-section was measured with mica track detectors, which were placed into another reaction chamber positioned directly next to the CORSET reaction chamber. For the two lowest bombarding energies (54 and 56 MeV) only the fission cross-section was measured since investigation of MED of fission fragments seems to be difficult due a very low cross-section.

3. The experimental results

The main difficulty in performing such investigations is connected with the necessity of production of nuclei with low excitation energies corresponding to the ion energy E_i near and below the Coulomb barrier and thus to a rapid decrease in the fission cross-section with decreasing E_i. Fig. 1 presents the cross section dependences on an ion energy for the $^{208}Pb(^{12}C,f)$ and $^{208}Pb(^{16}O,f)$ reactions.

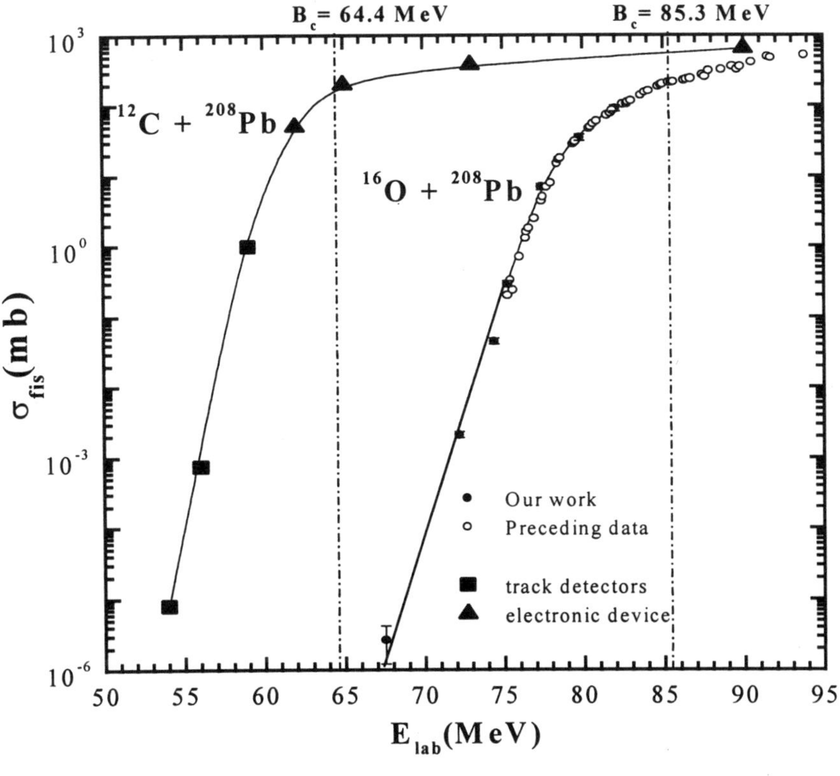

Fig. 1 *The fission cross-section on bombarding ion energy dependences.*

In the case of the ^{16}O ion we have an interaction between two magic nuclei. On the contrary, the ^{12}C ion is deformed. Thus it was expectable that the curve of the fusion cross-section for the $^{16}O + ^{208}Pb$ reaction had to be steeper. But the fission barriers for interactions of ^{208}Pb with ^{16}O and ^{12}C ions calculated by V.V.Pashkevich are 8 and 12.5 MeV correspondingly [16,17,18]. So, in the case of the ^{12}C ion we came closer to the fission barrier. Thus for the reaction with ^{16}O the fusion cross-section is practically equal to the fission cross-section, as opposed to the reaction with ^{12}C where at the lowest energies we cannot ignore the contribution of the evaporation residues into the fusion cross-section. In the immediate future we plan to measure the cross-section of the evaporation residues for the $^{12}C + {}^{208}Pb$ reaction. So, it will allow us to define the fission barrier experimentally for the first time.

Fig. 2 represents the mean total kinetic energy and its dispersion depending on the fragment mass.

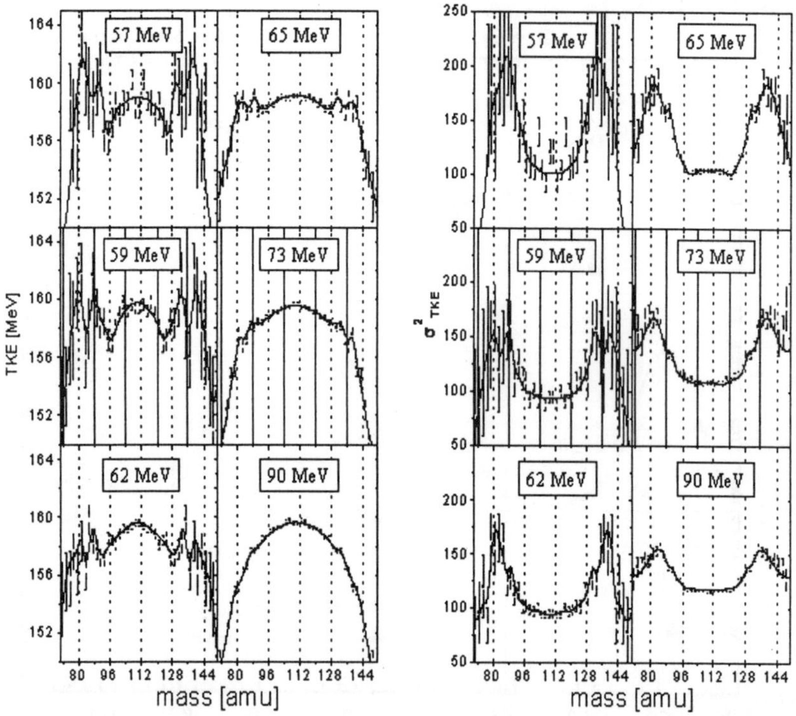

Fig. 2 *The mean total kinetic energy and its dispersion on fragment mass dependences*

It is seen that with a decrease in the excitation energy the TKE and σ^2_{TKE} values observed in the region of M > 125 increase. This raise is characteristic of the asymmetric fission. For the energy E_{lab}=90 MeV the properties of the fission fragments MED are close to the liquid drop model predictions [19,20] where the TKE(M) dependence is a parabola. However in the σ^2_{TKE}(M) dependence, which is very sensitive to the presence of different fission modes, the peak in the region of

M~135 is still present. This fact testifies that the asymmetric mode even for this energy is present, and it can be explained by the fact that the real excitation of the disintegrated nucleus is significantly lower than the initial one due to emitting of some pre-fission particles.

In Fig.3 the total mass distributions symmetrized with respect to M=A/2 and normalized to 200% are plotted. It is seen that on the slopes of the predominant symmetric fission there appears an asymmetric component at the lowest excitation energies. The solid lines are the gaussian fits of the symmetric component of the mass.distributions.

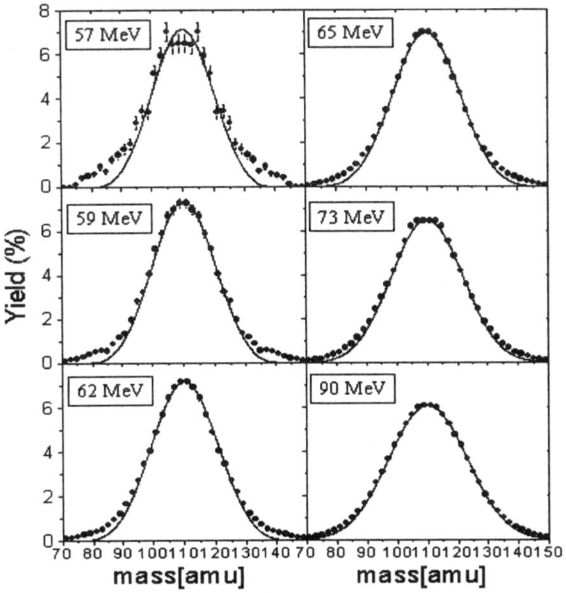

Fig. 3 *Total mass distributions*

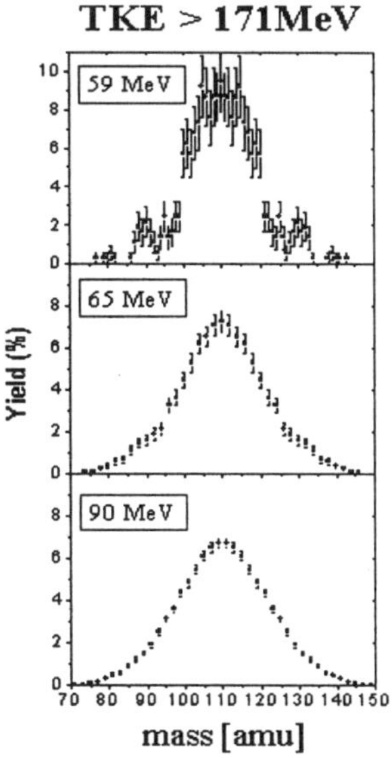

Fig. 4 *Mass distributions for the high TKE range*

Fig. 4 presents the fission fragment mass distribution involving Fig. 4 presents the fission fragment mass distribution involving only events with the total kinetic energy higher than 171 MeV. These distributions are plotted for the three energies of the bombarding ions, because E_{lab}= 90 MeV is the highest energy we measured, E_{lab}= 65 MeV is the closest to the Coulomb barrier and E_{lab}= 59 MeV is the lowest energy with good enough statistics. It is well known that the yield of the asymmetric component increases noticeably with a cooling of the disintegrating nucleus. In accordance with this remark in the selection of events with high kinetic energies there occurs enrichment of the mass distribution of fragments with asymmetric modes for $M_{a1}\approx$ 132 and $M_{a0}\approx$ 139. Concerning the highest energy the asymmetric component does not visually appear even in this TKE range.

In Fig. 5 the total kinetic energy distributions normalized to 200% are plotted in a logarithmic scale. The solid lines are the gaussian fits of the symmetric component yield into the total kinetic energy distribution. The dotted lines are the result of these fits subtraction from total TKE distributions.

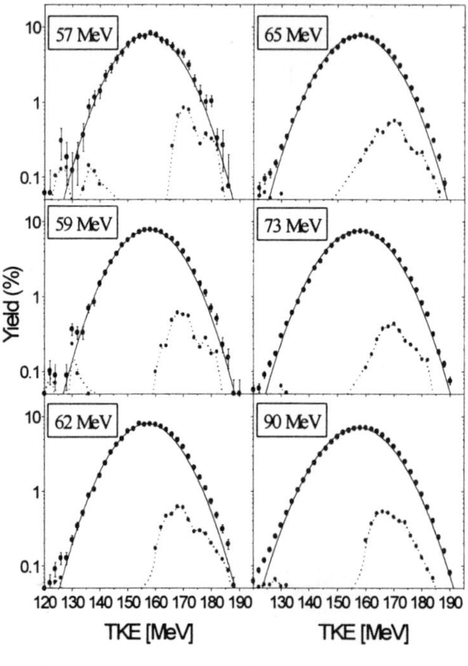

Fig. 5 *Total kinetic energy distributions*

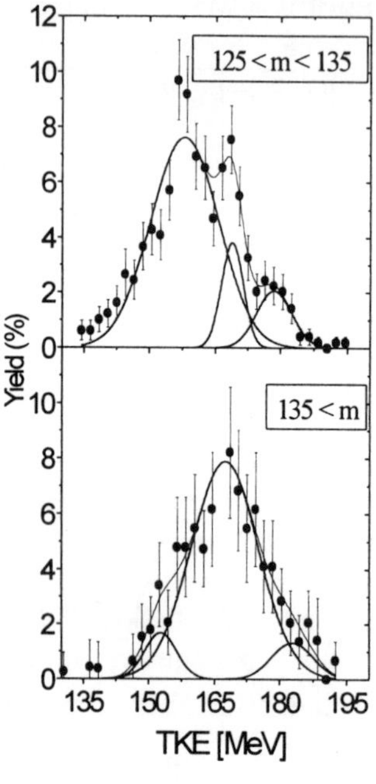

Fig. 6 *Total kinetic energy distributions for differen mass regions (E_{lab}=59 MeV)*

At the lowest energy (57 MeV) the raises on a right slope of the TKE distribution can be clearly seen. The presence in such a spectrum of several well-localized groups of fragments with different kinetic energies is one of the proofs of existence of different fission modes. The first raise in a region of 170 MeV corresponds to one of the asymmetric components and the second one in the region of 180 MeV corresponds to the other asymmetric component. To see more clearly the three-humped structure of the TKE distribution in the case of the higher excitation energies the TKE distributions of the fission fragments involving the events with asymmetric masses only can be investigated.

Fig. 6 is an example of such investigation for E_{lab}= **59** MeV. At the top of the figure the TKE distribution of the fragments involving the events with masses from 125 to 135 is plotted. The solid lines are the result of decomposition of this spectrum into three components. It is seen that in this mass region the symmetric component is still predominant but the two humps corresponding to the two asymmetric components are clearly seen. At the bottom of the figure the TKE distribution of the fragments involving the events with masses greater than 135 is plotted. In this case the predomination of one of the asymmetric modes is seen. Note that for the fragment with M=132 (which corresponds to the double-magic ^{132}Sn nucleus) in the case of the ^{220}Ra compound nucleus the complementary fragment is an ^{88}Sr nucleus, which has a magic number of neutrons (50). Thus in the case of disintegration of the ^{220}Ra compound nucleus a super-compact configuration with a very high TKE is formed. And it can be seen from Fig. 6 where the difference between the third and the first humps is more than 20 MeV. Basing on Fig. 3 - 6 some conclusion can be derived: for the ^{220}Ra compound-nucleus fission two asymmetric fission modes are observed; the first mode corresponds to $M_{a1}\approx132$ and TKE≈180 MeV; the second mode corresponds to $M_{a0}\approx139$ and TKE≈170 MeV. These are the fission modes called standard-1 and standard-2 in Brosa's and co-authors [21] terminology. Thus the generality of the fission fragment mass and energy distribution properties for ^{220}Ra with those of lighter nuclei as well as heavier ones is observed.

Fig. 7 shows the asymmetric component yields Y_a taken separately from the symmetric one. It were obtained by the gaussian fits of the symmetric component subtractions from the integral mass distributions. It is clearly seen that Y_a has a complex structure. The same structure is observed in a case of the $^{220\text{-}226}$Th compound-nuclei fission [6-9]. So, it was decomposed to the sum of two gaussians using the "least squares" method. The result of this decomposition gave the mean values for the fission mode masses M_{a1}=132 and M_{a0}=139 for the standard-1 and standard-2 modes respectively.

Fig. 8 shows the asymmetric component yields, which were obtained by the gaussian fits of the symmetric component subtractions from the total kinetic energy integral distributions. The result of the decomposition gave the mean values for the fission mode energies TKE_{a1}=180 MeV and TKE_{a0}=170 MeV for the standard-1 and standard-2 modes, respectively.

The circles in Fig.9 represent dependence of the ratio of symmetric to asymmetric fission probability Y_s/Y_a on the compound-nucleus excitation energy. It is seen that in the case of ^{220}Ra compound-nucleus, at the lowest excitation energies the yield of the asymmetric component decreases as opposed to the ^{226}Th

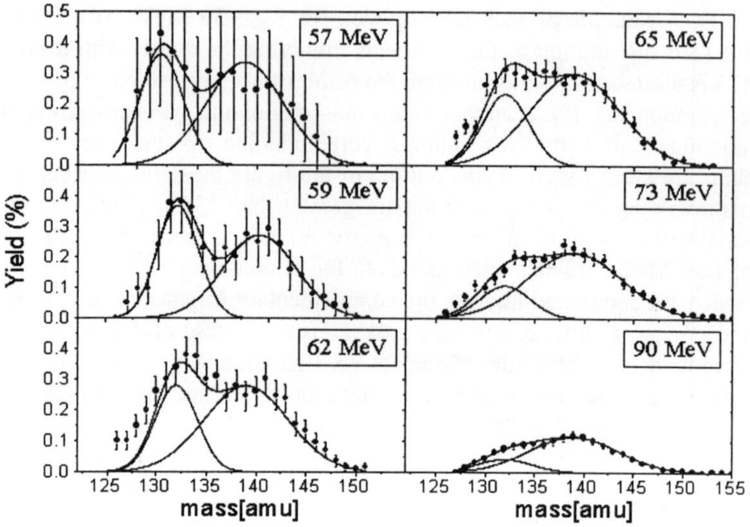

Fig. 7 *The yield of the asymmetric component in mass distributions*

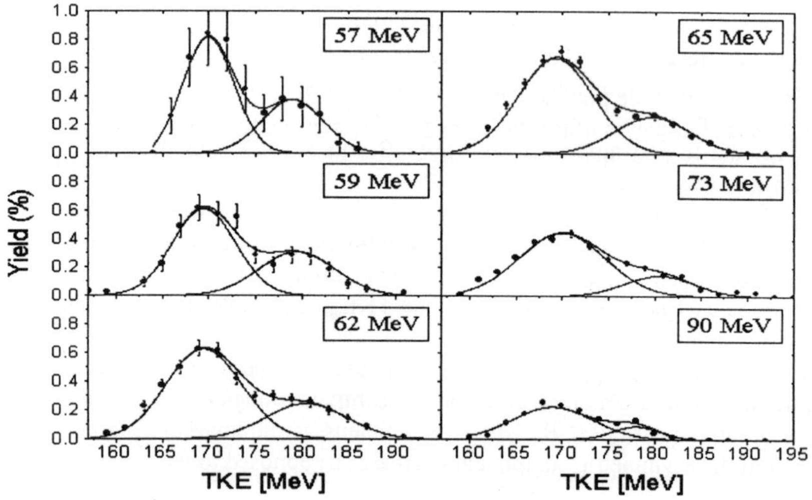

Fig.8 *The yield of the asymmetric component in TKE distributions*

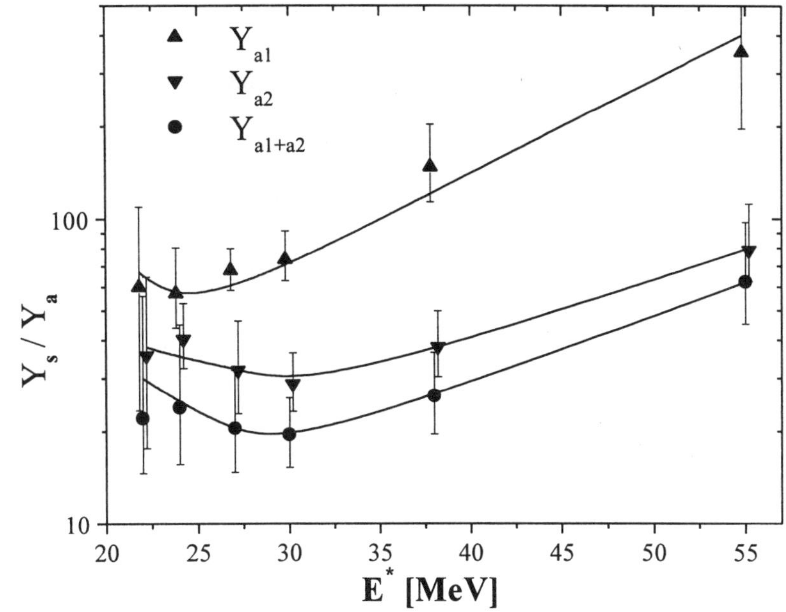

Fig. 9 *The symmetric to asymmetric fission probability ratio on an excitation energy dependence*

compound-nucleus. This fact is in good agreement with theoretical predictions made on the basis of the potential energy surface consideration [16-18]. It shows that on the potential energy surface there are two distinct well-separated symmetric and asymmetric valleys and the latter of them is also separated into two valleys. It shows also that for lighter isotopes the fission barrier for asymmetric fission is higher than for symmetric one ($E_f^a - E_f^s > 0$). For heavier isotopes the picture is the opposite ($E_f^a - E_f^s < 0$). Fig. 9 validates this remark. The triangles in Fig. 9 are the same dependences but presented separately for each of the fission modes. It is seen that the fission mode standard-1 is under much more greater influence of the excitation energy.

As a summary, we can confidently say that the three fission modes were observed in the fission of the ^{220}Ra compound-nucleus. They were clearly observed in the mass as well as energy distributions. The properties of the modes are in good agreement with the earlier known regularities for fission modes.

References

1. R. Vandenbosch and J. Huizenga, Nuclear Fission, Academic Press, New York, 1973.
2. F. Gönnenwein, in The Nuclear Fission Process, CRC Press Inc., Boca Raton, Florida, chap. 8 (1991) p.287.
3. M. G. Itkis et al., Z. Phys. **A 320**, 433 (1985); Sov. J. Part. Nucl. **19** (1988) 301; Nucl. Phys. **A 502** (1989) p.243.
4. H. J. Specht, Rev. Mod. Phys. **46** (1974) 733; Nukleonika, **20** (1975) 717.
5. B. D. Wilkins et al., Lect. Notes in Phys. **158** (1982) 150.
6. M. G. Itkis et al., Proc. of the Workshop on Nuclear Fission and Fission-product Spectroscopy (Seyssins, France, May 1994), H. Faust and G. Fioni (eds.), ILL, Grenoble (1994) p.77.
7. M. G. Itkis et al., Proc. of the XV EPS conf. Low Energy Nuclear Dynamics (LEND-95), (St. Petersburg, Russia, April 1995), Yu. Ts. Oganessian et al. (eds.), World Scientific, Singapore (1995) p.177.
8. M. G. Itkis et al., Tours Symp. On Nucl. Phys. III, (Tour, France, September 1997), M. Arnould et al. (eds.) AIP Conf. Proc. 425, Woodbury, N.Y., USA, (1998). P.189.
9. M. G. Itkis et al., Nucl. Phys. A (in press).
10. A. Kelic et al., Phys. Rev. C (in press).
11. G. G. Chubarian et al., JINR Rapid Communication No.4[90]-98, Dubna (1998) p.39; Proc. 2nd Int. Conf. On Exotic Nuclei and At. Masses (ENAM'98) (Shanty Creek Resort, Bellaire, Michigan, USA, June 1998), to be published.
12. K.-H. Schmidt et al., Phys. Lett. **B 325** (1994) 313.
13. S. Steinhäuzer et al., Proc. 3rd Int. Conf. On Dynamical Aspects of Nuclear Fission (Časta-Papiernička, Slovak Republic, August-September 1996), J. Kliman and B. Pustylnik (eds.), JINR, Dubna (1996) p.151.
14. K.-H. Schmidt et al., Nucl. Phys. **A 630** (1998) p.208.
15. E. M. Kozulin et al., in Heavy Ion Physics, Scientific Report 1995-1996, JINR, FLNR, Dubna (1997) p.215; N. A. Kondratiev et al., 4th Int. Conf. On Dynamical Aspects of Nuclear Fission (DANF'98) (Časta-Papiernička, Slovak Republic, October 1998), this volume.
16. V. V. Pashkevich, Int. School – Seminar on Heavy Ion Physics (Alushta, USSR, April 1983), Dubna, JINR (1983) p. 405.
17. V. V. Pashkevich, see Ref. 7, p.161.
18. V. V. Pashkevich, private communication.
19. J. R. Nix and W. J. Swiatecki, Nucl. Phys. **71** (1965) 1.
20. G. D. Adeev et al., Sov. J. Part. Nucl., **19** (1988) 529; Nucl. Phys. **A 502** (1989) 405.
21. U. Brosa et al., Phys. Reports **197** (1990) 167.

VARIABLE FRICTION AND THE EXTRA-PUSH EFFECT

Gérard Rudolf
Institut de Recherches Subatomiques, UMR7500, IN2P3-CNRS/Université Louis Pasteur,
BP28, F-67037 Strasbourg Cedex 2, France

This paper presents an investigation of the effect of shape-dependent or temperature-dependent friction on heavy ion trajectories and on the extra-push. The consequences of the nature of the friction on pre- and post-scission neutron multiplicities is examined in the frame of the Bass model. Temperature- or excitation-energy dependent friction allows to reproduce most experimental data on neutron multiplicities, to reconcile the results of the analysis of these multiplicities with those of theγ rays from the GDR and explains why cross sections for the synthesis of very heavy nuclei seem to ignore the extra-push effect.

1 Introduction

The discovery, some ten years ago, of the fission delay time has revealed that fission dynamics is determined by friction more than by barriers. Pre-scission neutron multiplicities analysed in the frame of the statistical model[1,2,3,4,5,6,7,8] suggested that fission lasts some 2-6 10^{-20}s. A more recent analysis of the same data came to the conclusion that this time may be as much as ten times longer [9]. Even longer times result from different techniques and an almost model independent analysis [10]. The situation is therefore rather confuse but the importance of nuclear friction in fission is clearly demonstrated.

The influence of friction on the dynamics of heavy ion collisions is, on the other hand, known since long. Indeed, when during the collision between two heavy ions the Coulomb and centrifugal forces overcome the nuclear ones, the two nuclei undergo a so-called dissipative collision. According to the balance between attactive and repulsive forces, they dissipate a more or less important part of their kinetic energy into excitation energy before reseparating. The relationship between the scattering angle and the total kinetic energy in the outgoing channel gives a measure of the friction. It is generally admitted that at low bombarding energies the one-body dissipation [11,12] gives a good description of the dynamics.

Fusion is only possible if friction exists, since the system must be trapped into the potential well. However for heavier systems where the contact occurs at a distance larger than the barrier elongation, friction hinders fusion. Indeed, some energy is lost already before the barrier is reached, so that the ingoing energy must exceed the barrier height by some amount called the extra-push.

Extra-push values can be calculated with reasonable accuracy with the one-body dissipation.

In the case of fission, on the contrary, the above mentionned experimental results show that friction must be much stronger. The large number of pre-scission neutrons, unexpected in the frame of the statistical model, lead to the conclusion that motion toward scission is overdamped. This can be intuitively understood by the fact that it corresponds to a large scale motion, implying a long sequence of rearrangements. This is the fondamental idea leading to shape-dependent friction [13].

However even this model predicts less pre-scission neutrons than is experimentally observed in the case of very heavy systems. This is due to the fact that the cross section for symmetric fragmentation is then dominated by fast fission, and not by fusion-fission. In this mechanism, a pocket exists in the asymmetric entrance channel, but not in the symmetric outgoing one. Thus, like in fusion, the system is first trapped behind the barrier of the entrance channel. It therefore has time to equilibrate the mass and energy degrees of freedom. This means that the lighter nucleus gains matter from the heavier one, until the pocket in the potential disappears. This mechanisms is generally considered as being 3 to 5 times faster than fusion, and thus producing a smaller number of pre-scission neutrons. Therefore, for the heaviest systems, it was necessary to admit that some neutrons are emitted by the fragments after scission but before full acceleration is attained, so that they appear as having been emitted by the composite system [7].

To explain the large number of pre-scission neutrons emitted in the ^{64}Ni + ^{238}U reaction whithout this hypothesis, Wilczynski et al. [14] have introduced another dependence of friction. If friction is supposed to be small in the entrance channel the nuclei interpenetrate deeply. If moreover it is supposed to be large in the exit channel, they take a long time to reseparate. By modifying the friction in the appropriate way, the reaction time becomes then much longer. This procedure allowed the authors of ref. [14] to account for the four pre-scission neutrons which are emitted in the ^{64}Ni + ^{238}U reaction although no fusion is supposed to occur. In fact, with such a picture fusion-fission becomes very difficult to distinguish from fast-fission. This is not astonishing if the dynamics of both mechanisms is dominated by friction more than by barriers.

Friction may also be temperature-dependent. This concept has been introduced in the analysis of γ rays from the GDR [15] and, from a theoretical point of view, by the linear response theory [16]. The two types of dependence of the friction have many consequences on the extra-push and the pre-scission multiplicities. In this paper, these consequences will be investigated in the frame of two very simple models.

2 Trajectory calculation

The aim of this calculation is to investigate qualitatively the influence of friction on reaction times, neutron pre-scission multiplicities and the extra-push effect. It can be considered as a very simplified version of the surface- friction model [17]. To remain as simple as possible, it will be one-dimensional, i.e. that only head-on collisions will be considered, and angular momentum will not be considered. The trajectories of two heavy ions are calculated step by step. At each step, one computes the energy E, the force f, the acceleration γ, the velocity v, the position r, the excitation energy E_{exc}, the evaporated energy E_{evap}, the neutron multiplicity ν:

$$E(r) = E_{coul}(r) + E_{nucl}(r) \tag{1}$$

$$f = \frac{dE(r)}{dr} + C_f f_f(r) v \tag{2}$$

$$\gamma = \frac{f}{\mu} \tag{3}$$

$$dv = \gamma dt \tag{4}$$

$$dx = vdt \tag{5}$$

$$E_{exc} = E_{cm} - TKE - E(r) - E_{evap} + (Q_{gg}) \tag{6}$$

$$E_{evap} = \Sigma dE_{evap} = \Sigma a_{evap} E_{exc} dt \tag{7}$$

$$d\nu = dE_{evap}/\Delta E \tag{8}$$

$$a_{evap} = 5.10^{-3} T^2 A^{2/3} (exp(E_{bind}/T))^{-1} 10^{22} s^{-1} \tag{9}$$

E_{nucl} is the Ngo potential [18]. The parameters have been slightly modified so that there is a plateau in the radial dependence of the potential in the exit channel. Thus it gives the border between fusion and fast-fission. T is the temperature and E_{bind} the mean value of the binding energy of the neutrons. f_f gives the radial variation of the friction force and is proportionnal to the surface of the window which is open between the two nuclei. The mass is supposed to be equilibrated at once at the smallest distance if this distance is lower than the elongation of the maximum of the potential in the entrance channel.

2.1 The friction coefficient and the extra-push

Figure 1 shows the result when C_f is taken equal to 220 $MeV fm^{-2} 10^{-22} s$. As soon as the nuclei come into contact, energy is lost (Fig.1a) so that the kinetic energy is too small to overcome the maximum of the potential. The two nuclei do not come closer than 12.5 fm, the total reaction time is about 20.10^{-22}s (Fig.1g) and the number of neutrons evaporated before reseparation is smaller than 0.5 (Fig.1d). Moreover, the source has not a well defined velocity (Fig.1h). This corresponds to a deep inelastic collision.

When C_f is slightly decreased to 190 $MeV fm^{-2} 10^{-22} s$ the energy loss is smaller and the system overcomes the barrier in the entrance channel. Thus the masses can equilibrate to form a new system for which there is no pocket in the potential (Fig.2e). The total time reaches some 150.10^{-22}s, and the neutron multiplicity atteins about 1.5. This corresponds to a fast fission reaction.

In Figs.1 and 2 the change from deep inelastic to fast fission reactions is due to a decrease of the friction coefficient, but for a fixed value of C_f it is achieved by increasing the bombarding energy. This energy corresponds precisely to the extra-push. Therefore, knowing the extra-push allows to determine C_f.

2.2 Shape dependent friction

A very simple form of shape dependence has been used to calculate the result presented in Fig.3. In fact C_f is taken equal to 190 $MeV fm^{-2} 10^{-22} s$ in the entrance channel, but jumps to 10000 $MeV fm^{-2} 10^{-22} s$ when the relative distance becomes larger than 12.8 fm in the exit channel.

The essential change when compared to Fig.2 is the reaction time. In this very simple model, the time to reach minimal distance can be considered as the formation time τ_{fo}, and the time needed later to reach scission as the fission time τ_f. Since the friction has not changed in the entrance channel, τ_{fo} has not changed and is equal to approximately 60.10^{-22}s (Fig.3c), but the very large friction from saddle to scission extends the total time to some 500.10^{-22}s (Fig.3g). About 4 neutrons are emitted mostly between the saddle and the scission points (Fig.3d).

With the behaviour of the friction assumed here, the extra-push is not modified with respect to the preceding case since the friction in the entrance channel has not changed.

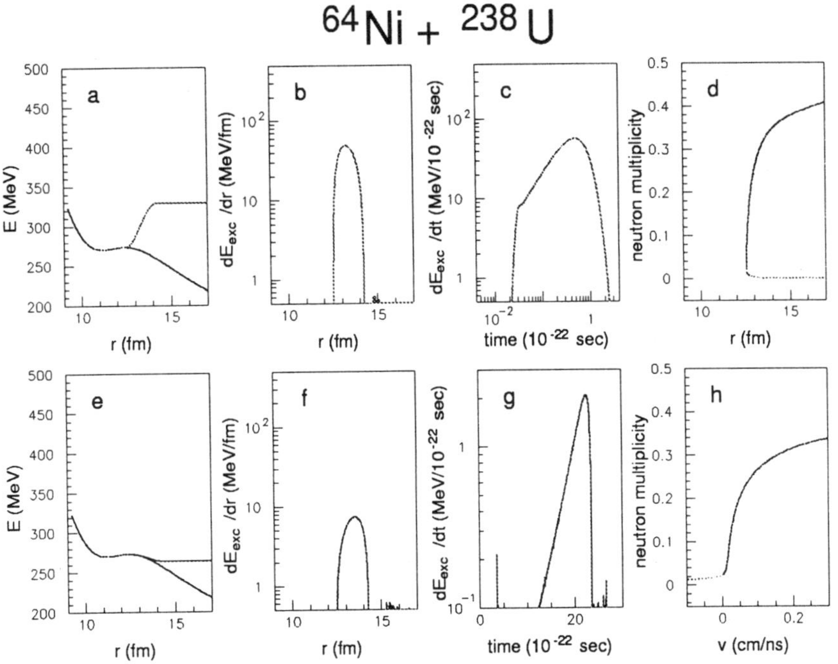

Figure 1: Result of the trajectory calculation when C_f is equal to 220 MeV 10^{-22}s fm^{-2}: a) potential energy and kinetic energy in the entrance channel; b) energy damping per relative distance unit in the entrance channel; c) energy damping per time unit in the entrance channel; d) neutron multiplicity as a function of the relative distance; e) potential energy and kinetic energy in the exit channel; f) energy damping per relative distance unit in the exit channel; g) energy damping per time unit in the exit channel; h) neutron multiplicity as a function of the relative velocity of the fragments.

2.3 Temperature dependent friction

A temperature dependent friction has been used to calculate the results shown in Fig.4. More precisely, C_f has been fixed to 50 $MeV fm^{-2} 10^{-22} s$ if the excitation energy is smaller than 25 MeV, and to 1000 $MeV fm^{-2} 10^{-22} s$ otherwise. This means that the jump occurs at a temperature of about 0.7 MeV. Because the friction becomes large already at an early stage of the reaction, τ_{fo} reaches now 200.10^{-22}s (Fig.4c), so that about 2 neutrons are already evaporated at this stage (Fig.4d). It takes some 300.10^{-22}s more to reach the saddle (Fig.4g), so that a total number of 4 neutrons are emitted before this point. The descent from saddle to scission brings an important amount of excitation

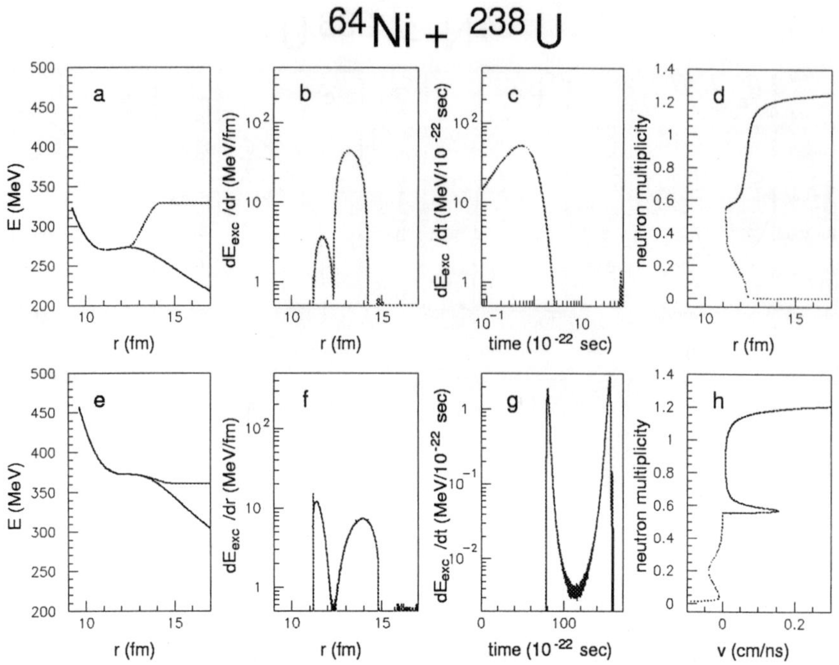

Figure 2: Same as fig. 1 when C_f is taken equal to 190 MeV 10^{-22}s fm^{-2}.

energy in the system, but it arrives after 500.10^{-22}s and lasts only 100.10^{-22}s so that this energy is evaporated mostly after scission (Fig.4d). One arrives thereby to the conclusion that most pre-scission neutrons are emitted before the saddle, in accordance with the results of the analysis of γ rays from the GDR [15]. This conclusion will be the starting point of the next section.

With a temperature dependence of the friction coefficient, the extra-push is of course modified. To make this point clear, let us come back first to the case of fixed C_f. When the bombarding energy just exceeds the extra-push, only the most central collisions lead to the interpenetration of the nuclei. For larger impact parameters, the angular momentum leads to deep inelastic collisions. When the bombarding energy is increased, the elongation at which the kinetic energy drops to 0 is more and more compact, (Fig.5a) which means that more and more partial waves contribute to fast fission. In other words, the cross section for fusion or fast fission is equal to 0 for energies below the extra-push, but increases continuously above.

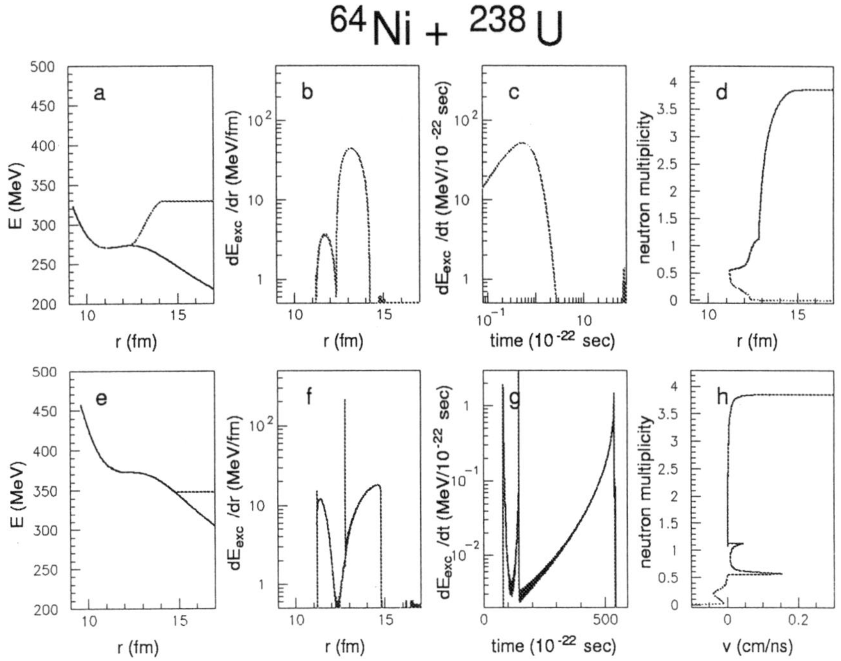

Figure 3: Same as fig. 1 when C_f is taken equal to 190 MeV 10^{-22}s fm^{-2} untill the relative distance is larger than 12.8 fm, and to 10000 MeV 10^{-22}s fm^{-2} after.

The situation is very different in Fig.5b. Here, an ad-hoc variation of C_f with temperature has been introduced. Because C_f is close to 0 at low temperature, the system can interpenetrate already at bombarding energies close to the barrier. But when the bombarding energy increases, the elongation of the point at which the kinetic energy is damped does not vary. This means that the excitation function of the cross section for fusion or fast-fission starts at smaller bombarding energy, but does not increase.

This may explain why the cross section for the synthesis of very heavy elements has a maximum which is close to the Bass barrier, and seems to ignore the extra-push [19]. Of course, this idea has to be checked quantitatively with a more realistic temperature dependence than that which has been assumed here.

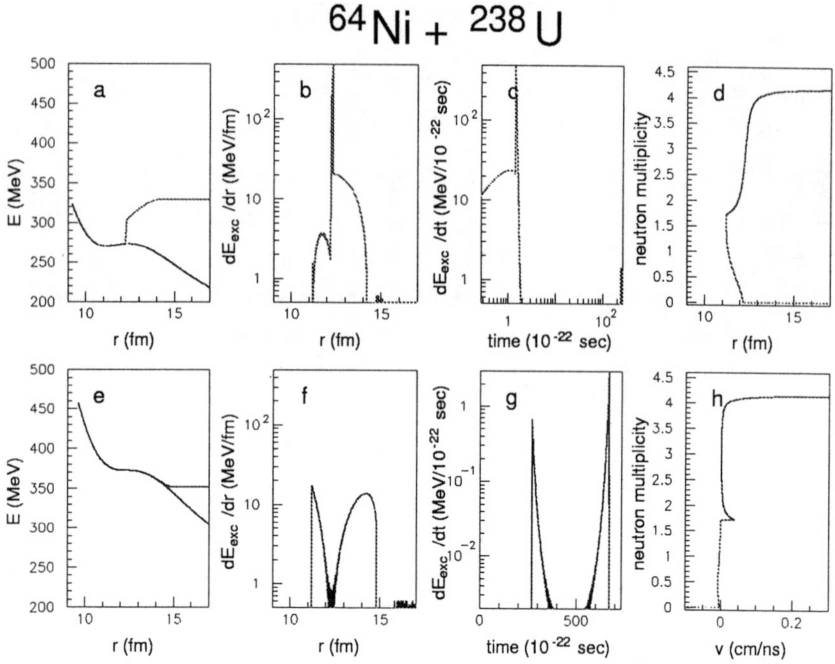

Figure 4: Same as fig. 1 when C_f is taken equal to 50 MeV 10^{-22}s fm^{-2} when the excitation energy is smaller than 25 MeV, and to 1000 MeV 10^{-22}s fm^{-2} when the excitation energy is larger than 25 MeV.

3 Neutron multiplicities in the frame of the Bass model

The model presented in the preceding section has been used to calculate neutron pre-scission multiplicities. However, its aim was only to give a qualitative description. To use it in a quantitative way would need to introduce the angular momentum, check the parameters of the potential with other observables, etc...

Instead, it is possible to use the Bass model [20] in a very simple way. The method is based on the knowledge, gained from statistical analysis, that fission is a slow process, and that the system evaporates almost all its energy before it fissions. This is confirmed by the results of the preceding section. Thus the time evolution of the system has not absolutely to be investigated. Indeed, if the delay time was infinite the excitation energy at the saddle point and at the scission point would be equal to 0. It would then be possible to deduce

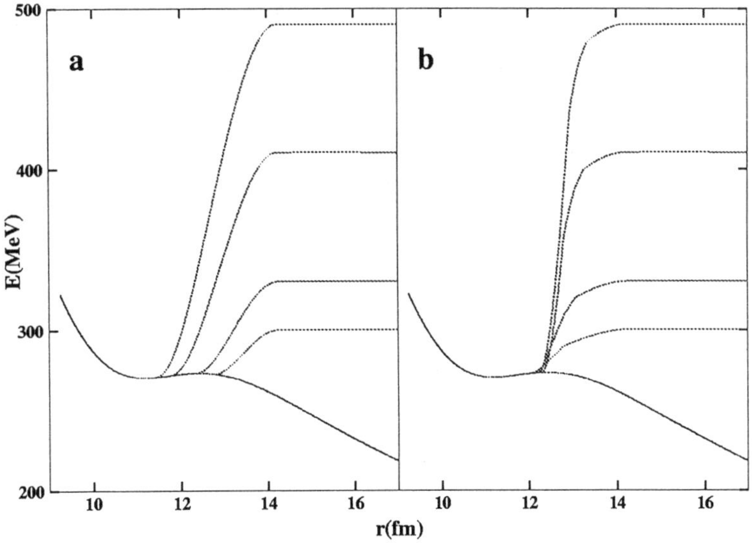

Figure 5: Trajectories at different bombarding energies when a) C_f is equal to 190 MeV 10^{-22}s fm^{-2}; b) C_f varies with temperature (see text).

pre-saddle, saddle-to-scission and post-scission multiplicities directly from barrier and scission energies. To correct for the finite delay time it is equivalent to correct these energies by an amount E_τ which can be a parameter to be adapted to the experimental data, exactly like the delay time is determined by comparing the predictions of the statistical model to the experiment.

This procedure can be used with elongation dependent friction[21] supposing that most pre-scission neutrons are emitted between saddle and scission, that fast-fission produces few pre-scission neutrons and that E_τ ranges between 0 and 60 MeV. But almost all published data could be reproduced by supposing that most pre-scission neutrons are emitted before saddle, that fast-fission produces nearly as many pre-scission neutrons as fusion and that E_τ is close to 25 MeV for all systems [22]. More precisely, Figs.5 and 6 show the result which is obtained when the following assumptions are made:

- all the pre-scission neutrons are emitted before the saddle is reached with an excitation energy E_τ. In other words, the energy which can be evaporated by pre-saddle particle emission is

$$E_{pre} = E_{cm} + Q_{gg} - E_{saddle} - E_\tau \qquad (10)$$

- For the partial waves which lead to fusion-fission, the saddle which has to be considered is the Bass barrier for the two equal mass nuclei formed in the exit channel if E_{cm} is large enough. The Bass barrier corresponds actually to the highest possible potential energy for two spheres, which may be a good approximation for the highest potential energy when the compound nucleus undergoes vibrations before fissioning.
- when E_{cm} is too small, the saddle is the usual deformed saddle given by the Sierk barrier.
- for the partial waves which lead to fast-fission, the pre-scission neutrons are emitted before the system atteins the deformed barrier, corrected by the same E_τ.

Some results of the calculation are shown in Figs. 6 and 7. More details can be found in ref. [22].

4 Discussion

The excitation energy E_τ was introduced as a parameter to be adjusted to the experiment. It turns out that a nearly constant value of E_τ can be used for all studied systems, and that this value corresponds approximately to the temperature above which friction becomes large. In the case of fusion, the excitation energy of the system decreases when it approaches the saddle, which is the undeformed one when the excitation energy is large enough and the deformed one if it is not. Moreover, because of the evaporation during the reaction process itself, the excitation energy decreases. The combination of the two effects makes that the excitation energy will become smaller than E_τ most probably in the vicinity of the first barrier encountered by the system. In the case of fast-fission, there is no barrier anymore in the exit channel but the slope of the potential is smallest at a given distance. This distance is not far from barrier systematics. It is at this distance that the system stays longest, and therefore evaporates most energy. Thus one may conclude that E_τ is explicitely related to the shape of the temperature or excitation energy dependence of the friction.

5 Conclusion

The calculations presented here do not pretend to yield really quantitative predictions, but a qualitative understanding of the dynamics of pre-scission evaporation. They open the way to the concept of two regimes in fission: a high energy regime in which friction is large and causes high pre-scission neutron multiplicities, and a low energy one where friction is small so that saddle to scission motion may be fairly fast. These two regimes seem to exist also for the entrance channel, which leads to reconsider the concept of the extra-push. If, moreover, friction is structure dependent[23], new paths to reach very heavy nuclei may exist.

References

1. D.J. Hinde et al., *Nucl. Phys.* **A452**, 550 (1986)
2. J.O. Newton et al., *Nucl. Phys.* **A483**, 126 (1988)
3. D.J. Hinde et al., *Phys. Rev.* **C39**, 2268 (1989)
4. D. Hilscher et al., *Phys. Rev. Lett.* **62**, 1099 (1989)
5. E. Mordhorst et al., *Phys. Rev.* **C43**, 716 (1991)
6. H. Rossner et al., *Phys. Rev.* **C45**, 719 (1992)
7. D.J. Hinde et al., *Phys. Rev.* **C45**, 1229 (1992)
8. K. Knoche et al., *Phys. Rev.* **C51**, 1908 (1995)
9. K. Siwek-Wilczynska et al., *Phys. Rev.* **C54**, 325 (19 96)
10. M. Morjean et al., contribution to this Conference
11. J. Blocki et al., *Ann. of Phys.* **113**, 330 (1980)
12. J. Randrup and W. Swiatecki, *Ann. of. Phys.* **125**, 193 (1980)
13. P. Froebrich and I.I. Gontchar, *Phys. Rep.* **292**, 131 (1998)
14. J. Wilczynski et al., *Phys. Rev.* **C54**, 325 (1996)
15. P. Paul and M. Thoennessen, *Annu. Rev. Nucl. Part. Sci.* **44**, 65 (1994)
16. H. Hofmann, *Phys. Rep.* **284**, 137 (1997)
17. P. Froebrich, *Phys. Rep.* **116**, 337 (1984)
18. T. Suomijarvi, R. Lucas, C. Ngo, E. Tomasi, C. Dalili and J. Matuszek, *Il Nuovo Cimento* **82A**, 51 (1984)
19. S. Hofmann et al., *Zeit. Phys.* **A358**, 377 (1997)
20. R. Bass, *Nucl. Phys.* **A231**, 45 (1974) **44**, 65 (1994)
21. G. Rudolf, *Nukleonica* **43**, 265 (1998)
22. G. Rudolf, invited talk at the Riken Symposium on Dynamics in Hot Nuclei, March13-14, 1998, and Preprint 98-19, IReS
23. A. Kelic et al., submitted to Phys. Rev., and G. Choubarian et al., contribution to this Conference.

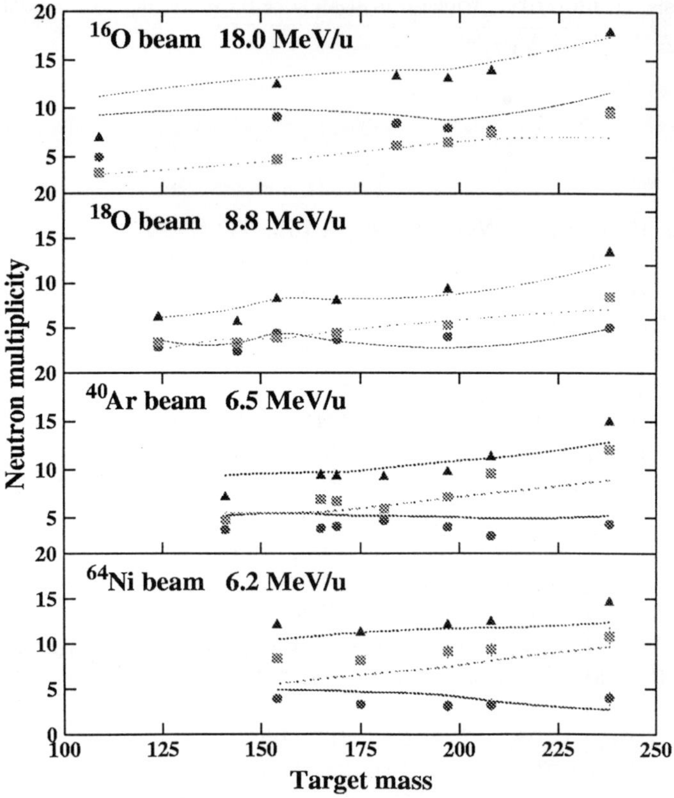

Figure 6: Experimental and calculated neutron multiplicities: ν_{tot} (triangles and dotted curve), ν_{pre} (circles and continuous curve), and ν_{post} (squares and dashed curve). The data are from ref. 7.

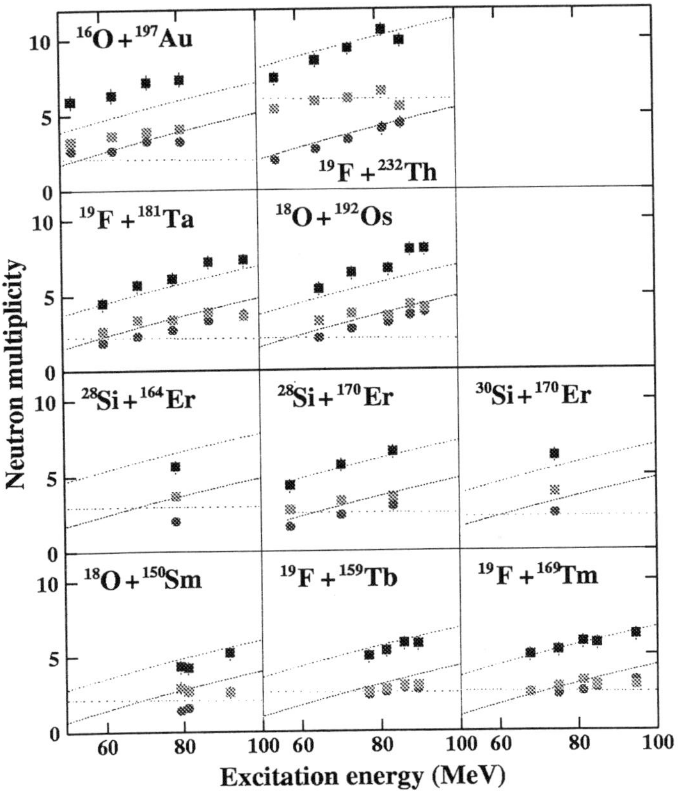

Figure 7: Same as Fig. 6. The data are from ref. 2.

LOW-ENERGY FISSION STUDIES WITH SECONDARY BEAMS

A. Heinz[a], C. Böckstiegel, H.-G. Clerc, A. Grewe, M. de Jong, J. Müller, S. Steinhäuser

Institut für Kernphysik, Technische Universität Darmstadt, Schloßgartenstraße 9, 64289 Darmstadt, Germany

K.-H. Schmidt, J. Benlliure, A. R. Junghans, B. Voss

Gesellschaft für Schwerionenforschung m.b.H., Planckstraße 1, 64291 Darmstadt, Germany

M. Pfützner

Institute of Experimental Physics, University of Warsaw, Ul Hoza 69, 00-381 Warszawa, Poland

Fission studies by use of secondary beams open the possibilty to extend the number of accessible fissioning systems and thus improve our understanding of low-energy fission. Nuclear fission from excitation energies around 11 MeV was studied in inverse kinematics at GSI Darmstadt for 70 neutron-deficient actinides and pre-actinides by use of relativistic secondary beams. The characteristics of multimodal fission of nuclei around ^{227}Th were systematically investigated. A systematic view on the large number of elemental yields measured revealed an even-odd structure also in the fission of odd-Z nuclei. This finding gives rise to a new interpretation of the enhanced production of even elements in nuclear fission of even-Z nuclei. Moreover, the yields of even-Z elements in near-symmetric charge splits were found to be enhanced by about 10% for long isotopic chains of uranium and thorium fissioning nuclei. In contrast to the actual understanding, the conclusion can be drawn that the even-odd effect is not a direct measure of the excitation energy at scission.

1 Introduction

Nuclear fission provides unique information on the reordering of nucleons in a large-scale collective motion. The signatures of shell effects and pairing correlations show up in fission from low excitation energies. Such data are of general importance for the role of shell structure and viscosity in nuclear dynamics. In a recent experiment, relativistic secondary beams were used to overcome restrictions of conventional experimental techniques which were limited to spontaneously fissioning nuclei and to nuclei in the vicinity of long-lived isotopes, used as target material. The new data systematically cover short-lived neutron-deficient actinides and pre-actinides. In this contribution,

[a]E-mail: A.Heinz@gsi.de

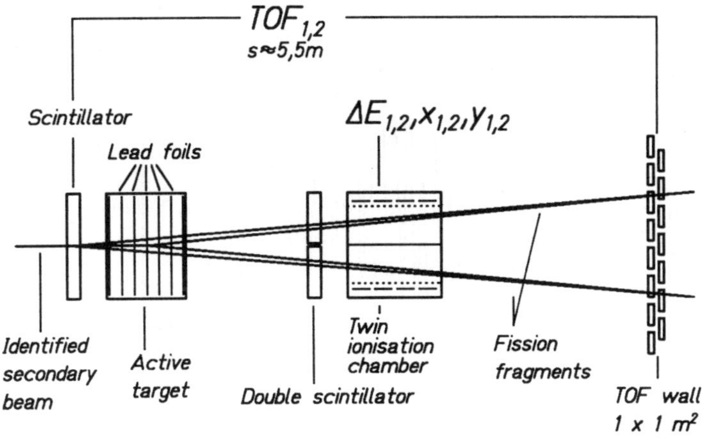

Figure 1: Schematical drawing of the set-up for the fission experiment mounted behind the fragment separator.

part of the large body of data acquired is presented and the resulting progress in the understanding of fission dynamics is sketched.

2 Experiment

The secondary-beam facility of GSI, Darmstadt, offers unique possibilities to provide secondary beams of neutron-deficient actinides and pre-actinides at relativistic energies produced by fragmentation of ^{238}U projectiles at $1 \cdot A$ GeV. Within the limits given by the primary-beam intensity and the fragmentation cross sections [1,2], nuclear charge and mass number of the secondary projectiles can freely be selected by tuning the fragment separator [3]. Intensities of typically 100/s were reached for the most abundantly produced isotopes. Fission from the desired excitation-energy range slightly above the fission barrier was induced in flight by electromagnetic excitation in a heavy target material. In the next chapter we give a detailed description of this excitation mechanism. The experimental set-up to study fission of secondary beams in inverse kinematics behind the fragment separator is sketched in Figure 1. As secondary target we used a stack of five lead foils with a total thickness of 3 g/cm^2 mounted in a gas-filled chamber which acts as a subdivided ionisation chamber (active target). With this device it is possible to discriminate fission induced in

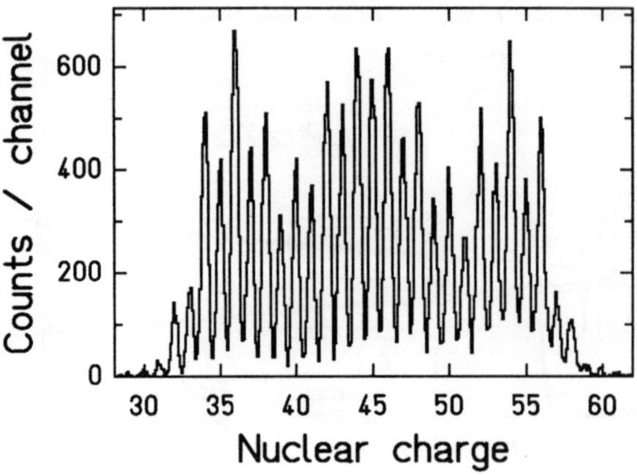

Figure 2: Nuclear-charge response of the double ionisation chamber with velocity correction applied as obtained for the fission fragments after electromagnetic excitation of ^{226}Th projectiles.

the lead foils against fission in other layers, e.g. the scintillator in front of the target. The average energy of the secondary projectiles in the lead target was about $430 \cdot A$ MeV. The differential energy loss of each fission fragment was measured separately with an horizontally subdivided twin ionisation chamber. In order to correct the energy loss for the velocity dependence, the time-of-flight of the fission fragments was measured by means of a scintillation detector placed in front of the lead target and a (1 m×1 m) scintillator wall located 5.5 m downstream.

Figure 2 shows the nuclear-charge response of the experimental set-up for fission fragments after electromagnetic-induced fission of ^{226}Th. Due to the high centre-of-mass velocities, an excellent charge resolution ($Z/\Delta Z = 120$) was achieved. Events stemming from reactions at lower impact parameters with nuclear contact were suppressed. For details of the analysis procedure see references[4,5].

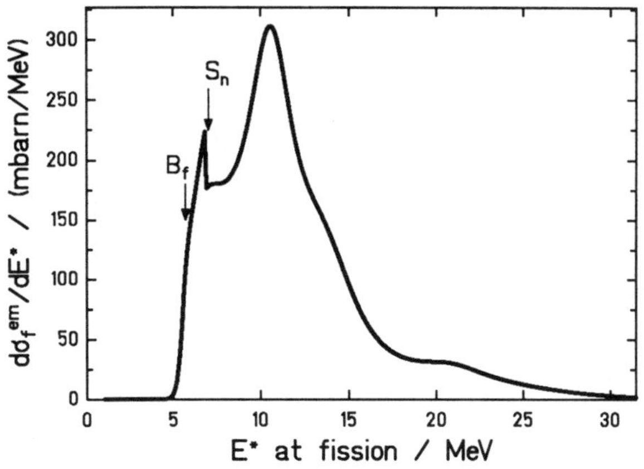

Figure 3: Calculated distribution of excitation energies at fission after electromagnetic excitation of ^{234}U projectiles at $430 \cdot A$ MeV in a lead target.

3 Excitation mechanism

The electromagnetic excitation in-flight in the secondary target is one of the most important ingredients of the experiment, ideally adapted to the kinematic properties and to the low intensities of the secondary beams. It populates states in the vicinity of the fission barrier with a large cross section of a few barns. Although the excitation energy acquired is not precisely known for a single event, the excitation-energy distribution can be calculated with rather good precision. The electromagnetic field of a lead target nucleus, seen by the projectile, can be represented by a flux of equivalent photons of different energies and multipolarities according to ref. [6]. The projectiles are excited according to the energy-dependent nuclear photo-absorption cross section which is dominated by the giant dipole resonance (GDR) with small contributions of the giant quadrupole resonances. First-chance fission represents the main source of fission, but also fission after evaporation of one or two particles (mostly neutrons) occurs with a probability of about 20%. The calculated excitation-energy distribution at fission after electromagnetic excitation of ^{234}U in the passage of a lead target at $430 \cdot A$ MeV is shown in Figure 3. Although in the deformed ^{234}U nucleus the GDR is split, the distribution shows only one peak at 11 MeV, because the high-energy component of the GDR is strongly suppressed by the decreasing equivalent photon spectrum. Multi-chance fission leads to a reduction of the excitation energy at fission. For details of the calculation see ref. [4]. By considering the small systematic variations of the photo-absorption cross sections as a function of mass number and deformation, the calculated excitation-energy distributions of the other nuclei investigated turned out to be rather similar. In all cases the peak of the excitation-energy distribution is predicted at less than 5 MeV above the fission barrier.

4 Results and discussion

In the present experiment, the elemental yields and the total kinetic energies of long isotopic chains from ^{205}At to ^{234}U have been determined. Part of the elemental yields are shown in Figure 4. The transition from symmetric fission in the lighter systems to asymmetric fission in the heavier systems is systematically covered for the first time. In the transitional region, around ^{227}Th, triple-humped distributions appear, revealing comparable intensities for symmetric and asymmetric fission. In particular for uranium and thorium isotopes strong even-odd effects are observed.

5 Fission channels

Turkevich and Niday [7] already noticed that different components which they named fission modes appear in the fission-fragment yields and in the kinetic-energy distributions. Later, models were proposed to deduce the fission characteristics from the properties of the scission configuration, e.g. ref. [8]. However, these do not explicitly consider the dynamical evolution of the system from saddle to scission which seems to be very important [9,10,11]. The concept of independent fission channels has been developed [12,13] according to which the fissioning system follows specific valleys in the potential energy in the direction of elongation. In this model, the yields of the fission channels are related to the properties of the fissioning nucleus near the fission barrier were the fission valleys appear. Several properties (e.g. average mass or charge split, mass or charge width, total kinetic energy) could be related to calculated properties of the highly deformed fissioning system near scission. Still, the mechanisms which determine the yields of the fission channels are not sufficiently well understood to allow for quantitative predictions. Therefore, they are usually deduced from experiment.

The present survey on fissioning systems with strongly varying charge distributions (see Figure 4) provides a systematic view on how the fission-fragment yields vary as a function of the nuclear composition. These data substantially

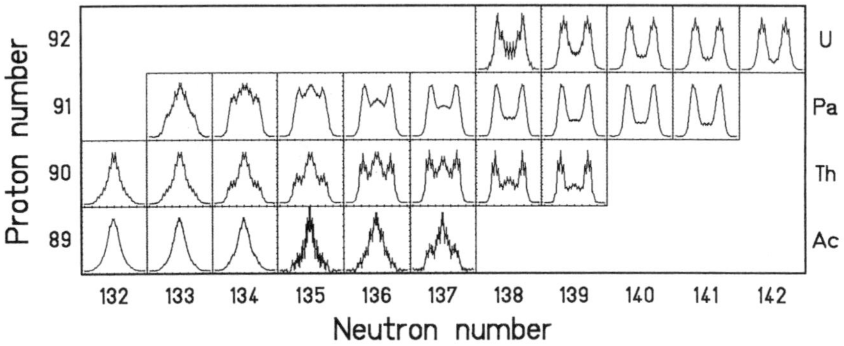

Figure 4: Measured fission-fragment charge distributions from ^{221}Ac to ^{234}U are depicted on a chart of the nuclides. (The figure shows only 28 out of the 70 systems investigated.)

increase the experimental knowledge relevant for a better theoretical understanding of the fission dynamics. At first glance, two fission components appear in the measured charge distributions, a symmetric and an asymmetric one. The transition is rather smooth, and the weights of the two fission components scale approximately with the mass of the fissioning nucleus, see Figure 5. The widths (standard deviation) of the symmetric and the asymmetric components were found to be close to 4.0 charge units and 2.1 charge units, respectively, for all nuclei for which they could be extracted.

In detail, the charge-yield distributions and the total kinetic energies of ^{228}Th, ^{226}Th, and ^{223}Th are shown in Figure 6. In order to consider the trivial variation of the total kinetic energy as a function of mass and charge split, the Coulomb repulsion V_C in the scission-point configuration was parametrized by

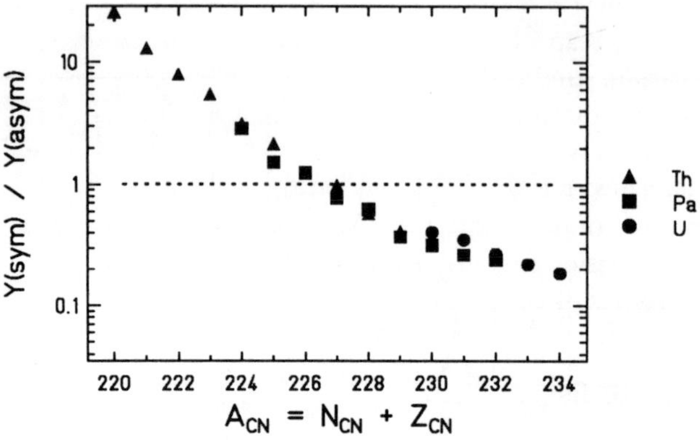

Figure 5: Ratio of the yields for symmetric fission divided by the yields for asymmetric fission for uranium, protactinium and throrium isotopes. The yields have been obtained by fitting Gaussians to the measured charge distributions.

the following expression, introduced in ref. [8]:

$$V_C = \frac{Z_1 \cdot Z_2 \cdot e^2}{r_0 \cdot \left(A_1^{1/3} \cdot \left[1 + \frac{2\beta_1}{3} \right] + A_2^{1/3} \cdot \left[1 + \frac{2\beta_2}{3} \right] \right) + d} \tag{1}$$

In this equation Z_i, A_i and β_i are nuclear-charge numbers, mass numbers and deformations of the fission fragments, $r_0 = 1.16$ fm is the nuclear-radius constant, e the elementary charge, and $d = 2$ fm the *tip distance* which effectively takes into account the neck formation at scission. The mass numbers were deduced from the measured charge numbers by the *unchanged charge density* (UCD) assumption. The deformation parameters were fixed at $\beta_i = 0.6$ for both fragments as predicted by the liquid-drop model [8]. Figure 6 reveals that the global structural effects observed in the charge yields strongly differ from those showing up in the total kinetic energies. From ^{228}Th to ^{223}Th, the weight of the asymmetric fission component decreases strongly, while the enhancement of the total kinetic energies at asymmetry with respect to the macroscopic model is nearly preserved. Moreover, the position of the enhanced kinetic energy rather seems to follow the position of the $N = 82$ shell in the heavy fragment. The $Z = 50$ shell does not have any noticeable influence neither on the yields nor on the kinetic energies.

Even for symmetric fission a deviation of the data on the mean total kinetic energies with respect to the prediction of the liquid-drop model can be observed. A careful analysis of our data for isotopic chains of uranium, protactinium, thorium, actinium and radium indicates that symmetric fission is at least not completely governd by macroscopic nuclear properties, but also an influence of deformed shells in the fission fragments seems to be important.

These findings impose important constraints on an improved dynamical description of the fission process which could provide a quantitative prediction of the characteristics of the fission products.

6 Dissipation in fission

Data on elemental fission-fragment yields were difficult to obtain in conventional fission experiments. The rather scarce data on total even-odd effects in elemental yields measured previously show a systematic variation with the fissility of the fissioning system (see Figure 7). From this variation, the intrinsic excitation energy acquired by dissipation up to scission was deduced to grow with increasing fissility [14]. The excellent nuclear-charge resolution of the present experiment allowed to determine the even-odd structure in the elemental yields for a large number of fissioning systems. The new results of the present work on total even-odd effects for thorium and uranium isotopes which are also shown in Figure 7 seem to be compatible with the previously measured data: The higher excitation energies present in electromagnetic-induced fission (see Figure 1) lead to a reduction by about 40%, but the dependence on the fissility parameter Z_{CN}^2/A_{CN} observed previously seems to be confirmed.

However, a closer look to the even-odd structure as performed in the following section will reveal that the situation is much more complex.

The present experiment for the first time provides data on even-odd structure in *symmetric* charge splits for long isotopic chains. Figure 8 shows the local proton even-odd effect as defined by Tracy et al. [15] for thorium isotopes both at symmetry and in the maximum of the asymmetric fission component. In contrast to the total even-odd effect shown in Figure 7, the local even-odd effect at symmetry remains constant over nine isotopes. The local even-odd effect at asymmetry is much larger but it also stays almost constant over the range investigated. We now understand the true reason for the strong variation of the even-odd effect for thorium isotopes (and also for uranium isotopes) investigated here: It is the strong variation of the relative weights of symmetric and asymmetric components in the fission yields, see Figures 4 and 5. With increasing neutron number (decreasing fissility), the relative weight of the asymmetric component grows, leading to a larger total even-odd effect.

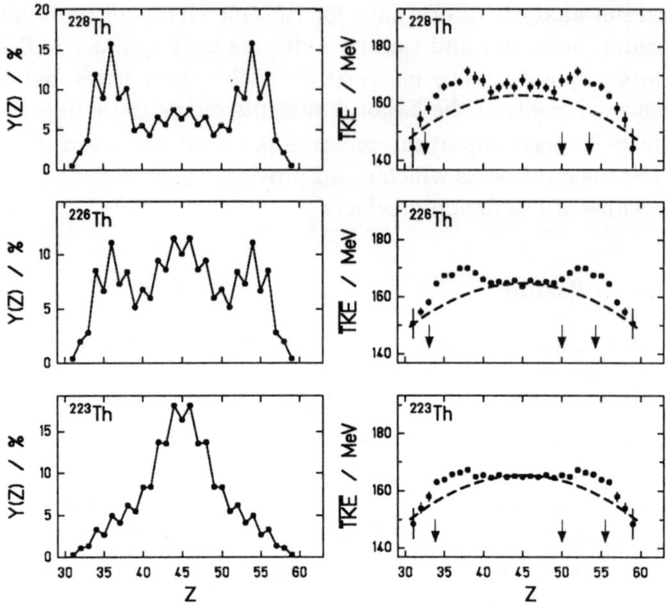

Figure 6: Measured elemental yields (left part) and average total kinetic energies (right part) as a function of the nuclear charge measured for fission fragments of several fissioning nuclei. Only statistical errors are given. The total kinetic energies are subject to an additional systematic uncertainty of 2%, common to all data (5). Arrows indicate the positions of neutron ($N = 50, 82$) and proton shells ($Z = 50$). The positions of the neutron shells are calculated from the proton numbers by assuming an unchanged charge density (UCD). The predictions of the liquid-drop model for the total kinetic energies are shown as dashed lines.

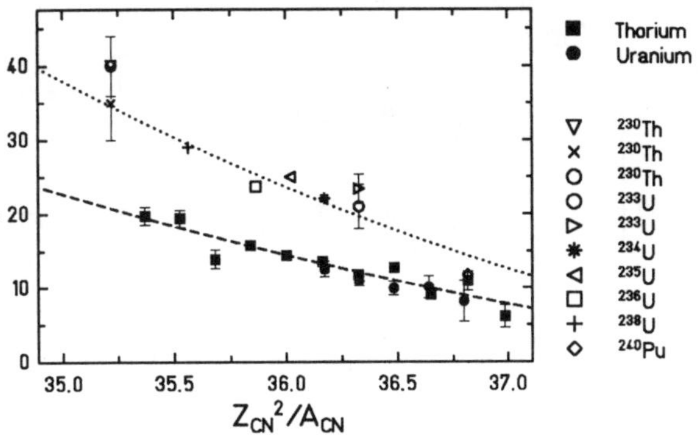

Figure 7: Total even-odd effect in the nuclear-charge yields observed in the fission of different nuclei. Data from thermal-neutron-induced fission (open symbols) from literature are compared to results of the present experiment for thorium and uranium isotopes (full symbols) as a function of the fissility parameter.

As already reported[16], the secondary-beam experiment also revealed strong local even-odd structures for all investigated *odd-Z* fissioning systems. A similar observation for a few other odd-Z systems was reported by Stumpf et al.[17]. We attributed this finding to an enhanced sticking probability of unpaired protons to the heavy fragment due to phase-space arguments in asymmetric charge splits[16]. This implies that an enhanced production of fission products with even proton numbers cannot directly be related to the probability of a fully paired proton configuration to survive up to the scission point as was done before[18]. Only for symmetric charge splits, the analysis used so far is valid. For asymmetric charge splits, however, part of the measured even-odd structure has to be attributed to the statistical contribution which explains great part of the higher even-odd effect found in asymmetry.

Let us now consider the global even-odd effects found for thermal-neutron-induced fission of the heavier systems shown in Figure 7. According to our new understanding, part of the measured total even-odd structure must be attributed to the statistical contribution. Here, symmetric fission is too weak to play any significant role. In these cases, another effect is expectedd to account for a part of the observed variation of the total even-odd structure: Due to the constant position of the heavy fission peak in mass number, the light and the heavy peak come closer to each other with increasing mass of the fissioning

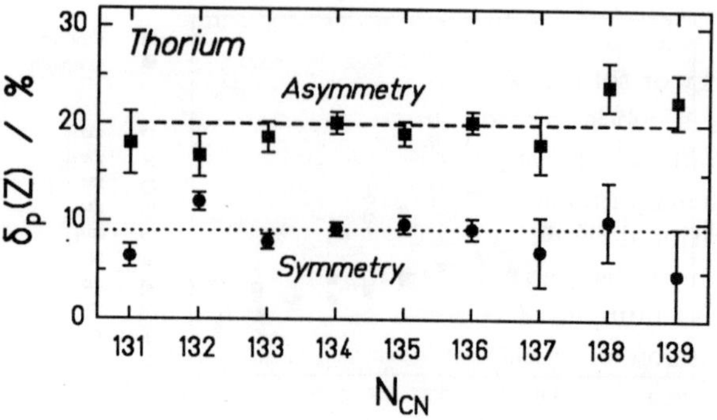

Figure 8: Local even-odd effect for symmetric and the most abundant asymmetric charge splits found in the fission of thorium isotopes after electromagnetic excitation.

system. Thus, the average asymmetry and consequently the statistical contribution to the even-odd structure decreases. Due to the restricted range of A_{CN}/Z_{CN} values, increasing mass goes in line with increasing fissilty.

¿From these findings we conclude that for all neutron-deficient thorium and uranium isotopes investigated in the secondary-beam experiment the dissipated energy from saddle to scission is about the same. In contrast to the first impression that we got from Figure 7, these systems do not follow the general trend deduced previously from the measured even-odd effect of heavier fissioning systems. Unfortunately, for those systems no information on local even-odd structure for near-symmetric charge splits is available due to the low yields at symmetry. Final conclusions on dissipation from the even-odd structure in asymmetric charge splits, the only information available in heavier systems, will require a far more detailed analysis and a full quantitative understanding of the statistical contribution to the even-odd structure in fission along the lines given in ref. [16], including the influence of shell structure.

7 Conclusion

A new experimental technique was applied to measure elemental yields and total kinetic energies after low-energy fission of short-lived radioactive nuclei. In contrast to most secondary-beam experiments of the first generation which are designed to determine one specific property of the secondary projectiles like binding energy or total interaction cross section, the present experiment is a quite elaborate secondary-reaction study where many characteristics of the fission process could be investigated. The beautiful data nicely demonstrate the decisive influence of nuclear structure on the fission process in a particular interesting transitional region around ^{227}Th. In contrast to the total kinetic energies, the element distributions were found to vary strongly, essentially as a function of mass number of the fissioning system. The relative weights of the asymmetric fission channels decrease with decreasing mass of the fissioning nucleus, but the behaviour of the total kinetic energies, which is related to the scission configurations, remains almost unchanged.

The new findings on even-odd effects in the nuclear-charge distributions after low-energy fission require a revised understanding of pair breaking in fission. Strong variations of the observed even-odd effect in the elemental yields as a function of fissility for different systems and as a function of asymmetry for specific systems are partly attributed to a statistical contribution to this quantity rather than to a variation of the intrinsic excitation energy at scission. Conclusions on the viscosity of cold nuclear matter have to be reconsidered. The detailed features of the fission fragments measured in the present work for a large number of fissioning systems in a continuously covered region of the chart of the nuclides provide an important test for elaborate models of low-energy nuclear dynamics.

Acknowledgments

We would like to thank K.-H. Behr, A. Brünnle and K. Burkhard and their team for their technical support, as well as H. Folger and his colleagues for manufacturing the targets. This work has been supported by the GSI Hochschulprogramm and by the Bundesminister für Bildung und Forschung (BMBF) under contract number 06 DA 473. The responsibility rests with the authors.

References

1. H.-G. Clerc *et al.* NPA **590**, 785 (1995)
2. A. R. Junghans *et al.* NPA **629**, 635 (1998)
3. H. Geissel *et al.* NIMB **70**, 286 (1992)
4. A. Grewe *et al.* NPA **614**, 400 (1997)
5. C. Böckstiegel *et al.* PLB **398**, 259 (1997)
6. B. Baur and C. A. Bertulani PRC **34**, 1654 (1986)
7. A. Turkevich and J. B. Niday PR **84**, 52 (1951)
8. B. D. Wilkins, E. P. Steinberg and R. R. Chasman PRC **14**, 1832 (1976)

9. P. Möller and S. G. Nilsson PLB **31**, 283 (1970)
10. G. A. Kudyaev, Yu. B. Ostapenko and G. N. Smirenkin SJN **45**, 951 (1987)
11. H. Kudo *et al.* PRC **25**, 3011 (1982)
12. V. V. Pashkevich NPA **169**, 275 (1971)
13. U. Brosa, S. Grossmann and A. Müller PRP **197**, 167 (1990)
14. F. Gönnenwein in *The Nuclear Fission Process*, ed. C. Wagemans (CRC Press, London, 1991), pp. 409
15. B. L. Tracy *et al.* PRC **5**, 222 (1972)
16. S. Steinhäuser *et al.* NPA **634**, 89 (1998)
17. P. Stumpf *et. al* in *Proc. Seminar on Fission* ed. C. Wagemans (Pont D'Oye, Belgium, 1995), pp. 196
18. H. Nifenecker *et al.* ZPA **308**, 39 (1982)

PROGRESS AND PRESENT STATUS OF THE INVESTIGATION OF THE FISSION FRAGMENT ANGULAR ANISOTROPY IN THE SLOW NEUTRON INDUCED FISSION OF ^{235}U

YU.N. KOPACH[1], A.B. POPOV[1], W.I.FURMAN[1], D.I. TAMBOVTSEV[2], L.K. KOZLOVSKY[2], N.N. GONIN[2], AND J. KLIMAN[1,3]

[1]*Joint Institute for Nuclear Research, 141980 Dubna, Russia*

[2]*SSC Institute of Physics and Power Engineering Obninsk, Russia*

[3]*Institute of Physics SAS Bratislava, Slovakia*

The progress and present status of the experimental and theoretical investigation of the fission fragment angular anisotropy in the slow neutron induced fission of a ^{235}U aligned target are presented. The upgraded experimental equipment has allowed us to gain more than a double increase of the accumulated statistics during the last two years. An improved data processing algorithm is developed. The obtained data are analyzed together with the total neutron, total fission and spin-separated fission cross sections in the neutron energy interval $0 - 20$ eV in order to obtain a new set of s-wave resonance parameters. The data are analyzed on the basis of a new theoretical description of fission process accounting for the interference between s-wave neutron resonances with different spins. Two fitting procedures were used, supplementary to each other. They are equivalent in the formalism of the description of cross sections, but different in the minimization algorithms. Three fission channels are taken into account for the spin group $J = 3$ ($K = 0, 1, 2$) and two for the spin group $J = 4$ ($K = 1, 2$). The obtained set of resonance parameters and K-channel contributions for the slow neutron induced fission of ^{235}U are discussed.

1 Introduction

Fission induced by slow neutrons is one of the unique tools for studying the quantum-mechanical aspects of the fission process. It gives a possibility to obtain information about the fission amplitudes $\gamma_{f\lambda}^{J^\pi K}$ for a given resonance λ. These parameters form a basis for a quantitative description of the fission process induced by resonance neutrons. Here, $J^\pi K$ are the spin, parity and the spin projection onto the symmetry axis of the fissioning nucleus.

It is known [1,2] that such amplitudes cannot be extracted unambiguously using only data on integral fission and neutron cross sections. A set of resonance parameters obtained by the fitting of the data using a multilevel two channel R-matrix parameterization is invariant to mirror reflection and rotation in the two-dimensional space of two channels. Additional information can be obtained from experiments on the interaction of polarized (or unpolarized) neutrons with a polarized (aligned) target.

We performed an experimental study of the energy dependence of the differential fission cross section (fission fragment angular anisotropy) of ^{235}U resonance neutron induced fission using an aligned target and unpolarized neutrons. In this case, even with s-wave neutrons one can observe a non-zero anisotropy of fission fragments which includes additional non-trivial dependence on the K quantum number. Thus, it gives an additional possibility for obtaining more complete and equivocal information about the s-wave fission amplitudes $\gamma_{f\lambda}^{J^\pi K}$. The feasibility of the extraction of such information from experimental data is determined by the completeness of the set of experimental data on different types of cross sections. For a well studied ^{235}U nucleus, there exists a unique set of rather accurate experimental data on total and spin-separated fission and total neutron cross sections. These data have to be analyzed together with our data on differential fission cross section in order to obtain a set of resonance parameters.

2 Experimental data

The experiment is being performed at the booster IBR-30 + LEA-40 in Dubna and is a development of the technique used by Pattenden and Postma. [3] The description of the experimental set-up used in the first stage of the experiment can be found in. [4] Since then, the experimental equipment has been substantially upgraded. The new $^3He-^4He$ dilution refrigerator is installed; one of the two ^{235}U mosaic single crystal samples is replaced by a new thinner one; and silicon surface barrier semiconductor detectors are replaced by more reliable silicon planar detectors specially designed for the operation at temperatures about 1 K. Such experimental modernization has resulted in an over double increase of the accumulated statistics during the past two years of measurements. An improved data processing algorithm is applied. [5] The statistical accuracy of the $A_2(E)$ data obtained in the neutron energy range $0.05-20$ eV with the energy bins 0.05 eV is about $2-3$ % for strong neutron resonances and $10-15$ % for inter-resonance intervals.

The energy dependence of the A_2 coefficient which characterizes the angular anisotropy of fission fragments is shown in fig. 1. The errors are the statistical ones. The systematical error of the whole data set is mainly due to the uncertainty of the electric hyperfine coupling constant P for ^{235}U nuclei in the uranyl rubidium crystal and is estimated as $\sim 20\%$. Our data are in good agreement with the data of Pattenden and Postma [3,6] obtained for separate resonances and broad inter-resonance intervals.

For a combined analysis, we also used the experimental cross sections for ^{235}U available from the National Nuclear Data Center (NNDC) at BNL.[7] The

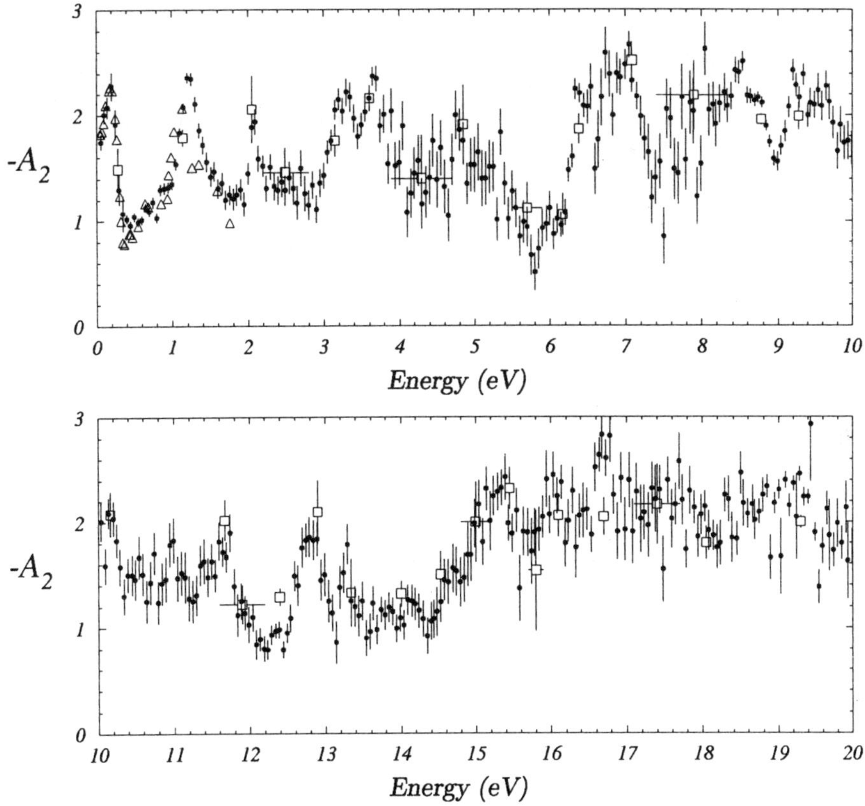

Figure 1. The $A_2(E)$ dependence. Circles – present work, squares [3] and triangles [6] – data of Pattenden and Postma.

data sets selected for the analysis are summarized in Table 1.

To simplify the analysis, we do not include the capture cross section in our fitting procedure. The $\sigma_{n\gamma}$ curve calculated on the basis of our final resonance parameters is however in reasonable agreement with the ENDF/B-VI evaluation. [7]

Table 1. The experimental data sets selected for the evaluation.

	Type of cross section	Author	Energy range
1.	$A_2(E)$	Present work	0.06 – 20 eV
2.	Fission (σ_{nf})	R.Gwin et al.[8]	0.0016 – 9.7 eV
3.	Fission (σ_{nf})	L.W.Weston et al.[9]	9.7 – 20 eV
4.	Spin-separated fission (σ^J_{nf})	V.L.Sailor et al.[10]	0.075 – 1.14 eV
5.	Spin-separated fission (σ^J_{nf})	M.S.Moore et al.[11]	1.6 – 20 eV
6.	Total (σ_{ntot})	F.D.Brooks et al.[12]	0.035 – 1.3 eV
7.	Total (σ_{ntot})	A.Michaudon et al.[13]	1.3 – 20 eV

3 Data analysis

For the analysis of the data we use the standard R-matrix formalism for the description of neutron cross sections. It gives the following expression for the reaction matrix (see [14]):

$$S^J_{nc} = e^{-i(\phi_n+\phi_c)}\left\{2\left((\mathbf{I}-\mathbf{K})^{-1}\right)_{nc} - \delta_{nc}\right\} \tag{1}$$

where

$$(\mathbf{I}-\mathbf{K})_{ij} = \delta_{ij} - \frac{i}{2}\sum_{\lambda}\frac{\sqrt{\Gamma_{i\lambda}\Gamma_{j\lambda}}}{E_\lambda - E - i\Gamma_{\gamma\lambda}/2} \tag{2}$$

Different types of cross sections can be expressed in terms of the S-matrix. The total neutron cross section is:

$$\sigma_{ntot} = \frac{2\pi}{k^2}\sum_J g_J Re(1 - S^J_{nn}) \tag{3}$$

As follows from,[15] the total and differential fission cross sections can be expressed via S-matrix (1) and (2), where the amplitudes of fission widths $\sqrt{\Gamma_{f\lambda}}$ characterize the Bohr fission channels $f = \{J^\pi K\}$. The total fission cross section is:

$$\sigma_{f0} = \frac{\pi}{k^2}\sum_J g_J \sum_K |S^J_{nf_K}|^2 \tag{4}$$

The spin separated fission cross sections are:

$$\sigma_{fJ} = \frac{\pi}{k^2} g_J \sum_K |S^J_{nf_K}|^2 \tag{5}$$

The angular anisotropy coefficient (see[15,16]) is:

$$A_2(E) = \frac{\sigma_{f2}(E)}{\sigma_{f0}(E)} , \tag{6}$$

where

$$\sigma_{f2} = G\frac{\pi}{k^2}\sum_{JJ'}\sqrt{g_J g_{J'}}U(\frac{1}{2}IJ'2;JI)\sum_K C^{J'K}_{JK20}S^{*J'}_{nfK}S^{J}_{nfK} , \tag{7}$$

and $G = \frac{15I^2}{\sqrt{(2I-1)I(I+1)(2I+3)}}$. Note that σ_{f2} contains terms responsible for the interference between resonances with different spins which were never taken into account in previous analyses.

Fitting of the experimental data sets listed in Table 1 was carried out using above formalism over the energy region $0 - 20$ eV. Two different methods were used. The first is the SAMMY code modified for the calculation of the $A_2(E)$ dependence. It uses Bayes' method and performs sequential analyses of different data sets retaining the covariance matrix from the previous fit. The second is the original code developed in FLNP, JINR. It uses the least-squares method implemented in the FUMILI subroutine and can fit several cross sections simultaneously. The two codes have identical formalisms of the description of cross sections and take into account the Doppler and resolution broadening. However, because of different minimization algorithms the results of fitting can be different. It should be noted that even different fits performed by the same code (SAMMY or FUMILI) may produce different sets of parameters. The energies of resonances, neutron and total fission widths are stable in most cases while partial fission widths may vary depending on the initial conditions. Combined use of the two methods allows us to obtain the most reliable set of resonance parameters and check its consistency.

According to Bohr hypothesis,[18] channels with lowest K values should be open in fission. From energy considerations, Lynn [19] concluded that for the spin group $J = 3$, channels with $K = 0, 1$ and possibly 2 should be open and for the spin group $J = 4$, channels with $K = 1$ and possibly 2 should be open. However, in previous works only two channels were taken into consideration for the two spin groups. Following the prediction of the theory, we included in our analysis three channels ($K = 0, 1, 2$) for $J = 3$ and two channels ($K = 1, 2$) for $J = 4$. For each spin group, three negative and two positive (> 20 eV) resonances were added. The radiative width Γ_γ was fixed and equals 0.039 eV for all resonances. All other parameters were varied. The results of the fit are shown in fig. 2. The dashed line in the lowest plot is calculated using a set of resonance parameters [17] obtained in the two channel approximation and without taking into account the information on $A_2(E)$.

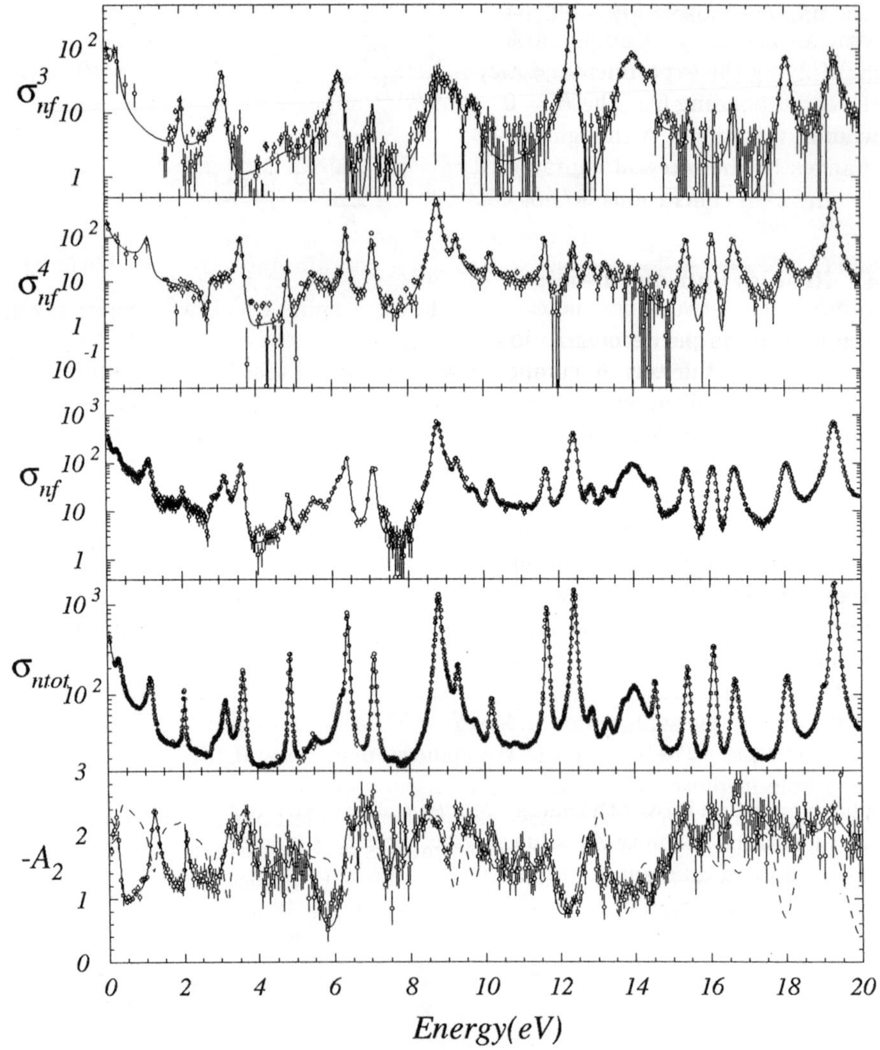

Figure 2. The comparison of the experimental data and fitting curves. Dashed line in the lowest plot calculated using resonance parameters from.[17]

We investigated the influence of the absolute normalization of A_2 (known with an accuracy of about 20%) on the variation of fitting parameters by multiplying the experimental data by a factor of 0.85 and 1.15 and performing the corresponding fit. The $K = 0$ component appears to be the most sensitive to such variations. Although the resulting parameters were slightly different, changes in the integral characteristics were insignificant (see next chapter). So our main conclusions do not change.

4 Results and discussion

Figure 3 shows the decomposition of the total and spin separated fission cross sections into different K-components. As is expected, the weights of the K-channels for different compound states λ are randomly distributed, and different K components have different energy behaviour. It can be noted that the contribution of the $K = 0$ component is significant both in spin-separated and total fission cross sections.

Such a small energy interval does not permit one to analyze the statistical and integral properties of resonances. Nevertheless, several definite conclusions can be drawn. Tables 2 and 3 represent the average contributions of different K components to A_2, spin-separated and total fission cross sections defined as

$$\frac{1}{N}\sum_i \frac{A_{2,K}(E_i)}{A_2(E_i)}, \qquad \frac{1}{N}\sum_i \frac{\sigma_{f0,K}(E_i)}{\sigma_{f0}(E_i)}, \qquad \frac{1}{N}\sum_i \frac{\sigma^J_{f0,K}(E_i)}{\sigma^J_{f0}(E_i)} \tag{8}$$

The summation is over the energy range 0–20 eV with a step 0.05 eV. One can see that as the absolute A_2 value changes by $\pm 15\%$ (coefficient C in the tables) the weights of different K components in A_2 do not change significantly. The average contribution of the JJ' interference term

$$\frac{1}{N}\sum_i \left| \frac{A_2^{JJ'}(E_i)}{A_2(E_i)} \right| \tag{9}$$

seems to be more sensitive (see Tab 2). Relative K contributions to cross sections vary much stronger than in the case of A_2. The $K = 0$ channel appears to be the most sensitive to such variations of the absolute A_2 value.

The contributions of different K components to fission cross sections do not give complete information about the degree of openness of different fission channels since they are influenced by the energy dependence of the cross section and by the fluctuations of the neutron widths. Thus, it is more consistent

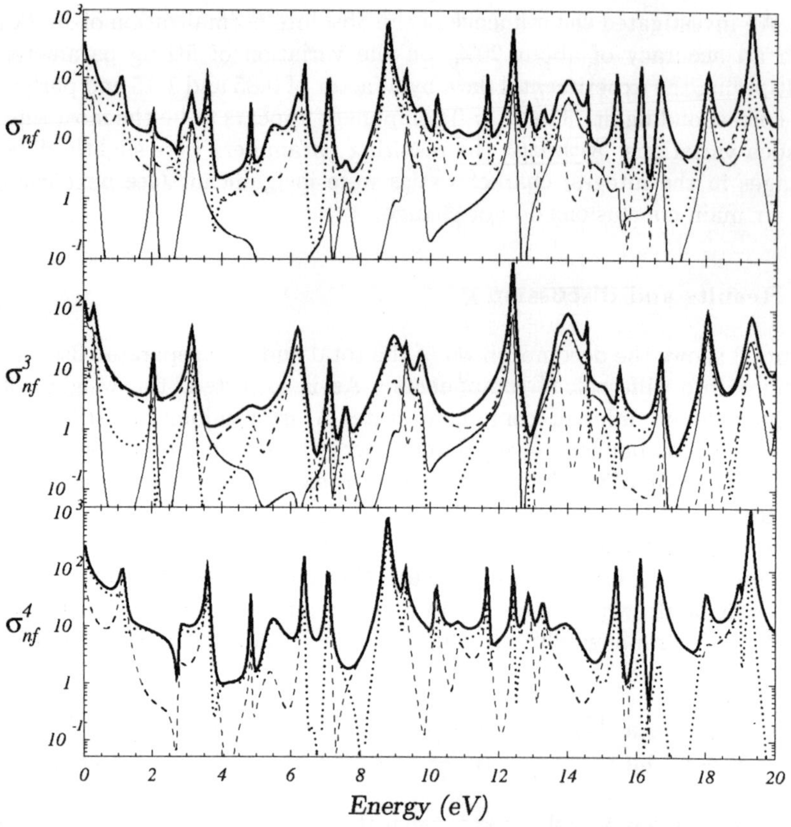

Figure 3. Decomposition of the total and spin-separated fission cross sections into K components. Solid line – K=0, dashed line – K=1, dotted line – K=2.

to define the average contribution as

$$W_K = \frac{1}{N_\lambda} \sum_\lambda \frac{\Gamma^\lambda_{f,K}}{\Gamma^\lambda_f} \tag{10}$$

where $\Gamma^\lambda_{f,K}$ and Γ^λ_f are the partial and total fission widths, respectively. In the limit of an infinite number of resonances it characterizes the degree of openness of a given K-channel. Table 4 shows the W_K values calculated separately for resonances from the spin groups $J = 3$ and $J = 4$. Although

Table 2. Relative contributions of different K components and JJ' interference term to A_2.

$K=0$	$K=1$	$K=2$	JJ'	SAMMY fit variant
0.14	0.65	0.21	0.22	C=0.85
0.15	0.64	0.21	0.15	C=1.00
0.15	0.63	0.22	0.12	C=1.15

Table 3. Relative contributions of different K components to the total and spin-separated fission cross sections.

σ	$K=0$	$K=1$	$K=2$	SAMMY fit variant
σ_{f0}	0.05	0.48	0.47	C=0.85
	0.07	0.51	0.42	C=1.00
	0.09	0.54	0.37	C=1.15
$\sigma_{f0}^{J=3}$	0.17	0.43	0.40	C=0.85
	0.23	0.42	0.35	C=1.00
	0.30	0.39	0.31	C=1.15
$\sigma_{f0}^{J=4}$		0.50	0.50	C=0.85
		0.55	0.45	C=1.00
		0.60	0.40	C=1.15

these values may fluctuate as larger numbers of resonances are included, it is obvious that the average weights of the $J^{\pi}K$ channels connected with the corresponding heights of fission barriers are comparable for all $J^{\pi}K$ channels with some prevalence of $K = 1$. It is interesting that our estimations of the contributions of $J^{\pi}K$ channels are close to the results obtained from the analysis of the total fission widths of neutron resonances with $J = 3$ and $J = 4$.[20] One has to note that these results were obtained in an indirect way and in a different energy interval.

Table 4. Degree of openness of different K channels for the spin groups $J = 3$ and $J = 4$.

σ	$K=0$	$K=1$	$K=2$	SAMMY fit variant
$\sigma_{f0}^{J=3}$	0.23	0.44	0.33	C=0.85
	0.32	0.39	0.29	C=1.00
	0.30	0.39	0.31	C=1.15
$\sigma_{f0}^{J=4}$		0.53	0.47	C=0.85
		0.54	0.46	C=1.00
		0.56	0.44	C=1.15

The integral distributions of the fission widths of resonances in the energy interval $0 - 20$ eV are given in fig 4. The experimental distributions for

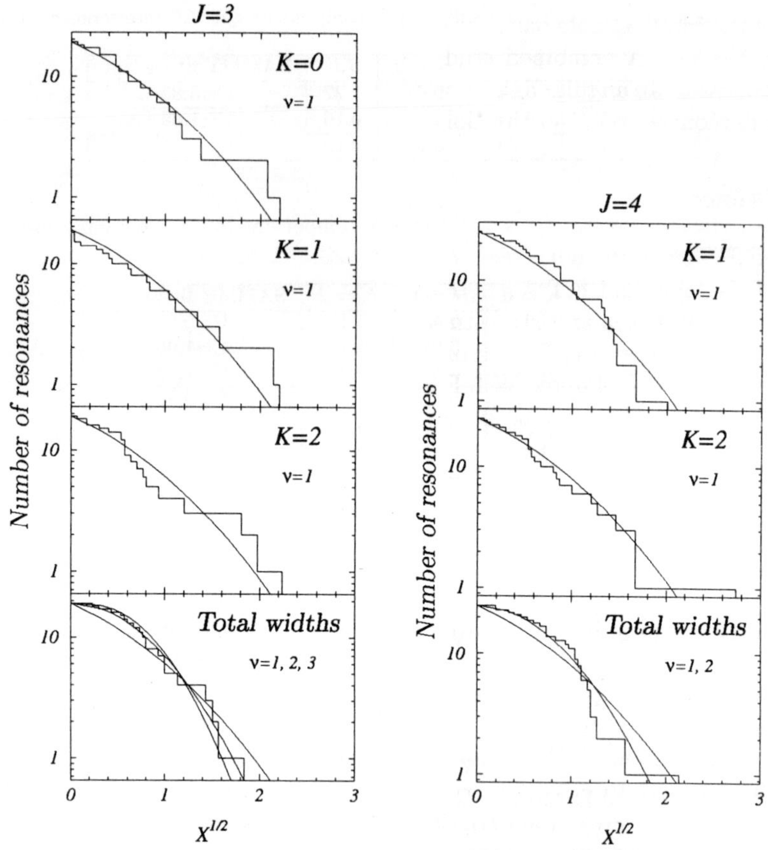

Figure 4. Integral distributions of partial and total fission widths (number of resonances with $\Gamma_{\lambda f}/\langle\Gamma_{\lambda f}\rangle > X$). Solid lines are the χ^2 distributions with ν degrees of freedom.

each separate $J^\pi K$ channel are in good agreement with the Porther-Thomas distribution with one degree of freedom. A small number of resonances does not allow us to analyze carefully the distributions of the total fission widths.

Finally, it should be noted that the obtained set of resonance parameters is not uniquely determined. It depends on the choice of negative resonances as well as on inclusion of resonances with small neutron and large total fission widths. However, since this parameter set includes all possible K channels, it forms the most reliable basis for quantitative analysis of $s-$ and $p-$resonance interference.[21] A combined study of fragment mass-energy distributions [22] together with an angular anisotropy coefficient can give new information about the interconnection [23] of the Bohr fission channels and fission modes.

References

1. G.F.Auchampaugh, *Nucl. Phys.* **A175**, 65 (1971).
2. D.B.Adler and F.T.Adler, *Phys. Rev.* **C6**, 985 (1972).
3. N.J.Pattenden and H.Postma, *Nucl. Phys.* **A167**, 225 (1971).
4. D.I. Tambovtsev, L.K. Kozlovsky, N.N. Gonin, N.S. Rabotnov, Yu.N. Kopach, A.B. Popov, W.I. Furman, J. Kliman, H. Postma, A.A. Bogdzel, and M.A. Guseinov, *Phys. At. Nucl.* **60/6**, 877 (1997).
5. Yu.N.Kopach, A.B.Popov, W.I.Furman, N.N.Gonin, L.K.Kozlovsky, D.I.Tambovtsev, and J.Kliman, *Phys. At. Nucl.* (1999), to be published.
6. H.Postma, in *Proc. 2nd Int. Symposium on Neutron Capture Gamma-ray Spectroscopy and Related Topics*, Petten, Netherlands, p. 619 (1974).
7. National Nuclear Data Center, Upton, NY, 11973-500, *http://www.nndc.bnl.gov.*
8. R.Gwin, R.R.Spencer, R.W.Ingle, J.H.Todd, and S.W.Scoles, *Nucl. Sci. Eng.* **88**, 37 (1984).
9. L.W.Weston and J.H.Todd, *Nucl. Sci. Eng.* **88**, 567 (1984).
10. R.I.Schermer, L.Parsell, G.Brunhart, C.A.Reynolds, V.L.Sailor, and F.J.Shore, *Phys. Rev.* **167**, 1121 (1968).
11. M.S.Moore, J.D.Moses, G.A.Keyworth, J.W.T.Dabbs, and N.W.Hill, *Phys. Rev.* **C18**, 1328 (1978).
12. F.D.Brooks, J.E.Jolly, M.G.Schomberg, and M.G.Sowerby, AERE-M-1670 (1966).
13. A.Michaudon, H.Derrien, P.Ribon, and M.Sanche, *Nucl. Phys.* **69**, 545 (1965).
14. ENDF/B-VI Summary Documentation, ENDF-201 (BNL-17541), 1991.
15. A.L.Barabanov and W.I.Furman, in *Proc. Int. Conf. on Nuclear Data for Science and Technology*, Gatlinburg, Tenessee, USA, May 9-13, p. 448 (1994).
16. A.L.Barabanov and W.I.Furman, *Z. Phys.* **A357**, 411 (1997).
17. L.C.Leal, H.Derrien, N.M.Larson, and R.Q.Wright, ORNL/TM-13516 (1997).
18. A.Bohr, in *Proc. 1st Int. Conf. Peaceful Uses At. Energy*, Geneva, Vol. 2, p. 151 (1956).
19. J.E.Lynn, "Theory of Neutron Resonance Reactions", (Claredon Press,

Oxford, 1968).
20. A.R.Musgrove *et al.*, *J. Phys.* **G7**, 549 (1981).
21. A.B.Popov, W.I.Furman, and L.Lason, in *Proc. of II Int. Workshop of Fission and Fission Fragment Spectroscopy*, edited by G.Fioni *et al.*, Seissyn, France, April 21-24, 1998, p. 349 (Woodbury, New York, 1998).
22. Sh.S.Zeinalov *et al.*, in *Proc. VI Int. Sem. on Int. Neutr. with Nucl.*, Dubna, Russia, May 13-16 1998, p. 274 (Dubna, 1998).
23. W.I.Furman and J.Kliman, in *Proc. 17th Int. Symp. on Nucl. Phys.*, Gaussig, 1987, p. 86 (Dresden, ZfK, 1988).

ANISOTROPY OF γ-RAY EMISSION IN BINARY AND TERNARY SPONTANEOUS FISSION OF ^{252}Cf

Yu.N. Kopach [1,2], M. Mutterer [1], P. Singer [1], M. Klemens [1], A. Hotzel [3], D. Schwalm [3], P. Thirolf [3], M. Hesse [4], and F. Gönnenwein [4]

[1] *Institut für Kernphysik, Technische Universität, 64289 Darmstadt, Germany*

[2] *Frank Laboratory of Neutron Physics, JINR, Dubna, Russia*

[3] *Max-Planck-Institut für Kernphysik, 69115 Heidelberg, Germany*

[4] *Physikalisches Institut der Universität, 72076 Tübingen, Germany*

Angular correlations of prompt fission γ-rays (energies from 0.1 to 1.5 MeV) from binary and light-charged-particle (LCP) accompanied spontaneous fission of ^{252}Cf were measured with a homogeneous 4π-crystal-ball $NaI(Tl)$ γ-ray spectrometer combined with a kinematic spectrometer for fission fragments and LCPs. The angular dependence of the γ-ray intensity was deduced both with respect to the fission direction and the direction of motion of the LCPs, and for various ternary fission modes accompanied by ^3H, ^4He, ^6He, Li and Be particles. The anisotropy of the γ-ray emission with respect to the fission axis was found to be equally pronounced in ternary and binary fission. This means that the emission of ternary particles at scission does not significantly influence the alignment of fragment spins contrary to conclusions from two previous experiments. In ternary fission, the correlation of γ-ray intensity and LCP motion was studied by projecting γ-ray emission angles on the plane perpendicular to the fission axis. These distributions were found to be isotropic relative to the projected LCP momentum, contrasting a suggested correlation between the ternary particle emission and a collective bending motion at scission.

1 Introduction

Fragments from nuclear fission are formed with rather high spins (6 to 8 $\hbar$, on average) which is attributed to a collective vibrational motion at scission (so-called bending or wriggling modes) [1,2,3], and to the Coulomb excitation (quadrupole torque interaction) in the early separation phase of the fragments [4,5]. However, recent data on fragment spin populations in cold (neutronless) fission from modern fine-grained γ-ray spectroscopy seem not to be consistent with theoretical calculations [6]. Generally, fragment spin vectors are mainly oriented perpendicular to the fission direction, in accordance both with the bending mode assumption and with post-scission Coulomb excitation. This alignment of the fragment spins causes the emission of the Yrast γ-rays to proceed anisotropically with respect to the fission axis, although fragment spin distributions are significantly influenced by neutron and statistical γ-ray emission [7,8].

Experimental access to intimate properties of the scissioning nucleus is are mainly oriented perpendicular to the fission direction, in accordance both with the bending mode assumption and with post-scission Coulomb excitation. This alignment of the fragment spins causes the emission of the Yrast γ-rays to proceed anisotropically with respect to the fission axis, although fragment spin distributions are significantly influenced by neutron and statistical γ-ray emission [7,8].

Experimental access to intimate properties of the scissioning nucleus is also acquired through the rare fission mode of light-charged-particle (LCP) accompanied fission ($\approx 1/260$ of the binary fission yield, for ^{252}Cf) [9]. It is, thus, interesting to study whether the emission of LCPs at scission influences the fragment spins and their alignment by comparing γ-ray angular correlations in ternary and binary fission, respectively. Only two experiments on γ-ray angular correlations in ternary fission were hitherto reported. They concern the angular distribution of γ-ray intensity in ^{252}Cf(sf) with respect to the fission direction, measured in coincidence with the relatively frequent ternary α-particles (about 90% of the ternary yield). In an early work, Ivanov [10] suggested that the γ-rays are emitted isotropically in ternary fission, and Pilz and Neubert [11] found the γ-ray-fragment correlation to be characteristically modified by the ternary α-particle release. Both experiments thus suggested a drastic change in the alignment of fragment spins. On the other hand, it was argued by Barabanov [12] that the results of Ref. [11] might be a signature of a mutual relation between the mechanism of ternary-particle emission and the excitation of collective modes during the saddle-to-scission stage. In the present paper we present an extended study on the γ-ray angular correlation in ternary fission of ^{252}Cf. In addition to ternary α-particles, emission of some much rarer LCP species, viz. ^{3}H, ^{6}He, Li and Be particles, was studied, and the angular dependence of the γ-ray intensity was not only determined with respect to the direction of fragment motion but also with respect to the emission direction of the LCPs. Brief account to the present work was given recently in Ref. [13]

2 Experiment

In the experiment, the Darmstadt-Heidelberg Crystal Ball (CB) [14] serving as a homogeneous 4π γ-ray spectrometer was combined with an efficient detection system for fission fragments and ternary particles. A detailed description of the experimental set-up and its performance is given in Refs. [15,16,17]

The γ-rays detected with high efficiency ($> 90\%$) are separated by time-of-flight from the prompt fission neutrons which are simultaneously registered with a rather high efficiency of about 60% . For avoiding in the coincidence time spectra any contamination in the prompt γ peak by the delayed pulses from the neutrons, which yield an about 20 times stronger anisotropy than the γ-rays (see Ref.[17]), a rather narrow time window of $\pm$ 2 ns was set onto the prompt γ-ray coincidence peak having a width of 3.5 ns fwhm. Energy calibration of the γ-ray pulse-height spectra was performed with various γ-ray sources, following standard procedures. Because of the dominance of full energy absorption in the large crystals the measured γ-ray spectra were not unfolded for the detector response. All crystals of the CB subtend equal solid angles ($\pi/40$, each), but relative γ-ray efficiencies varied up to $\approx$ 20% due to absorption in the charged-particle detectors mounted inside. The corresponding corrections were derived from the ^{252}Cf γ-ray spectra measured. Furthermore, an angular resolution correction accounting for the granularity of the CB was determined from a Monte-Carlo simulation (using the GEANT code), and the various γ-ray angular distributions presented below were unfolded accordingly.

3 Data analysis and results

3.1 Anisotropy of γ-ray emission with respect to fragment motion for binary and α-accompanied fission

The angular distribution $W'(E',\theta')$ of γ-rays with multipolarity ≤ 2 in the rest frame of a moving fragment is described by a series of even Legendre polynomials[7,8]

$$W'(E',\theta') = 1 + A_2(E')\ P_2(\cos\theta') + A_4(E')\ P_4(cos\theta') \tag{1}$$

where E' is the γ-ray energy and θ' is the angle with respect to the fission axis. The distribution $W(E,\theta)$ in the laboratory system is, due to the solid-angle aberration and Doppler shift, transformed[18] according to

$$\begin{aligned} W(E,\theta) \quad \simeq \quad & W'(E,\theta)\cdot(1+\beta\cdot\cos\theta) - \beta\cdot E\cdot\cos\theta\cdot\left.\frac{\partial W'}{\partial E'}\right|_{E,\cos\theta} \\ & - \beta\cdot(1-\cos^2\theta)\cdot\left.\frac{\partial W'}{\partial\cos\theta'}\right|_{E,\cos\theta} \end{aligned} \tag{2}$$

where β is the fragment velocity in units of c. Only first order terms in β are considered in Eq. (2) which is justified by the low β-values (≤ 0.05) involved.

Experimentally it is not possible to identify which of the two fragments is the source of an individual γ-ray event. Note that in our notation β is positive for the light mass group of fission fragments and negative for the heavy one. However, Eqs. (1, 2) remain valid when W denotes the common γ-ray angular

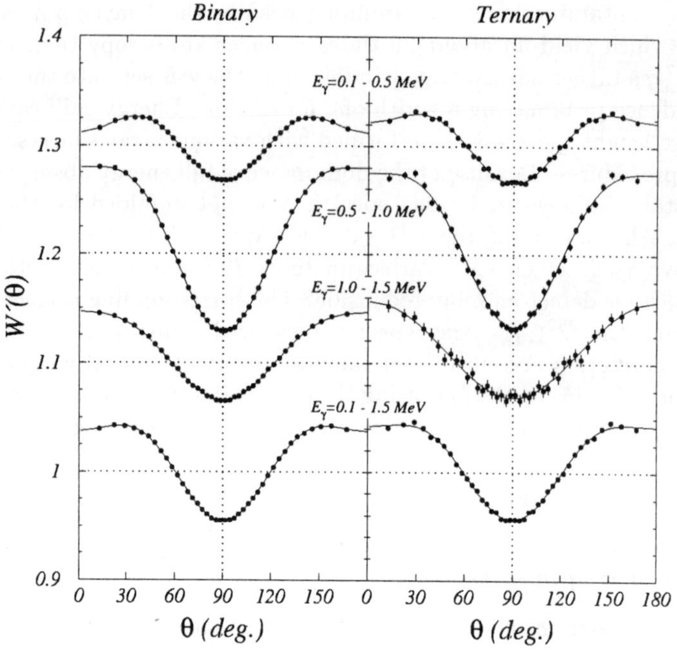

Figure 1: Rest-frame angular distributions $W'(\theta)$ of γ- rays with respect to the fragment motion for binary (left side) and α-accompanied (right side) fission of ^{252}Cf, for different γ-ray energy intervals. The scale is correct for the lowest plots; others are shifted consecutively by 0.1 units. Solid lines are fitted curves using Eq. (1). Error bars represent statistical errors.

distribution from the fragment pairs, and W' defines accordingly the average of the angular correlation functions (as in Eq. (1)) of the two fragments in their respective rest frames. The parameter β in Eq. (2) can than be interpreted as an "effective" velocity β_{eff} which will vary for different energy regions according to the relative contributions of the two fragments to the spectrum at these γ-ray energies. By grouping terms with the same power of $\cos\theta$ one can reduce Eq. (2) to:

$$\begin{aligned} W(E,\theta) &= 1 + A_2(E)\ P_2(\cos\theta) + A_4(E)\ P_4(cos\theta) \\ &\quad +B(E)\cdot[\cos\theta + C_3(E)\cos^3\theta + C_5(E)\cos^5\theta] \end{aligned} \tag{3}$$

where $A_2(E)$ and $A_4(E)$ are the average anisotropy coefficients for the two

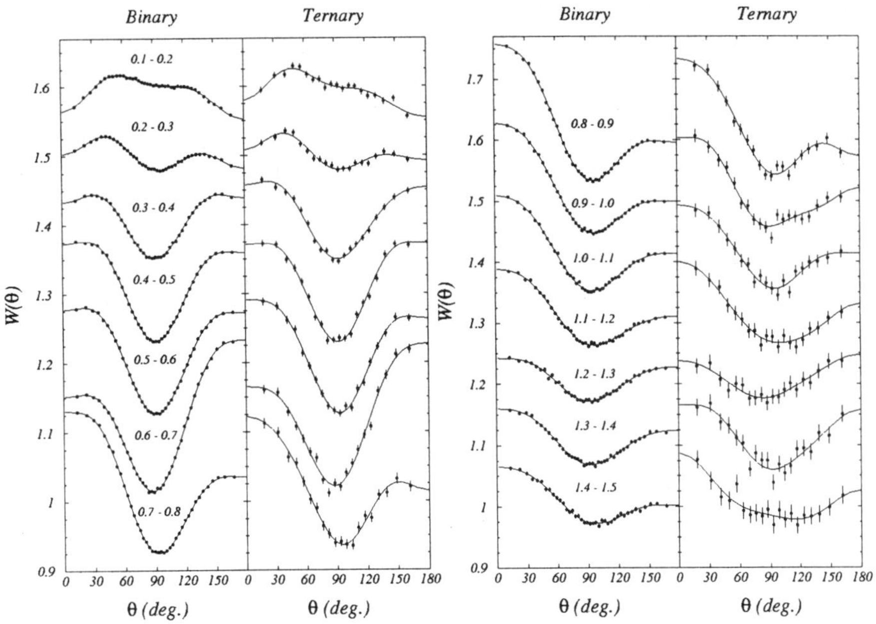

Figure 2: Angular distributions $W(\theta)$ of γ-rays with respect to the fragment motion in the laboratory system. Gamma-ray energy intervals are given in MeV. The scale is correct for the lowest plots; others are shifted consecutively by 0.1 units. Solid lines are fitted curves using Eq. (3).

fragments, $B(E)$ is proportional to β_{eff}, and C_i are fitting parameters.

As follows from Eq. (3), the center-of-mass distribution $W'(E,\theta)$ can be simply determined by averaging the γ-ray intensity $W(E,\theta)$ as measured per unit solid angle at an angle θ with respect to one fragment group and that at the respective complementary angle $\pi - \theta$:

$$W'(E,\theta) = 1/2[W(E,\theta) + W(E,\pi - \theta)] \tag{4}$$

Resulting symmetrized and integrated over 500 keV energy intervals angular distributions $W'(\theta)$ deduced from the present data according to Eq. (4) are compared in Fig. 1 for the binary and α-particle accompanied fission modes. The corresponding experimental angular distributions $W(\theta)$ integrated over finer energy bins of 100 keV are shown in Fig. 2. Solid lines in both patterns are the results of least-square fits to the data, according to Eqs. (1) and (3), respectively. Our data clearly show that there is no significant difference in the γ-ray anisotropy between the binary and α-particle accompanied fission modes.

Figure 3 displays the corresponding anisotropy ratios A, defined as $A = (W'(0°)/W'(90°)-1)$, for 100 keV energy intervals, in comparison with binary-fission data from Skarsvåg[8] which were deduced from an experiment with one fragment stopped in a thick backing. For binary fission of ^{252}Cf, both sets of data on $A(E_\gamma)$ are in good agreement yielding negative values at energies below 0.2 MeV, and maximum positive values in the γ-energy interval from 0.5 to 0.7 MeV. For the α- accompanied fission mode, all our data on $W'(\theta)$ (Fig. 1), $W(\theta)$ (Fig. 2) and $A(E_\gamma)$ (Fig. 3) do not reveal any significant difference compared to binary fission, in contrast to the conclusions from both former studies[10,11]. In Fig. 2, one can note drastic changes in the shapes of the curves between neighbouring energy bins (especially between 600 and 800 keV) which are obviously due to fluctuations of the multiplicities and multipolarities of the γ-rays coming from the light and the heavy group of fission fragments.

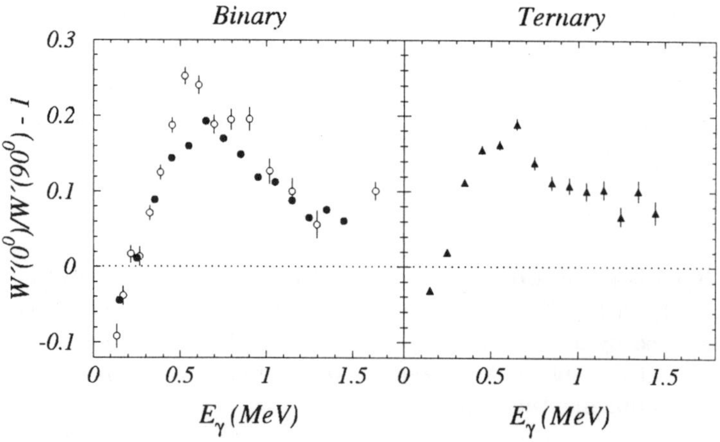

Figure 3: Anisotropy ratio A in binary (left side) and α-accompanied fission (right side) of ^{252}Cf, as a function of γ-ray energy. The present results (full symbols) are compared with binary fission data from Ref.[8] (circles).

Our A values, summarized in Tab. 1, show close agreement of our binary and α-accompanied fission data with $A = 0.092(5)$ as obtained by Skarsvåg[8] for binary fission, but the present result on α-accompanied fission is in obvious contradiction to the much smaller figures published previously. We also note

Table 1: Anisotropy ratio A and angular correlation parameters A_2 and A_4 for binary fission of ^{252}Cf, and ternary fission with emission of various LCPs. The γ-ray energy interval is from 0.1 to 1.5 MeV.

Fission mode:	Anisotropy parameters:		
	A	A_2	A_4
binary:	0.0872(3)	0.0683(1)	-0.0308(2)
	0.092(5) [8]		
	0.127(3) [10]		
α-accompanied:	0.0903(19)	0.0684(9)	-0.0261(12)
	0.015(22) [10]		
	0.03(2) [11]		
other LCPs:			
^{3}H	0.082(9)	0.062(4)	-0.023(6)
^{6}He	0.089(13)	0.068(6)	-0.027(8)
Li	0.02(4)	0.03(2)	-0.04(3)
Be	0.17(5)	0.12(2)	-0.03(3)

that, in the present experiment, we have observed the γ-ray multiplicity M_γ in the ternary cases to be lower by about 10% compared to binary fission [19], in agreement with [20] but again in conflict with [10] reporting $M_\gamma(\text{ter})/M_\gamma(\text{bin}) = 0.995(12)$. Our data in Tab. 1 reveal a slight (about 3%) increase in the A value for α-accompanied fission compared to the binary case, and, related to that, a somewhat lower A_4 coefficient. Such a small difference should not be surprising since the properties of fission fragments, such as the fragment mass distribution, the total excitation energy TXE, the average number of evaporated neutrons $\bar{\nu}$, and also the proton even-odd effect in the isotopic composition of the fragments [21] slightly change upon emission of ternary α-particles [9,22].

3.2 Anisotropy of γ-ray emission with respect to α-particle motion

For the analysis of the γ-ray-LCP correlation the experimental data for ternary fission were sorted into a coordinate system xyz (see Fig.1 in Ref. [17]) which allows us to deduce γ-ray angular correlations both with respect to the direction of fragment motion and the emission direction of the LCPs. For each event, the z axis was arbitrarily chosen as the direction of fission fragment from the light mass group. The x axis is in the plane made up by the three charged reaction products (the fragment pair and the LCP) and is close to the direction of the LCP which is emitted at the mean angle $\theta_{\alpha L} = 83^0$ (width $\approx 20^0$ fwhm) with respect to the light fragment motion [9,22]. The y axis is accordingly perpendicular to the xz plane making a right-handed coordinate system xyz. In this coordinate system one can analyze angular distribution of γ-rays with respect to the fission axis, $W(\theta)$, or angular correlations of γ-rays with the directions of the fragments and LCPs, $w(\phi_{ij})$, by projecting the 3-dimensional angular distribution on three complementary planes $(ij) = (xy), (zx), (zy)$, with ϕ_{ij} being the corresponding polar angles.

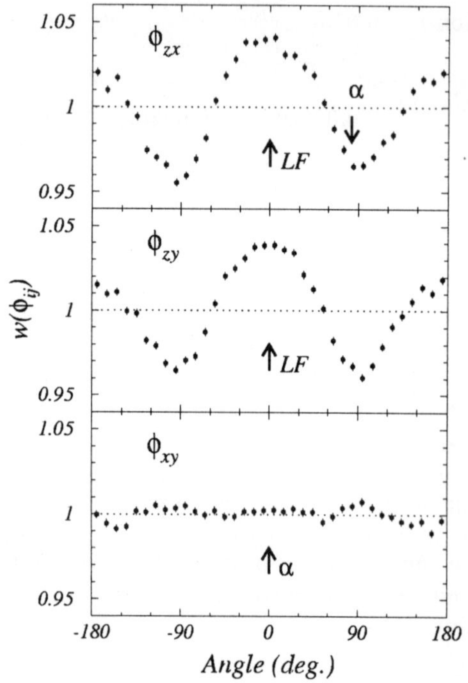

Figure 4: Projections of the measured γ-ray angular correlation in α-accompanied fission on different planes in the coordinate system xyz built-up by the momenta of the two fragments and the ternary particle (see text). The emission directions of the ternary α-particle and the light fission fragment are indicated by arrows.

For the case of α-accompanied fission and for γ-ray energies between 0.1 and 1.5 MeV, Fig. 4 displays the projections $w(\phi_{ij})$ of the LF-LCP-γ-angular correlation on the planes $(i,j) = (z,x), (z,y), (x,y)$, respectively, with ϕ_{ij} being the respective polar angle in the plane (i,j). The projections on the zx and

zy planes containing the fragment momenta (if we disregard the small recoil angle of 4°, on average, of the fragments caused by the emission of the α-particles) reflect the characteristic angular distribution of γ-rays peaked at the directions of the fragments. The projection onto the xy plane, i.e. perpendicular to the fission axis, reveals an isotropic distribution of γ-ray emission with respect to the α-particle direction ($\phi_{xy} = 0°$), within rather small experimental errors. As a consequence, there is no evidence for a significant correlation between the alignment of fragment spins and the direction of motion of ternary α-particles. Our result, thus, rules out an assumed pronounced correlation between the ternary particle emission and a collective bending motion at scission [12].

3.3 Gamma-ray angular correlations for other LCPs

Due to the high efficiency of our setup we were able to determine, for the first time, also the γ-ray angular correlations for ternary fission events accompanied by other LCPs such as ^3H, ^6He, Li and Be particles. The resulting spectra $W'(\theta)$ are summarized for all measured particles on the left-hand side in Fig. 5, while the right-hand side of the figure depicts the corresponding projections $w(\phi_{xy})$ on the xy plane perpendicular to the fission axis. The results deduced for the anisotropy ratio A and for the angular correlation parameters A_2 and A_4 are included in Table 1. The γ-ray angular distributions for ^3H and ^6He accompanied fission agree very well with that from binary fission. The emission of these particles shows similar to ^4He no effect on the γ-ray anisotropy. However, some deviations might be observed in coincidence with Li and Be particles, but the low counting statistics involved does not permit definite conclusions.

4 Summary and Conclusions

In summary, we have performed an experimental study with high statistical accuracy on the γ-ray angular correlations in the binary and ternary fission of ^{252}Cf using a homogeneous setup. No significant differences in the anisotropy of γ-ray emission relative to the fission axis were observed between binary fission and various ternary fission modes. This means that the emission of ternary particles at scission does not influence (or even destroy) the alignment of fragment spins, in contradiction to what has been concluded from previous work [10,11]. In ternary fission, the γ-ray angular distribution with respect to LCP motion, investigated for the first time, reveals no correlation between the directions of the γ-rays and the LCPs. The present study was thus able to clear up the experimental situation by demonstrating that the alignment of

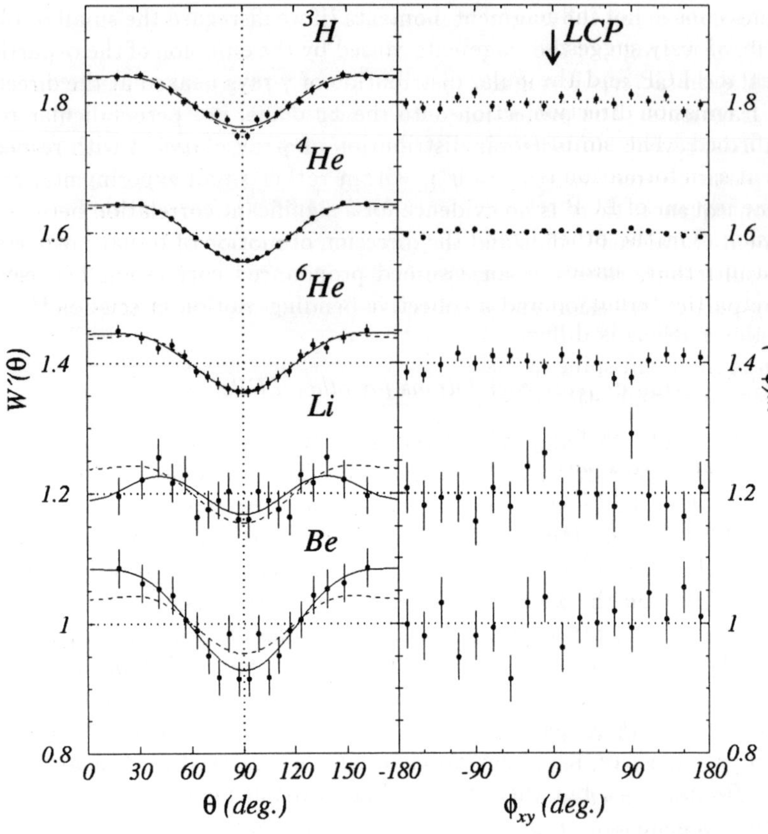

Figure 5: *Left side*: Angular correlations $W'(\theta)$ of γ-rays (0.1 - 1.5 MeV) in the center of mass system, for ternary fission events accompanied by ^3H, ^4He, ^6He, Li and Be particles. Solid lines are fitted curves using Eq. (1). The dashed line corresponds to the $W'(\theta)$ correlation for binary fission. *Right side*: Projections $w(\phi_{xy})$ of the γ-ray angular correlations on the xy plane perpendicular to the fission axis. The scales are correct for the lowest plots; others are shifted consecutively by 0.2 units.

fragment spins is not influenced by the ternary particle release at scission. Neither a previously suggested picture in terms of coupling of angular momentum between the LCP and the fragment spin [11] nor that of a correlation between the LCP emission direction and a collective bending motion at scission [12] could be confirmed. The simplest explanation for the present results could be that fragment spin formation is dominated by post-scission Coulomb excitation, i.e. after the instant of LCP release, rather than by a collective vibrational motion at scission. On the other hand, our present knowledge of the scission configuration in ternary fission is still incomplete, and the magnitude of a possible coupling of the ternary particle momentum and some collective motion of the scissioning nucleus is difficult to determine quantitatively. Evidently, the formation of fragment spins in fission is still a challenge to theory (see e.g. Ref. [6]), and even more sophisticated experimental studies might be needed for gaining further insight into the intimate details of the nuclear dynamics at scission.

This work forms part of the Ph.D. thesis of P. Singer (TU Darmstadt, 1997) and Yu. Kopach (JINR, Dubna, to be finished in 1999) and was supported in parts by the BMBF, Bonn (06DA461,06TU669). One of us (Yu.K.) would like to thank the DAAD, Bonn for a research grant. Helpful discussions with W.I. Furman, A.L. Barabanov, and E. Kankeleit are gratefully acknowledged.

References

1. J.R. Nix, and W. Swiatecki, Nucl.Phys.**71**, 1 (1965).
2. M. Zielinska-Pfabé, and K. Dietrich, Phys.Lett. **49B**, 123 (1974).
3. L.G. Moretto *et al.*, Nucl.Phys. **A502**, 453c (1989).
4. J.O. Rasmussen *et al.*, Nucl.Phys. **A136**, 465 (1969).
5. J.O. Rasmussen *et al.*, *Proc. Int. Conf. DANF96*, Častá Papiernička, Slovakia (1996), ed. J. Kliman, and B.I. Pustylnik, (JINR Dubna, 1996), p. 289.
6. J.O. Rasmussen, and R. Donangelo., *Proc. Int. Conf on Fission and Properties of Neutron-Rich Nuclei*, Sanibel Island, Fl., USA (1997), ed. J.H. Hamilton, and A.V. Ramayya, (World Scientific, Singapore, 1998), p. 150.
7. A. Wolf, and E. Cheifetz, Phys.Rev. **C13**, 1952 (1976).
8. K. Skarsvåg, Phys.Rev. **C22**, 638 (1980).
9. M. Mutterer and J.P. Theobald, in *Nuclear Decay Modes*, ed. D. Poenaru, (IOP Publ., Bristol, England, 1996), Chapt.12.
10. O.I. Ivanov, Sov.J.Nucl.Phys. **15**, 620 (1972).
11. W. Pilz, and W. Neubert, Z.Phys. **A338**, 75 (1991).
12. A.L. Barabanov, Phys.Atom.Nucl. **57**, 1157 (1994).

13. Yu.N. Kopach *et al.*, Phys.Rev.Lett. **82**, 303 (1999).
14. V. Metag *et al.*, Lecture Notes in Physics **178**, 163 (1983).
15. P. Singer *et al.*, Z.Phys. **A359**, 41 (1997).
16. M. Mutterer *et al.*, *Proc. Int. Conf on Fission and Properties of Neutron-Rich Nuclei*, Sanibel Island, Fl., USA (1997), ed. J.H. Hamilton, and A.V. Ramayya, (World Scientific, Singapore, 1998), p. 119.
17. M. Mutterer *et al.*, *Proc. Int. Conf. DANF98*, Častá Papiernička, Slovakia (1998), ed. J. Kliman *et al.*, (World Scientific, Singapore), in press.
18. A. Hotzel *et al.*, Z.Phys. **A356**, 299 (1996).
19. P. Singer *et al.*, *Proc. Int. Conf. DANF96*, Častá Papiernička, Slovakia (1996), ed. J. Kliman, and B.I. Pustylnik, (JINR Dubna, 1996), p. 262.
20. G.K. Metha *et al.*, Phys.Rev. **C7**, 371 (1973).
21. S.S. Kapoor *et al.*, Phys.Rev. **166**, 1190 (1968).
22. M. Mutterer *et al.*, *Proc. Int. Conf. DANF96*, Častá Papiernička, Slovakia (1996), ed. J. Kliman, and B.I. Pustylnik, (JINR Dubna, 1996), p. 250.

NEUTRON ENERGY DEPENDENCE OF FISSION FRAGMENT MASS & TKE DISTRIBUTIONS OF $^{235}U(n,f)$-REACTION BELOW 10 eV

Sh.S. Zeinalov[1], M.Florek[1], W.I.Furman[1], V. A. Kriachkov[2] and Yu.S.Zamyatnin[1]

1) JINR, 141980 Dubna, Russia.
2) IPPE, Obninsk, Russia

The first measurements [1] of the mass & TKE distributions of fission fragments (FF) from $^{235}U(n,f)$-reaction in the range of incident neutron energy from thermal up to few tens eV revealed small but important variations of fission mode weights in neutron resonances. Taking into account high density of ^{236}U compound states one may expect the interference between different states with the same spins in fission cross-section. The main goal of our investigation was to obtain experimental data necessary for multilevel, many fission mode analysis of resonance neutron induced fission of ^{235}U nuclear. It is clear that one needs enough large statistics for each neutron energy bin to reach this goal. This is very difficult task especially outside the strong neutron resonance areas. In present work the FF mass & TKE distributions were measured for neutron energy bins covering the incident energy range from thermal up to 10 eV using double Frisch-gridded ionization chamber and the intensive TOF facility IBR-30. Results of first analysis show a quality of the obtained data sufficient to accomplish partially the formulated above task. The R-matrix parameters of all neutron resonance inside the studied energy range necessary to carry out the analysis of our data are available from the investigation of FF angular anisotropy $A_2(E)$ measured in neutron induced fission of aligned ^{235}U target nuclei [2]. Preliminary analysis of our data shows rather complicated behavior of the TKE values and the fission mode contributions as function of neutron energy.

1. Experiment

The IBR-30 booster was used as pulsed neutron source with near 100 Hz pulse repetition frequency, 3.7 μs pulse width and 9.8 kW beam power. The experimental setup is schematically shown in fig.1. Neutron beam from reactor core was tailored by the collimator system to diameter of 5 cm on the target position. The target was mounted on the common cathode of the double ionization chamber with Frisch-grids at the flight distance of 14.54 m from the source. Between the exit of the tube and the target a lead collimator and cadmium filter was placed in order to reduce the thermal neutron flux at the target position. This gave us the possibility to carry out simultaneous measurements with approximately the same intensity for fission events induced by thermal and resonance neutrons if approximately the same width of different incident neutron energy bins was chosen. Distribution of fission events versus incident neutron energy is shown in fig. 2.

The construction and operation principles of the double ionization chamber for FF spectroscopy are similar to that described in [3]. For each fission event registered by the chamber the five parameters were recorded: two anode signal pulse heights proportional to FF kinetic energies, two grid signal pulse heights proportional to Cos(Θ) and the incident neutron time-of-flight code. Here Θ is the angle between FF velocity vector and the normal to the cathode plane. The code corresponding to time difference between reactor pulse and 'OR'ed anode signals was assigned to the incident neutron time-of-flight value. Data on Cos(Θ1) and Cos(Θ2) was used to correct FF pulse heights due to energy loses on the sample layer and backing material respectively. From corrected pulse heights of the signals the correlated kinetic energies of FF were obtained using a simple calibration procedure with two unknown parameters. These parameters were found using well known TKE and average light fragment kinetic energy values for thermal neutron induced fission of ^{235}U.

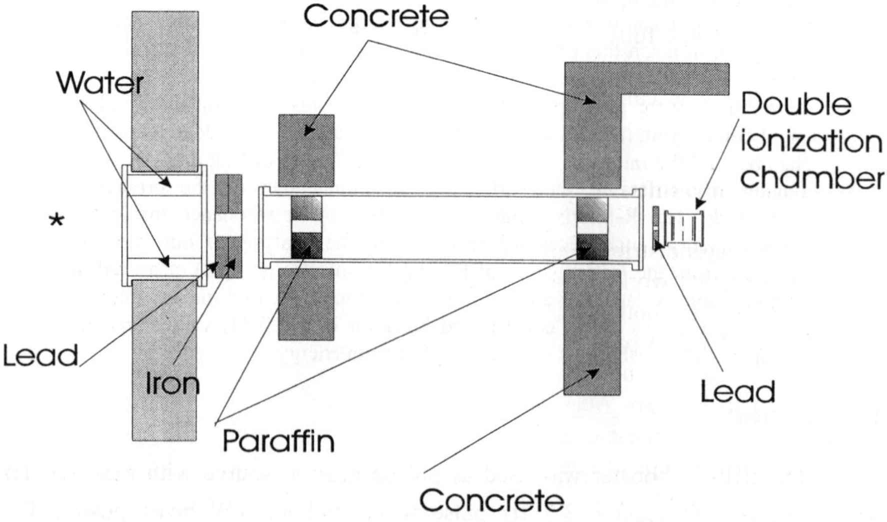

Figure 1. Schematic layout of experimental Setup.

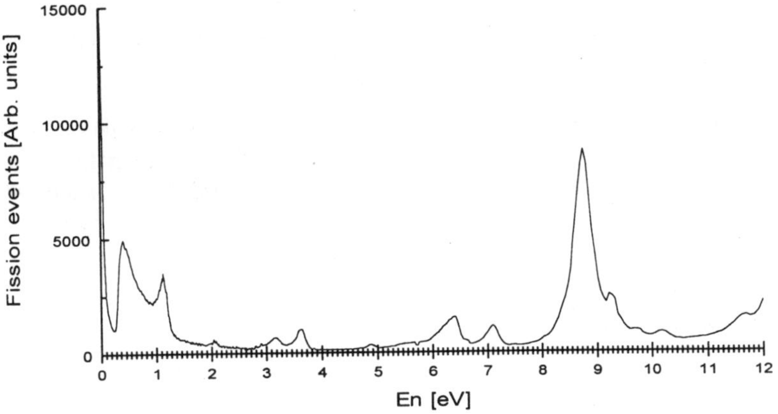

Figure 2. Measured distribution of fission events on incident neutron energy

The masses of correlated FF were obtained taking into account momentum and mass conservation laws as it usually is made in (2E) – experiments. The FF masses before neutron emission were calculated under the assumption that neutrons emit from fully accelerated fragments and the angular distribution of emitted neutrons is isotropic in the center-of-mass system. The number of neutrons emitted by FF was calculated using thermal neutron induced fission data from Apalin et al [6].

2. Discussion of results

The mass distribution of FF from thermal neutron induced fission of ^{235}U obtained in present work is shown in fig. 3a along with the data from [1]. A good agreement between both data (especially for heavy FF) is evident from Fig.3a. Following to the data presentation [1] the absolute mass yield difference between thermal and resonance neutron (averaged near En=8.77 eV) induced fission of ^{235}U is presented in Fig. 3b. A good agreement within experimental errors one could point out comparing these data with the same from [1]. A more detailed information on neutron energy dependence of FF mass yields in nearest vicinity of the resonance E_n=8.77 eV is represented in Fig. 4a-c. The each picture shows the difference between the mass yield at indicated neutron energy (inside 0.1 eV bin) and the same at E_n=8.55 eV. It should be noted that the mass dependence drawn in Figs. 4b and 4d are quite similar and have quantitative difference (in relative and absolute values) from the same presented in Figs. 4a and

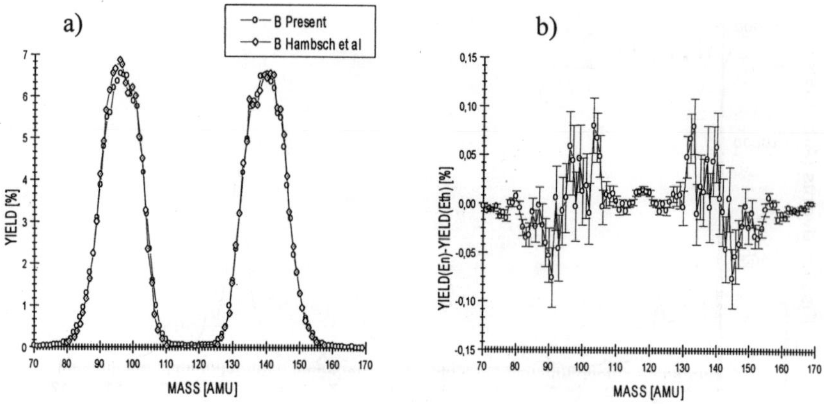

Figure 3. Mass yield for thermal neutron induced fission obtained in present work in comparison with the results [1] (a), and the absolute mass yield difference between resonance (En = 8.77 eV) and thermal neutron induced fission (b).

4c. The preliminary results on the similar mass and neutron energy dependence of the difference between FF mass yields for thermal and resonance neutron induced fission of ^{235}U target were presented in ref. [5]. These data were obtained without taking into account of the pulse height defect. But nevertheless overall qualitative agreement with the present data takes place.

The related data on neutron energy dependence of the mean TKE values near the resonance E_n=8.7 eV as well as TKE values for some low lying neutron resonances is shown in Fig. 5a. As it is clearly seen from comparison of Figs. 5a and 5b the energy dependence of the TKE does not coincide with the respective dependence of the spin separated fission cross-sections obtained in course of the data fit in [2]. This effect observed here for the first time needs more detailed further investigation and expanding it for wider neutron energy interval.

In previous studies of FF mass distribution variations for different neutron resonances the method of expansion by set of gaussians was used [1]. So whole mass distribution was composed of partial mass distributions corresponding to three fission modes predicted in ref. [6] for $^{235}U(n,f)$ process. Therefore in the present work the mass distributions of heavy FF group, for example, was fitted with following expression:

$$Y(M,E_n)=\sum_{j=1}^{3}\frac{W_i(E_n)}{\sqrt{2\pi\sigma_{Mi}^2}}\times\exp((M-M_i)^2/(2\times\sigma_{Mi}^2))$$

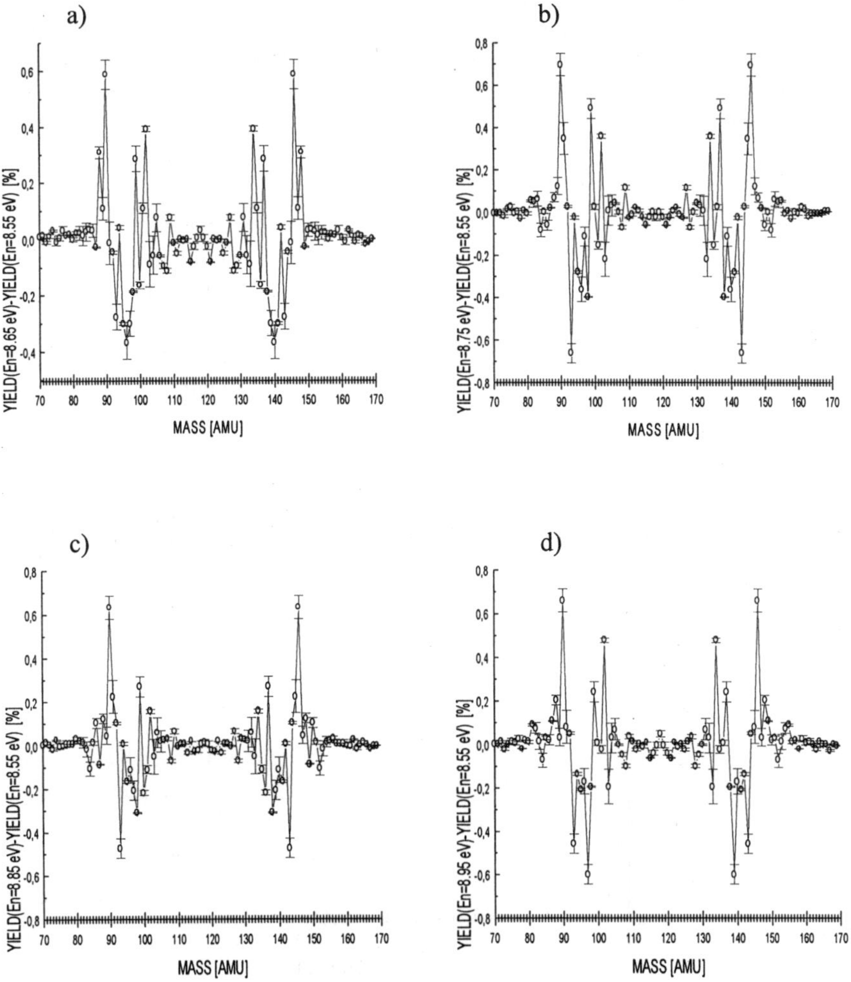

Figure 4. Absolute FF mass yield differences obtained at neutron energy bins of 0.1 eV near resonance at En = 8.77 eV in respect to the selected bin as indicated in each picture.

The W_i, M_i and σ_{Mi} are the weights of the gaussians, their mean masses and widths respectively. The weights of components was found as a result of the least squares fit of experimental mass distributions of $^{235}U(n_{th}, f)$ with the values of σ_{M1} and σ_{M2} taken from [1] and leaving free other parameters. The mean masses for two Brosa fission modes was found to be 133.7±0.1 for standard I and 140.7±0.1 for standard II respectively. These values are in a good agreement with theoretical calculations [7] and with the fitted values from [1].

Variations investigated in the work are very small (3-5%) in comparison with mean values of measured parameters. Time necessary to acquire valuable statistic using available resonance neutron beam is more than several hundred hours. In order to ensure

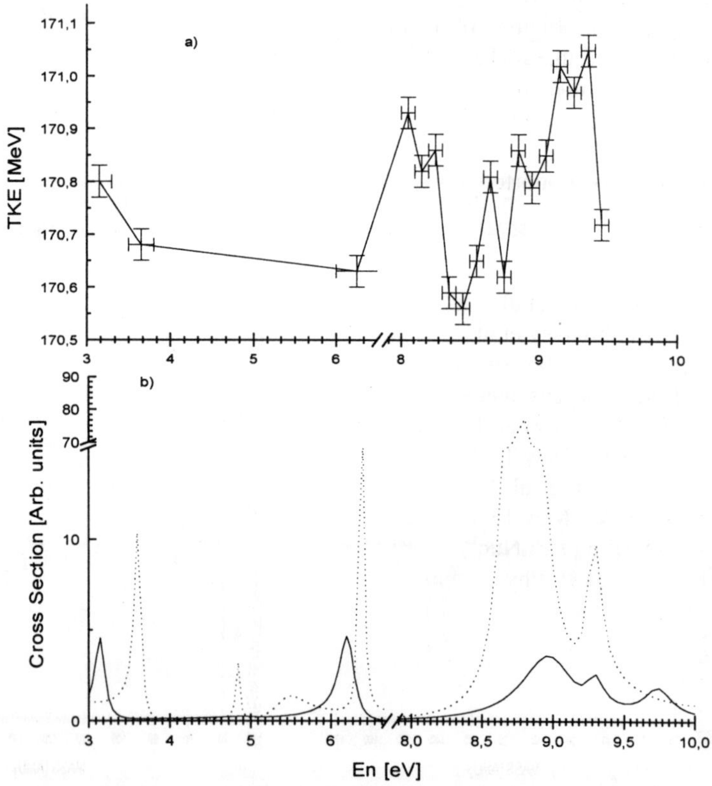

Figure 5. The dependence of TKE on incident neutron energy En (a), and (b) the calculated [5] spin separated cross-sections for $J^{\pi}=3^-$ (solid line) and for $J^{\pi}=4^-$ (dashed line)

acceptable level of stability during measurements collected data was tested. A grate share of collected in present experiment data doesn't pass the test and was rejected.

3. Conclusion.

In conclusion it is necessary to note that the new effects observed in the present paper are rather small. So its further investigation needs not only an essential increasing of statistics but also the crucial improvement of stability of operation of the detector and acquisition systems. For example the total statistics collected in present measurements with duration for about 1200 hours beam time was near 30 million events of resonance neutron induced fission. But more than 50% of collected data were rejected during off-line data selection and analysis procedure. In particular it was caused by instability of apparatus that was controlled using center of gravity position and FWHM of FF TKE peak. But if all experimental problems would be overcome the investigation of incident energy dependence of mass&TKE distributions of FF from resonance neutron induced fission allows one to clear out quantitatively the interconnection between fission modes and $J^{\pi}K$ fission channels in a realistic many level approach.

This work was supported in part by RFBR grant N 97-02-17590

References

1. F.-J. Hambsch et al, Nucl. Phys. A491 (1989) 56-90
2. D.I. Tambovtsev et al. International conference on Nuclear Data for Science and Technology, May 19-24, 1997, Trieste, Italy.
3 C. Budtz-Jorgensen et al, Nuclear Instr. And Meth. A258 (1987) 209-220
4 Sh. S. Zeinalov et al, ISIN-6 VI International Seminar on Interaction of Neutrons with Nuclei, May 13-16, 1998, Dubna, Russia, p. 274-278
5 W.I. Furman et al, ISIN-6 VI International Seminar on Interaction of Neutrons with Nuclei, May 13-16, 1998, Dubna, Russia, p. 260-273
6 V.F. Apalin et al. Nucl. Phys. 71 (1965) 553.
7 U. Brosa et al. Phys. Reports (Review Section of Phys. Lett.) 197, No. 4 (1990) 167-262.

THE TRANSPORT COEFFICIENTS FOR FUSION-FISSION REACTIONS

F.A. IVANYUK

Institute for Nuclear Research, Prospect Nauki 47, 252028 Kiev-28, Ukraine
E-mail: ivanyuk@kinr.kiev.ua

The collective friction and mass tensors are computed for the composite systems formed in the fusion of $^{18}O+^{208}Pb$, $^{20}Ne+^{249}Cf$ and $^{48}Ca+^{238}U$ (for the temperatures $0 \leq T \leq 2$ MeV) in the framework of the linear response theory and locally harmonic approximation. All components of friction and inertia tensors considered as functions of deformation show the peaked structure caused by avoided crossing of single-particle levels as well as by shell and pairing effects. The numerical values of friction and mass parameters differ considerably from the ones obtained with the wall formula for the friction and the Werner-Wheeler method for the collective mass.

1 Introduction

The synthesis of the superheavy elements is one of the most challenging problems of nuclear physics. Intensive experiments in this direction are carried out at JINR, Dubna and GSI, Darmstadt. A theoretical description of fusion-fission reaction is often based on the Langevin equation for collective variables Q_μ which parameterize the shape of the composite system. To solve the Langevin equation for $Q_\mu(t)$ one first of all needs to define the potential energy, friction $\gamma_{\mu\nu}$, mass $M_{\mu\nu}$ and diffusion $D_{\mu\nu}$ tensors.

The potential energy is commonly calculated within the microscopic-macro-scopic shell correction method [2] which describes the collective energy in the quasi-static picture very accurately. The diffusion tensor is often related to the friction tensor by the Einstein relation: $D_{\mu\nu} = T\gamma_{\mu\nu}$ and the tensors of friction $\gamma_{\mu\nu}$ and mass $M_{\mu\nu}$ are computed within the macroscopic approaches (the wall and window formula for friction [3] and the Werner-Wheeler (WW) method for inertia [4]). These methods provide rather simple expressions for the collective friction and mass coefficients. However the deformation and temperature dependence of macroscopic transport coefficients $\gamma_{\mu\nu}$ and $M_{\mu\nu}$ is not very reliable, at least at small excitation energies. For example, the observed [5] increase of the damping parameter $\eta = \gamma/2\sqrt{M|C|}$ with the excitation energy is difficult to explain either within the wall formula value for friction (where the friction parameter practically does not depend on the temperature T or excitation energy) or within the hydrodynamical model (friction parameter decreases as $1/T^2$). Also, the shell and pairing effects which are very important at low excitation energies are completely ignored in macroscopic models. Thus the necessity of having a microscopic theory for the collective transport coefficients is quite evident.

The possible solution of the problem can be based on an application of linear response theory (for review see [1]). The results of calculations presented below are obtained within an approach developed in [1] as well as in publications [6,7]. Because of limited space only few main quantities are defined below. For details, see [1, 6–8].

2 Collective dynamics in locally harmonic approximation

It is supposed that the nuclear many-body Hamiltonian can be approximated by [1]

$$\hat{H}(\hat{x}_i, \hat{p}_i, Q_\mu) = \hat{H}_{rmf}(\hat{x}_i, \hat{p}_i, Q_\mu) + \hat{V}^{(2)}_{res}(\hat{x}_i, \hat{p}_i), \tag{1}$$

where renormalized mean field Hamiltonian $\hat{H}_{rmf}$ depends explicitly on a few collective variables Q_μ and the residual two body interaction is assumed to be *independent* of Q_μ.

Considering the variation of Hamiltonian (1) due to the time dependence of $Q_\mu(t)$ as a small perturbation one can show that the slow collective motion in Q_μ can be described locally in terms of the so-called causal response function $\tilde{\chi}_{\mu\nu}(t-s)$

$$\tilde{\chi}_{\mu\nu}(t-s) = \Theta(t-s)\frac{i}{\hbar}\text{tr}\left(\hat{\rho}_{\text{qs}}(Q_0, T_0)[\hat{F}^I_\mu(t), \hat{F}^I_\nu(s)]\right) \equiv 2i\Theta(t-s)\chi''_{\mu\nu}(t-s). \tag{2}$$

Here, $\hat{F}^I_\mu(t)$ is the interaction representation of the derivatives of the Hamiltonian (1) with respect to Q_μ taken at an arbitrary fixed point Q^0_μ,

$$\hat{F}_\mu \equiv \left.\frac{\partial \hat{H}(\hat{x}_i, \hat{p}_i, Q_\mu)}{\partial Q_\mu}\right|_{Q_\mu = Q^0_\mu} \equiv \left.\frac{\partial \hat{H}_{rmf}(\hat{x}_i, \hat{p}_i, Q_\mu)}{\partial Q_\mu}\right|_{Q_\mu = Q^0_\mu}. \tag{3}$$

The $\hat{\rho}_{\text{qs}}$ represents the thermal equilibrium at Q^0_μ and is given by the canonical distribution $\rho_{\text{qs}}(Q^0_\mu, T) \propto \exp(-\hat{H}(Q^0_\mu)/T)$. To account for the pairing interaction the renormalized mean field Hamiltonian is chosen as that of independent quasiparticles (see [8]) $\hat{H}_{rmf} = U_0 + \sum_k E_k(\alpha^+_k \alpha_k + \alpha^+_{\bar{k}} \alpha_{\bar{k}})$, where U_0 is the ground state energy, E_k are the quasiparticle energies and α^+_k and α_k are creation and annihilation operators for quasiparticles.

The quantity of a special interest in the linear response theory is the Fourier transform $\chi_{\mu\nu}(\omega)$ of the response function (2). This can be found directly from (2)

$$\chi_{\mu\nu}(\omega) = \sum_{kj}{}' (n_k^T - n_j^T)\xi_{kj}^2 \frac{2(E_k - E_j)^2}{(\hbar\omega + i0)^2 - (E_k - E_j)^2} F_\mu^{kj} F_\nu^{jk}$$
$$+ \sum_{kj} (n_k^T + n_j^T - 1)\eta_{kj}^2 \frac{2(E_k + E_j)^2}{(\hbar\omega + i0)^2 - (E_k + E_j)^2} F_\mu^{kj} F_\nu^{jk}, \quad (4)$$

where $n_k^T = 1/(1+\exp(E_k/T))$, $\eta_{kj} \equiv u_k v_j + v_k u_j$, $\xi_{kj} \equiv u_k u_j - v_k v_j$ and u_k, v_k are coefficients of the Bogolyubov-Valatin transformation. The summation in (4) is over the single-particle states $|k\rangle$ and $|j\rangle$ and F_{kj}^μ are the matrix elements of $\hat{F}^\mu$ on these states.

To account for the residual interaction, $\hat{V}_{res}^{(2)}$, we first express the response function (4) in terms of the Green function $\mathcal{G}_{ks}^{(0)}(\omega \pm i0)$ for independent quasiparticles and then modify $\mathcal{G}_{ks}^{(0)}(\omega \pm i0)$ by introducing a spreading of the quasiparticle states. This is achieved by expressing the Green function in terms of the real and imaginary parts of the self-energy $\Sigma(\omega \pm i0, \Delta, T) = \Sigma'(\omega, \Delta, T) \mp i\Gamma(\omega, \Delta, T)/2$i,

$$\mathcal{G}_{ks}^{(0)}(\omega \pm i0) \to \mathcal{G}_{ks}(\omega \pm i0) = \frac{1}{\hbar\omega - E_{ks} - \Sigma'(\omega) \pm i\Gamma(\omega)/2}. \quad (5)$$

The calculation of Σ which includes the pairing is discussed in detail in [8].

The simplest expressions for the friction $\gamma_{\mu\nu}$ and mass $M_{\mu\nu}$ tensors within the linear response theory are offered by the so called zero frequency approximation,

$$\gamma_{\mu\nu}(0) = -i \frac{\partial \chi_{\mu\nu}(\omega)}{\partial \omega}\Big|_{\omega=0} \quad \text{and} \quad \mathrm{M}_{\mu\nu}(0) = \frac{1}{2}\frac{\partial^2 \chi_{\mu\nu}(\omega)}{\partial \omega^2}\Big|_{\omega=0} \quad (6)$$

which express the friction and mass tensors in terms of the derivatives of response function with respect to the frequency taken at $\omega = 0$.

3 Numerical results

In the present work the friction and mass tensors (6) are computed for the composite systems formed in the fusion reactions ^{18}O+^{208}Pb, ^{20}Ne+^{249}Cf and ^{48}Ca+^{238}U which were investigated recently at JINR, Dubna.

Like in [7], the independent particle Hamiltonian based on the deformed Woods-Saxon potential was used. The nuclear shape was parameterized in terms of Cassini ovaloids by two parameters ϵ and α_1 which describe the overall elongation and left-right asymmetry of the nucleus (see [9] and Fig. 1).

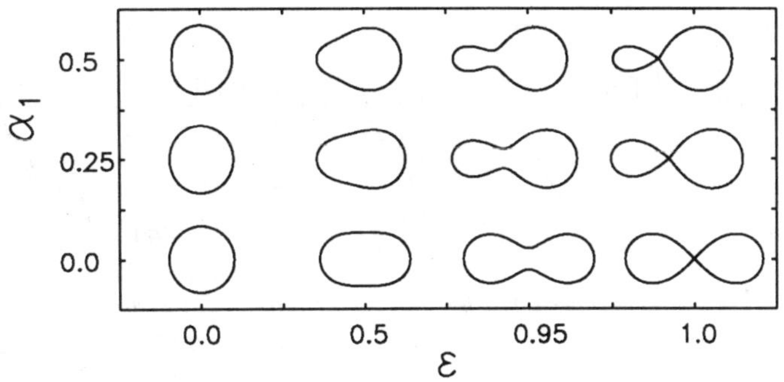

Figure 1. The two-dimensional family of shapes based on the parameterization of the nuclear surface in terms of Cassinian ovaloids (see [9]).

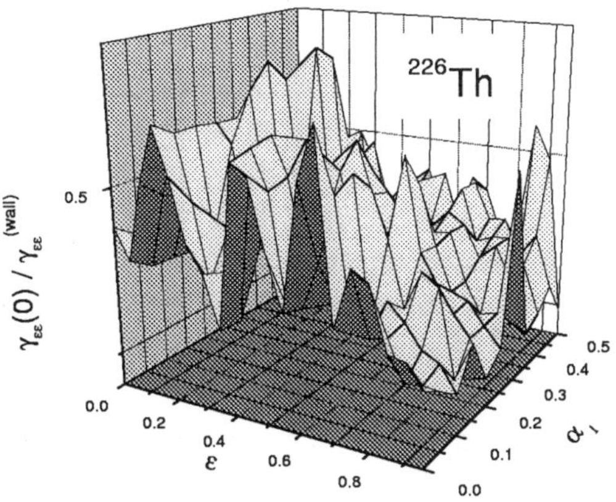

Figure 2. The $\epsilon\epsilon$-component of the friction tensor (6) (in units of wall formula value) for the ^{226}Th nucleus at the temperature $T = 1$ MeV.

The most interesting $\epsilon\epsilon$-component of the friction tensor (6), which corresponds to the elongation mode, is shown in Fig. 2. To remove the geometrical factors the friction coefficient $\gamma_{\epsilon\epsilon}$ was divided by the wall formula value [3]. First of all the rapid fluctuations against the deformation attract our attention. The analysis shows that the sharp peaks are mainly caused by the avoided crossings of single-particle levels. The fluctuations due to shell effects are much smoother.

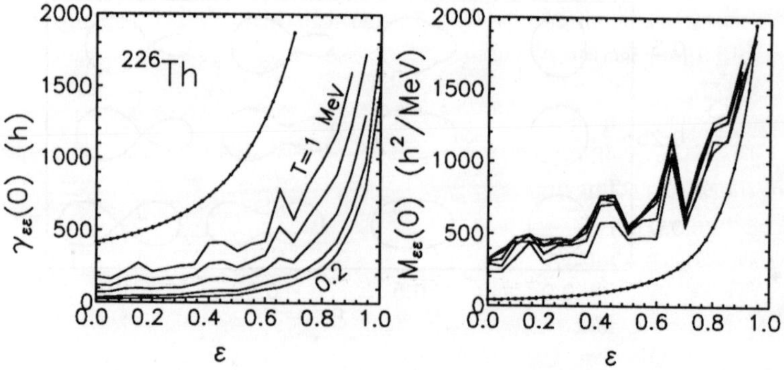

Figure 3. The $\epsilon\epsilon$-components of friction (left) and mass (right) tensors (6) for ^{226}Th nucleus versus the elongation parameter ϵ for temperatures $T = 0.2, 0.4, 0.6, 0.8, 1.0$ MeV. The lines with dots mark the wall formula friction value and the Werner-Wheeler mass parameter.

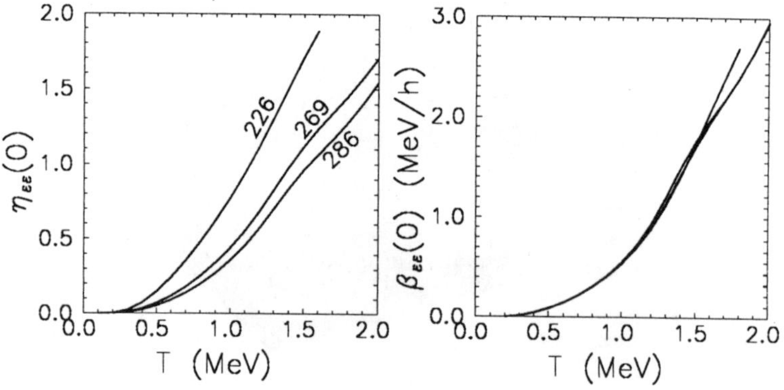

Figure 4. The damping factor $\eta_{\epsilon\epsilon} = \gamma_{\epsilon\epsilon}/2\sqrt{|C_{\epsilon\epsilon}|M_{\epsilon\epsilon}}$ (left) and the reduced friction coefficient $\beta_{\epsilon\epsilon} = \gamma_{\epsilon\epsilon}/M_{\epsilon\epsilon}$ (right) versus temperature. The numbers in Figure mark different mass numbers.

An example of the temperature dependence of the $\epsilon\epsilon$-component of friction and mass tensors (6) is shown in Fig. 3. It is seen that the friction parameter is a growing function of temperature. At $T = 1$ MeV $\gamma_{\epsilon\epsilon}(0)$ is smaller than the wall formula value still by a factor 2 approximately. As it was shown in [7] the wall formula value is reached at $T = 3 - 4$ MeV only.

The $\epsilon\epsilon$-component of the mass tensor is not very sensitive to the temperature below $T = 1$ MeV. At the same time $M_{\epsilon\epsilon}(0)$ is substantially larger than

$M_{\epsilon\epsilon}^{WW}$ computed within the WW method. The same relation between microscopic and macroscopic results for the friction and inertia holds for heavier systems, too.

Finally, Fig. 4 shows the damping factor $\eta_{\epsilon\epsilon} = \gamma_{\epsilon\epsilon}/2\sqrt{M_{\epsilon\epsilon}|C_{\epsilon\epsilon}|}$ and reduced friction coefficient $\beta_{\epsilon\epsilon} = \gamma_{\epsilon\epsilon}/M_{\epsilon\epsilon}$ at the saddle point as a function of the temperature. The damping factor reveals whether the collective motion is underdamped ($\eta \ll 1$) or overdamped ($\eta \gg 1$). As Fig. 4 shows, the collective motion changes from the underdamped to overdamped one at $T \approx 1 - 1.5$ MeV.

The damping factor can be deduced from the experimental data since it appears in the well known Kramers' formula of the decay rate: $\mathrm{R_K} = (\varpi_m/2\pi)\left(\sqrt{1+\eta^2} - \eta\right)\exp(-B/T)$. Here B is the fission barrier and ϖ_m determines the frequency of collective vibrations around the minimum of the potential.

Both η and β shown in Fig. 4 increase with the temperature. This behaviour is in qualitative agreement with the one found in [5] for ^{224}Th.

In conclusion, it is possible to affirm that the friction parameter is very small at low excitation energies. Both friction and mass tensors show complicated fluctuating structure caused by the avoided crossings of single-particle levels as well as by shell effects, and differ substantially from the ones obtained within macroscopic models in the $0.2 \leq T \leq 2$ MeV temperature range. Thus the results of dynamical computations based on the macroscopic friction and inertia parameters are not reliable at low excitation energies.

The derived transport coefficients will be used in the dynamical code for the description of fusion-fission reactions. The preliminary results of similar computations were reported in [10].

Acknowledgments

This work has been supported in part by the Deutsche Forschungsgemeinschaft. The author acknowledges greatly the use of the computer code by V.V. Pashkevich for the single-particle wave functions and matrix elements in the deformed Woods-Saxon potential.

References

1. H.Hofmann *Phys. Rep.* **284**, 1997 (137)
2. V.M.Strutinsky *Nucl. Phys. A* **95**, 1967 (420) *Nucl. Phys. A* **122**, 1968 (1)

3. J.Blocki, Y.Boneh, J.R.Nix et al. *Ann. Phys.* **113**, 1978 (330)
4. K.T.R.Davies, A.J.Sierk, J.R.Nix *Phys. Rev. C* **13**, 1976 (2385)
5. D.J.Hofman, B.B.Back, I.Diószegi et al *Phys. Rev. Lett.* **72**, 1994 (470); *see also* P.Paul, M.Thoennessen *Annu. Rev. Part. Nucl. Sci.* **44**, 1994 (65)
6. S.Yamaji, F.A.Ivanyuk, H. Hofmann *Nucl. Phys. A* **612**, 1997 (1)
7. F.A.Ivanyuk, H.Hofmann, V.V.Pashkevich, S.Yamaji *Phys.Rev. C* **55**, 1997 (1730)
8. F.A.Ivanyuk, H.Hofmann: *to be published*
9. V.V. Pashkevich *Nucl. Phys. A* **169**, 1971 (275)
10. F.A.Ivanyuk, G.I.Kosenko, Yu.Ts.Oganessian, V.V.Pashkevich: *Tours Symposium on Nuclear Physics III*, World Sci., Singapore 1997, p. 165

A DOUBLE-ARM TIME-OF-FLIGHT MICROCHANNEL PLATE SPECTROMETER CORSET FOR CORRELATION MEASUREMENT OF FISSION FRAGMENTS AND MEDIUM MASS HEAVY IONS

N.A.Kondratiev, E.M. Kozulin, I.V.Pokrovski, E.V.Prokhorova.

Joint Institute for Nuclear Research, 141980 Dubna, Moscow Region, Russia.

Studies of super heavy element fusion-fission processes by the use of multidetector correlation facilities for measurement of the neutron and gamma multiplicities dictate the need of creating a compact high-speed spectrometer for detecting fission fragments induced by heavy ions. For this purpose a double arm time-of- flight spectrometer CORSET was constructed using two wide-aperture microchannel plate (MCP) start detectors with an electrostatic mirror and two position sensitive module MCP stop detectors with an active area of 24 cm^2 for each module. The spectrometer was located in a thin wall spherical reaction chamber, 40 cm in diameter, with an additional container for radioactive targets inside. It was possible to replace the targets and detectors without deterioration of vacuum. DEMON neutron detectors and detectors with NaI(Tl) scintillators were used for registration of prompt neutron and gamma quanta emitted from compound nuclei and fission fragments. The spectrometer was tested and calibrated with fission fragments from a ^{252}Cf spontaneous fission source and with elastic scattering heavy ions during the experiments. A mass resolution of 3÷4 a.m.u. was obtained for the 10 cm time-of-flight path and M=48÷235. The position resolution was found to be 0.4 mm and it was possible, using folding angle information, to separate between fission fragments from spontaneous fission and fission fragments from ion induced reactions, and even those from some impurities in the target. The spectrometer had been previously used in the study of the ^{12}C, ^{18}O, ^{22}Ne, ^{24}Mg, ^{48}Ca and ^{64}Ni induced fission reactions during ~1000 hours without changing MCP in detectors.

1. Introduction

Investigation of fission fragment distributions, neutron and gamma-quanta emission in fusion-fission reactions induced by heavy ions using multidetector facilities is very useful for obtaining direct information on dynamics of atomic nuclei collective motion. Fission of compound nuclei at low excitation energies in the region Z~110 is very interesting in this aspect. A fission fragment spectrometer CORSET [1] using the time-of flight method (double velocity) was constructed for this kind of experiments. Our double-arm fission spectrometer measures two velocities and two-dimensional coordinates, allowing one to calculate mass, energy and folding angle of fragments. For such experiments, gas detectors with large areas are normally used [2,3] but in this case we used MCP detectors due to their best time resolution, small size and they are convenient in use.

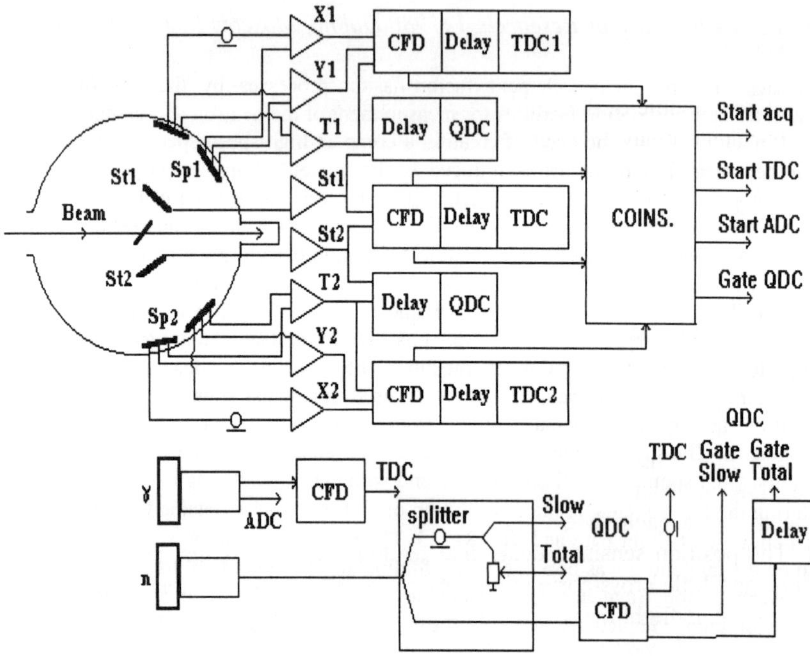

Figure 1 Schematic diagram of the double-arm fission spectrometer and electronics. The abbreviations are explained in the text.

2. Experimental arrangement.

2.1. Configuration

The correlation measurements were carried out using beams of the U-400 accelerator in Dubna, SARA in Grenoble, VIVITRON in Strasbourg and TANDEM in Catania. The experimental layout with electronic circuits is shown in Fig. 1. Each arm of the spectrometer consists of MCP start (St1,St2) detectors and two position sensitive MCP stop detectors (Sp1, Sp2). The start detector was positioned

4 cm away from the target, and the stop detectors were placed at a distance of 10 cm from the start detectors. The spectrometer was located in a thin-wall spherical reaction chamber 40 cm in diameter. There was a 44 mm diameter container for radioactive targets with thin windows for a beam and fission fragments. It is possible to simultaneously put up three different targets into the chamber. A glass window permits to check the position of the beam in the scintillator target. Replacement of the targets and change of the angle for each arm in the range ±30° are possible without deterioration of vacuum.

2.2. Detectors

Each module of the stop detector consists of thin entrance foil, two MCP's and a coordinate delay line readout [4,5]. The fission fragments and ions, whose position ant time are to be detected will pass through a 1 μm mylar foil with a 40 $\mu g/cm^2$ Au layer and the secondary electrons are accelerated toward the MCP with the size 43×63 mm. The number of electrons are amplified by two MCP's with an amplification of the order of 10^6 times of the incoming number of electrons, and transferred to a position readout device. The delay line readout consists of parallel-coupled wires in the x- and y-direction, one collecting and one non-collecting. Both have the length corresponding to 20 ns. The spacing between the 50 μm Cu-Be wires is 0.5 mm. We connected one end of the parallel par to the two inputs of a low noise differential amplifier and the other one to resistors whose impedance is equal to delay line impedance. Due to a differential amplifier, the final signal is not influenced by capacitive coupling, it corresponds only to the collected charge signal. The position sensitive delay line readout has high spatial resolution, small dead time and little cross-correlation between x- and y- coordinates. A fast "time" signal for time-of-flight measurement is extracted from the output side of the second MCP. Time and y-signals from two detector modules are connected with corresponding inputs of summed amplifiers directly and one of the x-signals after 25 ns delay. After this operation we can consider these two modules equivalent to one detector with correction to the x-coordinate.

The MCP start detectors with an electrostatic mirror are frequently used [6,8] in physical experiments. The detectors have been manufactured on the basis of the same 6×4 cm module-detector as used for the stop one and cover an angle of 70°×35°. An electrostatic mirror has been put on a horizontally situated MCP's. The mirror consists of a 1 μm mylar conversion foil with a 40 $\mu g/cm^2$. Au layer, an inner grid and a reflection grid made of 50 μm Cu-Be wire. The reflection grid and conversion foil were under a potential of about 1 kV with respect to the surface of the first MCP. The inner grid was under a zero potential. Due to the fast amplifier connected with the detector anode it is possible to decrees the MCP modal gain and to obtain a high counting rate of the detector.

2.3. Electronics and data acquisition system

A schematic diagram of the electrical connection is shown in Fig. 1. The outputs of spectrometer detector were connected to the fast amplifier placed inside the vacuum chamber. The outputs of all amplifiers were fed to constant fraction discriminators (CFD) in order to make measurements in the difference arrival time of the pulses from the start detectors, x- and y-delay line readout and "time" signal from the stop detector. We have further used a 100 ns cable delay line before the signals arrived in the time-to-digital converters (TDC, C414, CAEN). A rough selection of heavy ions and fission fragments can be carried out using charge signal from the detector time amplifier. A charge to digital converter (QDC) CAMAC module (2249SG, LeCroy) was used for this purpose. The same module was used for pulse shape discrimination of prompt neutrons from gamma-rays in neutron detectors with a liquid scintillator (NE213). A peak sensing analog digital converter (ADC, C420, CAEN) was utilized for gamma quanta energy determination in gamma detectors with NaI(Tl) scintillator. A start signal for data acquisition system was present if we had coincidence from two stop and one of the start detectors. The ADC, QDC and TDC's were all in a CAMAC crate and were controlled by a IBM PC 486 placed near the spectrometer. The second IBM computer for data collection was connected with the first by local network. The data acquisition system can record up to 64 parameters per events with counting rate of 10^3 per second.

3. Results

Prior to experiments the spectrometer was tested and calibrated with a ^{252}Cf source. The obtained total kinetic energy (TKE) and mass yield for heavy fission fragments is shown in Fig. 2. They are compared with the data from the work by Schmitt [7], obtained using silicon detectors (2E-method) with and without correction for neutron emission. A rather wide mass distribution of fission fragment in the present work is mainly due to the thickness of the source. The ratio between the peak and the valley in this distribution of ~20 can be taken as a measure of thespectrometer mass resolution and we estimate it as 3÷5 a.m.u.

A system for correlation measurements was applied to the ^{12}C, ^{18}O, ^{22}Ne, ^{24}Mg, ^{48}Ca and ^{64}Ni ions induced experiments. A two dimensional plot of the fragment mass versus total kinetic energy of both fragments for the 230 MeV ^{48}Ca beam is displayed in Fig. 3. The

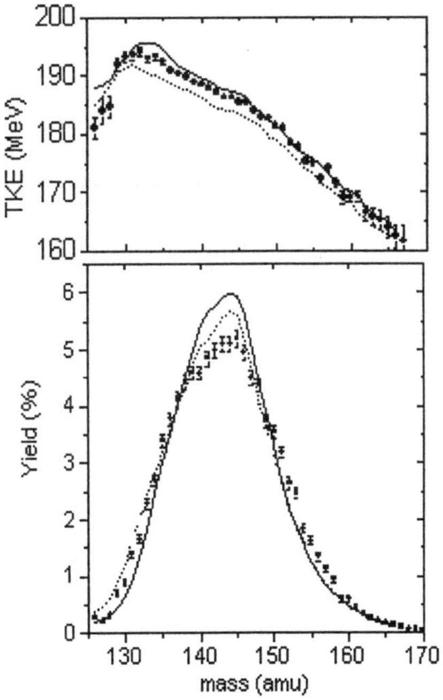

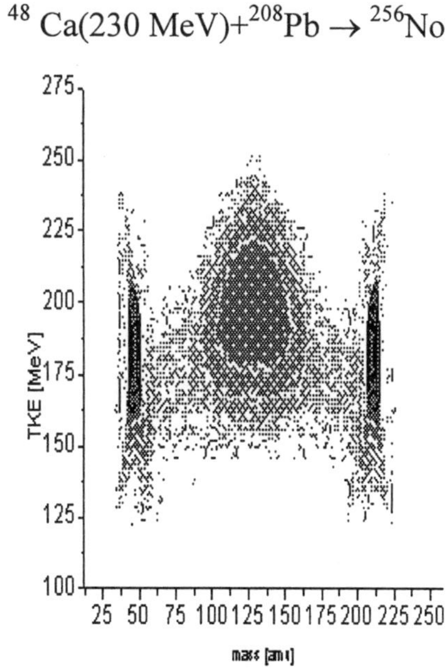

Figure 2 Total kinetic energy and mass yield of the ^{252}Cf(s,f) fission fragments calculated by us (points) are compared with those by Schmitt [7] with taking into account correction for neutrons (line)

Fig.ure 3 TKE vs. reaction product mass two-dimensional plot for ^{208}Pb(^{48}Ca,f) reaction at E_{lab}=230MeV

^{208}Pb nuclei from the target and elastically scattered ^{48}Ca ions were used for mass and energy calibration of the spectrometer and for obtaining its real mass resolution. The mass distribution of these ions with separation on a small surface area of the detectors was presented in Fig. 4. The obtained mass resolution FWHM~4 is in good agreement with our estimation for ^{252}Cf measurements.

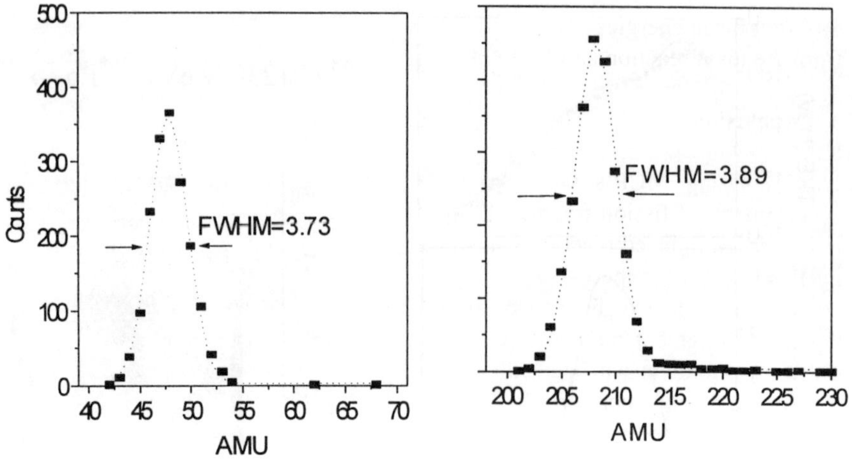

Figure 4. The mass spectrum of elastic scattering ^{48}Ca and ^{208}Pb ions from a 200 μg/cm^2 target with separation on a small surface area of the detectors.

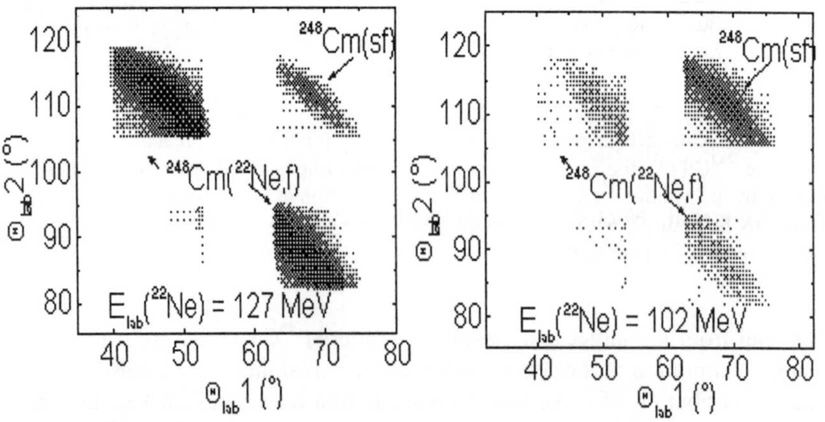

Figure 5. Two-dimensional plots of fission fragments in the lab angles ($\theta_{lab}1$, $\theta_{lab}2$) for the reaction $^{248}Cm(^{22}Ne,f)$ and spontaneous fission from the target.

Coordinate resolution for the 226Ra alpha-particles was found in work [1] as 0.4 mm. This is more than enough for separation of the reaction products in experiment using a folding angle. Fig. 5 shows two different kinds of events measured during the same time, one is spontaneous fission fragments from the ^{248}Cm target and the other - fission fragments from the $^{248}Cm(^{22}Ne,f)$ reaction at two different beam energies . Good spatial resolution allows us to separate fission events from the main reaction and fission fragments from some impurities in the target.

4. Conclusion

Using MCP's we have built a very small and simple time-of-flight spectrometer of fission fragments for neutron and gamma correlation measurements in heavy ion induced reactions. Mass resolution of 4 a.m.u. is quite enough because in this kind of experiment the used thickness of the target has comparable contribution. The excellent spatial resolution allows a more precise determination of the linear fragment moments and separation of different kinds of fission. Charge pulse measurement with detectors allows to select false events with big ion energy loss and to receive more precise data.

This work was supported in part by INTAS Foundation, grant No. 97-1929.

References

1. E.M.Kozulin et al., Heavy Ion Physics JINR Dubna 1997 p.215
2. M.G.Itkis et al., Heavy Ion Physics. JINR Dubna 1997 p.179
3. The 4-fragment spectrometer FOBS. JINR Preprint E13-97-118 Dubna 1997.
4. S.E.Soboltka et al., IEEEE Transaction on Nucleus Science, v.35 No.1 1998, p.384
5. O.H.Odland et al., Nucl. Instr. and Meth., v.A378, 1996, p.149
6. W.Starzecki et al., Nucl. Instr. and Meth. v.193, 1982, p.499
7. H.W.Schmitt et al., Physics review. v.141, 1966, p.1146.
8. Y.S. Kim, et al., Nucl. Instr. and Meth., v.A329, 1993, p.403

EFFICIENT ALGORITHMS OF MULTIPARAMETER γ-RAY SPECTRA PROCESSING

M. MORHÁČ, J. KLIMAN, V. MATOUŠEK, I. TURZO, M. VESELSKÝ
Institute of Physics, Slovak Academy of Sciences
Dúbravská cesta 9, 842 28 Bratislava, Slovak Republic
E-mail: morhac@savba.sk

In the paper, the efficient algorithms to process multiparameter γ-ray spectra are presented. They allow to search for peaks, to separate useful information from background and to improve resolution in 1-, 2- and 3-parameter γ-ray spectra. The algorithms are generalized to n-parameter data.

1 Introduction

Experimental progress in the understanding of nuclear structure depends critically on the ability to analyze coincidence data in multidimensional space. The requirements for using higher dimensions and higher resolutions are continuously increasing. The classical techniques developed to analyze one-parameter spectra cannot be directly extended to higher-fold data. Some operations are extremely time consuming. This problem is becoming relevant for big sizes and for multidimensional spectra, where the number of operations grows exponentially with the sizes. The implementations of some methods, mainly on PC computers, without optimization would be impossible. In the paper we present algorithm of multidimensional peak searching, multidimensional background elimination and deconvolution.

2 Peak searching

Without being interested in details we outline the algorithm of peak searching for 2-parameter spectra. We assume that the peak can be described by Gauss function and that the background can be approximated by a linear function within short interval. Then the number of counts as a function of the channel position x, y is

$$N(x,y) = G_1(x,y) + G_2(x) + G_3(y) + B + Cx + Dy \tag{1}$$

where $G_1(x,y)$ represents a 2-dimensional Gaussian, $G_2(x)$ and $G_3(y)$ represent 1-dimensional Gaussian ridges and $B+Cx+Dy$ is linear background. The second partial derivative of 1 in both variables is independent of $G_2(x), G_3(y)$ and background. It implies that the method is not sensitive to either of

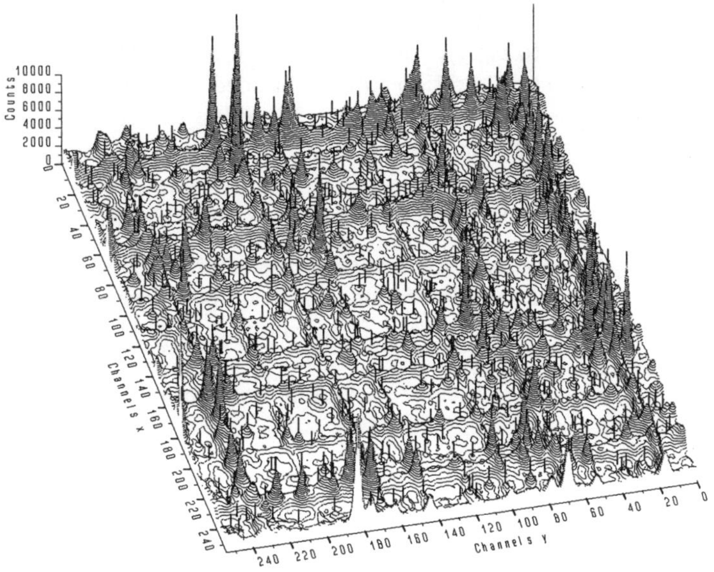

Figure 1. Example of 2-parameter spectrum with peaks found.

both ridges or their sum. Then maxima in the second derivative of a spectrum correspond to positions of a 2-dimensional peaks. The slices in the maxima in both directions are compared to their standard deviations. Applying conditions for the existence and the shape of peaks [1] the positions of 2-dimensional Gaussians can be determined. The method was extended also to 3-dimensional spectra. An example of Gaussians found in a 2-parameter spectrum is presented in Fig. 1. Positions of found peaks can be used as initial estimates to fit 2-parameter peaks in spectrum. An alternative way is to eliminate background from spectrum and then improve resolution sufficiently enough to obtain separated peaks.

3 Background estimation algorithm

The aim of algorithm is to separate useless information (background, combinations of coincidences of the background with peaks - ridges) from useful information contained in n-fold coincidence peaks of n-parameter spectrum.

Let us suppose that we estimate baseline spectrum of one-parameter spectrum stored in the vector $v(i)$. From $v(i)$ we shall calculate step by step the vectors $v_1(i), v_2(i)$ up to $v_m(i)$, where m is a free parameter. The new value in the channel i in the pth iteration step is obtained by comparison of average of the values $v_{p-1}(i-p), v_{p-1}(i+p)$ with the value $v_{p-1}(i)$. We accept the minimum of these values, i.e.,

$$v_p(i) = \min\left\{ v_{p-1}(i),\ \frac{1}{2}\left[v_{p-1}(i+p) + v_{p-1}(i-p)\right]\right\}. \tag{2}$$

Analogously to one dimensional background estimation one can proceed in derivation of an algorithm to determine background of 2-parameter spectrum. From the 2-parameter spectrum $v(i,j)$ we shall successively create matrices $v_1(i,j), v_2(i,j)$ up to $v_m(i,j)$.

The matrix elements that enter background estimation in the pth iteration step are denoted

$$\begin{aligned} P_1 &= v_{p-1}(i+p, j+p); & P_2 &= v_{p-1}(i-p, j+p), \\ P_3 &= v_{p-1}(i+p, j-p); & P_4 &= v_{p-1}(i-p, j-p), \\ S_1 &= v_{p-1}(i+p, j); & S_2 &= v_{p-1}(i, j+p), \\ S_3 &= v_{p-1}(i, j-p); & S_4 &= v_{p-1}(i-p, j). \end{aligned} \tag{3}$$

Our aim is that the peak ridges in both directions (coincidences of peaks with linear background) as well as their sum were incorporated into the background estimate. Therefore we compare the values in the elements S with the average value of neighbouring elements P. Then we replace the new values S by

$$\begin{aligned} S_1 &= \max\{S_1,\ (P_1+P_3)/2\} - (P_1+P_3)/2; \\ S_2 &= \max\{S_2,\ (P_1+P_2)/2\} - (P_1+P_2)/2; \\ S_3 &= \max\{S_3,\ (P_3+P_4)/2\} - (P_3+P_4)/2; \\ S_4 &= \max\{S_4,\ (P_2+P_4)/2\} - (P_2+P_4)/2. \end{aligned} \tag{4}$$

Then the new value in the matrix element i, j in the pth iteration step is

$$v_p(i,j) = \min\{v_{p-1}(i,j),\ (S_1+S_4)/2 + (S_2+S_3)/2 + (P_1+P_2+P_3+P_4)/4\}. \tag{5}$$

Background elimination algorithm can be generalized to three and multidimensional data [2]. An example of 2-parameter spectrum before and after background elimination is given in Fig. 2.

4 Deconvolution

The Gold method of deconvolution [3] [4] proved to be the most efficient for nuclear spectra deconvolution. In this section we outline the algorithm of Gold

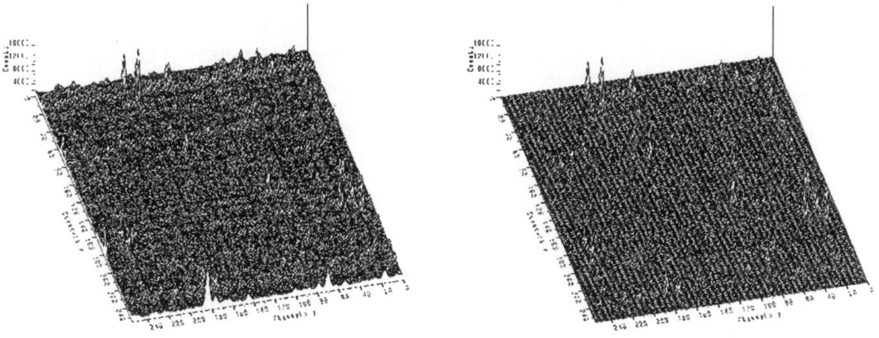

Figure 2. Two-parameter spectrum before (a) and after (b) background suppression.

deconvolution. For a discrete system the relationship between a measured value $x(i)$ and the raw result of measurement $y(i)$ can be expressed as

$$y(i) = \sum_{k=o}^{N-1} h(i-k)x(k) \qquad i = 0, 1, \ldots, N-1 \tag{6}$$

where $h(i)$ is an impulse response and N is the number of samples of vectors h, x. The knowledge of the instrumental function $h(i)$ is required. Eq. 6 can be written in matrix form as

$$\bar{y} = H\bar{x} \tag{7}$$

where columns of H are represented by the vectors h shifted by one position. Multiplying both sides of 7 by H^T yields

$$H^T\bar{y} = H^T H\bar{x} \tag{8}$$

wherefrom one gets

$$\bar{y_1} = H_1\bar{x}. \tag{9}$$

The direct inversion of H_1 cannot lead to a stable solution. Therefore the method of regularization must be employed.

The basic form of iterative method of deconvolution for a general linear discrete system is

$$\bar{x}^{k+1} = \bar{x}^k + \mu(\bar{y} - A \cdot \bar{x}^k) \tag{10}$$

where A is system matrix, k represents the number of iterations and μ is the relaxation factor. In our case the matrix $H^T H$ in 8 is symmetric, so

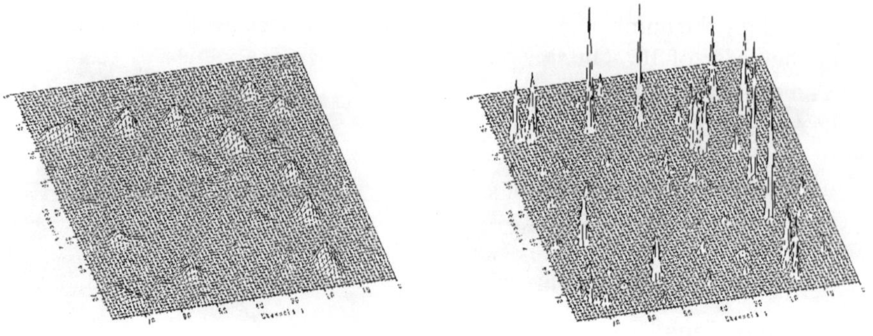

Figure 3. Two-parameter spectrum before (a) and after (b) deconvolution.

its eigenvalues are real. The eigenvalues of matrix $(H^T H) \cdot (H^T H)$ are the squares of eigenvalues of matrix $H^T H$ and therefore must be positive. The real positive eigenvalues of the system matrix is basic condition of stable solution of 9. Then 10 becomes

$$\bar{x}^{k+1} = \bar{x}^k + \mu[H^T H H^T \bar{y} - H^T H H^T H \bar{x}^k] \tag{11}$$

wherefrom one gets

$$\bar{x}^{k+1} = \bar{x}^k + \mu[\bar{y}' - H' \cdot \bar{x}^k] \tag{12}$$

By choosing relaxation factor μ according to [3] we obtain the Gold algorithm of deconvolution

$$x_i^{k+1} = \frac{y_i'}{\sum_{m=0}^{N-1} H_{im}' \cdot x_m^k} \cdot x_i^k \tag{13}$$

The advantage of Gold method in nuclear spectra deconvolution is the positivity of solution $\bar{x}$. The method can be generalized to 2-parameter spectra [4]. The example of 2-parameter spectrum before and after the deconvolution using the Gold method is given in Fig. 3.

The direct extension of the computational algorithm of one dimensional non-optimized Gold deconvolution to two-, three- and multidimensional system leads to exponential growth of numerical operations. Given a rather slow convergence speed of iterative Gold method one needs considerable number of iteration steps ($\approx$ 1000). Therefore in practical cases the method without optimization is not directly applicable in multidimensional deconvolution.Optimization can be achieved according to two facts. Fist, in γ-ray spectroscopy the number of channels of impulse response with non-zero counts is of the order of 10. Outside this interval the counts vanish to zeros. This can be exploited in matrix vector multiplication in the denominator of 13. It is useless to multiply with zeros. Second, in 13 one can observe that if for an i, y_i or x_i^k equals zero then for all following iteration steps $j > k$ x_i^j will remain zero. It means that for such a point one can stop corrections in the iteration step k. Both these facts allow to decrease dramatically computing time of 13.

5 Conclusions

All the algorithms described so far were implemented in "Integrated Multiparameter Nuclear Data Analysis Package" [5]. It includes standard conventional procedures and routines as well as new developed algorithms. The package is a tool that integrates multiparameter data taking, processing and display. It was used to process data from multidetector system Compact Pocket Ball (ORNL Oak Ridge) and Gammasphere (LBL) Berkeley with the aim to obtain information on high excited states of neutron-rich nuclei from spontaneous fission. It is also utilized in the field of relativistic nuclear physics in experiments on NUCLOTRON accelerator (JINR Dubna) in the frame of programme SPHERE.

Acknowledgments

The work was partially supported by the Grant Agency of Slovak Republic through contract GAV 2/5121/98. The authors would like to thank the GAMMASPHERE collaboration, J.H. Hamilton, J.O. Rasmussen, G. M. Ter-Akopian and A. V. Daniel for the help in data preparation and valuable discussions.

References

1. M.A. Mariscotti *Nucl. Instr. Meth. Phys. Res. A* **50**, 309 (1967)
2. M. Morháč, J. Kliman, V. Matoušek, M. Veselský, I. Turzo *Nucl. Instr. Meth. Phys. Res. A* **401**, 113 (1997)
3. R. Gold: *An iterative unfolding method for matrices. Technical Report ANL-6984.* Argonne National Laboratory, Argonne, Illinois 1964
4. M. Morháč, J. Kliman, V. Matoušek, M. Veselský, I. Turzo *Nucl. Instr. Meth. Phys. Res. A* **401**, 385 (1997)

5. M. Morháč, J. Kliman, V. Matoušek, I. Turzo *Nucl. Instr. Meth. Phys. Res. A* **389**, 89 (1997)

COMPARATIVE ANALYSIS OF THE SYNTHESIS OF Z=114 ELEMENT IN THE COLD AND WARM FUSION REACTIONS

G.G.ADAMIAN[1,2,3], N.V.ANTONENKO[1,2], E.A.CHEREPANOV[1], A.K.NASIROV[1,3], W.SCHEID[2] AND V.V.VOLKOV[1]

[1] *Joint Institute for Nuclear Research, 141980 Dubna, Russia*
[2] *Institut für Theoretische Physik der Justus-Liebig-Universität, D-35392 Giessen, Germany*
[3] *Institute of Nuclear Physics, 702132 Tashkent, Uzbekistan*

Within the dinuclear system concept, the preference of the reaction ^{48}Ca+^{244}Pu is shown in comparison with the cold fusion reaction for the synthesis of Z=114 element.

1 Introduction

In order to reach the superheavy $Z = 114$ element, the experimentalists plan to use both Pb-based (GSI, Darmstadt) and actinide-based (FLNR, Dubna) fusion reactions [1,2]. With last warm fusion reactions one may obtain the compound nucleus close to the predicted center of stability island. The theoretical investigations are directing to reveal the optimal experimental conditions for the formation of the superheavy elements. However, the experimental results on fusion of heavy nuclei can not be explained satisfactory in the known macroscopic models [3]. In these models the fusion cross sections are overestimated by few orders of magnitude and the isotopic trends [3,4] are found to be opposite to the experimental ones. In the synthesis of superheavy nuclei with cold fusion reactions [2] the optimal excitation energy of a compound nucleus is much smaller than the energy predicted in the macroscopic models. These facts demonstrate an importance of the correct choice of relevant collective coordinates which are mainly responsible for the complete fusion. The effects of nuclear structure in the fusion process are quite strong to change the found macroscopic trends.

In the macroscopical models the fusion is mainly considered as the motion in relative distance between the centers of nuclei R (or the elongation of system). In opposite to these models in the dinuclear system (DNS) concept [6,7,8,9] the complete fusion occurs through the mass asymmetry coordinate $\eta = (A_1 - A_2)/A$ (A_1 and A_2 are the mass numbers of the DNS nuclei and $A = A_1 + A_2$). Starting with the basic idea that DNS configurations do not dissolve in the R coordinate (which is explained by the effect of structural forbiddenness [5]), the DNS fusion concept gives impressive agreement with the experiment [7,8,9]. In the present paper we analysed with the DNS concept the cold ^{74,76}Ge+^{208}Pb $\rightarrow^{281,283}$114+1n and warm ^{48}Ca+^{244}Pu$\rightarrow^{292-xn}$114+xn fusion reactions using for the synthesis of the element $Z = 114$. For the warm fusion reactions, the fusion probability is larger than in Pb-based reactions due to smaller losses (quasi-fission) in the entrance channel. However, the large fusion probability can be compensated by decreasing survival probability of the compound nucleus due to higher excitation energy which corresponds to the 3n-4n channels. Therefore, without detail calculations all the factors determining the evaporation residue cross section it is difficult to say which reaction (cold or warm) is preferable.

2 DNS concept of fusion

In accordance with the DNS approach the evaporation residue cross section is written as [7,8,9]

$$\sigma_{ER}(E_{\rm c.m.}) = \sum_{J=0}^{J_{cr}} \sigma_c(E_{\rm c.m.}, J) P_{CN}(E_{\rm c.m.}, J) W_{sur}(E_{\rm c.m.} + Q, J). \quad (1)$$

The capture cross section $\sigma_c(E_{\rm c.m.}, J)$ defines the transition of the colliding nuclei over the Coulomb barrier and the formation of the DNS when the kinetic energy $E_{\rm c.m.}$ and angular momentum J of the relative motion are transformed into the excitation energy and angular momentum of the DNS. The good agreement between the our calculations (using optical model [11]) and experimental data for $\sigma_c(E_{\rm c.m.})$ was obtained for reactions on the ^{208}Pb and ^{238}U targets. The surviving probability $W_{sur}(E^*, J)$ of the formed compound nucleus with excitation energy $E^* = E_{\rm c.m.} + Q$ (Q is Q-value of the reaction) was considered with the statistical approach [11]. For the xn evaporation channel W_{sur} is expressed as

$$W_{sur}(E^*, J) = P_{xn}(E^*, J) \prod_{i=1}^{x} \left[\frac{\Gamma_n(E^*, J)}{\Gamma_n(E^*, J) + \Gamma_f(E^*, J)} \right]_i, \quad (2)$$

where P_{xn} is the probability of evaporation of x neutrons and $[\Gamma_n/(\Gamma_n+\Gamma_f)]_i$ is the probability of the neutron evaporation in competition with fission channel at the i step of the de-excitation cascade. For the partial widths of neutron emission and fission the following expressions are used:

$$\Gamma_n(E_m^*, J) = \frac{2\mu_n}{(\pi\hbar)^2 \rho_m(E_m^*)} \int_0^{E_m^* - B_n} \sigma_{inv}(E_n)\rho_d(E_m^* - B_n - E_n)E_n dE_n, \quad (3)$$

$$\Gamma_f(E^*, J) = \frac{1}{2\pi \rho_m(E_m^*)} \int_0^{E_m^{s*} - B_n} \rho_m(E_m^{s*} - B_f - \varepsilon)d\varepsilon \quad (4)$$

Here E_m^* (E_m^{s*}), μ_n and σ_{inv} are the thermal energy of the mother nucleus in ground–state (saddle–point configuration), the reduced mass of the system (neutron plus daughter nucleus) and the inverse cross section [11], respectively. The ρ_m (ρ_d) is Fermi-gas expression for the level density of mother (daughter) nucleus. The shell effects are damped exponentially $f = \exp(-\gamma E^*)$ with γ=0.061 in the expressions for the fission barrier $B_f = fB_f(E_{CN}^* = 0)$ and for the parameter of level density $a(E^*) = A(0.134 - 1.21 \times 10^{-4}A)[1 + (1-f)\Delta W/E^*]$ (ΔW is the shell correction in nuclear mass formula) [11]. For calculations, we used the value of $B_f(E^* = 0) = B_f^{stat}$ from Ref. [12]. The probability of complete fusion $P_{CN}(E_{\text{c.m.}}, J)$ depends on the degrees of freedom which are responsible for the complete fusion and quasi-fission processes after the capture stage. The fusion channel in the DNS model is the evolution of the DNS to compound nucleus by nucleon transfer from a light nucleus to a heavy one (diffusion in mass asymmetry η degree of freedom). The quasi-fission (decay of the DNS) channel is considered as motion to larger R and strongly reduces the fusion cross section in the reactions with heavy nuclei. To find the complete fusion probability [7,8]

$$P_{CN} = \frac{\lambda_\eta^{Kr}}{\lambda_R^{Kr} + \lambda_\eta^{Kr}} - \frac{\lambda_\eta^{Kr}\lambda_R^{Kr}}{\lambda_R^{Kr} + \lambda_\eta^{Kr}} \frac{\tau_\eta - \tau_R}{1.72}, \quad (5)$$

we can use a two-dimensional the Kramers-type expression (quasi-stationary solution of the Fokker-Planck equation) for the rates of fusion λ_η^{Kr} and quasi-fission λ_R^{Kr} through the fusion (B_{fus}^*) in the η and quasi-fission (B_{qf}) in the R barriers. The first term in (5) is the contributions from the quasi-stationary rates. The second term is related to the transient times τ_R and τ_η. We can neglect this term for $\tau_k \ll 1/\lambda_k^{Kr}$ ($k = R, \eta$) ($Z_1 Z_2 < 2500$) [7].

3 Potential energy

The bulk of the fusion characteristics (for example, B_{fus}^* and B_{qf}) is given by potential energy of the DNS which is calculated as the sum of binding energies of the nuclei B_i (i=1,2) and nucleus-nucleus potential $V(R, \eta, J)$ [7,8]:

$$U(R, \eta, J) = B_1 + B_2 + V(R, \eta, J) - [B_{12} + V'_{rot}(J)]. \quad (6)$$

The shell effects are included in these binding energies. The isotopic composition of the nuclei forming the DNS is chosen with the condition of a N/Z-equilibrium in the system. The value of $U(R, \eta, J)$ is normalized to the energy of the rotating compound nucleus by $B_{12} + V'_{rot}$. Deformations of the DNS nuclei taken into account in the calculation of the potential energy surface [7,8]. For the heavy nuclei in the DNS, which are deformed in the ground state, the parameters of deformation are taken from Ref. [10]. The light nuclei of the DNS are assumed to be deformed only if the energies of their 2^+ states are smaller than 1.5 MeV. As known from experiments on subbarrier fusion of lighter nuclei, these states are easily populated. For the collision energies considered, the relative orientation of the nuclei in the DNS follows the minimum of the potential energy during the evolution in η. Taking the deformation of the nuclei in the DNS as a function of η, we choose the way for the DNS evolution on the potential energy surface calculated in the DNS concept. As the result, the inner fusion barrier takes the value as indicated in Fig.1.

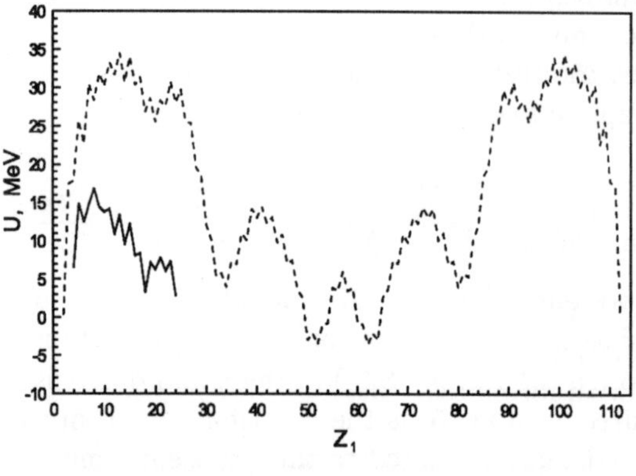

Figure 1. Dependence of the potential energy of the DNS on the charge number Z_1 of one nucleus in the DNS for the $^{48}Ca+^{244}Pu$ reaction. The calculated results with and without deformation of the DNS nuclei are presented by solid and dashed lines, respectively. The result with deformation is presented for Z_1-values which are of interest to calculate B^*_{fus}.

The U gives us the information about the energy threshold for fusion which determines the optimal bombarding energy. In the DNS model, the inner fusion barrier B^*_{fus} in mass asymmetry supplies the hindrance for complete fusion. The top of this barrier coincides with the maximum of the DNS potential energy as a function of η. If the excitation energy of the DNS is sufficient to overcome this barrier, then the fusion occurs.

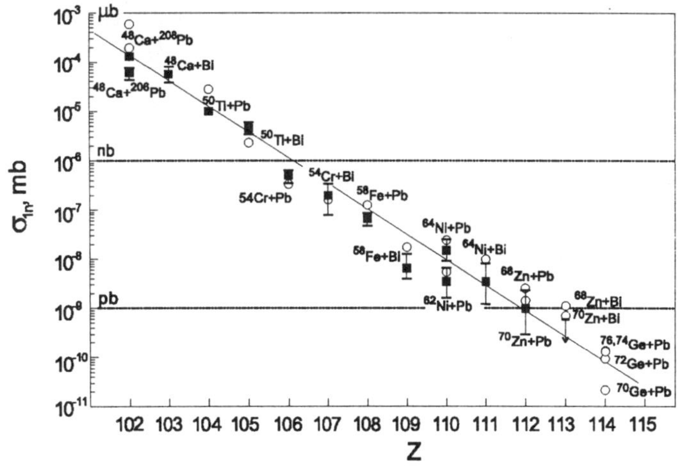

Figure 2. Experimental data (references see in review [2])(black squares) and theoretical calculations (open circles) for synthesis of elements from 102 to 114 in cold fusion reactions $(HI, 1n)$. The combinations are noted in the figure. The line has been drawn to guide the eyes.

4 Cold fusion reactions

The calculations for Pb-based reactions were performed with one set of parameters and with the same assumptions. In Fig.3 [13] (left panel) one can see that our calculations with formula (1) describe quite well the experimental σ_{ER} [14] for the ^{48}Ca+ ^{208}Pb $\rightarrow^{256-xn}$114 reaction (left panel). In the recent experiments the limit of the heaviest elements production in the cold fusion has been reached. As one can see in Fig.2 [13], σ_{1n} of reaction ^{70}Zn+^{209}Bi$\rightarrow^{279}$113 is less than 1 pb. For the ^{74,76}Ge+^{208}Pb $\rightarrow^{281,283+1n}$114 reactions, we obtained σ_{1n}=0.1pb, which are close to the estimations for the same reactions in Ref. [8]. The value of σ_{1n} decreases by about 5 orders of magnitude from the nucleus with Z=102 to the nucleus Z=112 mainly due to the strong decrease of P_{CN} [8,9]. Theoretical estimations of σ_{1n} (Fig.2) for cold fusion reactions are in an agreement with known experimental data (references see in review [2]).

5 Warm fusion reactions

The use of more asymmetric combinations of the colliding nuclei than in the Pb-based reactions may help to extend the possibility of the superheavy elements production. Therefore, we analyzed the hot fusion reactions $^{48}Ca+$ $^{232}Th, ^{238}U, ^{244}Pu$ with the DNS fusion concept.

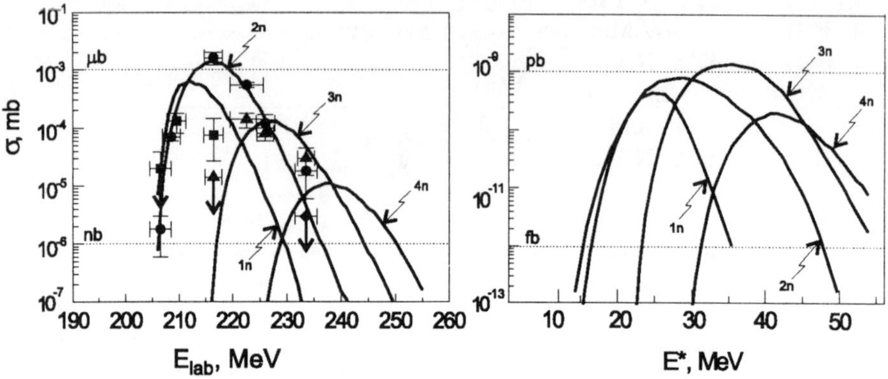

Figure 3. Theoretical (solid lines) and experimental [14] (solid point) cross sections for xn–channels in the collisions $^{48}Ca+^{208}Pb$ (left panel) and $^{48}Ca+^{244}Pu$ (right panel), as a function of the bombarding energy E_{lab} and excitation energy of compound nucleus E^*, respectively.

Fig.3 [13](right panel) shows that the maximal calculated yield is about 1.5 pb for the hot fusion reaction $^{48}Ca+^{244}Pu \rightarrow {}^{289+3n}114$. This corresponds to the $3n$ evaporation channel at an excitation energy 35 MeV of compound nucleus.

6 Summary

The strong decrease in cold fusion cross section for synthesis of heaviest elements is mainly caused by a decrease of fusion probability P_{CN} due to strong competition between the complete fusion and quasi-fission in the DNS. For the $^{48}Ca+^{244}Pu \rightarrow {}^{289+3n}114$ reactions, the P_{CN} is larger about 10^5 time than in the $^{76}Ge+^{208}Pb \rightarrow {}^{283+1n}114$. The gain in fusion and capture probabilities for actinide-based reaction relative to cold fusion reaction is not compensated by the relative loss in survival probability of the compound nucleus. Therefore, comparison of the formation cross section of Z=114 element in Pb– and actinide–based reactions shows that second one is preferable.

References

1. Oganessian Yu.Ts. in book of Proc. Int. Conf. Structure of Vacuum and Elementary Matter, Wilderness, South Africa, March 10-16, 1996, p.11-24; A.V.Yeremin, V.K.Utyonkov and Yu.Ts.Oganessian *in* AIP Conf. Proc. 425 (1998) 16.
2. S.Hofmann, Rep. Prog. Phys. 61 (1998) 570.
3. K.H.Schmidt, W.Morawek, Rep. Prog. Phys. 54 (1991) 949.
4. G.G.Adamian, N.V.Antonenko, S.P.Ivanova and W.Scheid, Nucl. Phys. A 646 (1999) 29.
5. Yu.F.Smirnov and Yu.M.Tchuvil'sky, Phys.Lett.B 134 (1984) 25; G.G.Adamian, N.V.Antonenko, Yu.M.Tchuvil'sky, Phys.Lett. B (1999) (in press).
6. V.V.Volkov, *in* Proc. Int. School-Seminar on Heavy Ion Physics, Dubna, 1986 (JINR, Dubna, 1987) p.528; Izv. AN SSSR ser. fiz. 50 (1986) 1879; *in* Proc. Int. Conf. on Nuclear Reaction Mechanisms, Varenna, 1991, ed. E.Gadioli (Ricerca Scientifica, 1991) p.39.
7. N.V.Antonenko, E.A.Cherepanov, A.K.Nasirov, V.B.Permjakov and V.V.Volkov, Phys. Lett. B 319 (1993) 425; Phys. Rev. C 51 (1995) 2635; G.G.Adamian, N.V.Antonenko and W.Scheid, Nucl. Phys. A 618 (1997) 176; G.G.Adamian, N.V.Antonenko, W.Scheid and V.V.Volkov, Nucl. Phys. A 627 (1997) 332; R.V.Jolos, A.K.Nasirov and A.I.Muminov, Euroup. J. Phys. A (1999) (in press).
8. G.G.Adamian, N.V.Antonenko, W.Scheid and V.V.Volkov, Nucl. Phys. A 633 (1998) 154.
9. V.V.Volkov, in the present proceedings.
10. S.Raman et al., At. Data Nucl. Data Tables 36 (1987) 1.
11. E.A.Cherepanov, *in* Proc. Int. Symp. on In Beam Nuclear Spectropscopy, Debrecen, 1984, p.499; E.A. Cherepanov and A.S. Iljinov, Nucleonika 25 (1980) 611.
12. R.Smolanczuk, J.Skalski, A.Sobiczewski, Phys.Rev.C 52 (1995) 1871.
13. E.A.Cherepanov, Int. Workshop on Rare Nuclear Process in Low Energy Heavy Ion Physics, New Delhi India, November 16-20, 1998 (in press).
14. A.V.Yeremin, et al. Rapid Comm. JINR 6(92)-98.

LIST OF PARTICIPANTS

Y. Aritomo
Flerov Laboratory of Nuclear Reactions
Joint Institute for Nuclear Research
Dubna, 141980, Moscow region, Russia
aritomo@yukawa.kyoto-u.ac.jp

P. Armbruster
GSI
Postfach 12 05 52
Max-Planck-Str. 1
D-64220 Darmstadt, Germany
P.Armbruster@gsi.de

P. Bandžuch
Institute of Physics
Slovak Academy of Sciences
Dúbravská cesta 9
SK-84228 Bratislava, Slovak Republic
bandzuch@savba.sk

J. Benlliure
GSI
Max-Planck-Str. 1
D-64291 Darmstadt, Germany
J.Benlliure@gsi.de

M. Beňovič
Institute of Physics
Slovak Academy of Sciences
Dúbravská cesta 9
SK-84228 Bratislava, Slovak Republic
benovic@savba.sk

E. Běták
Institute of Physics
Slovak Academy of Sciences
Dúbravská cesta 9
SK-84228 Bratislava, Slovak Republic
betak@savba.sk

T. Bolognese
CEA/SACLAY
DSM/DAPNIA/SPP
F-91191 Gif sur Yvette, France
bolognese@dapnia.cea.fr

N. Carjan
University of Bordeaux
Centre d'Etudes Nucleaires
Le Haut Vigneau
33175 Gradignan Cedex, France
carjan@in2p3.fr

G. G. Chubarian
Texas A&M University
Cyclotron Institute
College Station, TX 11843-3366, USA
chubarian@leper.tamu.edu

L. Dematté
EC-JRC, IRMM
Retiesweg
B-2440 Geel, Belgium
dematte@irmm.jrc.be

T. von Egidy
Technische Universität München
Fakultät für Physik
Physik-Department E18
James-Franck-Str.
D-85748 Garching, Germany
egidy@e18.physik.tu-muenchen.de

M. Florek
Department of Nuclear Physics
Faculty of Mathemathics and Physics
Comenius University, Mlynská Dolina
SK-84215 Bratislava, Slovak Republic
florek@fmph.uniba.sk

G. Giardina
Instituto Nazionale di Fisica Nucleare
Dipartimento di Fisica
Universita di Messina
I-98166 Messina, Italy
giardina@nucleo.unime.it

Š. Gmuca
Institute of Physics
Slovak Academy of Sciences
Dúbravská cesta 9
SK-84228 Bratislava, Slovak Republic
gmuca@savba.sk

F. Gönnenwein
Universität Tübingen
Physikaliches Institut
Auf der Morgenstelle 14
D-72076 Tübingen, Germany
goennenwein@uni-tuebingen.de

W. Greiner
Institut für Theoretische Physik
Universität Frankfurt am Main
Robert-Mayer-Strasse 8-10
P. O. BOX 111932
D-60054 Frankfurt am Main, Germany
greiner@th.physik.uni-frankfurt.de

G. G. Gulbekian
Flerov Laboratory of Nuclear Reactions
Joint Institute for Nuclear Research
Dubna, 141980 Moscow region, Russia
gulbekian@flnr.jinr.ru

J. H. Hamilton
Department of Physics and Astronomy
Vanderbilt University
Nashville, TN 37235, USA
hamiltj1@atrvax.vanderbilt.edu

A. Heinz
G S I
Max-Planck-Str. 1
D-64291 Darmstadt, Germany.
heinz@axp602.gsi.de

F. P. Heßberger
G S I
Max-Planck-Str. 1
D-64291 Darmstadt, Germany
hess@rzri6f.gsi.de

M. G. Itkis
Flerov Laboratory of Nuclear Reactions
Joint Institute for Nuclear Research
Dubna, 141980 Moscow region, Russia
itkis@flnr.jinr.ru

F. A. Ivanyuk
Institute for Nuclear Research
Prospect Nauki 47
252028 Kiev-28, Ukraine
ivanyuk@kinr.kiev.ua

M. Jandel
Institute of Physics
Slovak Academy of Sciences
Dúbravská cesta 9
SK-84228 Bratislava, Slovak Republic
jandel@savba.sk

S. A. Karamian
Flerov Laboratory of Nuclear Reactions
Joint Institute for Nuclear Research
Dubna, 141980 Moscow region, Russia
karamian@flnr.jinr.ru

J. Kliman
Flerov Laboratory of Nuclear Reactions
Joint Institute for Nuclear Research
Dubna, 141980 Moscow region, Russia
kliman@flnr.jinr.ru

N. A. Kondratiev
Flerov Laboratory of Nuclear Reactions
Joint Institute for Nuclear Research
Dubna, 141980 Moscow region, Russia
kondratiev@flnr.jinr.ru

Yu. N. Kopach
Frank Laboratory of Neutron Physics
Joint Institute for Nuclear Research
Dubna, 141980 Moscow region, Russia
kopach@nfsun1.jinr.ru

A. Kötzle
Universität Tübingen
Physikaliches Institut
Auf der Morgenstelle 14
D-72076 Tübingen, Germany
achim.koetzle@uni-tuebingen.de

L. Krupa
Flerov Laboratory of Nuclear Reactions
Joint Institute for Nuclear Research
Dubna, 141980 Moscow region, Russia
krupa@flnr.jinr.ru

J. Leja
Department of Nuclear Physics
Faculty of Mathemathics and Physics
Comenius University, Mlynská Dolina,
SK-84215 Bratislava, Slovak Republic
leja@center.fmph.uniba.sk

M. Matoš
Department of Nuclear Physics
Faculty of Mathemathics and Physics
Comenius University, Mlynská Dolina,
SK-84215 Bratislava, Slovak Republic
matos@center.fmph.uniba.sk

V. Matoušek
Institute of Physics
Slovak Academy of Sciences
Dúbravská cesta 9
SK-84228 Bratislava, Slovak Republic
matousek@savba.sk

M. Morjean
GANIL
BP 5027
F-14076 CAEN Cedex 5, France
morjean@ganil.fr

M. Mutterer
Institut für Kernfysik
Technische Hochschule
D-64289 Darmstadt, Germany
mutterer@hrz1.hrz.th-darmstadt.de

J. B. Natowitz
Cyclotron Institute
Texas A&M University
College Station, TX 77843, USA
natowitz@comp.tamu.edu

Yu. Ts. Oganessian
Flerov Laboratory of Nuclear Reactions
Joint Institute for Nuclear Research
Dubna, 141980 Moscow region, Russia
oganessian@flnr.jinr.ru

I. V. Pokrovski
Flerov Laboratory of Nuclear Reactions
Joint Institute for Nuclear Research
Dubna, 141980 Moscow region, Russia
pokrovski@flnr.jinr.ru

G. Pollarolo
Dipartimento di Fisica Teorica
Universita Torino
Via Pietro Giuria 1
I-10125 Torino, Italy
nanni@to.infn.it

G. Rudolf
IReS
BP 28
F-67037 Strasbourg, France
gerard.rudolf@ires.in2p3.fr

E. Rurarz
Soltan Institute for Nuclear Studies
Department PII
PL-05400 Swierk near Otwock, Poland
rurarz@india.cyf.gov.pl

A. Sobiczewski
Soltan Institute for Nuclear Studies
Warsaw University
ul. Hoza 69
00-681 Warszawa, Poland
sobicz@fuw.edu.pl

Š. Šáro
Department of Nuclear Physics
Faculty of Mathemathics and Physics
Comenius University, Mlynská Dolina,
SK-84215 Bratislava, Slovak Republic
saro@fmph.uniba.sk

W. H. Trzaska
University of Jyväskylä
Department of Physics
P.O. Box 35
40351 Jyväskylä,Finland
trzaska@jyfl.jyu.fi

V. K. Utyonkov
Flerov Laboratory of Nuclear Reactions
Joint Institute for Nuclear Research
Dubna, 141980 Moscow region, Russia
utyonkov@flnr.jinr.ru

E. Vardaci
Universtá di Napoli and INFN
Dipartimento Scienze FisicheVia
CinthiaI-80126 Napoli, Italy
Emanuel.Vardaci@na.infn.it

M. Veselský
Institute of Physics
Slovak Academy of Sciences
Dúbravská cesta 9
SK-84228 Bratislava, Slovak Republic
fyzimarv@savba.sk

T. Wada
Department of Physics,
Konan University,
8-9-1 Okamoto, Higoshinada
kobe 658-8501, Japan
wada@konan-u.ac.jp

M. Wöstheinrich
Universität Tübingen
Physikalisches Institut
Auf der Morgenstelle 14
D-72076 Tübingen, Germany
marcus.woestheinrich@uni-tuebingen.de

V. V. Volkov
Flerov Laboratory of Nuclear Reactions
Joint Institute for Nuclear Research
Dubna, 141980 Moscow region, Russia
volkov@flnr.jinr.ru

A. N. Yeremin
Flerov Laboratory of Nuclear Reactions
Joint Institute for Nuclear Research
Dubna, 141980 Moscow region, Russia
yeremin@flnr.jinr.ru

AUTHOR INDEX

V

W

X

Y

Z